Technology Platforms for 3D Cell Culture

Registered Office
John Wiley & Sons Ltd., The Atrium, Southern Gate, Chichester, West Sussex, PO19 8SQ, UK

Editorial Offices
9600 Garsington Road, Oxford, OX4 2DQ, UK
The Atrium, Southern Gate, Chichester, West Sussex, PO19 8SQ, UK
111 River Street, Hoboken, NJ 07030-5774, USA

For details of our global editorial offices, for customer services and for information about how to apply for permission to reuse the copyright material in this book please see our website at www.wiley.com/wiley-blackwell.

Library of Congress Cataloging-in-Publication data are available

ISBN: 9781118851500

A catalogue record for this book is available from the British Library.

Wiley also publishes its books in a variety of electronic formats. Some content that appears in print may not be available in electronic books.

Cover image: Courtesy of Stefan Przyborski

Set in 9.5/13pt Meridien by SPi Global, Pondicherry, India

1 2017

MIX
Paper from responsible sources
FSC www.fsc.org FSC® C013604

Contents

List of contributors

Brigitte M. Angres
Cellendes GmbH, Reutlingen, Germany

Ayla Annac
InvivoSciences, Inc., Madison, Wisconsin, USA

Kazuya Arai
SCIVAX Life Sciences, Inc., Kawasaki, Kanagawa, Japan

Andrew M. Blakely
Department of Molecular Pharmacology, Physiology and Biotechnology, Center for Biomedical Engineering, Rhode Island Hospital, Department of Surgery, Brown University, Providence, Rhode Island, USA

John J. S. Cadwell
FiberCell Systems Inc., Frederick, Maryland, USA

D.D. Carson
Department of BioSciences, Rice University, Houston, Texas, USA
Department of Genetics, M. D. Anderson Cancer Center, Houston, Texas, USA

Igor Charvet
Hepia/HES-SO Campus Biotech, Geneva, Switzerland

P. E. Constantinou
Department of BioSciences, Rice University, Houston, Texas, USA

Michael Conway
InvivoSciences, Inc., Madison, Wisconsin, USA

Sean Curran
Department of Molecular Pharmacology, Physiology and Biotechnology, Center for Biomedical Engineering, Brown University, Providence, Rhode Island, USA

B. P. Danysh
Department of BioSciences, Rice University, Houston, Texas, USA

N. A. Delk
Department of Biological Sciences, The University of Texas at Dallas, Richardson, Texas, USA

N. J. Doty
BioTime, Inc., Alameda, California, USA

B. J. Engel
Department of BioSciences, Rice University, Houston, Texas, USA

I. E. Erickson
BioTime, Inc., Alameda, California, USA

M. C. Farach-Carson
Department of BioSciences and Department of Bioengineering, Rice University, Houston, Texas, USA

Andreia Fernandez
InSphero AG, Schlieren, Switzerland

David Fluri
InSphero AG, Schlieren, Switzerland

E. L. S. Fong
Department of Bioengineering, Rice University, Houston, Texas, USA

Markus Furter
InSphero AG, Schlieren, Switzerland

D. A. Harrington
Department of BioSciences, Rice University, Houston, Texas, USA

Johannes Haugstetter
InSphero AG, Schlieren, Switzerland

Elena Heister
The Electrospinning Company Ltd, Rutherford Appleton Laboratory, Harwell Oxford, Didcot, UK

Manabu Itoh
SCIVAX Life Sciences, Inc., Kawasaki, Kanagawa, Japan

Jens M. Kelm
InSphero AG, Schlieren, Switzerland

Eleanor Knight
Department of Biosciences, Durham University, Durham, UK

Marshall Kosovsky
Corning Life Sciences, Tewksbury, Massachusetts, USA

MaryAnn Labant
3D Biomatrix, Inc., Ann Arbor, Michigan, USA

Elizabeth Leary
Department of Molecular Pharmacology, Physiology and Biotechnology, Center for Biomedical Engineering, Brown University, Providence, Rhode Island, USA

Kali L. Manning
Department of Molecular Pharmacology, Physiology and Biotechnology, Center for Biomedical Engineering, Brown University, Providence, Rhode Island, USA

M. Martinez
Department of BioSciences, Rice University, Houston, Texas, USA

Robert J. McKean
The Electrospinning Company Ltd, Rutherford Appleton Laboratory, Harwell Oxford, Didcot, UK

R. L. Milton
Department of BioSciences, Rice University, Houston, Texas, USA

Hiromi Miura
SCIVAX USA, Inc., Woburn, Massachusetts, USA

Jeffrey R. Morgan
Department of Molecular Pharmacology, Physiology and Biotechnology, Center for Biomedical Engineering, Brown University, Providence, Rhode Island, USA

Stephen S. Navran
Synthecon, Inc., Houston, Texas, USA

Bramasta Nugraha
Department of Biosystem Science and Engineering, Swiss Federal Institute of Technology (ETH Zürich), Basel, Switzerland
Roche Pharmaceutical Research and Early Development, Basel, Switzerland

M. V. Onorato
BioTime, Inc., Alameda, California, USA

Stefan Przyborski
Department of Biosciences, Durham University, Durham, UK
ReproCELL Europe, NETPark, Sedgefield, UK

M. Mamunur Rahman
SCIVAX USA, Inc., Woburn, Massachusetts, USA

Antonio Romo-Morales
Department of Biosciences, Durham University, Durham, UK

Suparna Sanyal
Corning Life Sciences, Tewksbury, Massachusetts, USA

Nicole A. Slawny
3D Biomatrix, Inc., Ann Arbor, Michigan, USA

Luc Stoppini
Hepia/HES-SO Campus Biotech, Geneva, Switzerland

Randy Strube
InSphero AG, Schlieren, Switzerland

Lars E. Sundstrom
Elizabeth Blackwell Institute for Health Research, School of Clinical Sciences, University of Bristol, Bristol, UK

Michael Susienka
Department of Molecular Pharmacology, Physiology and Biotechnology, Center for Biomedical Engineering, Brown University, Providence, Rhode Island, USA

Tetsuro Wakatsuki
InvivoSciences, Inc., Madison, Wisconsin, USA

William G. Whitford
GE Healthcare, Life Sciences, Cell Culture, Logan, Utah, USA

J. Malcolm Wilkinson
Kirkstall Ltd, Rotherham, UK

Helmut Wurst
Cellendes GmbH, Reutlingen, Germany

T. I. Zarembinski
BioTime, Inc., Alameda, California, USA

Preface

Understanding basic cellular biology relies on research involving cell-based *in vitro* assays that are also used to model disease, test and screen compounds and assess the safety of chemicals. Most often, the investigation of such biological processes is based on studying homogeneous populations of mammalian cells cultured as monolayers on flat, two-dimensional (2D) polystyrene substrates. However, cells naturally exist within a complex three-dimensional (3D) tissue microenvironment composed of mixed cell populations, extracellular proteins and both physical and chemical signals from multiple sources.

It should be recognised that many important discoveries have been made from conventional 2D culture approaches. This reductionist view to understanding basic biological processes has value but is limited since cells growing on 2D substrates do not always reflect the true physiological behaviour of their native counterparts in real tissues. When forced to grow on a flat 2D substrate, cells adapt and radically change their shape, proliferate in an aberrant fashion and often lose their differentiated phenotype, resulting in abnormal cellular behaviour. It is widely appreciated that such structural modifications to the physical environment can result in changes to gene transcription and protein translation, remodelling of the cytoskeleton and irregular cell signaling.

Accordingly, the anatomy of a cell, i.e. its structure and form, are inextricably linked to its physiological function. Technologies are now becoming available that enable researchers to culture cells in 3D that in turn enhance the value of such cell-based assays and the generation of more accurate and physiologically relevant results. The growth of tissue-like structures in 3D in combination with media perfusion/circulation creates a dynamic system which advances cell-based models still further towards the recreation of more '*in vivo*-like' conditions.

Maintaining the natural 3D architecture of a cell is therefore considered one of the fundamental steps toward enhancing the value of cell-based assays. There are now several technologies available that promote solutions for 3D cell culture. In general, these fall into one of three categories: hydrogels (e.g. Matrigel™, collagen gels); cell aggregates (e.g. hanging drop methods, low adherence plates); and scaffolds (e.g. porous physical supports). There is no panacea and no one solution is suitable for all 3D culture needs. The availability of these technologies from commercial sources now allows the investigator to select the most appropriate method suitable for their experimental requirements. Moreover, 3D cell culture technology is often combined with new developments in dynamic media perfusion, to further enhance cell growth and physiological relevance of the model. There are numerous advanced technologies that enable media perfusion encompassing cleverly designed devices and mini-benchtop bioreactors.

The aim of this text is provide the reader with a review of the types of technology available and examples of where such methods may be applied. This has been divided into descriptions of 3D cell culture technologies by certain commercial developers representative of these key areas and the users of such methods. The reader will learn about the options available to perform 3D culture, explore the options for media perfusion, from where such 3D technologies can be acquired, how they work and how they can be used. Any cell biologist considering 3D cell culture and aiming to enhance the physiological relevance of their cell-based assay should consult this text as a guide to getting started with such methods.

Key features

- A review of the current state-of-the-art technologies available for 3D cell culture and media perfusion models.
- Contributions from leading developers and researchers active in 3D cell technology and advanced cell-based assays.
- Instruction and guidance on performing specific 3D culture methods and media circulation systems.
- Examples of where such technologies have been successfully applied.
- Guidance on resources and technical support to help get started using 3D culture methods with options of dynamic media circulation.
- Relevance to multiple fields including stem cells, tissue engineering, cell-based screening assays, etc.
- Examples of advanced physiologically relevant *in vitro* models, including use of 3D culture and perfusion technology, organotypic models, co-cultures, etc.

Primary readership

Interest in 3D cell culture and the ability to create more tissue-like constructs are developing rapidly in the scientific community. Researchers recognise its value and are keen to apply such technology to their experimental systems. This book will be especially topical given the drive to improve the value of *in vitro* cell-based assays and generate more physiologically relevant data. A quick scan of the scientific literature will indicate that interest in this sector is developing rapidly. While many different approaches that enable 3D culture have been developed, most are based on research in academic labs and are published in scientific journals. While of value, they are not always technologies readily available to the majority of investigators interested in 3D cell culture. Several approaches to culture cells in 3D have now been commercialised. Numerous small and large companies have undertaken the process of translating research into the creation of marketable products. These organisations are the pioneers of a new era in cell culture methods and have made such technology available to the greater scientific community through the development of bespoke products and applications. This was not possible until now and the time is right for this book to bring together the different approaches that are readily available and to support this rapidly growing sector of 3D cell culture.

Any cell biologist practising conventional 2D cell culture will be interested in this book as an opportunity to enhance the value of their cell-based assays and perform 3D culture methods. Specific sectors of interest include cancer cell biology, stem cells, tissue engineering, *in vitro* alternatives to animal use, liver toxicology, neuroscience and those requiring specialised cell culture models (e.g. co-culture, organotypic models). Cell culture as a technique is very general and is performed in academic, industrial and government laboratories worldwide. It is anticipated that many will benefit from this text as a comprehensive guide to technologies that are readily available, to act as a reference and assist in the selection of technology and guidance for use. Therefore the primary readership will be the practising bench scientists (postgraduate and postdoctoral students, research fellows, research scientists and the like). The secondary readership may be managers of cell culture facilities/departments, R&D heads of section, technical supervisors, advisors, etc. who would recommend methods to perform. Fringe interests include parties interested in cell biology in general and methods associated with the field.

The pharmaceutical industry is particularly interested in developing new approaches to enhance its ability to discover new compounds and has identified 3D culture as a priority area. Similarly, contract research organisations are now using 3D culture methods to provide additional information to their clients. Collectively, these are substantial opportunities. Societies interested in alternatives to animal research, cell biology, tissue engineering, cancer biology, etc. should all find this text of value.

Stefan Przyborski
Professor of Cell Technology
Durham University, UK
January 2017

List of abbreviations

2D	Two-dimensional
3D	Three-dimensional
5-FU	5-Fluorouracil
A1AT	α1 Antitrypsin
ADME Tox	Absorption, distribution, metabolism, excretion and toxicity
ALI	Air-liquid interface
ASC	Adipose-derived stem cells
ATP	Adenosine-5′-triphosphate
ATRA	All trans-retinoic acid
BBB	Blood–brain barrier
bFGF	Basic fibroblast growth factor
BME	Basement membrane extracts
BNDF	Brain-derived neurotrophic factor
CAF	Cancer-associated fibroblast
CD	Cytochalasin D; cluster of differentiation
CDI	1,1′-Carbonyldiimidazole
CDM-HD	Chemically defined medium for high-density cell culture
C/EBPa	CCAAT-enhancer-binding proteins
CHO	Chinese hamster ovary
CLL	Chronic lymphocytic leukaemia
CNS	Central nervous system
CP	Cortical plate
CTG	CellTracker green; CellTiter-Glo
CV	Coefficient of variation
CYP1A1	Cytochrome P450, family 1, subfamily A, polypeptide 1
CYP2E1	Cytochrome P450 2E1
CYP3A4	Cytochrome P450 3A4
d4T	2′,3′-Didehydro-3′deoxythymidine
DNA	Deoxyribonucleic acid
DCIS	Ductal carcinoma *in situ*
DLS	Dynamic light scattering
DMEM	Dulbecco's modified eagle medium
DMSO	Dimethyl sulfoxide
DNP	Dinitrophenol
DOX	Doxorubicin
DPMK	Drug metabolism and pharmacokinetics
DRG	Dorsal root ganglia
DTT	Dithiothreitol
EthD	Ethidium homodimer-1

EB	Embryoid body
ECM	Extracellular matrix
ECS	Extracapillary space
EDTA	Ethylenediamine tetra-acetic acid
EEF1	Elongation factor 1 complex
EFP	Evoked field potential
EGF	Epidermal growth factor
EGFR	Epidermal growth factor receptor
EHS	Engelbreth-Holm-Swarm
EHT	Engineered heart tissue
ELISA	Enzyme-linked immunosorbent assay
EMT	Epithelial-mesenchymal transition
ESC	Embryonic stem cell
ET	Engineered tissue
EVS	Extravascular space
FA	Focal adhesion
FACS	Fluorescence activated cell sorting
FBS	Fetal bovine serum
FDA	Food and Drug Administration; fluorescein diacetate
GABA	γ-Aminobutyric acid
GAG	Glycosaminoglycan
GFAP	Glial fibrillary acid protein
GFP	Green fluorescent protein
GMP	Good Manufacturing Practice
GSH	Glutathione synthase
GSSG	Oxidised glutathione
HA	Hyaluronic acid
HARV	High aspect ratio vessel
HCA	High-content analysis
HCV	Hepatitis C virus
HCVpp	Hepatitis C virus pseudoparticles
HD	Hanging drop
HDMEC	Human dermal microvascular endothelial cell
HDP	Hanging drop plate
H&E	Haematoxylin and eosin
HF	Hollow fibre
HFBR	Hollow fibre bioreactor
HFIM	Hollow fibre infection model
HIPE	High internal phase emulsion
HiPSC	Human induced pluripotent stem cell
HIV	Human immunodeficiency virus
HMEC	Human microvascular endothelial cell
hMSC	Human mesenchymal stem cell
HNF4α	Hepatocyte nuclear factor 4 α
HPC	Hydroxypropylcellulose
HSC	Haematopoietic stem cell
HS	Heparan sulfate

HTS	High-throughput screening
HUVEC	Human umbilical vein endothelial cell
ICM	Inner cell mass
ID	Internal diameter
IDC	Invasive ductal carcinoma
IgG	Immunoglobulin G
iPSC	Induced pluripotent stem cell
LC-MS	Liquid chromatography-mass spectrometry
LCST	Liquid crystal solution temperature
LDH	Lactate dehydrogenase
NF	Neurofilament
M-dPEG-NHS	Maleimide-dPEG8-N-hydroxysuccinimide ester
Mal-PVA	Maleimide-functionalised polyvinyl alcohol
MAP	Microtubule-associated protein
MCR	Multicellular resistance
MCTS	Multicellular tumour spheroid
MDCK	Madin–Darby canine kidney
MEA	Microelectrode array
MeHg	Methyl-mercury
mESC	Murine embryonic stem cell
MRP2	Multidrug-resistant associated protein 2
MPLSM	Multiphoton laser scanning microscopy
MSC	Mesenchymal stem cell
MMP	Matrix metalloproteinase
mRNA	Messenger ribonucleic acid
MWCO	Molecular weight cut-off
NAPQI	N-acetyl-p-benzoquinone imine
NASA	National Aeronautics and Space Administration
NCP	NanoCulture Plate
NE	Neuroepithelium
NGF	Nerve growth factor
NIH	National Institutes of Health
NMR	Nuclear magnetic resonance
NPC	Neural precursor cells
PA	Polyacrylamide
PBS	Phosphate buffered saline
PCL	Poly-ε-caprolactone
PCNA	Proliferating cell nuclear antigen
PCR	Polymerase chain reaction
PDMS	Poly-dimethyl-siloxane
PEG	Polyethylene glycol
PEGDA	Polyethylene glycol diacrylate
PEGSSDA	Polyethylene glycol diacrylate with internal disulfide bonds
PEGnor/I2959	Polyethylene glycol norbornene/Irgacure 2959
PET	Polyethylene terephthalate
PFA	Paraformaldehyde
PGA	Polyglycolic acid

PGLA	Poly-D, L-lactide-co-glycolide
Pgp	P-glycoprotein
PHA	Polyhydroxyl alkanoate
PI	Propidium iodide
PK/PD	Pharmacokinetic/pharmacodynamic
PLA	Polylactic acid
PLLA	Poly-L-lactide
PolyHEMA	Polyhydroxyethylmethacrylate
PPAR	Peroxisome proliferator-activated receptor
PPI	Paired pulse inhibition
PrC	Prostate cancer
PS	Polysulfone
PSC	Pluripotent stem cell
PTFE	Polytetrafluoroethylene
PVA	Polyvinyl alcohol
PVDF	Polyvinylidene fluoride
QV	Quasi Vivo
RCCS	Rotary Cell Culture System
RFP	Red fluorescent protein
RGD	Arginine-glycine-aspartate
RGDS	Arginine-glycine-aspartate-serine
RIG-I	Retinoic acid-inducible gene 1
RLU	Relative luminescence unit
ROCK	Rho-associated coiled-coil protein kinase
RPM	Revolutions per minute
RT	Room temperature
RTK	Receptor tyrosine kinase
RT-PCR	Real-time polymerase chain reaction
RWV	Rotating wall vessel
SCARB1	Scavenger receptor class B, member 1
SEM	Scanning electron microscopy
shRNA	Short hairpin RNA
STA	Staurosporine
STLV	Slow turning lateral vessel
SU	Single use
TAX	Taxol
TCEP	Tris (2-carboxyethyl) phosphine
TCP	Tissue culture polystyrene
TEER	Transepithelial electrical resistance
TGF-β2	Transforming growth factor-β2
TMRE	Tetramethylrhodamine ethyl ester
TMTC	Trimethyltin chloride
TPZ	Tirapazamine
ULA	Ultra-low attachment
UV	Ultraviolet
XPS	X-ray photoelectron microscopy

CHAPTER 1

An introduction to the third dimension for routine cell culture

Antonio Romo-Morales[1] and Stefan Przyborski[1,2]
[1] *Department of Biosciences, Durham University, Durham, UK*
[2] *ReproCELL Europe, NETPark, Sedgefield, UK*

Introduction

In recent years, the advent of three-dimensional (3D) cell culture technologies has led to a paradigm shift in our understanding of eukaryotic cell culture. The challenge of reproducing the complexity of whole tissues *in vitro* is being addressed through various approaches incorporating biological parameters known to influence cellular behaviour. As such, the increasing number of publications utilising these technology platforms is evidence of the transition into 3D cell culture. This book juxtaposes these efforts and successes with the shortcomings of culturing mammalian cells with conventional methods. However, full adoption of these techniques for routine mammalian cell biology research will require their validation. This book therefore serves as a guiding tool for researchers who seek to shift towards more advanced cellular assays that recreate *in vivo*-like conditions, compiling readily available techniques for 3D cell culture.

Two-dimensional (2D) *in vitro* models have been vital to understand biological processes and mechanisms in cellular biology. For decades, cellular monolayers have been used to model disease, screen and assess the efficacy and toxicity of chemical compounds and develop anticancer treatments. Although valuable, it should be recognised that these conventional cell culture approaches are a simplistic method, overlooking important biological parameters that influence cellular behaviour. 2D cell culture does not provide an *in vivo*-like environment where physical cues, cell-cell and cell-matrix communication and the interplay of different cell types can be reproduced. This results in a poor reflection of physiological cellular behaviour, as well as limited potential to form more complex tissue-like structures. These disadvantages become more significant in the context of drug testing, where monolayers of cultured cells fall short in reflecting how drugs interact with target molecules *in vivo*. The lack of inclusion

Technology Platforms for 3D Cell Culture: A User's Guide, First Edition. Edited by Stefan Przyborski.

of the signalling context as part of the cell culture system hinders the predictive value of traditional cell-based drug screening methods (Bhadriraju & Chen, 2002; Sun *et al.*, 2006).

Cells naturally exist within a complex 3D tissue environment composed of heterogeneous cell populations, extracellular proteins, forming an intricate system of physical and chemical cues that impact the natural response of cells. Replicating the native environment is a fundamental step towards making these models more physiologically accurate and enhancing the value of the results drawn from these culturing systems. Here, we examine specific areas where 2D cell culture fails and anticipate the areas of improvement that 3D cell culture seeks to tackle.

Structure and cell adhesion

Culturing cells in 2D imposes physical constraints that impede cells from organising naturally and spreading vertically (Figure 1.1). For cells to adopt their native morphology, they need to form integrin-mediated adhesions with the extracellular matrix (ECM). Flat polystyrene or glass substrates cannot faithfully capture the topographically complex extracellular environments, and therefore they tend to force an apical-basal cell polarity on all cells. This characteristic polarity, seen in monolayer-cultured cells, may be relevant to epithelial cells but it impedes mesenchymal cells in acquiring their characteristic stellate morphology. In turn, cell shape and tissue architecture will probably affect the growth of 2D-cultured mesenchymal cells and thus their differentiation. Similarly, apical-basal polarity and the formation of 3D structures are means by which tumour cells develop resistance to apoptosis (Weaver *et al.*, 2002). In this way, the inability of cells to establish integrin-induced cell polarity in monolayers prevents the

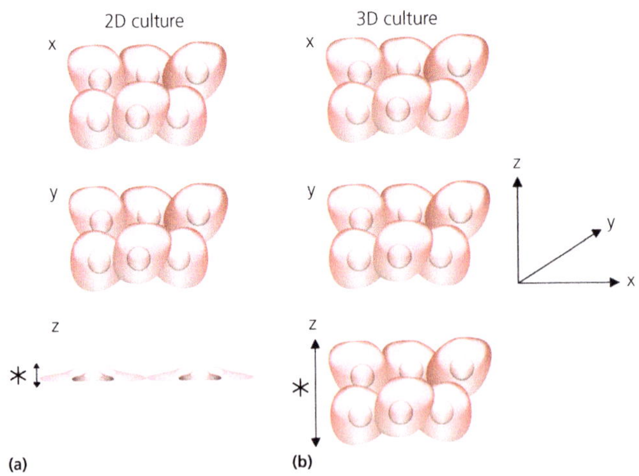

Figure 1.1 Cell flattening. This schematic shows how cells remodel in a flat, 2D environment (a). 3D cell culture (b) ensures cell integrity is preserved maintaining a more physiological shape and form.

study of the mechanisms by which tumour cells can escape extrinsic control. 3D models that support growth in the vertical dimension will be able to recapitulate such mechanisms and study tumourigenesis accordingly. Amongst other features, such models must incorporate the tumour microenvironment, as it has been identified as a key component driving tumour progression (Castelló-Cros & Cukierman, 2009).

Finally, when only 5% of anticancer candidate compounds in preclinical development are licensed after undergoing successful phase III testing (Hutchinson & Kirk, 2011), it is evident that there is an urgent need for more robust and higher quality models to assess these agents. Even if these 3D models delay the time it will take for drugs to reach phase III trials, ultimately it will be a more cost-effective approach to deliver more predictable results.

Mechanotransduction

The patterning of cell-adhesive ligands on more complex substrates and the development of 3D platforms, ranging from solid scaffolds to the manipulation of fluids at a microscale with microfluidics, have become popular avenues of advanced cell culture. These examples corroborate that cell adhesion and structure are two salient features of 3D cell culture (Baker & Chen, 2012). Three-dimensionality, however, has become a generalised statement for all discrepancies between traditional cell culturing systems and newer technology platforms for 3D cell culture. As such, there are other important features of advanced culturing systems, which reside in mechanotransduction and the impact of cells adapting to their surroundings through mechanosensing.

Cells are naturally exposed to mechanical stresses that can influence biological processes such as mitosis, cell migration, stem cell differentiation and self-renewal (Eyckmans *et al.*, 2011). This occurs via adhesion-mediated signalling, which is the mechanism whereby the cells' contractile ability and response to these pressures are transduced into biochemical signals, modifying their behaviour. The machinery behind mechanotransduction involves several cytoskeletal proteins, spanning long distances enabling mechanical continuity and acting as mediators of force transmission (Wang *et al.*, 1993). Whilst intermediate filaments, made of vimentin, keratin and laminin monomers, establish the intracellular structure, actin and myosin form contractile filaments that bind to a cluster of proteins connecting the cytoskeleton to the ECM through transmembrane integrin receptors (Eyckmans *et al.*, 2011). Focal adhesions (FAs) are found amongst this group of proteins and are a key and well-documented unit in cell-ECM adhesion (Kuo, 2014). When force is applied to this unit and cells undergo mechanical deformation, the intracellular structure and organelle positioning in the cell will be disrupted because of the interconnectedness of the cytoskeleton with the cell membrane. Force transmission is also reciprocal; in normal circumstances, cells can also exert forces towards the extracellular space. The continuous polymerisation and depolymerisation of microtubules coupled with the engagement of myosin II pulling actin filaments during contraction creates mechanical forces that are transmitted to focal adhesions (Eyckmans *et al.*, 2011). In turn, this force can remodel the ECM, depending on intracellular activity.

Knowing that these forces are constantly reshaping cells and their exterior, the question then becomes: how do these forces transduce into biochemical signals? One mechanism is through restructuring of the ECM resulting in the exposure of new sites for signalling molecules to engage with or release of growth factors bound to the matrix. Mechanical forces are known to release and activate transforming growth factor (TGF)-β1, which in turn can induce the transdifferentiation of fibroblasts into myofibroblasts and affect developmental processes, wound healing and tumourigenesis (Wipff *et al.*, 2007).

In this way, flat polystyrene or glass substrates for cell culture will inherently lead to the remodelling of cellular architecture (Vergani *et al.*, 2004), providing an inexact representation of native tissue. Along with the flattening of the cell, force transmission through focal adhesions will alter the shape of the cell nucleus, modifying gene expression and therefore protein synthesis (Thomas *et al.*, 2002). Moreover, the rigidity of the substrate where cells reside can enhance cell proliferation but inhibit cell differentiation due to limited cell-cell and cell-matrix interactions (Cukierman *et al.* 2002). To successfully model physiological responses, *in vitro* experiments have to embrace these variables to choose a suitable platform that acknowledges cell integrity, tissue organisation and the impact of mechanotransduction on cell behaviour.

Crosstalk and effector transport

3D cell culture is an enabling technology, bringing the possibility of studying the intricate developmental processes occurring in early embryogenesis as well as the instances when these go awry (Yamada & Cukierman, 2007). The study of branching morphogenesis entails being able to reproduce tubular structures (Fata *et al.*, 2004). Along with structural support, it is necessary to create an information-rich environment with the necessary signalling molecules to facilitate development and differentiation into more elaborate structures. By having a more natural spread of receptors and adhesion molecules distributed across the cell surface, cells not only can achieve this but also engage in a dialogue with neighbouring cells and the supporting stroma (Cukierman *et al.*, 2002). The increase in these interactions enhances intercellular signalling and preserves the transmission of instructive signals for tissue homeostasis.

The ECM is also responsible for laying out the compartments for dispersal of nutrients. It establishes the tissue architecture where gradients of nutrients, oxygen, pH and waste products can manifest. Biological gradients are essential in exerting pressures that can stimulate or inhibit cellular activities, and thus regulate processes such as cell migration and homing (Baker & Chen, 2012). Accurate modelling of the events occurring *in vivo* will have to consider the ECM as a spatial organiser. For example, recapitulating the full picture of cancer means that we need to abandon the reductionist approach of monolayers. To understand how multicellular drug resistance arises, the topography of the tissue or organ needs to be acknowledged. Often, therapeutic agents fail to target all cancerous cells because they lie in inaccessible or deeper areas of the tissue. Similarly, drug resistance can also be attributed to hypoxia, which is why it must be considered as an important factor of the microenvironment when emulating *in vivo* conditions (Asthana & Kisaalita, 2012). Deficient early cell-based models cannot reproduce different oxygen concentrations, and therefore it is no surprise that therapeutic agents slip through the screening, ultimately failing at later stages of the drug development pipeline.

For these reasons and the shortcomings of 2D cell culture, 3D cell technologies seek to fix the discrepancy between the events occurring *in vivo* and the conventional methods used in tissue culture.

Technology platforms for 3D cell culture are predominantly categorised into scaffold-based and scaffold-free systems. Scaffold-based technologies provide physical support in the form of matrices made from natural or synthetic materials to create a suitable microenvironment for optimal cell growth, differentiation and function. Hydrogels are a popular 3D culture method that falls into this category; they work on the same principle of preserving native cellular shape and tissue architecture, enabling a more physiologically relevant function through multiple applications. Conversely, scaffold-free culture systems do not rely on an exogenous input acting as a cellular framework. These technologies encourage the formation of multicellular masses, often referred to as aggregates or spheroids. In this way, spheroids can form their own ECM and then assemble into 3D microtissues.

Finally, these different technologies can be combined in a cleverly designed manner to create another set of platforms for 3D cell culture. Seemingly complex and robust, mini-bioreactors with perfusion flow are mainly concerned with maintaining a constant or controlled supply of biochemical and mechanical cues to improve the quality of engineered tissues.

Aggregate-based technologies

Aggregate-based technologies consist in coaxing cells into forming 3D tissue-like masses or spheroids, by exploiting the biophysical variables acting on the media in which they are grown. Aggregates have the advantage of secreting their own ECM and self-organising into microtissues with multiple cell-cell interactions. Along with self-assembly, other virtues of these technologies include maintaining a consistent spheroid size, not requiring additional materials for culturing, and being compatible with high-throughput screening (HTS). Overall, spheroids offer a simple transition into 3D cell culture and are rapidly becoming an attractive tool in tissue engineering for developing banks of mini-organoids to be used in personalised medicine, drug screening and regenerative therapy (Barker, 2014; van de Wetering *et al.*, 2015; Yui *et al.*, 2012).

Different methods have been developed to generate this type of cell culture for routine use. Cells can be initially cultured in a drop of medium, which is then suspended on the lid of a cell culture dish (Figure 1.2). The lack of surface to attach to encourages cells to aggregate at the apex of these hanging droplets to then form spheroids. Hanging drop plates have multiple concave wells where the cell suspension is distributed, reproducing this phenomenon and maximising the production of tissue-like masses. Moreover, hanging drop plates are covered with a lid that prevents evaporation, maintaining a humid and sterile environment. These suspension cultures are adequate for cells that can proliferate in a non-adhesive environment where aggregation is favoured (Jo & Park, 2000).

An alternative to the hanging drop technique involves using attachment-resistant cell surfaces, which are coated with hydrophilic polymers that inhibit cell adherence (Jo & Park, 2000). This mechanism forces cells to float in the medium, stimulating them to coalesce into spheroids. The surface of the bottom of the plates can also be modified to control spheroid shape. Whilst flat bottoms result in irregular morphology and size, U- or V-shaped surfaces have been optimised to promote formation of single clusters of cells for use in high-throughput studies. Still, there are other constraints that can limit the use of low-adherence substrates. For example, coating the substrate is a time-consuming procedure that can delay the cell seeding procedure. Similarly, the production of 3D spheroids is cell type dependent, which can limit the applications of all scaffold-free

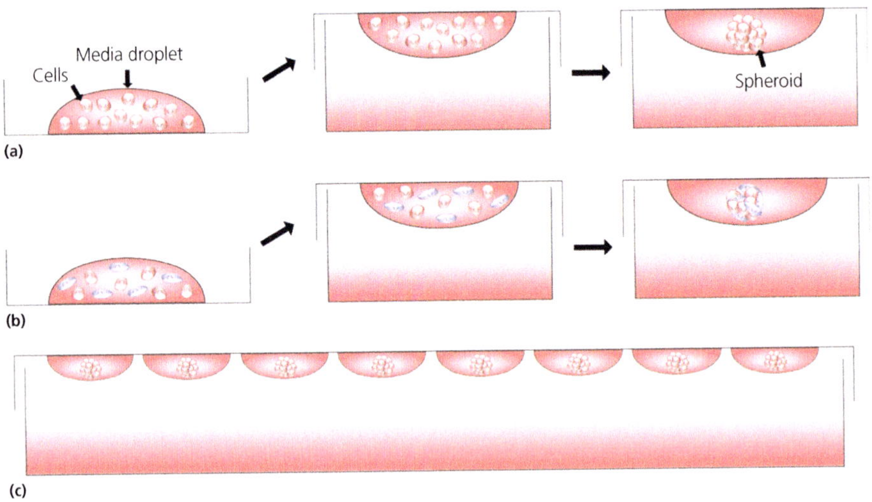

Figure 1.2 3D cell aggregates. (a) Formation of 3D microtissues using the hanging drop technique. Droplets of cell suspension are placed on the lid of a Petri dish, which is gently inverted and placed on top of the dish containing medium to maintain a humid atmosphere. Suspended cells come together in the apex of the droplet, forming a compact 3D aggregate. (b) Co-culture alternative cell types within each technology. (c) Multiple 3D cell aggregates can be produced in a single dish.

3D cell culture technologies. Furthermore, a third technique to encourage spheroid formation by reproducing the native microenvironment of cells focuses on micropatterned surfaces. Microcontact printing methods can engineer surfaces with defined simple tessellations such as square or honeycomb patterns that can also generate spheroids (Yoshii *et al.*, 2011). Although obtaining uniform size is difficult, this method is another interesting prospect in this area of 3D cell culture technologies.

Regardless of how spheroid formation is achieved, these methods make it possible to scale down experiments and work with smaller volumes. This can reduce the cost of exogenous molecules used when studying the influence of growth factors on cellular function, for example. Similarly, suspension cultures are also advantageous since they keep a high local concentration of endogenous factors improving tissue function (Szczepny *et al.*, 2009). In this way, aggregate-based cultures are highly suited for building more realistic models that permit co-culturing methods, where different cell types are grown in the same droplet. Co-culture with other cell types can establish a signal-rich environment which can be used to study the effect of paracrine signalling in real tissue, as well as cellular interactions. Varying configurations of co-cultures allow for adjusting the ratio of cell types to accurately model their native context or merging two spheroids to form a 'Janus spheroid' (Hsiao *et al.*, 2012; Torisawa *et al.*, 2009).

By working in conjunction with stem cell biology, the *in vitro* differentiation of pluripotent stem cells (PSCs) using this type of culture results in cell aggregates referred to as embryoid bodies (EBs). Regarding morphology, these 3D masses can resemble morula-like structures or they may develop into cystic EBs akin to embryos in the blastula stage (Abe *et al.*, 1996). These spheroids have the potential to form tissues from different germ layers within one single EB (Ader & Tanaka, 2014). The lack of available human tissue of this kind and the need for a 3D model to study early developmental processes have favoured EBs as a platform to study organogenesis and test inductive factors and lineage decisions. The size of these 3D masses is known to affect their potential for differentiation (Bratt-Leal *et al.*, 2009) and often the steps in the generation of these aggregates can cause cell loss and size variation. These difficulties have prompted the use of microcarriers to propagate human pluripotent stem cells (hPSCs), avoiding the manual cutting of the monolayer to induce spheroid formation and scaling up the production of evenly sized EBs (Lam *et al.*, 2015).

Spheroids are particularly useful to simulate low nutrient conditions such as hypoxia, but this has also obstructed their use in tissue engineering. This is because, as the size of the spheroid increases, oxygenation becomes problematic due to poor vasculature and thus the centres of these 3D tissue-like masses develop necrosis. Oxygen is known to diffuse across 100–200 μm of tissue thickness (Griffith & Swartz, 2006), which is why if an organoid exceeds these measurements, it can be rendered unviable for implantation. Regenerative medicine has striven to bypass this barrier and engineer larger tissues by maintaining an optimal size. Conversely, there are instances where the risk of hypoxia is in fact welcomed. Low oxygen concentrations are physiologically relevant when modelling embryogenesis and tumour progression. Hypoxia is known to induce angiogenesis and instruct the release of growth factors from the tumour stroma and infiltrating immune cells, which ultimately play a role in tumour development (Cukierman & Yamada, 2007). In this way, the interdependence between microenvironment factors, such as size and oxygen levels, should be included in the experimental design and in the selection of a platform to more closely represent the *in vivo* environment of cells (Ashtana & Kisaalita, 2012).

Spheroid culture has served as an instrument to grow rudimentary structures that imitate the anatomy and physiology of real organs. In recent years, these organoids have garnered significant attention as they offer a wide spectrum of opportunities, from disease models and drug screening tools to grafts with therapeutic potential. Part of their success stems from exploiting cells' biological ability to self-organise into structures of higher complexity. These, however, are not perfect; organoids suffer from batch-to-batch variation, may lack certain cell types or may not fully mimic all stages of organ development. Inclusion of native signalling cues in the culture system has improved the outcomes of these organotypic cultures. For example, culturing intestinal stem cell Lgr5+ cells to induce greater levels of Wnt signalling, Noggin and epidermal growth factor (EGF) signalling has resulted in enhanced intestinal crypt physiology (Sato *et al.* 2009, 2011).

Building on this optimised method to culture intestinal organoids, a biobank was developed from colorectal carcinoma patients as a strategy to delve further into the genetic alterations found in this epithelial cancer (van de Wetering *et al.*, 2015). Aggregate-based technologies offer a uniform and reproducible tool to analyse the genotype-phenotype correlations in intestinal carcinoma. Likewise, these 'miniguts' can be tested against the available anticancer drugs and push forward the case for personalised medicine and cancer genetics. These possibilities exhibit the versatility and vast potential of this platform for 3D cell culture. As with the rest of these innovative systems, their strengths lie in their specific approach to solving the lack of three-dimensionality and restricted portrayal of living tissue through traditional culturing techniques.

Scaffold-based technologies

Ranging from hydrogels and microcarriers to microfluidic surfaces and solid scaffolds, this category encompasses the broadest spectrum of platforms for 3D cell culture. The unifying characteristic is that the platform serves as an artificial matrix that allows cell growth in a new dimension in order to escape the geometrical limitations of monolayer cultures. Based on this principle, different technologies have been developed to satisfy various niches in biological research.

Hydrogels

Hydrogels are moderately different from solid scaffolds. A first evident distinction is in the strength of the physical support they give. Hydrogels are loose scaffolds consisting of crosslinked natural or synthetic materials for cell encapsulation (Figure 1.3). In this way, these superabsorbent matrices are better suited to modelling soft tissue because of their tissue-like flexibility and viscoelasticity (Tibbitt & Anseth, 2009). Recreating the stem cell niche with only hyaluronic acid as a matrix supporting the growth of hESCs does not reflect the natural complexity of the ECM (Gerecht *et al.*, 2007). The addition of specific proteins, however, can significantly influence the differentiation of hESCs in 3D models. Coating hydrogels with ECM molecules enables the cultured cells to engage in the cross-talk of *in vivo*-like cues. Success with these platforms hinges on a combination of signalling via chemical and molecular pathways and biomechanical properties.

In a similar fashion to scaffolds, these gels can have a porous architecture facilitating the mass transfer of drugs, nutrients and oxygen to reach all areas. The idea of the choice of material determining the application and benefits of the technology is also very present in these platforms. Hydrogels can be derived from a wide variety of sources that in turn affect their compatibility and properties. For example, animal-derived hydrogels mainly use collagen, which is the most abundant protein in the ECM, making it a natural biological ligand for integrin attachment. Matrigel® is a popular commercially available hydrogel composed of tumour extract derived from Engelbreth-Holm-Swarm (EHS)

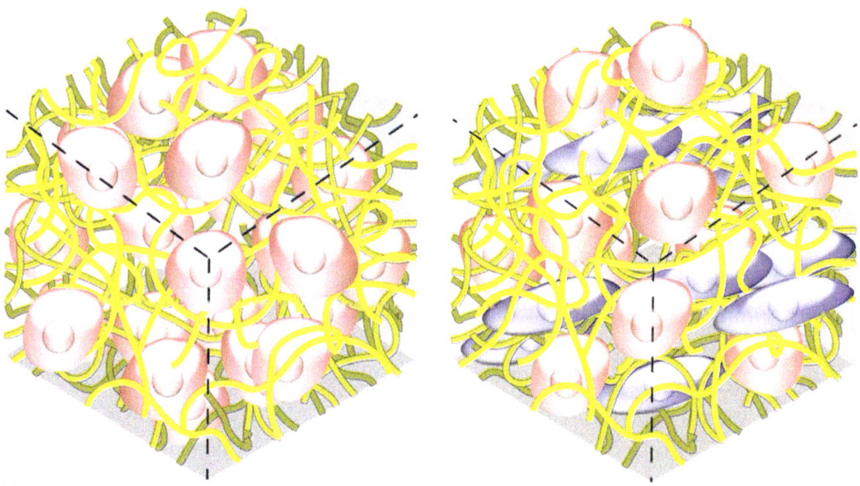

(a) (b)

Figure 1.3 Hydrogels. 3D culture using hydrogel technology. (a) The cartoon shows cells within a matrix of protein molecules that create a nano-scale microenvironment mimicking the structure of the extracellular matrix. Cells are embedded within the proteinaceous 3D framework within an aqueous-based gel. (b) Co-culture of alternative cell types using hydrogel technology.

mouse sarcoma. It is known to contain various growth factors, a rich protein mix including collagen IV, laminin and enactin and other undefined constituents (Vukicevic *et al.*, 1992). Though these may result in batch-to-batch variation, hindering reproducibility, Matrigel can promote cellular functions that would otherwise remain unseen by providing the necessary endogenous factors (Benton *et al.*, 2014). Moreover, plant-derived hydrogels have been developed by crosslinking alginate monomers (Zimmermann *et al.*, 2007). Despite having no adulteration with animal proteins, they cannot escape from biological variation, rendering them unviable for HTS. Synthetic hydrogels solve these issues by using inert materials, whilst still being able to be supplemented with bioactive molecules to enhance their use.

Exploiting synthetic hydrogels with careful manipulation of their properties has resulted in the creation of injectable hydrogels. By controlling the gelation time and degradation of these materials, synthetic hydrogels can be utilised as a delivery mechanism of cultured cells to sites that would otherwise require an invasive procedure (Temenoff & Mikos, 2000). Other practical uses have seen these matrices employed to investigate developmental processes such as branching and vascular morphogenesis (Lo *et al.*, 2012).

Microcarriers are another system that can incorporate hydrogels that use these matrices as the basis to build microscopic spheres (90–500 μm in diameter) for culturing entrapped cells in 3D. With a high surface area to volume ratio, this technology also allows the culturing of anchorage-dependent cells (van Wezel, 1967). The main application of this system has been as a high-yield culture for the production of biologics in industry (Wu *et al.*, 2004). These spheres usually have a magnetic core, allowing control during media changes (Justice *et al.*, 2009).

Overall, synthetic hydrogels have vast potential as a culturing technique that can have research, therapy and industrial applications. Notwithstanding, the general obstacles faced by hydrogels include short culture periods due to diffusion of nutrients across the hydrogel. Also, using ultraviolet (UV) light to cure the gel is believed to be damaging to cells (Nicodemus & Bryant, 2008).

Solid scaffolds

Solid scaffolds were originally devised for transplantation applications in wound healing. Seeding cells in biodegradable scaffolds enabled creation of 3D cellular structures, which could be incorporated into living tissue where the exogenous framework would eventually degrade and be replaced by healthy natural tissue. In recent years, however, there has been a growing interest in introducing *in vitro* scaffolds for routine use in cell culture.

The materials used in the fabrication process are very important in shaping the purpose of the scaffold-based technology. Components of the native ECM including collagen, fibrin and hyaluronic acid (HA) (Gerecht *et al.*, 2007; Matsiko *et al.*, 2012) have been used effectively to create 3D matrices to support cell growth. These constituents have the benefit of being biocompatible and possessing readily available adhesion sites that can increase the complexity of the tissue.

Decellularised scaffolds are an example of a natural matrix where the native composition and architecture of tissue are fully preserved. Organs and tissue sections can undergo physical, chemical and enzymatic treatment to remove all cellular antigens whilst still preserving the ECM (Song & Ott, 2011). The preparation of such scaffolds can involve ionic detergents, which circumvents the problems of enzymatic treatment and collagen degradation (Gilbert *et al.*, 2006). Ensuring collagen remains intact also conserves its bioactive sites, facilitating the culture of cells in this decellularised matrix. Depending on the purposes of this type of scaffold, the recovery and processing techniques will vary to achieve optimum use.

Similarly, scaffolds can be produced from naturally derived materials such as alginate, silk and gelatin (Zimmermann *et al.*, 2007). Both decellularised matrices and these types of scaffold share the advantage of being biodegradable, which makes them suitable for growing grafts or to lay the foundations for new functioning cells to repopulate damaged tissue.

Despite being beneficial in the context of tissue engineering, working with biological materials in the laboratory affects consistency. A partial solution has been to use biodegradable polymers such as polyglycolic acid, polylactic acid and their co-polymer polylactic-*co*-glycolic acid (Mikos *et al.*, 1993). This is not ideal because their degradation results in the release of unwanted by-products that can alter cell behaviour. The build-up of lactic acid, for example, is known to cause suboptimal culturing conditions for embryonic stem cells (ESCs), decreasing pluripotency markers and inducing spontaneous differentiation (Ouyang *et al.*, 2006). The added variability coupled with short shelf-life and problematic storage make biodegradable materials unsuitable for standard use in 3D cell culture. In light of these shortcomings, synthetic scaffolds with defined composition have risen as a more consistent alternative. Inert and non-degradable materials such as polymers, titanium and ceramic-based platforms can be carefully tweaked to capture the cellular niche, creating scaffolds suitable for cell culture (Boccaccini & Blaker, 2005; van den Dolder *et al.* 2003).

The methodology behind the making of these matrices separates them into two categories: fibrous and porous scaffolds. One example of how fibrous scaffolds are manufactured is through electrospinning, a technique by which polymer jets are passed through an electric field (Reneker & Chun, 1996). The electrospun fibres that accumulate in the collector plate are then used to form interlaced structures or aligned patterns in which cell positioning can be regulated. This technique is highly flexible, allowing a variety of substances, from biologically active to synthetic polymers, to be used as jetting materials. In fact, it is possible to use two or more materials to produce heterogeneous scaffolds (Yang *et al.*, 2005).

Porous scaffolds, on the other hand, make available a controllable 3D space where cells can enter and grow, forming contacts and interactions with adjacent cells (Figure 1.4). The dimensions of the voids are known to affect cell seeding as well as how cells behave and grow in the scaffold (Knight & Przyborski, 2014).

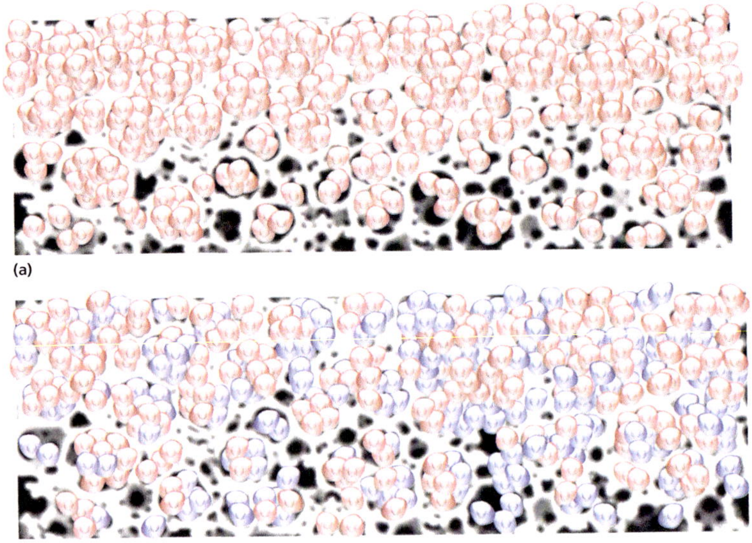

(a)

(b)

Figure 1.4 **Solid scaffolds.** (a) Porous solid scaffold supporting 3D cultured cells. Cells enter the porous framework of the solid scaffold where they do not flatten, they maintain their 3D structure and they bind to one another forming 3D tissue-like masses. (b) Co-culture of alternative cell types using scaffold technology.

Voids are interconnected by small pores, which prevent cells from being isolated within the 3D microenvironment and also contribute to greater cell infiltration. Achieving pore formation is an elaborate process that can be carried out through different techniques. Particulate leaching is a physical process in which a polymer is cast around soluble beads known as porogens (Reignier & Huneault, 2006). Popular porogens include sugar, salt and paraffin wax. Although this method has the advantage of tight control over pore size, it has limited connectivity amongst these spaces, which may result in heterogeneous cultures. An alternative procedure ensuring greater interconnectivity through multiple pores is emulsion templating. This method for fabricating solid scaffolds incorporates polymerisation by high internal phase emulsion (HIPE). This biphasic emulsion consists of an aqueous and a non-aqueous monomer/surfactant phase, which results in a highly porous scaffold linked by interconnecting pores (Barbetta *et al.*, 2000). A third method is gas foaming technology, which can generate large internal volume and 3D spaces by agitating polymers to create foam. Phase separation ensues from these conditions, causing the dissolved gas to split from the polymer. The free gas molecules then join to reduce free energy forming clusters and in turn leave porous structures, suitable for cell growth (Harris, 1998). Regulating the agitation and the use of high-pressure gases facilitates control of the porosity of the scaffold, although the low pore interconnectivity still remains a problem (Salerno *et al.*, 2009).

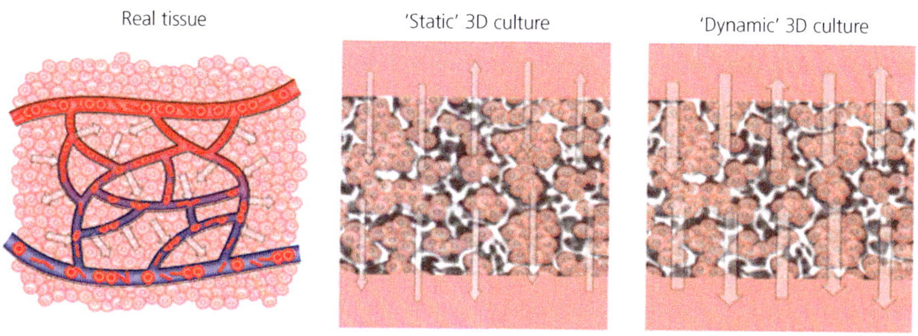

Figure 1.5 Perfusion model. Unlike real tissue, 3D cell culture models lack a vascular and capillary bed. Exchange of gases, nutrients and waste products occurs by diffusion, most often in a static culture where unstirred layers can build up in stagnant media. Dynamic 3D culture involves perfusion and movement of the media to reduce unstirred layers and increase exchange..

The lack of biological activity and natural cell adhesion sites can be overcome by coating these substrates with ECM proteins such as laminin and fibronectin (Knight & Przyborski, 2014). Despite providing physical support in the form of 3D spaces where cells can proliferate, these voids have poor mass transfer since these cultures are static systems. For these reasons, scaffolds are usually engineered as thin membranes (200 μm) that permit sufficient exchange of nutrients and waste products. This in turn enriches the physiological accuracy of these models, allowing researchers to study *in vivo* phenomena in a controlled *in vitro* setting.

3D bioreactors

Perfusion flow culturing systems can be identified as another division in 3D cell culture. These systems focus on replicating continuous circulation of nutrients and waste in cells and tissues (Figure 1.5). In addition, microfluidic culture systems and 3D bioreactors serve to model dynamic biological processes and the consequences of *in vivo* forces such as shear stress and fluid turbulence. Pulsating blood flow, for example, causes a mechanical stretch on endothelial and smooth muscle cells, which in turn can trigger cell signalling pathways, altering their behaviour (Tzima *et al.*, 2005). Similarly, flow rates are known to favour certain developmental decisions such as an arterial versus a venous phenotype in vasculogenesis (Le Noble *et al.*, 2004).

Regarding the culture type, microfluidics are often recognised as a scaffold-based platform in the literature whereas 3D bioreactors would largely fall under scaffold-free technologies since they generally produce suspension cultures aided by a constant agitation maintaining cells in suspension. This classification overlooks the fact that the concept behind these platforms can be implemented on both scaffold-based and scaffold-free technologies. Therefore, it is also possible to consider them as a separate category of 3D culture systems, borrowing aspects from both, and thus exemplifying how this field of science is in fact multidisciplinary.

Directional flow type technologies that primarily involve pumping of media over cultured cells can be performed on a variety of scales, including large bioreactors composed of complex tubing arrangements, smaller scale bench-top versions and micro-scale fluid control devices. Microfluidics consist of the engineered manipulation of fluids at a micro-scale (Sackmann *et al.*, 2014). These lab-on-a-chip microtechnologies exploit fluid behaviour at the submillimetre scale because the rules controlling forces such as laminar versus turbulent flow, surface tension and capillary forces are vastly different compared to the macro scale (Sackmann *et al.*, 2014). The fabrication of these intricate systems uses processes such as microcontact printing, photolithography and replica moulding (Ito *et al.*, 1997; Park & Shuler, 2003; Sun *et al.*, 2012). The mechanism behind this culturing system involves an array of pillars that support the growth of cells. These micropillars also immobilise cells, preventing fluids from displacing them and ensuring a controlled transient or continuous flow of media circulating through the culture system. Passing a collagen matrix creates a thin layer surrounding the cell, which establishes cell-matrix interactions, introducing more complexity to the model. These upgrades contribute to an inexpensive and efficient model for drug screening, compatible with automation where single cell manipulation is feasible. The potential to quantitatively and qualitatively examine the impact of fluid forces acting on cells, while minimising reagent volume, makes this platform very attractive in medical and biological research.

Also referred to as agitation-based approaches, these culturing systems' principal purpose is to recreate biophysical cues experienced by cells in live tissue. By ensuring constant movement in the culture system, cells are prevented from adhering to the walls of the container and instead they are encouraged to form cell-cell interactions (Breslin & O'Driscoll, 2013). These systems, such as rotational and spinner flask bioreactors, form spheroids by continuous rotation or stirring, respectively. Constant motion and perfusion flow allow for transfer of nutrients and waste to and from the suspension culture. These bioreactors are well equipped for large production and long-term maintenance of cell aggregates, aided by easy media changes to suit these purposes (Rodday *et al.*, 2011). Disadvantages of bioreactors include larger media requirements since the culture system operates with greater volumes. Whilst the shear force can affect cell physiology (Lin & Chang, 2008), it can also exert pressures that are constantly occurring in an *in vivo* context. Moreover, size variation and poor uniformity in morphology are problems that can be addressed by combining this approach with other aggregate-based technologies more suitable for culturing multiple spheroids with consistent dimensions. In this way, suspension cultures can be initially generated through this technique but later transferred to rotational culture systems or spinner flask bioreactors. The enhanced environment of bioreactors will complement the model and facilitate long-term culturing. However, using these culture systems for drug screening would require replating spheroids into suitable plates that can ensure one spheroid per well and maintain a uniform size for HTS (Breslin & O'Driscoll, 2013).

Barriers to adoption and future directions

Amongst the obstacles discussed in each section, this rapidly growing multidisciplinary field faces the difficulty of trying to incorporate the various biological parameters into one single platform. The ideal 3D cell culture platform is imagined as a system comprising multiple cell types in a chamber that recreates the *in vivo* forces acting on cells. These would include structure and surface modifications, cellular interactions between adjacent cells and the ECM, mechanical and fluid flow forces. The problem, however, is that tissues are widely diverse and have a variable set of needs, which obstructs the efforts of designing an all-encompassing technology that meets every biological requirement. Experimental models need to show the different facets of the same tissue in a reproducible, measurable and reliable manner and in certain cases this is not easy, straightforward or possible. For these reasons, scientists have moved away from this approach. Rather than a panacea for culturing cells *in vitro*, this is a matter of utilising various aspects of 3D cell technology, depending on the biological question to be explored.

An anatomical or histological view of disease would argue that pathological conditions normally reflect an alteration of the tissue organisation or an insult to the cellular structure. In order to fully grasp the progression from one state into another, it is necessary to replicate such modifications. At times, mimicking the *in vivo* forces influencing cell behaviour can be conflicting. For example, static cultures may be adequate to establish gradients that in turn can allow for a close study of avascular tissues, such as tumours. On the other hand, lack of perfusion flow and circulation of nutrients makes it difficult to build 3D systems with vascularised tissue. This example illustrates the issues researchers encounter and the importance of weighing these factors when planning an experiment to address their proposed biological inquiries. Commercialised 3D cell culture technologies provide an array of accessible solutions that are flexible and easily adaptable to different experimental set-ups. These platforms can enhance cell-based models by bridging the gap between traditional monolayer cultures and animal models, ultimately expanding our understanding of cellular biology. By working in concert with other modern resources in cell biology, such as human stem cells, these technologies can create robust tissue mimetics by introducing a 3D component that enables cell differentiation into more complex structures. Creating human tissue *in vitro* offers exciting possibilities for advancing drug discovery with an enhanced predictive accuracy of drug candidate compounds, as well as furthering regenerative medicine. The success of platforms for 3D cell culture technology explored hereafter will depend on overcoming the barriers to adoption and validating their potential for routine use.

3D culture systems face various challenges before full adoption becomes a reality. Even when their worth is undeniable, conclusions from one system may not be true for another. Whilst using synthetic materials and platforms compatible with HTS can mitigate the problem of variable results, these methods may fail to mimic some characteristic of an *in vivo* setting or be unsuited to carrying out downstream analyses. These drawbacks hint at a wider issue in 3D cell culture. There are multiple answers to the question of reproducing a more *in vivo*-like setting. These technologies will invariably impact cell culture in different ways, making it difficult to see the path towards advanced cell culture as a single step to improve the biological relevance of cell-based assays. For these reasons, culturing systems are shifting their focus towards investigating the interdependence of factors known to enrich the representation of physiological phenomena. This, however, is often prohibited by the nature of the 3D culturing system that cannot be subjected to other conditions, such as oxygen concentrations and mechanical forces. Even though these platforms may lack flexibility, 3D cell culture will innovate and seek to examine the synergy of microenvironment factors, offering a higher degree of complexity and moving a step closer to reflecting true physiological behaviour.

References

Abe, K., Niwa, H., Iwase, K. *et al.* (1996) Endoderm-specific gene expression in embryonic stem cells differentiated to embryoid bodies. *Experimental Cell Research*, **229**, 27–34.

Ader, M. & Tanaka, E. M. (2014) Modeling human development in 3D culture. *Current Opinion in Cell Biology*, **31**, 23–28.

Ashtana, A. & Kisaalita, W. S. (2012) Microtissue size and hypoxia in HTS with 3D cultures. *Drug Discovery Today*, **17**, 810–817.

Baker, B. M. & Chen, C. S. (2012) Deconstructing the third dimension – how 3D culture microenvironments alter cellular cues. *Journal of Cell Science*, **125**, 3015–3024.

Barbetta, A., Cameron, N. R. & Cooper, S. J. (2000) High internal phase emulsions (HIPEs) containing divinylbenzene and 4-vinylbenzyl chloride and the morphology of the resulting PolyHIPE materials. *Chemical Communications*, **3**, 221–222.

Barker, N. (2014) Adult intestinal stem cells: critical drivers of epithelial homeostasis and regeneration. *Nature Reviews Molecular Cell Biology*, **15**, 19–33.

Benton, G., Arnaoutova, I., George, J. *et al.* (2014) Matrigel: from discovery and ECM mimicry to assays and models for cancer research. *Advanced Drug Delivery Reviews*, **79–80**, 3–18.

Bhadriraju, K. & Chen, C. S. (2002) Engineering cellular microenvironments to improve cell-based drug testing. *Drug Discovery Today*, **7**, 612–620.

Boccaccini, A. R. & Blaker, J. J. (2005) Bioactive composite materials for tissue engineering scaffolds. *Expert Review of Medical Devices*, **2**, 303–317.

Bratt-Leal, A. M., Capenedo, R. L. & McDevitt, T. C. (2009) Engineering the embryoid body microenvironment to direct embryonic stem cell differentiation. *Biotechnology Progress*, **25**, 43–51.

Breslin, S. & O'Driscoll, L. (2013) Three-dimensional cell culture: the missing link in drug discovery. *Drug Discovery Today*, **18**, 240–249.

Castelló-Cros, R. & Cukierman, E. (2009) Stromagenesis during tumorigenesis. *Methods in Molecular Biology*, **522**, 275–305.

Cukierman, E., Pankov, R. & Yamada, K. M. (2002) Cell interactions with three-dimensional matrices. *Current Opinion in Cell Biology*, **14**, 633–640.

Eyckmans, J., Boudou, T., Yu, X. & Chen, C. S. (2011) A hitchhiker's guide to mechanobiology. *Developmental Cell*, **21**, 35–47.

Fata, J. E., Werb, Z. & Bissell, M. J. (2004) Regulation of mammary gland branching morphogenesis by the extracellular matrix and its remodeling enzymes. *Breast Cancer Research*, **6**, 1–11.

Gerecht S., Burdick J. A., Ferreira L. S. *et al.* (2007) Hyaluronic acid hydrogel for controlled self-renewal and differentiation of human embryonic stem cells. *Proceedings of the National Academy of Sciences USA*, **104**, 11298–11303.

Gilbert, T. W., Sellaro, T. L. & Badylak, S. F. (2006) Decellularization of tissue and organs. *Biomaterials*, **27**, 3675–3683.

Griffith, L. G. & Swartz, M. A. (2006) Capturing complex 3D tissue physiology in vitro. *Nature Reviews Molecular Cell Biology*, **7**, 211–224.

Harris, L. D., Kim, B.-Y. & Mooney, D. J. (1998) Open pore biodegradable matrices formed with gas foaming. *Journal of Biomedical Materials Research*, **42**, 396–402.

Hsiao, A. Y., Tung, Y.-C., Qu, X. *et al.* (2012) 384 Hanging drop arrays give excellent Z-factors and allow versatile formation of co-culture spheroids. *Biotechnology and Bioengineering*, **109**, 1293–1304.

Hutchinson, L. & Kirk, R. (2011) High drug attrition rates – where are we going wrong? *Nature Reviews Clinical Oncology*, **8**, 189–190.

Ito, Y., Chen, G., Guan, Y. & Imanishi, Y. (1997) Patterned immobilization of thermoresponsive polymer. *Langmuir*, **3**, 2756–2759.

Jo, S. & Park, K. (2000) Surface modification using silanated poly(ethylene glycol)s. *Biomaterials*, **21**, 605–615.

Justice, B. A., Badr, N. A. & Felder, R. A. (2009) 3D cell culture opens new dimensions in cell-based assays. *Drug Discovery Today*, **14**, 102–107.

Knight, E. & Przyborski, S. A. (2014) Advances in 3D cell culture technologies enabling tissue-like structures to be created in vitro. *Journal of Anatomy*, **227**(6), 746–756.

Kuo, J.-C. (2014) Focal adhesions function as a mechanosensor, in *Mechanotransduction* (eds A. J. Engler & S. Kumar), Elsevier Academic Press, Amsterdam, pp. 55–73.

Lam, A. T., Chen, A. K., Ting, S. Q. *et al.* (2015) Integrated processes for expansion and differentiation of human pluripotent stem cells in suspended microcarriers cultures. *Biochemical and Biophysical Research Communications*, **473**(3), 764–768. Le Noble, F., Moyon, D., Pardanaud, L. *et al.* (2004) Flow regulates arterial-venous differentiation in the chick embryo yolk sac. *Development*, **131**, 361–375.

Lin, R.-Z. & Chang, H.-Y. (2008) Recent advances in three-dimensional multicellular spheroid culture for biomedical research. *Biotechnology Journal*, **3**, 1172–1184.

Lo, A. T., Mori, H., Mott, J. & Bissell, M. J. (2012) Constructing three-dimensional models to study mammary gland branching morphogenesis and functional differentiation. *Journal of Mammary Gland Biology and Neoplasia*, **17**, 103–110.

Matsiko, A., Levingstone, T. J., O'Brien, F. J. & Gleeson, J. P. (2012) Addition of hyaluronic acid improves cellular infiltration and promotes early-stage chondrogenesis in a collagen-based scaffold for cartilage tissue engineering. *Journal of the Mechanical Behavior of Biomedical Materials*, **11**, 41–52.

Mikos, A. G., Sarakinos, G., Leite, S. M. *et al.* (1993) Laminated three-dimensional biodegradable foams for use in tissue engineering. *Biomaterials*, **14**, 323–330.

Nicodemus, G. D. & Bryant, S. J. (2008) Cell encapsulation in biodegradable hydrogels for tissue engineering applications. *Tissue Engineering.Part B, Reviews*, **14**, 149–165.

Ouyang, A., Robin, N. & Yang, S.-T. (2006) Long-term culturing of undifferentiated embryonic stem cells in conditioned media and three-dimensional fibrous matrices without extracellular matrix coating. *Stem Cells*, **25**, 447–454.

Park, H. T. & Shuler, M. L. (2003) Integration of cell culture and microfabrication technology. *Biotechnology Progress*, **19**, 243–253.

Reignier, J. & Huneault, M. A. (2006) Preparation of interconnected poly(3-caprolactone) porous scaffolds by a combination of polymer and salt particulate leaching. *Polymer*, **47**, 4703–4717.

Reneker, D.H. & Chun, I. (1996) Nanometre diameter fibres of polymer, produced by electrospinning. *Nanotechnology*, **7**, 216–223.

Rodday, B., Hirschhaeuser, F., Walenta, S. & Mueller-Klieser, W. (2011) Semiautomatic growth analysis of multicellular tumor spheroids. *Journal of Biomolecular Screening*, **16**, 1119–1124.

Sackmann, E. K., Fulton, A. L. & Beebe, D. J. (2014) The present and future role of microfluidics in biomedical research. *Nature*, **507**, 181–189.

Salerno, A., Oliviero, M., Di Maio, E. *et al.* (2009) Design of porous polymeric scaffolds by gas foaming of heterogeneous blends. *Journal of Material Sciences: Materials in Medicine*, **20**, 2043–2051.

Sato, T., Vries, R. G., Snippert, H. J. *et al.* (2009) Single Lgr5 stem cells build crypt-villus structures in vitro without a mesenchymal niche. *Nature*, **459**, 262–265.

Sato, T., Stange, D. E., Ferrante, M. *et al.* (2011) Long-term expansion of epithelial organoids from human colon, adenoma, adenocarcinoma, and Barrett's epithelium. *Gastroenterology*, **141**, 1762–1772.

Song, J. J. & Ott, H. C. (2011) Organ engineering based on decellularized matrix scaffolds. *Trends in Molecular Medicine*, **17**, 424–432.

Sun, T., Jackson, S., Haycock, J. W. & MacNeil, S. (2006) Culture of skin cells in 3D rather than 2D improves their ability to survive exposure to cytotoxic agents. *Journal of Biotechnology*, **122**, 372–381.

Sun, Y., Weng, S. & Fu, J. (2012) Microengineered synthetic cellular microenvironment for stem cells. *Wiley Interdisciplinary Reviews. Nanomedicine and Nanobiotechnology*, **4**, 414–427.

Szczepny, A., Hogarth, C. A., Young, J. & Loveland, K. L. (2009) Identification of hedgehog signaling outcomes in mouse testis development using a hanging drop-culture system. *Biology of Reproduction*, **80**, 258–263.

Temenoff, J. S. & Mikos, A. G. (2000) Injectable biodegradable materials for orthopaedic tissue engineering. *Biomaterials*, **21**, 2405–2412.

Thomas, C. H., Collier, J. H., Sfeir, C. S. & Healy, K. E. (2002) Engineering gene expression and protein synthesis by modulation of nuclear shape. *Proceedings of the National Academy of Sciences USA*, **99**, 1972–1977.

Tibbitt, M. W. & Anseth, K. S. (2009) Hydrogels as extracellular matrix mimics for 3D cell culture. *Biotechnology and Bioengineering*, **103**, 655–663.

Torisawa, Y.-S., Mosadegh, B., Luker, G. D. *et al.* (2009) Microfluidic hydrodynamic cellular patterning for systematic formation of co-culture spheroids. *Integrative Biology*, **1**, 649–645.

Tzima, E., Irani-Tehrani, M., Kiosses, W. B. *et al.* (2005) A mechanosensory complex that mediates the endothelial cell response to fluid shear stress. *Nature*, **437**, 426–431.

Van den Dolder, J., Spauwen, P. H. M. & Jansen, J. A. (2003) Evaluation of various seeding techniques for culturing osteogenic cells on titanium fiber mesh. *Tissue Engineering*, **9**, 315–325.

Van de Wetering, M., Francies, H. E., Francis, J. M. *et al.* (2015) Prospective derivation of a living organoid biobank of colorectal cancer patients. *Cell*, **161**, 933–945.

Van Wezel, A. L. (1967) Growth of cell-strains and primary cells on micro-carriers in homogeneous culture. *Nature*, **216**, 64–65.

Vergani, L., Grattarola, M. & Nicolini, C. (2004) Modifications of chromatin structure and gene expression following induced alterations of cellular shape. *International Journal of Biochemistry & Cell Biology*, **36**, 1447–1461.

Vukicevic, S., Kleinman, H. K., Luyten, F. P. *et al.* (1992) Identification of multiple active growth factors in basement membrane matrigel suggests caution in interpretation of cellular activity related to extracellular matrix components. *Experimental Cell Research*, **202**, 1–8.

Wang, N., Butler, J. P. & Ingber, D. E. (1993) Mechanotransduction across the cell surface and through the cytoskeleton. *Science*, **260**, 1124–1127.

Weaver, V. M., Lelièvre, S., Lakins, J. N. *et al.* (2002) β4 integrin-dependent formation of polarized three-dimensional architecture confers resistance to apoptosis in normal and malignant mammary epithelium. *Cancer Cell*, **2**, 205–216.

Wipff, P. J., Rifkin, D. B., Meister, J. J. & Hinz, B. (2007) Myofibroblast contraction activates latent TGF-beta1 from the extracellular matrix. *Journal of Cell Biology*, **176**, 1311–1323.

Wu, S.-C., Liu, C.-C. & Lian, W.-C. (2004) Optimization of microcarrier cell culture process for the inactivated enterovirus type 71 vaccine development. *Vaccine*, **22**, 3858–3864.

Yamada, K.M. & Cukierman, E. (2007) Modelling tissue morphogenesis and cancer in 3D. *Cell*, **130**, 601–610.

Yang, F., Maurugan, R., Wang, S. & Ramakrishna, S. (2005) Electrospinning of nano/micro scale poly (L-lactic acid) aligned fibers and their potential in neural tissue engineering. *Biomaterials*, **26**, 2603–2610.

Yoshii, Y., Waki, A., Yoshida, K. *et al.* (2011) The use of nanoimprinted scaffolds as 3D culture models to facilitate spontaneous tumor cell migration and well-regulated spheroid formation. *Biomaterials*, **32**, 6052–6058.

Yui, S., Nakamura, T., Sato, T. *et al.* (2012) Functional engraftment of colon epithelium expanded in vitro from a single adult Lgr5+ stem cell. *Nature Medicine*, **18**, 618–623.

Zimmermann, H., Shirley, S. G. & Zimmermann, U. (2007) Alginate-based encapsulation of cells: past, present, and future. *Current Diabetes Report*, **7**, 314–320.

PART I
Aggregate-based technologies

CHAPTER 2

Gravity-enforced microtissue engineering

Randy Strube, Johannes Haugstetter, Markus Furter, Andreia Fernandez, David Fluri and Jens M. Kelm

InSphero AG, Schlieren, Switzerland

Introduction

The hanging drop technology

The hanging drop (HD) technology is used to culture cells and tissues originated from embryology as a means to investigate basic developmental processes that would be otherwise restricted to growing embryonic stem cells on flat culture surfaces (Keller, 1995). The classic hanging drop culture comprises a small droplet of cell culture medium placed on a sterile surface such as the lid of a bacterial dish which is then inverted to generate the hanging droplet. The hanging drop is stabilised by its surface tension, preventing dispersion across the surface (Kelm *et al.*, 2003). This allows contact-free tissue culture as well as the reassembly of microtissues or embryoid bodies without the use of supporting materials.

The inventor of the HD method was Ross Granville Harrison who adapted HD technology to grow frog neuronal tissue. Utilising this technology, he was the first to observe the development of growth cones in developing neurons and provided the basis for the discovery of nerve growth factor (NGF) by Rita Levi-Montalcini, for which she received the 1986 Nobel Prize for Physiology or Medicine (Levi-Montalcini, 1964). Currently, HD technology is not only extensively used in stem cell biology and growing whole embryos, but it has become one of the major technologies used to reassemble spherical tissue structures (Kelm *et al.*, 2005). The ability to resolve three-dimensional (3D) structures, endogenous extracellular matrix and cell-cell contacts is an important advance that made the hanging drop a widely used tissue culture method. However, besides improving the cell's biology, a decisive step towards industrial application is the development of automation-compatible technologies enabling assay automation with a high throughput. Moreover, the lack of any supporting materials allows compatibility with a vast variety of currently established endpoints.

Technology Platforms for 3D Cell Culture: A User's Guide, First Edition. Edited by Stefan Przyborski.

An intriguing characteristic of self-assembled cell populations is their capacity to self-organise into tissue-like structures when multiple cell types are mixed together. For example, it was shown that mixing human umbilical vein endothelial cells (HUVECs) together with arterial-derived fibroblasts always resulted in a peripheral endothelial layer creating a barrier between connective tissue and the surrounding medium (Kelm *et al.*, 2005). Even more astonishing was the self-sorting of sensory neurons in combination with connective fibroblasts. Sensory neurons derived from either embryonic mouse or rat dorsal root ganglia (DRG) segregated and accumulated over seven days at the periphery of a connective microtissue, generating DRG-like morphology including an organised outgrowth of the axons. Even Schwann cells were observed aligning the axons initiating myelination (Kelm *et al.*, 2006). Recently, a whole-brain model 4 mm in diameter was generated applying the concept of cellular self-assembly reflecting in large part the development of the brain (Lancaster *et al.*, 2013).

Industrial implementation of HD technology has been impeded by its highly manual and cumbersome nature. Therefore, InSphero developed a HD system enabling automation-compatible production in hanging drops (GravityPLUS™, see product description) and assaying them in specially designed assay plates (GravityTRAP™, see product description). The standard multiwell design of the GravityPLUS technology in combination with the SureDrop™ inlet enables seamless integration into an automated production process. The microcapillary connecting the inlet with the cell culture compartment allows for regular medium exchange of up to 70% and readdition of either cells or substances. An automated production process using a 96-channel pipette enables fast dispersal throughout the plate, leading to a homogeneous distribution of cells. As a result, size variation as measured by diameter can be kept below 5% using dispersed cell lines (Drewitz *et al.*, 2011). The HD production plate is complemented by the microtissue assay plate to allow for high compatibility with analytical and liquid handling instrumentation.

Microtissue models for efficacy testing

Drug sensitivity is not only a result of the genetics but is rather altered by the microenvironment as highlighted by Arno Ostmann (Ostmann, 2012). Therefore, in oncology the multicellular tumour spheroid (MCTS) model has become the standard 3D model to test antitumour drugs with more organotypic models. Multicellular resistance (MCR) towards anticancer drugs and radiation was one of the early observed differences in the response of cells grown either in monolayer or spheroid format, exemplifying that a 3D environment can have a severe impact on drug sensitivity. For example, vinblastine exhibited an IC_{50} value of 0.008 µmol/L in monolayer cultures and 53 µmol/L in spheroid cultures of A549 cells (Desoize & Jardillier 2000a,b). One major determinant which impacts drug sensitivity is hypoxia, which occurs in areas of the tumour which are distant from the capillary network. Low oxygen partial pressure within the tumour can lead to either higher sensitivity or resistance, depending on the drug mechanism and structure (Thoma *et al.*, 2014). Extracellular acidification is another example which affects the competence of basic (e.g. doxorubicin) and acidic (e.g. chlorambucil) drugs. Whereas the uptake of basic drugs is decreased (higher resistance), the uptake of acidic drugs is increased (higher sensitivity) (Hall *et al.*, 2004; Wilson *et al.*, 2011).

The HD method is a very versatile production tool that can generate a wide variety of tumour microtissue models from either single cell type (Drewitz *et al.*, 2011; Kelm *et al.*, 2003, 2005) or multicell type models (Amann *et al.*, 2014). Interestingly, certain fibroblast cell lines exhibit a capacity to transform their phenotype in cancer cell line co-cultures towards cancer-associated fibroblasts (CAFs). This phenomenon was described by Amann and co-workers demonstrating transformation of SV80 fibroblasts to CAFs in a microtissue co-culture with an A549 lung cancer cell line (Amann *et al.*, 2014).

Tumour microtissues have also been applied for drug target discovery and validation. The simple fact that cancer cell phenotype and metabolism change in a 3D set-up will enable the discovery of novel targets which cannot be seen in a monolayer model (Thoma *et al.*, 2013). The potential to generate tumour microtissues with increasing complexity depending on the purpose and/or rationale of the testing concept makes the tumour microtissue model a versatile tool to foster drug discovery and target validation.

Microtissue models for safety testing

The toxicologist's toolbox for assessing toxicological effects is huge and ranges from simple enzymatic tests to the use of animals and humans (Tuschl *et al.*, 2008). Safety testing, for either early safety or mechanistic toxicology, requires biological function and structure to be maintained as well as being possible *in vitro*. The spheroid model is a good balance between biological complexity and ease of use. Over the past decades, a number of different microtissue types based on either cell lines or primary cells have been created, including liver (Kelm *et al.*, 2005; Messner *et al.*, 2013), heart (Kelm *et al.*, 2004), kidney, brain (Urich *et al.*, 2013) and vascular models (Kelm *et al.*, 2005).

One of the most prominent examples where 3D configuration is essential is the liver. To maintain liver-specific functionality, the hepatocytes need to maintain their 3D polarised structure and cell-cell contacts (Berthiaume *et al.*, 1996). Reaggregation of hepatocytes to hepatospheres (3D InSight™ liver microtissues) re-establishes hepatocyte structure, leading to improved liver-specific metabolic activity and functionality as compared to monolayer counterparts. Hepatocyte polarisation is also essential for the expression and correct localisation of cell membrane receptors and transporter proteins. For example, hepatocytes express a distinct set of transporter proteins on their sinusoidal, basolateral and apical (canalicular) membrane which is lost in cultures, which do not maintain cell polarisation (Berthiaume *et al.*, 1996). This spatial 3D organisation has to be recapitulated *in vitro* to maintain liver-specific functionality. Aside from scaffold- and hydrogel-based technologies, hepatocytes do reform liver-specific 3D tissues by self-assembly in hanging drops in a similar way to cancer cells (Messner *et al.*, 2013). In comparison to the majority of other methods, cellular spheroids retain most of the cell-cell contacts and the contacts between the cell and the extracellular matrix (ECM), since the cells are in close contact and they produce their own ECM (Godoy *et al.*, 2013).

Product description

Founded in 2009, InSphero AG provides industry-leading 3D cell culture products and services to the life sciences marketplace, counting all the top 15 global pharmaceutical and cosmetics companies as its clients. Built upon a strong and growing portfolio of technology and process patents, and by fostering collaboration with leading assay and imaging technology partners, the company offers a complete set of solutions for 3D cell culture workflows (Figure 2.1). Products include 3D InSight assay-ready microtissues manufactured using the patented GravityPLUS hanging drop system, which is also available commercially. The company also offers a line of 3D InSight culture media optimised for maintenance of 3D microtissues, along with 3D InSight assay kits, featuring optimised protocols compatible with InSphero 3D microtissues and media. Partnering with SCREEN Holdings, InSphero offers a rapid, bright-field imager designed to streamline label-free assessment of spheroid size and morphology. This range of products not only makes InSphero a convenient single-source provider of 3D cell culture technologies to users in the marketplace, but also serves as the basis of an extensive service portfolio, making InSphero the world's largest 3D-focused contract research organisation. Each of these products is discussed in more detail below.

Assay-ready 3D InSight microtissue models

As previously described, various methods (including hanging drop) and related cell culture plates and materials are readily available to enable production of 3D microtissues. However, procurement and maintenance of critical cell lines and donor materials, and variability introduced by lab- and technician-specific handling, can be burdensome and costly, and compromise the quality and consistency of results obtained using 3D models. The commercial availability of off-the-shelf, standardised 3D microtissues provided in an assay-ready format is therefore of benefit not only to ensure consistency and reproducibility within a specific laboratory or institution, but also to reliably translate results between independent labs and drug development programmes. For example, standardised assays to quantify cellular adenosine-5′-triphosphate (ATP) content or lactate dehydrogenase (LDH) release have been adopted within the compound derisking industry as 'gold standard' endpoint assays with which to measure cell viability.

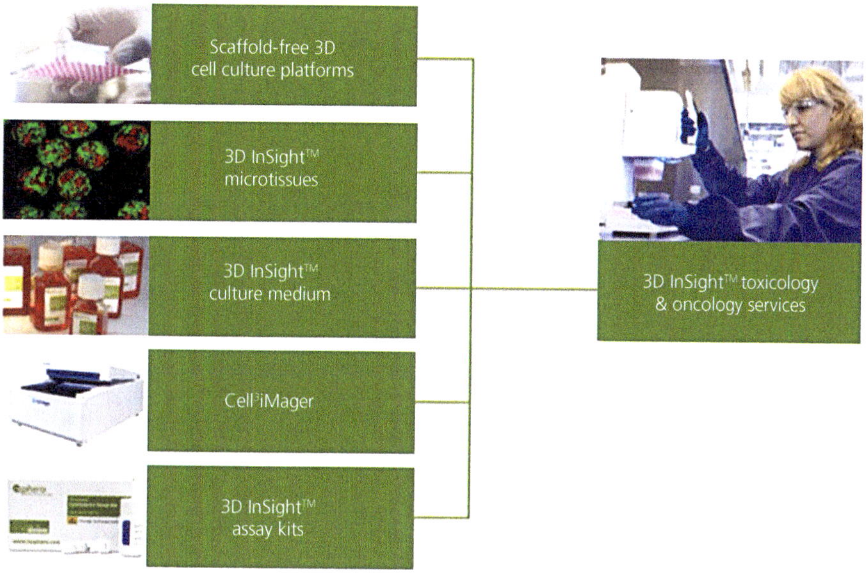

Figure 2.1 **InSphero 3D InSight portfolio.** The InSphero 3D InSight portfolio includes products for the production, cultivation and assessment of 3D microtissues, as well as services utilising gold standard microtissues and 3D-optimised protocols. Source: © 2015, InSphero AG.

With a similar goal in mind, InSphero has developed a portfolio of 3D InSight microtissue models, designed to both standardise and streamline microtissue production and assay workflow (Figure 2.2). These assay-ready microtissues reduce the man-hours required within development groups for cell culture preparation, freeing up valuable resources to focus more on screening objectives using a standardised, readily available cell model. Each microtissue is designed and manufactured to provide an organotypic microtissue morphologically and functionally superior to 2D cell culture. Microtissues are manufactured using InSphero's scaffold-free GravityPLUS hanging drop system, removing assay artifacts and microtissue processing obstacles presented by other scaffold-based 3D cell culture materials. Assay-ready microtissues are provided in a 96-well GravityTRAP plate, with a single microtissue per well, allowing simple medium exchanges, compound dosing and endpoint processing for long-term culture and repeat dosing protocols. Plates of microtissues are delivered using overnight shipping services to labs around the world.

Currently available 3D InSight liver microtissue models include primary human, rat and dog liver microtissues, consisting of either hepatocyte/Kupffer cell co-cultures or hepatocyte-only monocultures. These liver co-culture models enable *in vitro* assessment of inflammation-mediated cytotoxicity such as that seen with the drug trovafloxacin. Cell line-derived liver microtissues established from HepG2 and HepaRG hepatoma cell lines provide convenient early screening models with robust liver-specific metabolic enzyme profiles.

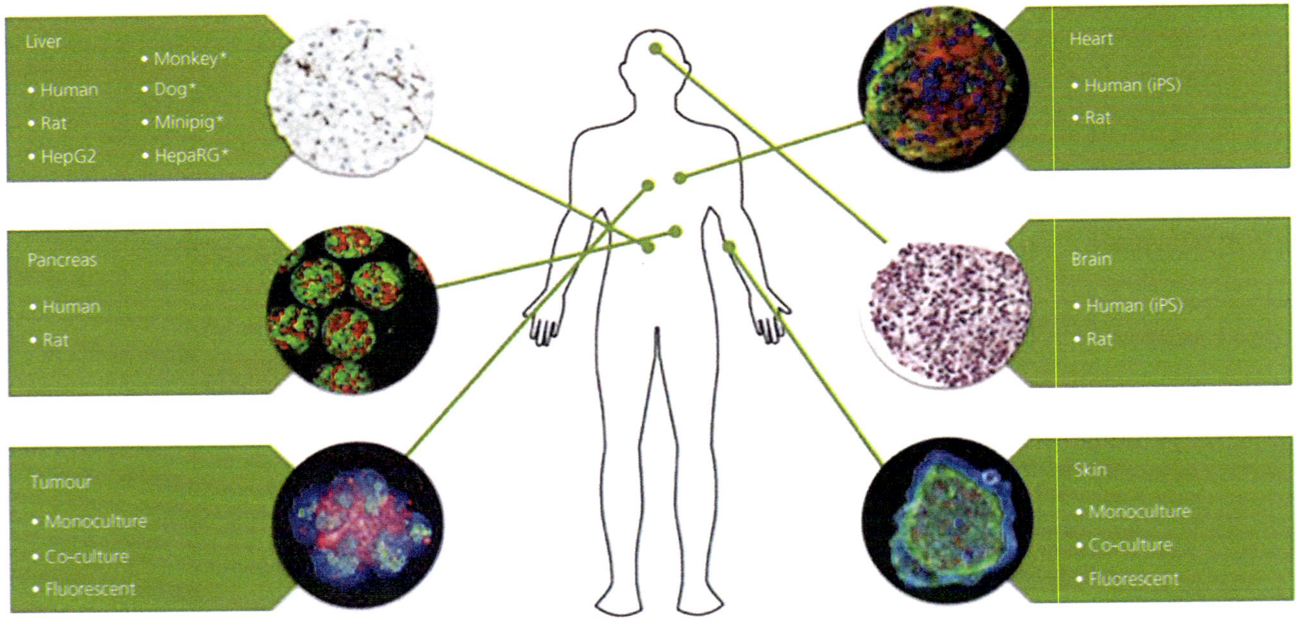

Figure 2.2 **InSphero 3D InSight microtissue model.** InSphero is continuously expanding its 3D InSight microtissue model portfolio. Available models comprise more than 100 homo- and heterotypic tumour microtissue models, liver microtissues from various species, cardiac, brain, pancreas and skin models. Source: © 2015, InSphero AG.

More than 100 3D InSight tumour microtissues are also readily available, either established from monoculture tumour cell lines or as co-cultures with a selection of fibroblast and endothelial cell lines. Genetically engineered tumour/stromal co-cultures are also available, allowing simultaneous monitoring of individual cell populations within the microtissues.

Additional organotypic models include long-lived 3D InSight human and rat pancreatic microislets that display glucose-responsive insulin secretion for over four weeks *in vitro*, serving as a convenient model for diabetes and metabolism research. Rat- and iPS-derived human cardiac and brain microtissues, as well as human dermal microtissues, are in development at the time of writing, and hold promise as models for *in vitro* toxicity and disease modelling. Custom microtissue production protocols can also be confidentially developed and/or manufactured using proprietary or modified cell lines through InSphero's 3D InSight services.

Scaffold-free 3D cell culture platforms: GravityPLUS hanging drop system and GravityTRAP ULA plates

Until recently, scaffold-free reaggregation of single cells into functional 3D microtissues by the hanging drop method was a conceptually simple but technically challenging task. The GravityPLUS hanging drop system (Figure 2.3), introduced by InSphero in 2009, facilitates reliable, automation-compatible and affordable 3D cell culture in hanging drops. The ease of use and versatility of the platform have been reinforced by its adoption in laboratories ranging from academic research to high-throughput screening.

The patented design of the GravityPLUS plate simplifies formation of hanging drop spheroids. Up to 50 µL of cell suspension is loaded from the top via the SureDrop inlet with either a manual pipette or a 96-tip robotic liquid handling device, forming a stable hanging drop. Within 2–4 days, a single microtissue forms in each drop as the cells settle to the bottom of the drop and begin self-aggregating. The extremely stable drop allows removal and replacement of up to 70% of the drop medium during the aggregation process, as may be necessary for long-term culture or differentiation protocols. The improved oxygenation resulting from the proximity of the forming microtissue to the surrounding air-liquid interface of the hanging drop makes it a reliable method for self-assembly driven production of microtissues from most cell types, including primary cells, immortalised tumor cell lines and more complex co-culture microtissue models such as tumor-stromal co-cultures.

Once formed, microtissues are easily harvested into the sister GravityTRAP plate by simple addition of media sufficient to exceed the surface tension capacity of the hanging drop. The proprietary non-adhesive coating of the GravityTRAP plate allows for culturing over weeks without attachment and decomposition of the microtissues. The unique tapered design of the GravityTRAP wells allows reliable medium exchange without microtissue aspiration, and a flat 1.0 mm window at the well bottom facilitates spheroid visualisation and image capture by traditional microscopy or more complex high-content analysis platforms.

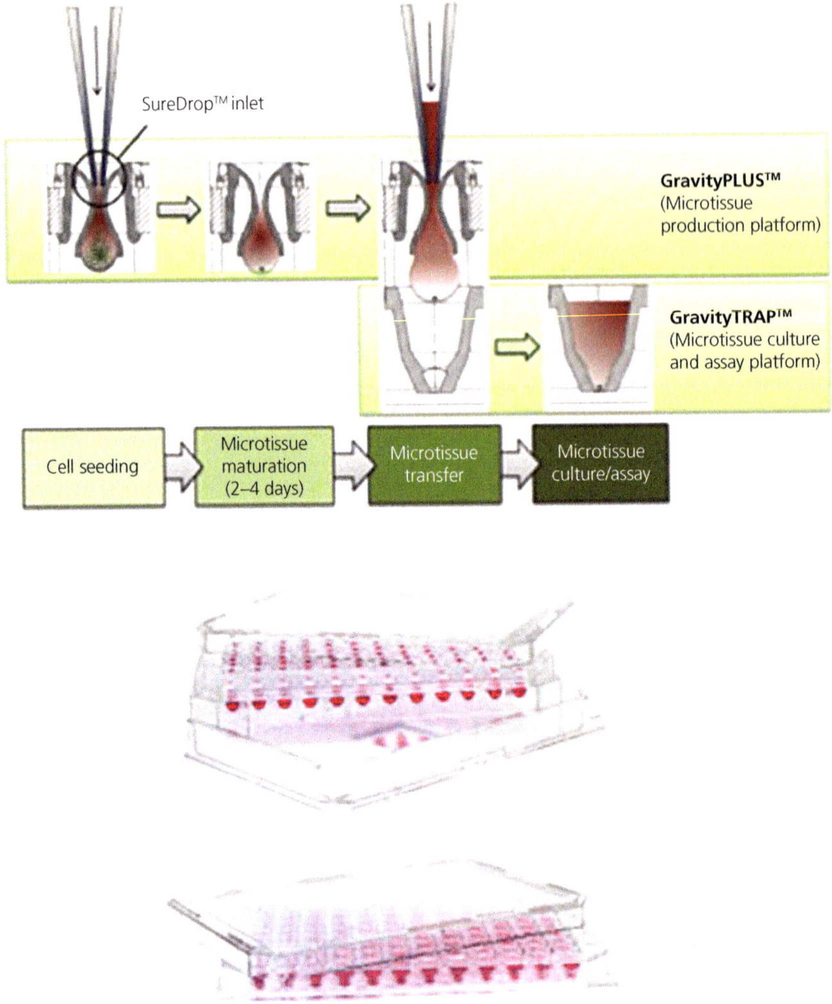

Figure 2.3 **GravityPLU hanging drop system.** This system is composed of a GravityPLUS hanging drop plate (for microtissue production) and complementary GravityTRAP multiwell assay plate (for long-term cultivation and assay of microtissues). Source: © 2015, InSphero AG.

Production of consistently sized microtissues is essential for producing reliable 3D cell-based assays, an obstacle difficult to overcome using traditional hanging drop techniques. The SureDrop inlet in the GravityPLUS plate allows for precise dispensing into and aspirating from hanging drops, and is the only design of its kind on the market. Wells are elastically suspended and mechanically adjust to the pipette tip to assure a tight seal between tip and inlet, reliably eliminating operator and instrument pipetting variations. As the drop volume corresponds to the seeding cell number, microtissues produced in GravityPLUS plates display outstandingly low diameter variations, down to 5% and less.

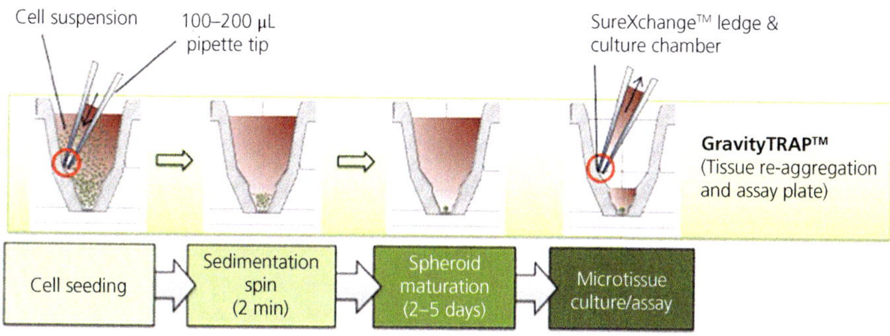

Figure 2.4 **GravityTRAP.** Reliable, reproducible microtissue production in the GravityTRAP ULA plate is possible with many immortalised tumour cell lines. Source: © 2015, InSphero AG.

Many cell types, particularly immortalised tumor cell lines, are capable of forming spheroid microtissues via self-aggregation without the use of the hanging drop method, for instance when plated on non-adherent treated (e.g. ultra-low attachment (ULA) treated) round-, flat- or V-bottom multiwell cell culture plates. For such instances, the InSphero GravityTRAP ULA plate is offered as a stand-alone enhanced ultra-low attachment plate, where cells in suspension are simply seeded into the ULA plate and allowed to settle by gravity or brief centrifugation into the bottom of each well (Figure 2.4). Microtissue maturation in the GravityTRAP ULA plate can then occur in a similarly reproducible manner as in the hanging drop, but is less conducive to production of primary cell-derived or co-culture microtissues. The tapered SureXchange™ well design and narrow culture chamber remove the risk of disturbing the microtissue during medium exchange, while the flat bottom improves imaging and spheroid localisation within the well.

3D InSight assay kits

The 3D InSight cytotoxicity assay kit is a homogeneous biochemical assay for measuring total cellular ATP as an endpoint for cell viability. The assay kit incorporates proprietary luminescent technology from Promega, and is validated for use with InSphero 3D InSight microtissues and culture media. The assay is designed for use with multiwell formats, making it ideal for automated high-throughput screening (HTS), cell proliferation and viability assays for assessing toxicity, drug efficacy and more. The proven lytic capacity of the assay reagent allows efficient lysis of small microtissues, making it ideal for use with 3D InSight liver microtissues and 3D InSight pancreatic microislets.

3D InSight cell culture media

Cells cultured in scaffold-free spheroids, such as InSphero's assay-ready 3D InSight microtissues, maintain their tissue-specific phenotype over long periods in culture. This longevity is key to making them one of the most powerful *in vitro* systems for predictive testing of compound efficacy and toxicity. To ensure consistent and successful maintenance of 3D microtissues, InSphero offers a range of cell culture media optimised for 3D spheroids, including liver, pancreatic islet and tumour microtissue formulations. Developed by InSphero scientists, each medium is enhanced to maintain microtissue functionality over long cultivation periods. The media are freshly prepared at InSphero and shipped ready to use at 4 °C.

Cell³iMager for microtissue imaging

Imaging spheroids using conventional or even automated microscopes is a slow, low-throughput process. High-content imaging systems overcome some of these throughput issues but are comparatively expensive and still suffer from slow image capture/processing and image analysis software that has not been optimised for 3D spheroids. As a solution for rapid, robust imaging to monitor spheroid growth and morphology, InSphero offers the SCREEN Cell³iMager™.

The Cell³iMager facilitates analysis of spheroids by fast, parallel scanning in a bright field, allowing determination of spheroid count per well, diameter, area, pseudo volume and loss of circularity in individual spheroids. Image capture is rapid, requiring less than 1 minute per plate at 2400 dpi (10.6 µm/pixel). The four-plate stage can accommodate 6-well to 384-well plates per scan run, with resolution up to 9600 dpi (2.6 µm/pixel). The reagent-free, label-free system allows faster sample processing (30 × 384-well plates/hour versus 2.5 plates/hour with conventional systems), and convenient measurement of spheroid growth over time in a non-destructive way. The user-friendly 3D software package provided with the Cell³iMager offers multiple analysis options, and provides flexibility to customise scanning recipes for automated removal of debris (e.g. dust, fibres) and compensation for undesired well effects such as shadowing. The rapid, flexible spheroid imager provides a convenient tool for phenotypic drug discovery, drug sensitivity testing, co-culture associated loss of spheroid volume, combinatorial drug testing, drug-target discovery and validation, and as a quality control mechanism when growing spheroids.

3D InSight services

The rapidly emerging demand for industry and academic labs to build 3D technology into development and research protocols often requires costly adaptation and validation of 3D models prior to full implementation into a standardised workflow. The broad range of InSphero's 3D cell culture products described herein is accompanied by an established and well-respected scientific support staff that can offer fast turnaround on 3D cell-based screening services using scaffold-free 3D InSight microtissue models or custom-developed 3D models. These services make InSphero the world's largest 3D-focused contract research organisation, offering toxicity-focused compound derisking, oncology-focused efficacy screening, and the easyEST™ service for determining compound embryotoxicity and teratogenicity.

Toxicity testing services can be performed on any number of compounds, using 3D InSight microtissues and a panel of toxicity and DMPK mechanistic indicators including hepatotoxicity, Kupffer cell function, mitochondrial impairment, reactive metabolites, metabolic competence, apoptosis, steatosis, cholestasis, proliferation and 'omic-level' mRNA and protein expression profiling. Oncology-focused services include the cytotoxicity endpoints described above, as well as label-free growth profiling, siRNA-mediated target validation, antibody penetration analysis and immunomodulatory antibody testing, all available with 3D InSight tumour microtissues or co-cultures. The easyEST embryonic stem cell test is a standardised, multicomponent testing service for evaluating compound embryotoxicity and teratogenicity, following the ECVAM Invittox 113 protocol. The service utilises embryoid bodies (EB) produced using the GravityPLUS hanging drop system, improving throughput and consistency of the EST at a fraction of the cost of using the classic hanging drop EB production protocol.

Tumour microtissue size profiling for drug testing

Materials and methods

3D microtissue production

HCT-116 (DSMZ no. ACC 581) cells were expanded as a monolayer in cell culture flasks using VLE RPMI 1640 medium (Biochrom GmbH, Germany) supplemented with 10% fetal bovine serum (FBS Superior, Biochrom GmbH, Germany), 100 unit/mL penicillin and 100 µg/mL streptomycin at 37 °C and 5% CO_2. To produce HCT-116 3D InSight colorectal cancer microtissues, cells were detached from the 2D culture flask using trypsin/EDTA and seeded into 96-well hanging drop GravityPLUS plates. After microtissue formation, spheroids were transferred into GravityTRAP plates for further experiments. Medium was refreshed at days 4, 6 and 8 post transfer to assay plates.

Compound treatment of HCT-116 3D InSight microtissues

Compounds (staurosporine, taxol, doxorubicin) to treat HCT-116 microtissues in GravityTRAP plates were purchased from Enzo Life Sciences (Lausen, Switzerland) and dissolved in DMSO. From the stock solutions, twofold serial dilutions in DMSO were prepared for each compound, leading to 200× concentrated working dilutions. The assayed concentration range of staurosporine was 2 to 1000 nM, for taxol it was 1.6 nM to 800 nM and for doxorubicin the range was between 0.8 nM and 400 nM.

Microtissues were treated in GravityTRAP plates with the indicated compound dilutions (n=4), leading to 0.5% DMSO in the cell culture medium. As a control, microtissues were exposed to 0.5% DMSO without a compound. Upon medium exchange, compounds were readded where indicated. The treatment was performed over 10 days but for two tissues per group, the compound exposure was discontinued after five days and they were thereafter incubated in medium without compound (n=2).

3D InSight microtissue size profiling

GravityTRAP plates containing treated microtissues were scanned with Cell³iMager with a resolution of 4800 dpi. Microtissue size was assessed with the software provided by the manufacturer. The measurement parameters used are indicated in Table 2.1.

Air bubbles interfering with the size analysis were excluded manually, either by adjusting the image section to be analysed by the software or by cropping the area designated to a microtissue after size measurement.

The acquired data was normalised to the non-treated control:

$$Average_{size,norm} = Average\left(\frac{S_1}{S_{1;control}};\frac{S_2}{S_{2;control}};\frac{S_3}{S_{3;control}}\right) \quad (1)$$

S: size of microtissue.

Table 2.1 Settings used for microtissue size analysis.

Scanner settings	
Resolution	4800 dpi
Well width	1000 µm
Border width	15 µm
Well ROI width	100%
Allowable object's maximum area	85%
Noise reduction filter size	2
Debris threshold	35%
Object size upper limit	958 µm
Compactness upper limit	97
Straight fibreness upper limit	6
Object detection	
Smoothing size	969 µm
Object threshold base	9
Object threshold all area	29
Object edge smoothing	Yes
Alive or dead	
Area	10418–2400000 µm^2
Circularity lower limit	30%
Spheroid density	26–300

ROI, region of interest.

Determination of cell viability

In order to determine the number of viable cells per microtissue, the amount of intracellular ATP as a measure for metabolic activity was determined by using CellTiter-Glo® luminescent cell viability assay (Promega, Madison, USA). The assay was performed according to the manufacturer's protocol with slight adaptations specific to microtissues. In brief, supernatant was removed and lysis of microtissues initiated by adding 40 µL diluted CellTiter-Glo reagent (1:2 in PBS). The cell lysis was enhanced by pipetting up and down prior to transfer into an opaque white assay plate. Relative luminescence (RLU) was recorded with a TECAN™ Infinite 200 microplate reader, with sensitivity set to 1000 ms.

The acquired data was normalised to the non-treated control:

$$Average_{ATP,norm} = Average\left(\frac{R_1}{R_{1;control}};\frac{R_2}{R_{2;control}};\frac{R_3}{R_{3;control}}\right) \quad (2)$$

Results

It is commonly accepted that tumour sensitivity or resistance to chemotherapeutic agents is not only genetically determined but also driven by the tumour microenvironment. Metabolic gradients and the extracellular matrix can influence nutrient availability to various segments of a solid tumour, and also limit the ability to administer an effective dose of chemotherapeutic agent. These biological and therapeutic gradients result in the formation of phenotypically different tumour cell subpopulations exposed to a variety of therapeutic to subtherapeutic drug doses *in vivo*. 3D tumour microtissues, or multicellular tumour spheroids, are considered a more representative, organotypic model for assessment of tumour growth. They contain layers of cells that exhibit more *in vivo*-like size- and gradient-dependent proliferation and viability profiles. Tumour size is the most frequently used *in vivo* endpoint when assessing antitumour efficacy in animal xenograft models, whereas proliferation is the more typically evaluated growth endpoint *in vitro* using 2D monolayer cultures. Such 2D *in vitro* assays frequently fail to correlate with *in vivo* observations, owing to the inability of 2D cultures to recapitulate the native tumour microenvironment described above, and making measurement of 3D spheroid growth (i.e. size) over time a more desirable endpoint.

We have tested three reference drugs and monitored growth over time by non-disruptive quantification of size and confirmed the size-derived IC_{50} values by quantification of intratissue ATP content. Size profiling after 10 days treated with taxol (TAX), doxorubicin (DOX) and staurosporine (STA) displayed dose-dependent size differences. Even with a high degree of tissue debris and dissolving of clear spherical appearance, the core tissue borderline was detected (red line around the microtissues) (Figure 2.5). The growth profiles over time displayed different characteristics of the drugs. Whereas taxol displayed a very

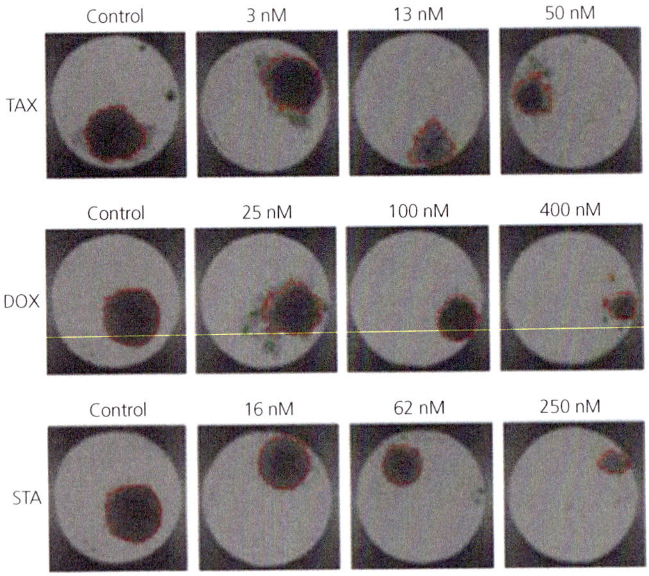

Figure 2.5 Size profiling over time with HCT-116 microtissues. The HCT-116 microtissues treated with different concentrations of taxol (TAX), doxorubicin (DOX) and staurosporine (STA) after a 10-day treatment with repeated dosing. Dose response is clearly visible for all drugs tested.

narrow concentration window, having either no effect (3 nM) or a strong impact on growth (13 nM), a more gradual response to staurosporine was observed between control and 1000 nM (Figure 2.6).

Differences between the drugs were also observed after removing the compound after a five-day treatment to assess recurrence of tumour growth. Whereas no regrowth was observed using the highest drug concentrations, tumour regrowth was observed for lower drug concentrations, at 6 nM for taxol, 50 nM for doxorubicin and 125 nM for staurosporine (Figure 2.7). Whereas the doxorubicin and staurosporine-treated microtissues regrew symmetrically, the taxol-treated microtissues regrew side-specific (see Figure 2.7). The growth curves of continuously treated tumour microtissues and the single-dosed groups only confirm the qualitative analysis.

High concentrations of the drugs led to nearly complete tumour removal after five days for taxol whereas at lower concentrations, tumour started to regrow (Figures 2.8 and 2.9). For doxorubicin, 50 and 100 nM treated groups were compared. Whereas both concentrations affected growth of the tumour microtissues in a similar way, the 50 nM group demonstrated a stronger capacity to regrow after drug removal (see Figures 2.8 and 2.9). In the case of staurosporine, tissue which displayed similar size reduction at 125 nM and 500 nM after five days regrew substantially compared to the continuous treated counterparts (see Figures 2.8 and 2.9).

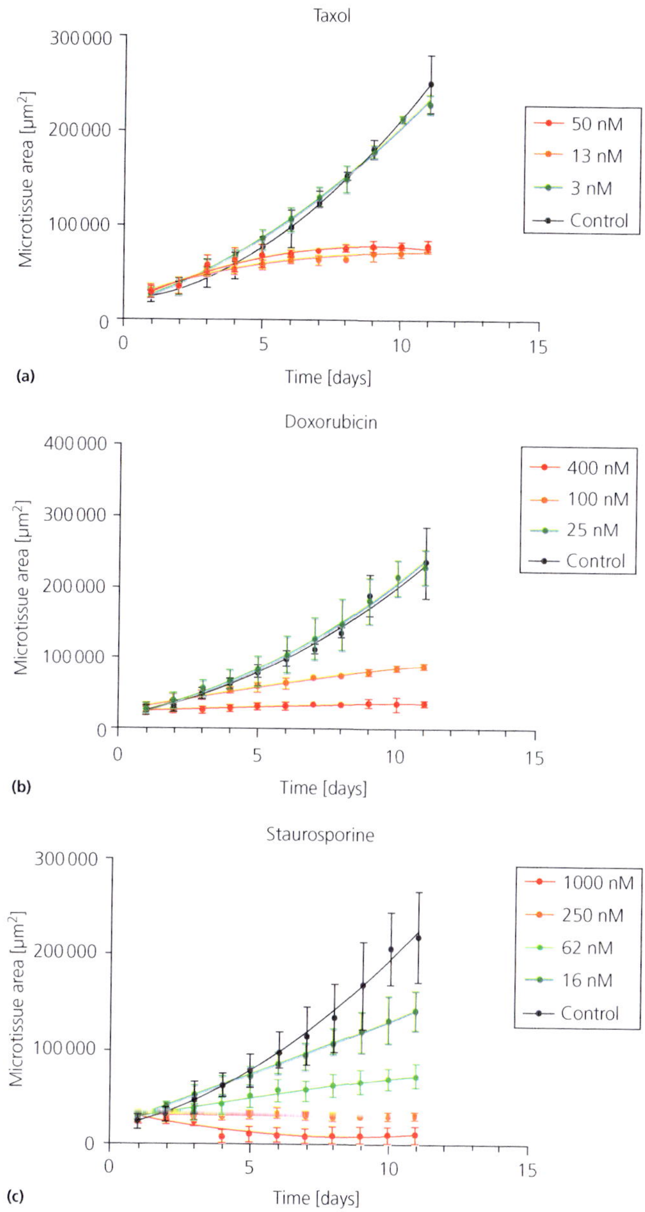

Figure 2.6 **Growth profiling over time with HCT-116 microtissues.** HCT-116 microtissues treated with different concentrations of (a) taxol (TAX), (b) doxorubicin (DOX) and (c) staurosporine (STA) over 10 days. Growth is calculated via the area over time in [μm²].

Comparing dose response curves from days 3, 5, 7 and 10, we observed in general decreasing IC_{50} values for taxol which were even more pronounced for doxorubicin (Figures 2.10 and 2.11). Whereas after three days no dose response was observed for taxol, the IC_{50} values did not change after five days, remaining

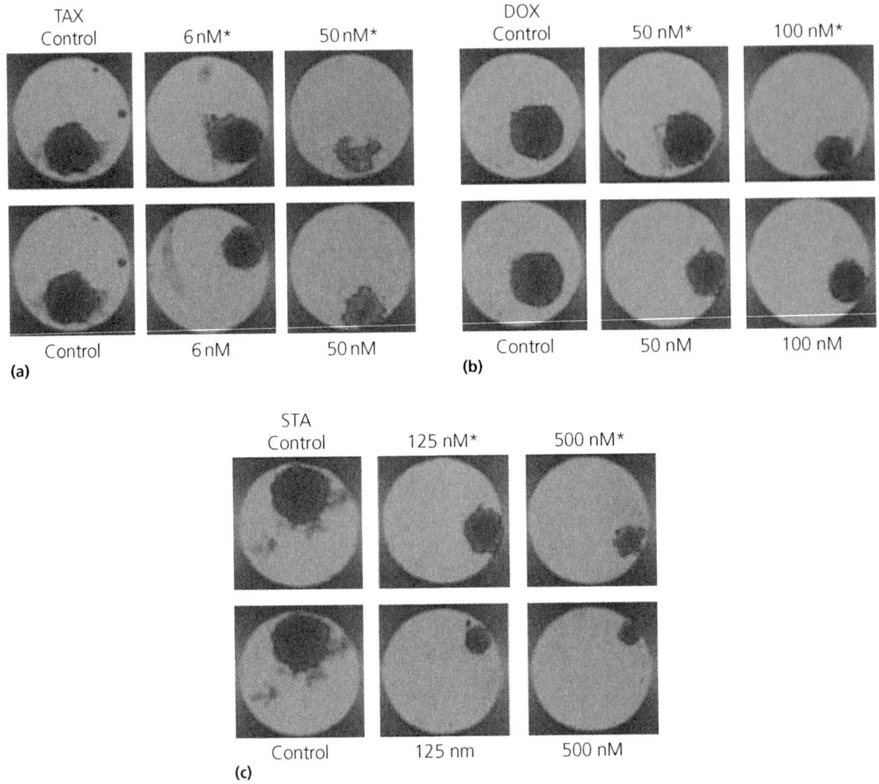

Figure 2.7 Size comparison of HCT-116 microtissues. HCT-116 after 10 days' treatment with (a) taxol (TAX), (b) doxorubicin (DOX) and (c) staurosporine (STA) compared to cultures where drug exposure was stopped after five days. Growth is calculated via the area over time in [μm^2].

constant between 5 and 6 nM. However, the sigmoidal nature of the dose response became stronger with later time points (see Figure 2.10). Doxorubicin displayed a clearly reduced IC_{50} value over the 10-day dosing period, from 159.7 nM after three days to 46.22 nM after 10 days, highlighting that longer incubation time with repeated dosing can have a significant effect on the IC_{50} value (see Figure 2.11).

For staurosporine, the calculated IC_{50} values (approximately 40 nM on average) did not alter significantly over the 10-day period, but similar to the dose response curves of taxol, the sigmoidal character became more distinct (Figure 2.12). The size-derived IC_{50} values were benchmarked at day 10 against an ATP-derived IC_{50} determination. ATP quantification resulted in similar IC_{50} values substantiating the size-based analysis (Table 2.2).

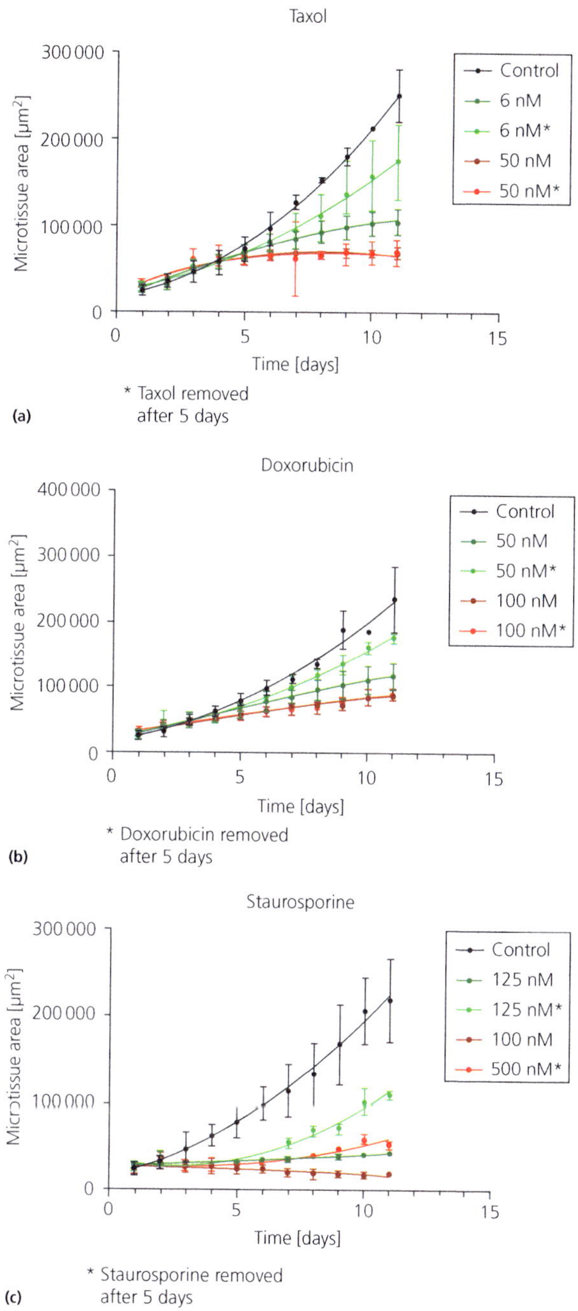

Figure 2.8 Growth profiling over time with HCT-116 microtissues. HCT-116 treated with different concentrations of (a) taxol (TAX), (b) doxorubicin (DOX) and (c) staurosporine (STA) over 10 days compared to cultures where drug exposure was stopped after five days. Growth is calculated via the area over time in [μm²].

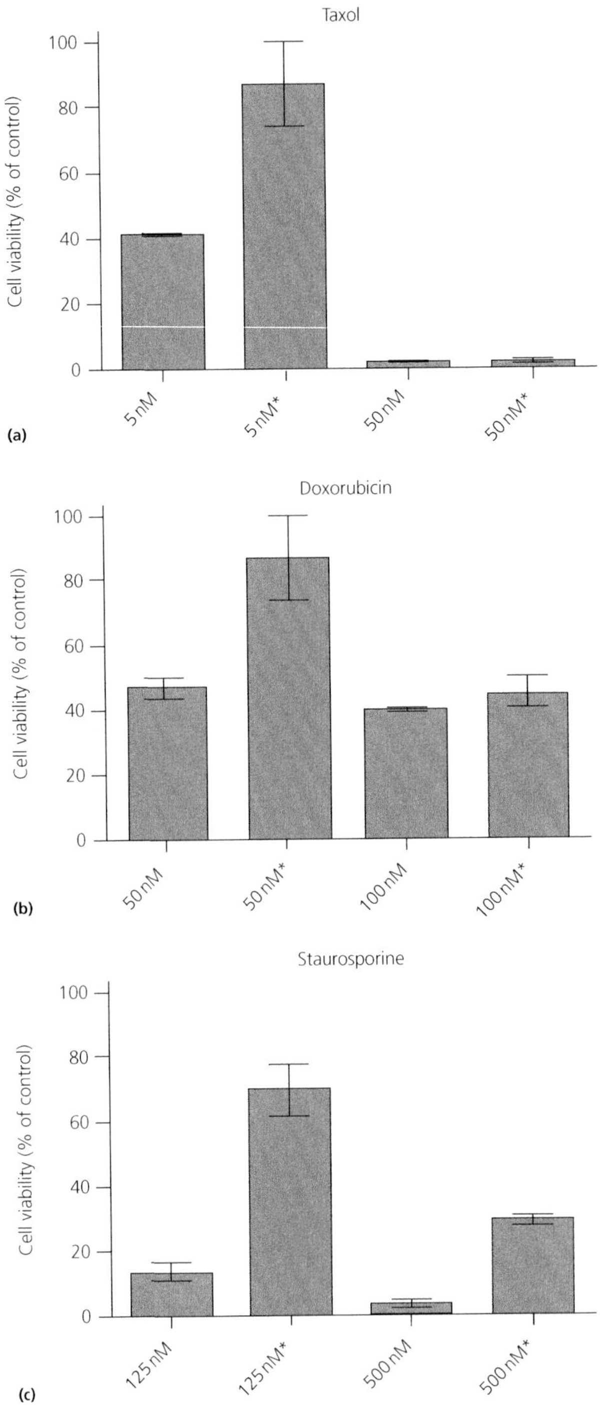

Figure 2.9 ATP-content of HCT-116 microtissues. The ATP content of HCT-116 microtissues after drug exposure over 10 days with (a) taxol, (b) doxorubicin and (c) staurosporine compared with the ATP content of microtissues which were not treated with the drug after day 5*.

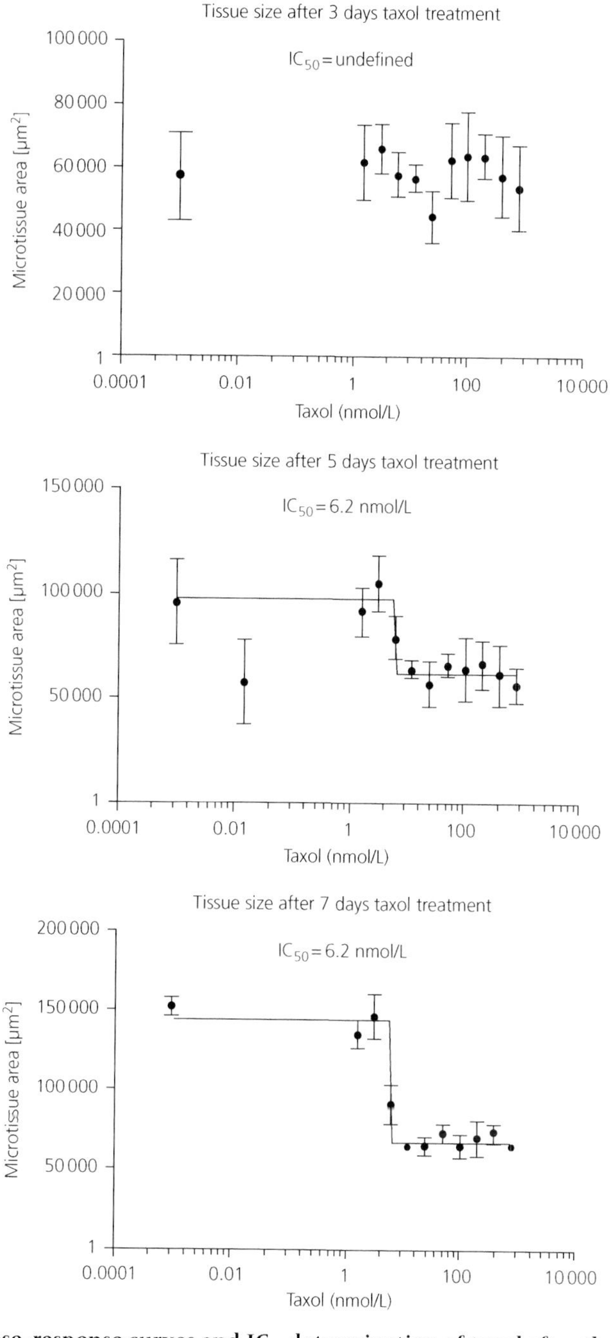

Figure 2.10 Dose–response curves and IC_{50} determination of taxol after three, five, seven and 10 days based on size profiling. To benchmark the area-based IC_{50} calculations, dose–response was additionally measured by the intratissue ATP content after 10 days.

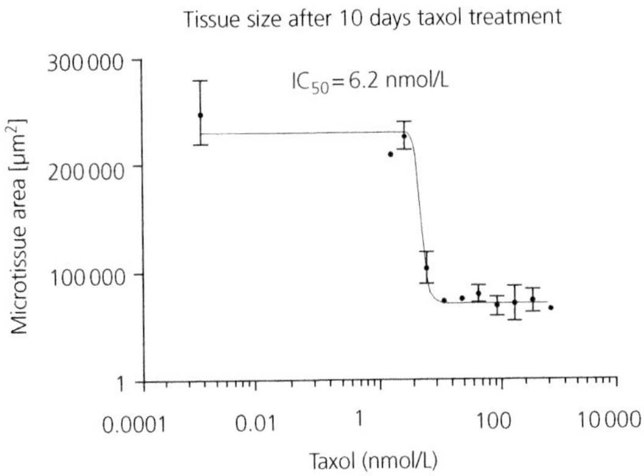

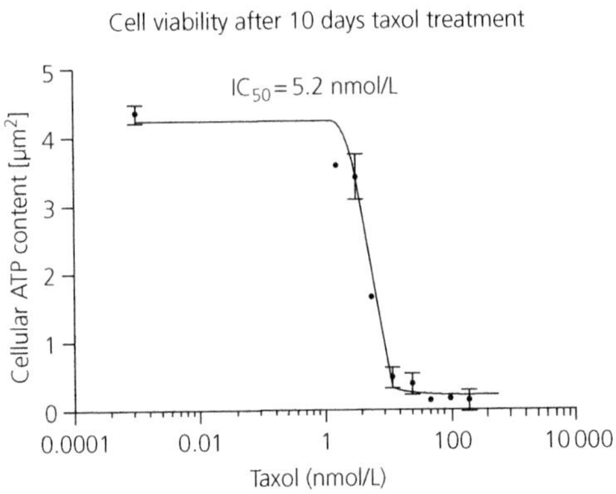

Figure 2.10 (Continued)

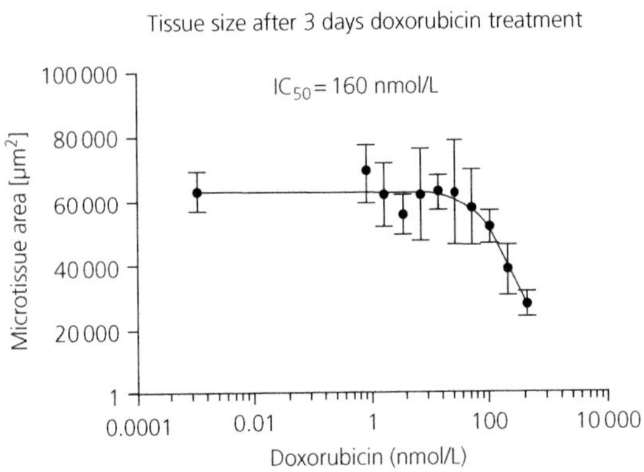

Figure 2.11 Dose–response curves and IC_{50} determination of doxorubicin after three, five, seven and 10 days based on size profiling. To benchmark the area-based IC_{50} calculations, dose–response was additionally measured by the intratissue ATP content after 10 days.

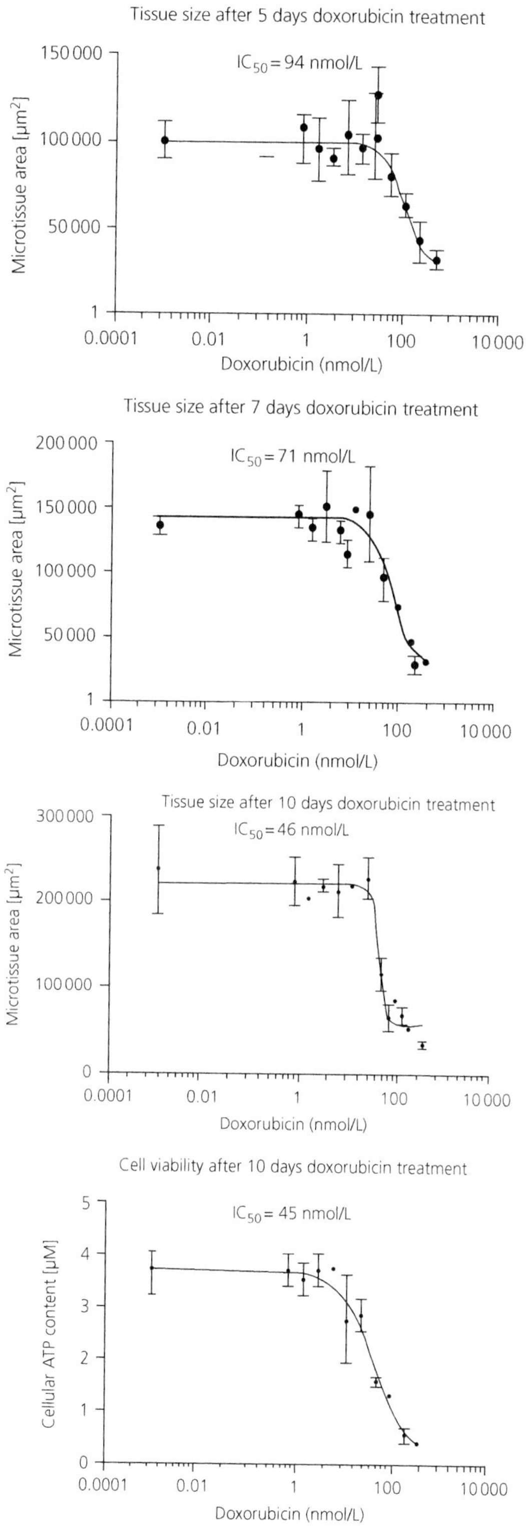

Figure 2.11 (Continued)

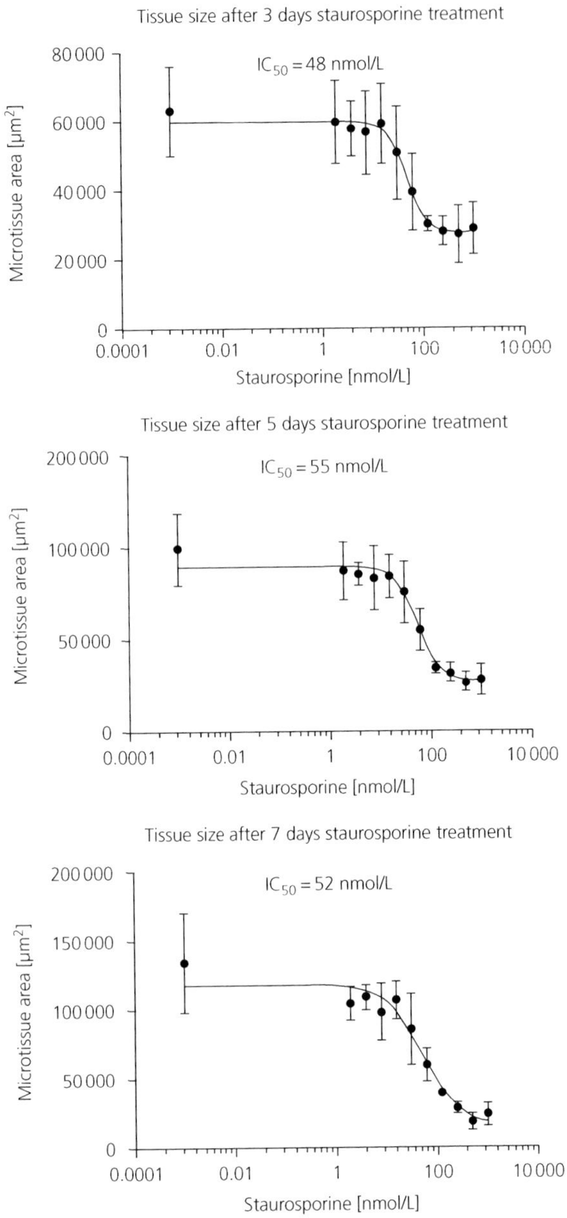

Figure 2.12 **Dose–response curves and IC_{50} determination of staurosporine after three, five, seven and 10 days based on size profiling.** To benchmark the area-based IC_{50} calculations, dose–response was additionally measured by the intratissue ATP content after 10 days.

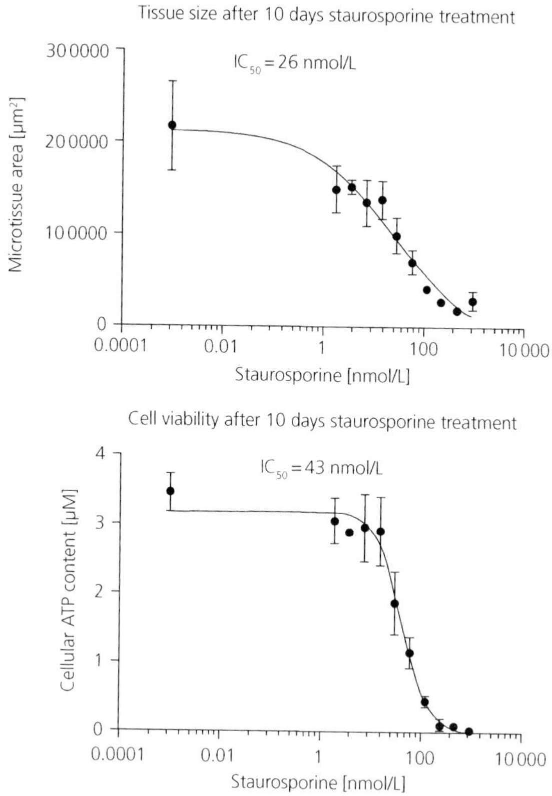

Figure 2.12 (Continued)

Conclusions

In vivo efficacy testing is mainly based on size quantification over time. The tumour microtissue model enables the use of size as a translational endpoint to assess *in vitro* drug efficacy. Size has been shown to be a robust and reliable endpoint also for *in vitro* drug efficacy studies using microtissue models. Different growth profiles were monitored as a consequence of different drug mechanisms. Even though pharmacokinetics are not reflected in static culture conditions, drug sustainability determined by assessing tumour recurrence provides first insights into how a dosing scheme might impact compound efficacy. Overall, size analysis is a powerful tool for either phenotypic drug discovery or a more in-depth assessment of antitumour drug efficacy.

Table 2.2 IC_{50} values.

Time (days)	IC_{50} taxol (area)		IC_{50} taxol (ATP)	
	Mean	*95% CL*	*Mean*	*95% CL*
3	n.a.	n.a.	-	
5	6.2 nM	n.a.	-	
7	6.2 nM	n.a.	-	
10	5.2 nM	3.7–7.3 nM	5.2 nM	4.3–6.4 nM

Time (days)	IC_{50} doxorubicin (area)		IC_{50} doxorubicin (ATP)	
	Mean	*95% CL*	*Mean*	*95% CL*
3	160 nM	60–420 nM	-	
5	94 nM	52–170 nM	-	
7	71 nM	44–110 nM	-	
10	46 nM	35–62 nM	45 nM	24–83 nM

Time (days)	IC_{50} staurosporine (area)		IC_{50} staurosporine (ATP)	
	Mean	*95% CL*	*Mean*	*95% CL*
3	48 nM	38–61 nM	-	
5	55 nM	39–77 nM	-	
7	52 nM	25–110 nM	-	
10	26 nM	5–140 nM	42 nM	32–57 nM

ATP, adenosine triphosphate; CL, confidence limits; n.a., not applicable.

Troubleshooting and notes (Table 2.3)

Medium exchange in the GravityTRAP plate

Remove the medium (maximal volume 90 μL per well) by placing the tip of the pipette at the ledge as shown in Figure 2.13a. Pipette at low speed if using an automated multichannel pipette (<30 μL/sec). Add 50 μL of prewarmed tissue-specific medium by placing the pipette tip at the ledge (see Figure 2.13a).

Microtissue retrieval

For certain analytical endpoints such as histology, microtissues need to be recovered from the GravityTRAP wells. Translocation of microtissues from the GravityTRAP is performed by standard liquid handling processes as shown in Figure 2.14.

Table 2.3 Troubleshooting overview.

Problem	Cause	Solution
Uneven drop volume due to evaporation	Low humidity in incubator needs to be compensated	Use humidifier pad, add 15 mL of 0.75× PBS. At high evaporation rates, reduce ionic strength of humidifier pad as low as 0.25× PBS
Transfer rate of microtissues is below 100%	Microtissues not detectable due to air bubbles	Prewetting of the wells will prevent inclusion of air bubbles. Prewet GravityTRAP as described. Transfer MT by applying medium from the top at low pipetting speed (<10 uL/sec)
	Microtissue(s) did not settle at the bottom of the well(s)	Centrifuge GravityTRAP plate at 250 RCF for 2′ to deposit MT at the bottom of the cavity. Alternatively transfer MT by aspiration of entire drop volume from the top (MT size should be less than 500 μm in diameter)
Drops fall off	Too low ionic strength of humidifier solution in the bottom tray may result in growing drops, which eventually fall off	Increase ionic strength of the humidifier pad solution
Cells do not form a compact spheroid	Certain cell types need additional measures to improve microtissue formation	Medium and process optimisation required

MT, microtissue; PBS, phosphate buffered saline; RCF, relative centrifugal force.
Source: © 2015, InSphero AG.

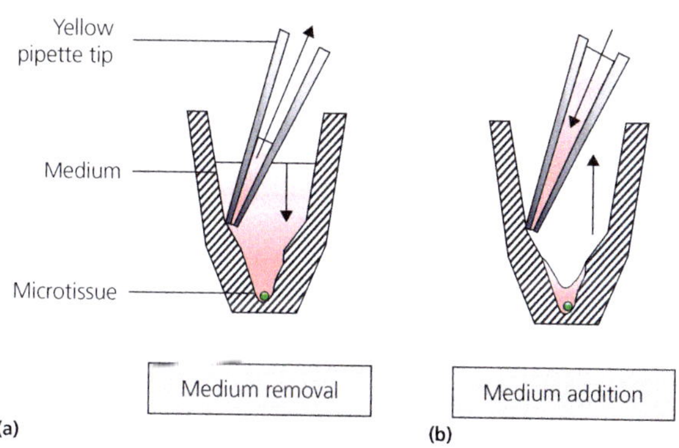

Figure 2.13 **Medium exchange in the GravityTRAP plate.** Pipette tip position during liquid removal (a) and addition (b) into GravityTRAP wells. Source: © 2015, InSphero AG.

Summary

The microtissue or spheroid model has been shown to be a versatile format to create 3D cell culture models of cell lines, primary cells and stem cells. The hanging drop technology enables efficient and scalable production processes for most

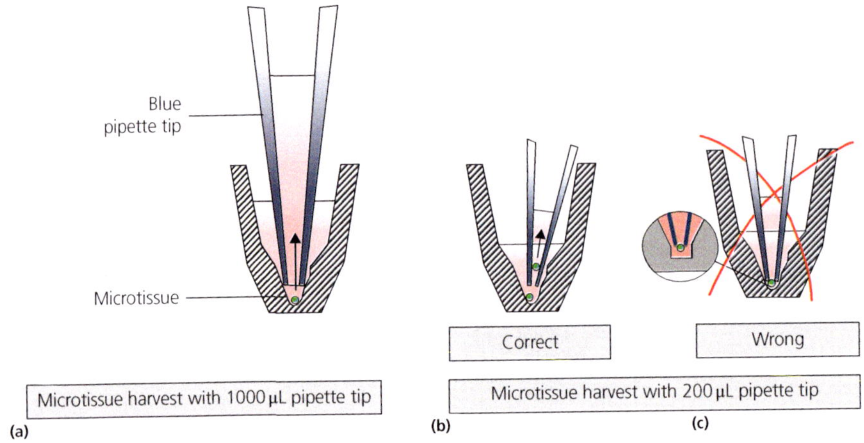

Figure 2.14 **Microtissue retrieval from GravityTRAP plates.** Harvest is recommended with either 1 mL pipette tips (a) or 200 µL pipette tips (b,c). Tip positioning slightly above the bottom of the plate avoids microtissue damage. During aspiration, slowly lower the tip until the bottom is reached. Source: © 2015, InSphero AG.

of the cell sources and origins. They are perfectly suited for phenotypic drug discovery concepts as well as for safety testing at early and later stages. Enabling the use of similar assay endpoints throughout the whole drug discovery and development process, the microtissue model has become the new cell culture standard to foster drug discovery, development and safety testing.

References

Amann, A., Zwierzina, M., Gamerith, G. *et al.* (2014) Development of an innovative 3D cell culture system to study tumour-stroma interactions in non-small cell lung cancer cells. *PLos One*, **9**, e92511.

Berthiaume, F., Moghe, P. V., Toner, M. and Yarmush, M. L. (1996) Effect of extracellular matrix topology on cell structure, function, and physiological responsiveness: hepatocytes cultured in a sandwich configuration. *FASEB Journal*, **10**, 1471–1484.

Buzhor, E., Harari-Steinberg, O., Omer, D. *et al.* (2011) Kidney spheroids recapitulate tubular organoids leading to enhanced tubulogenic potency of human kidney-derived cells. *Tissue Engineering Part A*, **17**, 2305–2319.

Desoize, B. & Jardillier, J. (2000a) Contribution of three-dimensional culture to cancer research. *Critical Reviews in Oncology/Hematology*, **36**, 59–60.

Desoize B. & Jardillier J. (2000b) Multicellular resistance: a paradigm for clinical resistance? *Critical Reviews in Oncology/Hematology*, **36**, 193–207.

Drewitz, M., Helbling, M., Fried, N. *et al.* (2011) Towards automated production and drug sensitivity testing using scaffold-free spherical tumour microtissues. *Biotechnology Journal*, **6**, 1488–1496.

Godoy, P., Hewitt, N. J., Albrecht, U. *et al.* (2013) Recent advances in 2D and 3D in vitro systems using primary hepatocytes, alternative hepatocyte sources and non-parenchymal liver cells and their use in investigating mechanisms of hepatotoxicity, cell signaling and ADME. *Archives of Toxicology*, **87**, 1315–1530.

Hall, M. D., Martin, C., Ferguson, D. J. P. *et al.* (2004) Comparative efficacy of novel platinum (IV) compounds with established chemotherapeutic drugs in solid tumour models. *Biochemical Pharmacology*, **67**, 17–30.

Keller, G. M. (1995) *In vitro* differentiation of embryonic stem cells. *Current Opinion in Cell Biology*, **7**, 862–869.

Kelm, JM. & Fussenegger, M. (2004) Microscale tissue engineering using gravity-enforced cell assembly. *Trends in Biotechnology*, **22**, 195–202.

Kelm, J. M., Timmins, N. E., Brown, C.J. *et al.* (2003) Method for generation of homogeneous multicellular tumour spheroids applicable to a wide variety of cell types. *Biotechnology and Bioengineering*, **83**, 173–180.

Kelm, J. M., Diaz Sanchez-Bustamante, C., Ehler, E. *et al.* (2005) VEGF profiling and angiogenesis in human microtissues. *Journal of Biotechnology*, **118**, 213–229.

Kelm, J. M., Ittner, L. M., Born, W. *et al.* (2006) Self-assembly of sensory neurons into ganglia-like microtissues. *Journal of Biotechnology*, **121**, 86–101.

Lancaster, M. A., Renner, M., Martin, C. A. *et al.* (2013) Cerebral organoids model human brain development and microcephaly. *Nature*, **501**, 373–379.

Levi-Montalcini, R. (1964) The nerve growth factor. *Annals of the New York Academy of Sciences*, **118**, 149–170.

Messner, S., Agarkova, I., Moritz, W. *et al.* (2013) Multi-cell type human liver microtissues for hepatotoxicity testing. *Archives of Toxicology*, **87**, 209–213.

Ostmann, A. (2012) The tumour microenvironment controls drug sensitivity. *Nature Medicine*, **18**, 1332–1334.

Thoma, C. R., Stroebel, S., Rösch N. *et al.* (2013) A high-throughput-compatible 3D microtissue co-culture system for phenotypic RNAi screening applications. *Journal of Biomolecular Screening*, **18**, 1330–1337.

Thoma, C. R., Zimmermann, M., Agarkova, I. *et al.* (2014) 3D cell culture systems modeling tumour growth determinants in cancer target discovery. *Advanced Drug Deliveries Reviews*, **69–70**, 29–41.

Tuschl, G., Lauer, B. & Mueller, S. O. (2008) Primary hepatocytes as a model to analyze species-specific toxicity and drug metabolism. *Expert Opinion on Drug Metabolism & Toxicology*, **4**, 855–870.

Urich, E., Patsch, C., Aigner, S. *et al.* (2013) Multicellular self-assembled spheroidal model of the blood brain barrier. *Scientific Reports*, **3**, 1500.

Wilson, W. & Hay, M. P. (2011) Targeting hypoxia in cancer therapy. *Nature Reviews Cancer*, **11**, 393–410.

Trademark Disclosure

GravityPLUS™, Gravity TRAP™, SureDrop™, SureXchange™, 3D InSight™ and easyEST™ are trademarks of InSphero AG. CellTiter-Glo® is a registered trademark of Promega Corp. TECAN™ is a trademark of Tecan Trading AG.

CHAPTER 3

Physiologically relevant spheroid models for three-dimensional cell culture

Nicole A. Slawny and MaryAnn Labant
3D Biomatrix, Inc., Ann Arbor, Michigan, USA

Introduction

The basic steps for implementing three-dimensional (3D) cultures are:
1 determination of a culture format and assay
2 literature searches for information on the specific cells in the chosen format
3 preliminary experimentation to optimise culture conditions.

Prior to using a 3D cell culture format, consideration needs to be given to the proposed analytical methods. The chosen format and the measured output parameters must be compatible. For example, washing samples can be challenging in some 3D formats; therefore, whether cell samples will be analysed directly or transferred into a receiving plate must be determined in advance. Sample size is another consideration; samples may need to be pooled for analysis. Assay time is critical, especially if the cells will be removed from the 3D culture system and transferred to a two-dimensional (2D) system. In this circumstance, assays need to be relatively short. Once a sample is transferred to a 2D plate, eventually the 3D cellular construct will start to adhere, and the biology will begin to change.

A singular 3D cell culture format does not fit all needs. The choice of format depends on experimental design. Table 3.1 provides a high-level comparison of the Perfecta3D® hanging drop plates to other available 3D cell culture formats.

History of the Perfecta3D hanging drop plates

The Perfecta3D hanging drop plates (HDPs) were developed by 3D Biomatrix to enable faster efficient discovery by making *in vitro* research more physiologically relevant. The HDPs were developed after careful consideration of well-known obstacles in the *in vitro* discovery process, such as:

- missed discoveries due to existing *in vitro* early preclinical testing processes allowing too many false-positives to move forward to animal testing, thus contributing to high drug attrition rates

---

Technology Platforms for 3D Cell Culture: A User's Guide, First Edition. Edited by Stefan Przyborski.

Table 3.1 A high-level comparison of 3D cell culture formats.

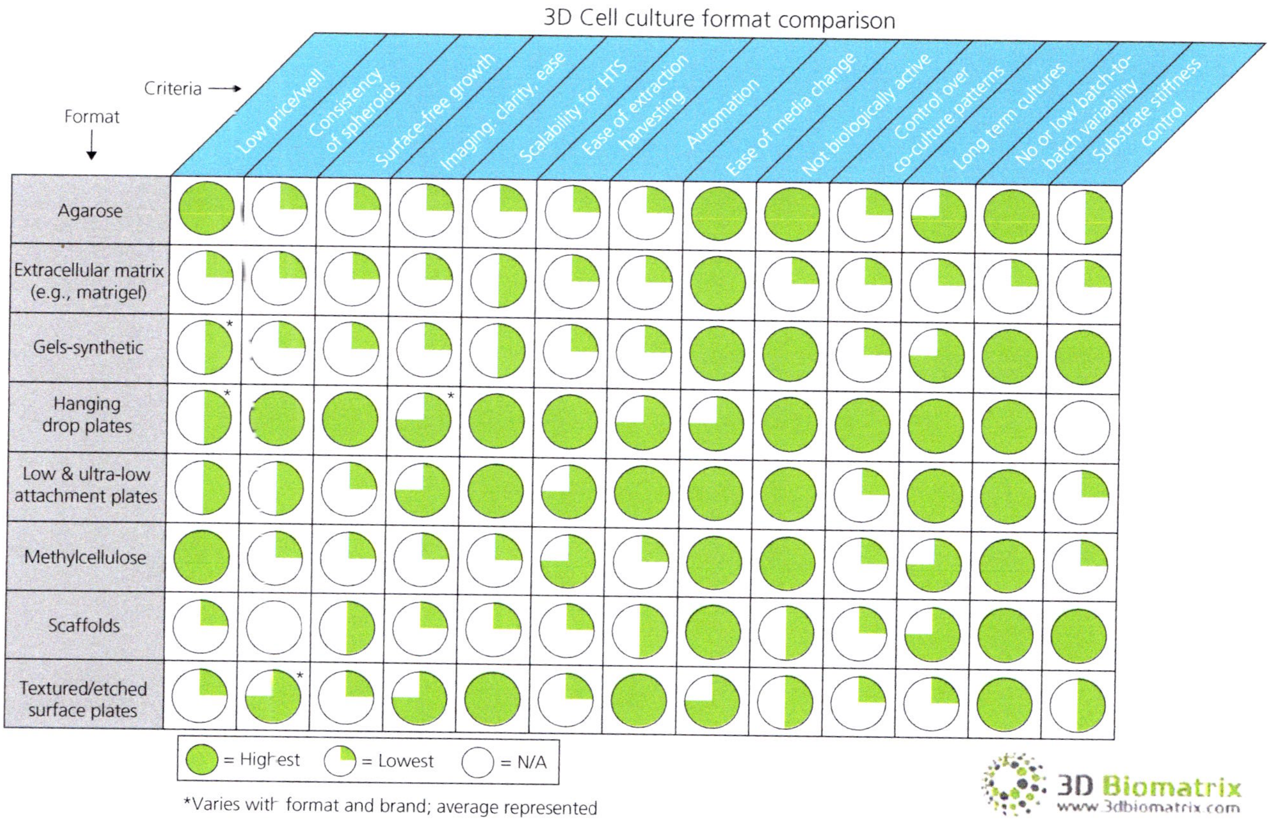

- a lack of consistent and more physiologically relevant preclinical model systems
- cost of failure from early false-positive or false-negative results.

To address such obstacles successfully, it was clear that a solution had to target improving the early stages of research. In discovery as an example, scientists seeking new therapies to treat disease know that if a compound is going to fail, it is best that it fails early; more physiologically relevant models are needed earlier in the research process. Different methods to achieve a 3D cell culture environment had to be considered such as gels, scaffolds and other matrices, coated or textured 2D plates and scaffold-free methods.

Because spheroids have long been considered to be well-characterised models for 3D culture and screening due to their similarity to physiological tissues compared to other models (del Duca *et al.*, 2004; Friedrich *et al.*, 2007, 2009; Kubbies, 2006; Kunz-Schughart *et al.*, 2004), developing a spheroidal environment became the focus, with the aim of addressing the overall challenges of adopting 3D cell culture in drug discovery and research laboratories. 3D Biomatrix's goals were to:

- create a system that provided an efficient, inexpensive 3D environment for consistent, reproducible, spheroid formation and culture that was easy to use and worked with existing laboratory supplies and equipment
- design a standard-size tool that would be applicable to laboratories of any size, supporting both small research projects as well as high-throughput screening (HTS) and high-content analysis (HCA) applications.

Spheroids, which are self-assembled multicellular aggregates that produce their own extracellular matrix (ECM), encompass the complex cell-matrix and cell-cell interactions that mimic functional properties of the corresponding tissue *in vivo* (Friedrich *et al.*, 2009) (Figure 3.1). Spheroids naturally mimic avascular tumours with inherent metabolic (oxygen, carbon dioxide, nutrients, wastes) and proliferative gradients; thus, they serve as excellent physiological tumour models known to provide more reliable and meaningful therapeutic results compared to 2D cell culture tests (Friedrich *et al.*, 2009; Kunz-Schughart *et al.*, 2004). Co-culture of cancerous and non-cancerous cells (i.e. endothelial, stromal and epithelial cells) extends the system's predictive cytotoxicity capabilities.

The hanging drop technique has evolved

Using the hanging drop technique to attain a 3D environment to grow spheroids is not a new concept. In fact, it has been around since at least the early 1900s. More recent adaptations include placing drops of cell suspension on the underside of culture plate lids and inverting them quickly so that the drops cling to the lid. While this method allows for cellular self-organisation not influenced by artificial substrates such as plastic or synthetic polymer-covered surfaces or scaffolds, it tests even the most patient researcher. The drops can slide into each other and often require stringent temperature control while producing variably sized spheroids. Further, the spheroids created in this manner are difficult to monitor, feed, maintain and harvest for further testing.

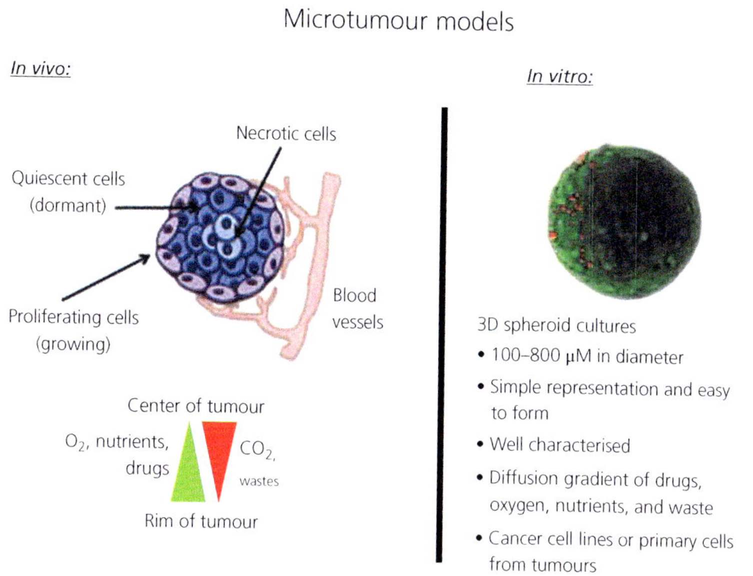

Figure 3.1 **3D Biomatrix microtumour model.** Spheroids encompass the complex cell-matrix and cell-cell interactions that mimic functional properties of the corresponding tissue *in vivo*.

The Perfecta3D HDP is one of the simplest tools available today to achieve a 3D cell culture and co-culture environment for spheroid development. As illustrated later in this chapter, a drop of cell suspension (media and cells) is pipetted into the top of each well, and the plate geometry causes it to hang stably below the well via surface tension. Without contacting any surfaces or matrices, the cells aggregate into a single spheroid. The hanging drop confines spheroid formation to one per well, allowing scientists to control the spheroid diameter with the type and number of cells in each well. This also allows the user to create or avoid a necrotic core in the spheroid, as experimentally required. The access holes on the top of the plate allow for media exchange, the addition of compounds, reagents, different cell lines, ECM or additional cell types or spheroids for co-cultures.

Unlike older hanging drop culture methods previously mentioned, HDPs do not require any flipping or inversion (Figure 3.2). The plates are used for long- or short-term cultures and co-cultures. The Perfecta3D HDPs meet industry standards and do not require any further investment in instrumentation, allowing HTS and HCA to be conducted with existing instrumentation and automation set-ups.

As *in vivo* as *in vitro* gets

The HDPs allow cell cultures and co-cultures to grow as a 3D spheroid 'microtissue' suspended in media. Researchers can test drugs on cells, such as stem cells, and solid tumours such as breast, colon, lung, liver, prostate, pancreatic and other

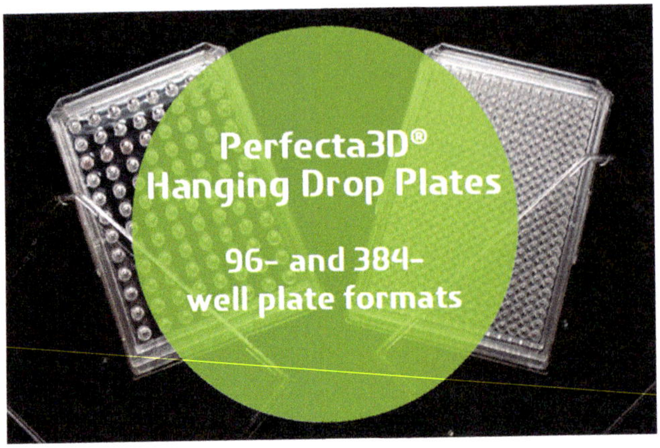

Figure 3.2 **Perfecta3D hanging drop plates.** The Perfecta3D HDPs are available in 96- and 384-well plate formats.

Table 3.2 An overview of the features of HDPs.

Pros	Cons
• No added materials • Easy to use • Consistent spheroid formation; control over size • Ability to add media and test compounds • Conduct assays directly in the hanging drop • Short- or long-term cultures • Co-culture development • Ability to harvest spheroids • Transparent • HTS and HCA capable; compatible with liquid handling tools • Inexpensive	• No support or porosity • Limited physical flexibility • Size of spheroid limited to 800 μm

HCA, high-content analysis; HTS, high-throughput screening.

cancer and avascular tumours, to see how they would react in a 3D environment similar to the body. The 3D environment of the HDP contains only media and cells; there are no other variables introduced to the spheroid culture such as gels, scaffolds or contact with plastic. The opportunity to identify and move only the most promising therapies forward to the animal and clinical phases is a compelling proposition for researchers, who want to increase efficiency while reducing costly later-stage failures.

The 3D Biomatrix products are designed to remove the biggest hurdles in finding the most promising therapies as early as possible using a 3D cell culture solution (Table 3.2).

- HDPs are comparatively low in price, and are easy to learn how to handle and use.
- HDPs do not need to be inverted to create the hanging drop.

- Media can be changed easily and test compounds added.
- Assays can be conducted directly in the hanging drop.
- HDPs can be used for both short- and long-term cultures.
- HDPs allow control of spheroid size with one spheroid per well.
- HDPs can be used to grow different forms of co-cultures and embryoid bodies.
- Spheroids can be transferred to any standard well plate.
- HDPs integrate with existing laboratory processes and instruments.
- HDPs can be used in HTS and HCA applications.

Laboratory automation equipment and 3D assays

Perfecta3D HDPs work with most brands of liquid handling instruments, plate readers and imagers. Liquid handlers, such as the Beckman Coulter Biomek® FX, BioTek MultiFlo™ FX, Eppendorf epMotion®, Agilent Bravo, Integra Viaflo, and CyBio CyBi®-Well, all work with the 96- and 384-well HDPs. Microplate imagers and plate readers, such as the BioTek Synergy and Cytation3™, Nexcelom Bioscience Celigo®, Tecan Infinite® readers,and TTP Labtech acumen® high-content imager, also work with the HDPs. Standard and inverted microscopes, both bright field and fluorescence, can be used for viewing and imaging. In addition, spheroids can be transferred from the HDP to a standard well plate, opening up the opportunity to use any automation equipment or confocal microscope for further spheroid analysis.

While most existing assays can be optimised for use in 3D spheroid culture, 3D Biomatrix also provides easy-to-use kit formats for a 3D cell viability assay, a 3D cytotoxicity assay and an invasion assay, all of which include HDPs.

Description of technology

One of the most exciting prospects among recent advances in 3D cell culture is standardised mass production of 3D spheroids for HTS applications. 3D Biomatrix's 96- and 384-well Perfecta3D HDPs are designed to enable consistent formation of spheroids using conventional liquid handling tools. This scaffold-free platform is simple to use, and generates spheroids with consistent sizes and shapes for controllable and reliable testing. Adjusting the seeding density, from as few as 50 cells to as many as 500 000 cells per well, allows production of varying spheroid sizes.

The plate consists of the main hanging drop culture plate and a complementary lid and tray, which serve to maintain sterility and reduce evaporation. Access holes in the culture plate allow manipulation of fluids and spheroids from the topside. Two liquid reservoirs constructed around the periphery of the culture plate and tray help to alleviate evaporation (Figure 3.3), a common problem in culturing hanging drops. The platform's dimensions meet ANSI/SBS standards.

Hanging drops are created by dispensing small volumes of cell suspensions, using standard pipette tips, into the access holes on the top of the plate.

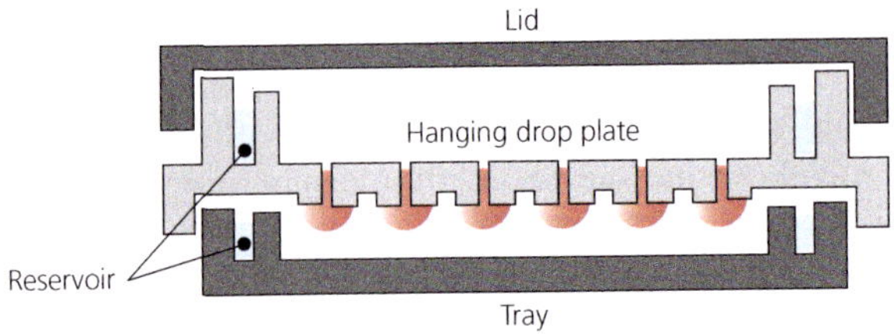

Figure 3.3 **Schematic of the Perfecta3D HDP.**

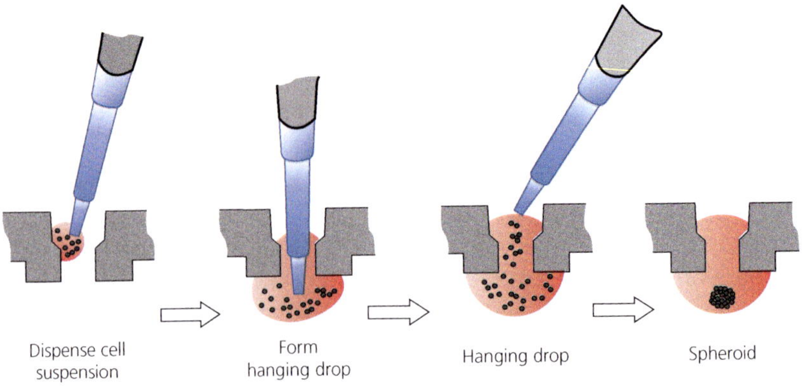

Figure 3.4 **Cell seeding into Perfecta3D HDP.** Spheroids are created by dispensing cell suspensions into the access holes of the Perfecta3D HDP, just like pipetting into conventional multiwell plates.

In a similar fashion, additional media can be added for long-term culture growth, and experimental reagents and drugs can be added to or removed from each hanging drop. A plateau structure on the bottom of each well stabilises the hanging drops (Figure 3.4).

It is surprising to some at first that the Perfecta3D HDPs do not have a bottom surface that cells would eventually attach to, therefore cells in the suspended drop aggregate into a spheroid.

Spheroids can be harvested and analysed using absorbance, fluorescence and luminescence assays measured with a plate reader. Microscopic imaging of spheroids can be performed directly with the transparent plate, lid and tray assembled. The platform also offers simplified liquid handling procedures and compatibility with HTS instruments and high-content imagers, as mentioned previously (Figure 3.5).

Cancer researchers have come to realise the critical importance of including stromal cells in their model systems for drug testing. Tumour cells are in communication with many other cell types, such as immune cells and fibroblasts that send and receive signals to and from the tumour. This constant communication is believed to be important in cancer progression and needs to be taken into consideration during experimental set-up (Junttila & de Sauvage, 2013).

Perfecta3D® hanging drops plates new user quick guide

Seeding

- Watch this short video illustrating drop formation: http://www.3ldbiomatrix.com/manual-pipetting-video
- Volume: 96-well plates: up to 50 μL per well; 384-well plates: up to 30 μL per well.
- If a drop fails to hang, simply insert a pipette tip into the well from the top to disrupt the surface tension. Do not add more liquid or push the pipette plunger up or down.
- The number of cells that can be grown in the Perfecta3D hanging Drop Plates varies widely (50–500,000 cells/drop) with cell type.

Feeding

- Some cell types can take up to 5 days to form a single spheroid; you may see multiple spheroids/well at first before they merge.
- Depending on the incubator's evaporation rate and the cells' metabolic activity, you may need to change media every day or every other day.
  - 96-well plates: Remove 10 μL and replace with 14 μL fresh media *every day* **or** remove 14 μL and replace with 20 μL fresh media *every other day*
  - 384-well plates: Remove 5 μL and replace with 7 μL fresh media *every day* **or** remove 7 μL and replace with 10 μL fresh media *every other day*
- Micro-tissues or co-cultures can be developed by adding different cell types or by adding one spheroid to another.

Treatment

- Compounds, small molecules, or even viral transduction reagents can be added directly to the hanging drops.
- Many assays that do not require lysis or washing can be performed directly in the perfecta3D hanging drop plates.
- If you transfer the spheroids before experimentation, they may stick to the surface of standard well plates and flatten. Conduct assays within an hour of transfer.
- Transfer spheroids to standard well plates by adding excess media or centrifuging (spin for 1 minute at 300 × g).

Analysis

- Easily analyze spheroids by microscopy using 4X and 10X objectives or transfer to a standard well plate for high magnification imaging.
- Anticipate that readouts will be different from what you observe in two-dimensional (2D) cultures, because three-dimensional (3D) spheroids recapitulate more realistic cell behavior, such as differential zones of proliferation.

Figure 3.5 Perfecta3D HDP user guide.

Such co-cultured spheroids can be easily manipulated using the Perfecta3D HDPs by varying the timing and order of seeding the different cell types into the hanging drops (Figure 3.6), and the ability to form different types of co-cultures has been demonstrated (Hsiao *et al.*, 2012).

PC-3_DsRed human prostate cancer cells, human umbilical vein endothelial (HUVEC) cells and MC3T3-E1 mouse preosteoblast cells were seeded together at a 1:50:50 ratio to form mixed co-culture spheroids (Figure 3.7). PC-3_DsRed cells, which were transfected with DsRed lentivirus to express red fluorescent protein, can be seen randomly distributed within the spheroid.

Concentric layer patterning was achieved by initially forming a spheroid of one cell type as the inner core. A cell suspension of the second cell type was subsequently added to the existing hanging drop to form an exterior coating around the inner core.

PC-3_DsRed human prostate cancer cells and MC3T3-E1 mouse preosteoblast cells were co-cultured at a 1:100 ratio, with PC-3_DsRed cells preferentially patterned in the centre core of the spheroid, randomly distributed within the spheroid, or as the exterior coating of the spheroid (Figure 3.8).

In another example, CellTracker™ Green-labelled COS7 African green monkey kidney fibroblast cells were cultured to form spheroids. On day 3, CellTracker

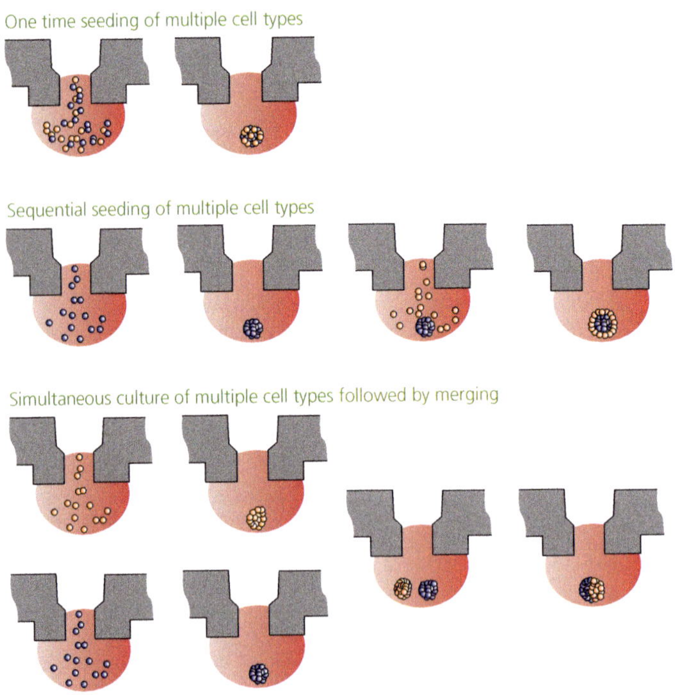

Figure 3.6 Co-culture variations in the Perfecta3D HDPs. Different culturing modalities allow for mixed co-culture spheroids to study cellular interactions.

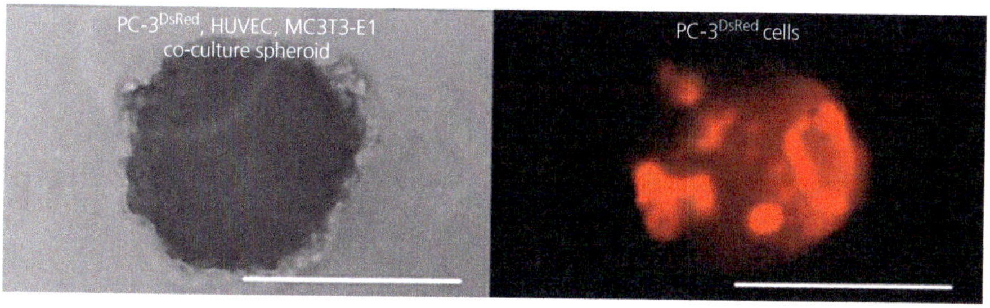

Figure 3.7 **Example of a mixed co-culture spheroid.** This example comprises PC-3_DsRed human prostate cancer cells, human umbilical vein endothelial (HUVEC) cells and MC3T3-E1 mouse preosteoblast cells seeded together at a 1:50:50 ratio. PC-3_DsRed cells, which were transfected with DsRed lentivirus to express red fluorescent protein, can be seen randomly distributed within the spheroid. Source: Hsiao 2012. Reproduced with permission of John Wiley & Sons.

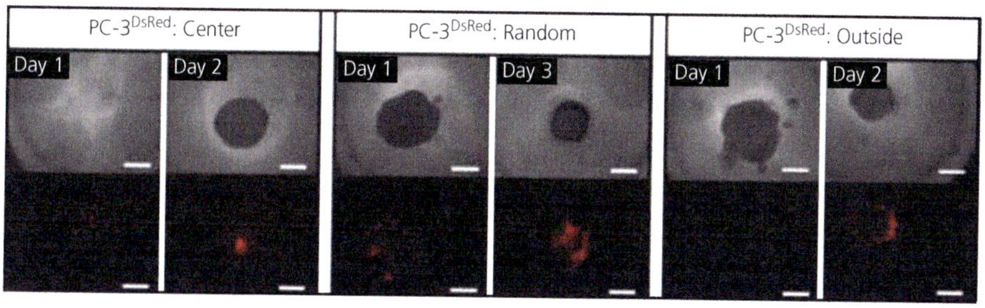

Figure 3.8 **PC-3_DsRed human prostate cancer cells and MC3T3-E1 mouse preosteoblast cells spheroids.** These cells were co-cultured at a 1:100 ratio to show a variety of patterned co-cultures achieved using different seeding techniques. Source: Hsiao 2012. Reproduced with permission of John Wiley & Sons.

Red-labelled COS7 cells were added to each hanging drop containing a CellTracker Green-labelled COS7 spheroid. The red cells attached to the periphery of the existing green spheroid and subsequently formed into a single spheroid with the green cells (Figure 3.9).

Janus spheroids are spheroids composed of two groups of cells juxtaposed next to each other so that each group essentially forms a hemisphere. This cellular patterning method opens new possibilities for understanding and manipulating interactions between different cell types in 3D cell culture. A Janus spheroid can be formed easily in the Perfecta3D HDPs by retrieving a spheroid from one well and then gently pipetting it into another well containing a spheroid, similar to the removal and addition of liquid.

The formation of Janus spheroids is unique to hanging drop plates, such as the Perfecta3D HDPs, which have an access hole large enough to allow the transfer of an additional spheroid with a pipette.

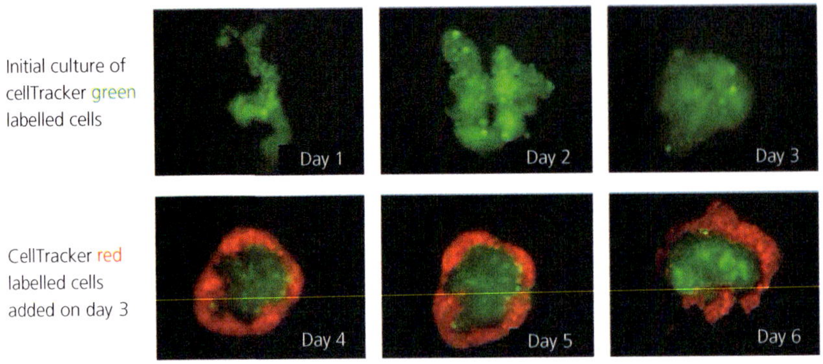

Figure 3.9 Patterned co-cultures. In another example of patterned co-cultures, CellTracker Green-labelled COS7 African green monkey kidney fibroblast cells were cultured to form spheroids, then on day 3, CellTracker Red-labelled COS7 cells were added to each hanging drop. The red cells attached to the periphery of the existing green spheroid and subsequently formed into a single spheroid with the green cells. Source: Hsiao 2012. Reproduced with permission of John Wiley & Sons.

CellTracker Green-labelled and Red-labelled COS7 African green monkey kidney fibroblast cells were cultured separately in different wells to form spheroids. On day 3, a green spheroid was transferred into each hanging drop containing a red spheroid. Over the next few days, the two spheroids slowly aggregated together to form a single Janus spheroid with half green cells and half red cells (Figure 3.10).

Experimental results obtained using the Perfecta3D hanging drop plates

Experimental results obtained using the Perfecta3D HDPs were first published in 2011 (Tung *et al.*, 2011). In this study, it was demonstrated that drugs with different modes of action produce distinct responses in the physiological 3D spheroids compared to conventional 2D cell monolayers.

COS7 African green monkey kidney fibroblasts, ES-D3 murine embryonic stem cells (mESCs) and A431.H9 human epithelial carcinoma cells were cultured in the Perfecta3D HDPs to form spheroids. mESCs readily formed spheroids on day 1, whereas COS7 cells formed smaller aggregates initially that later assembled into a larger spheroid. A live/dead assay indicated that more than 90% of the cells remained alive after 12 days of culture, demonstrating that the Perfecta3D HDPs are suitable for longer-term spheroid cultures. In some circumstances, spheroids have been maintained in the HDPs for up to 40 days.

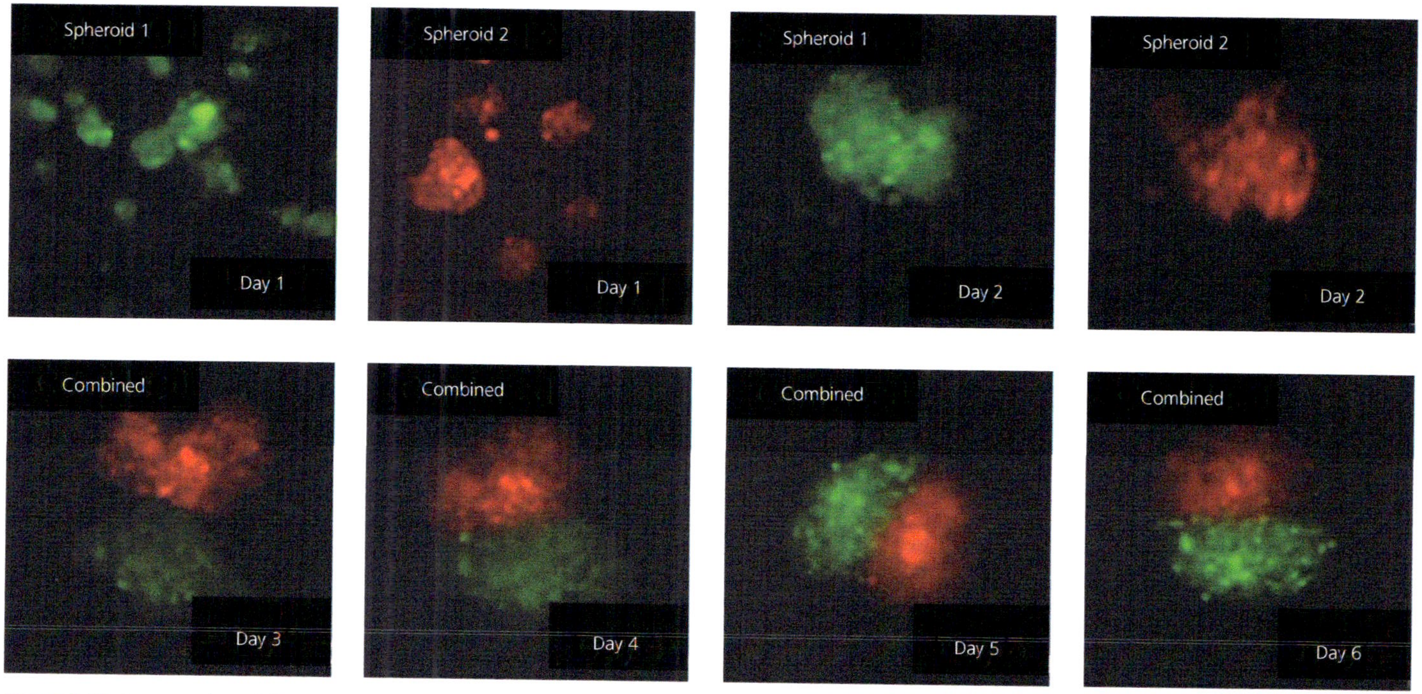

Figure 3.10 **Janus spheroids.** To illustrate the formation of Janus spheroids, CellTracker Green-labelled and Red-labelled COS7 African green monkey kidney fibroblast cells were cultured separately in different wells to form spheroids. On day 3, a green spheroid was transferred into each hanging drop containing a red spheroid. The two spheroids slowly aggregated together to form a single Janus spheroid with half green and half red cells. Source: Hsiao 2012. Reproduced with permission of John Wiley & Sons.

A431.H9 cells were seeded at various cell densities and the size of the spheroids was recorded over a seven-day culture period. The data demonstrated that cells remained proliferative by the end of the culture period. Moreover, the data showed that spheroid size can be fine-tuned by altering the cell seeding density and the length of culture period. This is important since the size of cell colonies has been reported to affect the differentiation of stem cells (Hwanga *et al.*, 2009).

Two anticancer drugs with distinctly different activity profiles were used to assess chemosensitivity: 5-fluorouracil (5-FU), a conventional compound that inhibits cellular proliferation, and tirapazamine (TPZ), a hypoxia-triggered cytotoxin that causes DNA damage. Data showed that A431.H9 cells are more resistant to 5-FU under 3D spheroid than 2D culture conditions. However, the opposite is true for treatment with TPZ. Data indicated that A431.H9 cells are more resistant to TPZ when cultured under 2D rather than 3D spheroid culture conditions.

5-Fluorouracil primarily targets proliferating cells. Therefore, it is more effective against the rapidly proliferating cells in 2D monolayer cultures and does not kill the quiescent cells in the spheroids. In contrast, TPZ is a hypoxia-activated cytotoxin, so it is more effective in the spheroids, where active oxygen consumption by cells and limits in diffusive oxygen transport create a hypoxic core similar to that of solid tumours (Sutherland, 1988).

The IC_{50} of A431.H9 cells treated with 5-FU was about 0.1 μM in 2D condition and 1 to 100 μM, depending on cell density, in 3D spheroid cultures. For the TPZ-treated cells, the IC_{50} was about 50 μM when cultured in 2D and 8 μM for all spheroid sizes. Such distinctly and dramatically different responses from the same cells to the same drugs tested under different culture conditions illustrated the importance and need of using 3D models in drug screening and testing.

The combined treatment of 5-FU and TPZ was more effective at killing the A431.H9 cells. At a drug concentration of 10 μM, cell viability decreased from 75% and 40% when treated with 5-FU and TPZ, respectively, to only 20% when the spheroids received the combined treatment. By combining two drugs of distinctly different mechanisms, both the proliferating cells in the peripheral layer and cells in the hypoxic core of the spheroids can be targeted effectively.

Hanging drop plates and high-throughput screening

In a 2013 study, spheroid cultures generated in the Perfecta3D HDPs were used to create a colorectal cancer model for a proof-of-concept drug validation experiment (Horman *et al.*, 2013a). HCT-116 human colorectal carcinoma cells were used to generate spheroids by seeding 20 000 cells in 96-well HDPs and 5000 cells in 384-well HDPs. Cells were maintained in hanging drops for three days to permit spheroid formation. The resulting spheroids were uniformly sized and displayed characteristic morphologies as observed by light microscopy.

Spheroids were transferred into 384-well, black-sided, clear-bottomed, high-content imaging assay plates (Greiner) by the addition of supplementary media,

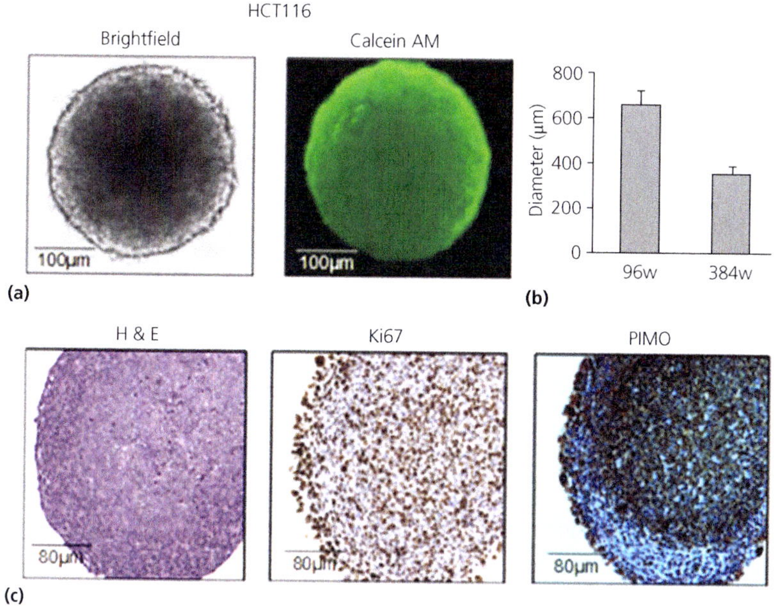

Figure 3.11 96- and 384-well spheroid formation, imaging, drug treatment and analysis. (a) HCT-116 human colorectal carcinoma spheroids from 384-well HDPs display characteristic morphology under bright-field microscopy and can be stained with calcein AM for fluorescence-based size and viability analysis. (b) 384-well spheroids were on average 350 μm, and 96-well spheroids measured approximately 650 μm. Errors bars represent standard deviations. (c) Immunohistochemistry sections of 384-well HCT-116 colorectal carcinoma spheroids were stained with H&E, Ki67 (proliferation) and PIMO (hypoxic adducts). Source: Horman 2013. Reproduced with permission of Nature Publishing Group.

and viable cells were stained with calcein AM. Spheroid size analysis was performed by capturing images via confocal microscopy (Figure 3.11).

The resulting spheroids had an average diameter of 650 μm from 96-well HDPs and 350 μm from 384-well HDPs, demonstrating excellent reproducibility of spheroid size relative to initial cell plating number. Sectioning and immunohistochemistry were performed to further characterise the tumour-like morphologies (haematoxylin and eosin stain, H&E), proliferative zones (Ki67 antibody) and presence of hypoxic adducts (pimonidazole, PIMO) of 384-well tumour spheroids. Although the spheroids were less than 500 μm in diameter, there was evidence of hypoxic cores, and proliferation was detected throughout the spheroids, particularly along the outermost layer (see Figure 3.11). Based on these observations, HDP-derived spheroids constitute a robust 3D tumour model for high-throughput drug discovery.

The previous work was extended to investigate the potential of conducting RNAi screens in 3D (Horman *et al.*, 2013b). In this work, a HTS spheroid model of colorectal cancer tumour formation was developed to test whether the silencing of specific integrin receptor and receptor tyrosine kinase (RTK) genes inhibited spheroid formation.

Prior to initiation of the RNAi screen, HCT-116 colorectal carcinoma cells demonstrated reproducible spheroid formation with a consistent mean diameter of 350 μm ± 25 μm in the 384-well Perfecta3D HDPs. Two methods of viral infection were assessed. For the EGFP–reporter gene method, 5000 HCT-116 cells in 25 μL of McCoy's 5A medium were dispensed per well into a 384-well Perfecta3D HDP. Immediately after, 5 μL of viral supernatant containing a gene-specific short hairpin RNA (shRNA)-EGFP-lentivirus or a non-targeting shRNA-EGFP-lentivirus control was dispensed into the HDP. Infection and spheroid formation occurred for three days under standard growth conditions.

For the puromycin (puro)-selection method, 50 000 HCT-116 cells in standard, flat-bottom, 96-well plates were incubated overnight in medium with viral supernatant containing a gene-specific shRNA-puro-lentivirus or a non-targeting shRNA-puro-lentivirus control. After 24 hours the transduced cells were selected in 2 μg/mL puromycin for three days. Cells were then trypsinised, washed and transferred to the 384-well Perfecta3D HDPs (30 μL per well), and spheroids were allowed to form for three days.

The shRNA-EGFP-lentivirus spheroids were transferred into 384-well, black-sided, clear-bottom imaging plates (Greiner) and imaged via the TTP Labtech acumen eX3 high-content imager. The non-targeting shRNA-EGFP-lentivirus and the ensuing viral infection did not noticeably alter spheroid formation, indicating that this is an efficient system to deliver shRNAs to spheroids.

The puro-selection method was utilised to infect cells with shRNA-expressing lentiviruses against platelet-derived growth factor receptor-α (PDGFRα), PDGFR-β, integrin β1 (ITG-β1) or a non-targeting shRNA–puro-lentivirus control (NT control). After three days, spheroids were transferred into 384-well assay plates, stained with calcein AM and imaged on the acumen eX3. Relative spheroid formation was calculated by dividing the average total intensity measurement of the shRNA-infected spheroids by the average total intensity measurement of the NT control spheroids. Genetic knockdown of PDGFR-α, PDGFR-β or ITG-β1 led to significantly less spheroid formation than seen with the NT control (Figure 3.12).

Lowering PDGFR-α, PDGFR-β or ITG-β1 expression significantly reduced spheroid formation consistent with what has been reported in the literature, and described later in this chapter. These results demonstrate that viral genomic reagents can be used in conjunction with 3D spheroid culture in Perfecta3D HDPs and high-content analysis with the acumen eX3 to screen RNAi libraries in an HTS-compatible format.

In addition, compatibility of the Perfecta3D hanging drop plates with existing automation and analytical instrumentation has been further demonstrated using the BioTek MultiFlo FX and BioTek Cytation 3 Cell Imaging Multi-Mode

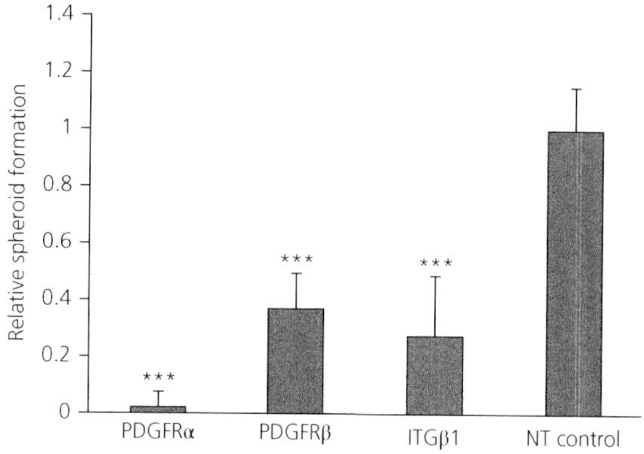

Figure 3.12 Relative spheroid formation of HCT-116 colon tumour cells. Relative spheroid formation after infection with shRNA-expressing lentiviruses against PDGDFR-α, PDGFR-β, ITG-β1 or NT control. Relative spheroid formation was calculated by dividing total calcein AM-positive cytokine-specific shRNA or NT control area by calcein AM-positive NT control area. Errors bars represent standard deviation. $***P<0.001$. Source: Horman 2013. Reproduced with permission of Nature Publishing Group.

Reader to dispense cell suspensions and to track spheroid formation using digital wide-field imaging.

Automating each of these processes provides a 3D solution that is less labour intensive and more reproducible than previous methods, and further promotes the use of this spheroid formation method for low- or high-throughput purposes.

Two important steps must be accomplished to facilitate spheroid formation for 3D culture in the HDPs: dispensing cells and medium, and tracking spheroid formation. Automated dispensing is faster than manual pipetting, and is the preferred method, especially with the 384-well HDP. Cell imaging in the hanging drop is possible with a 4× or 10× objective, or a long working distance 20× objective. Tracking simultaneous formation of multiple spheroids can be accomplished with the Cytation 3.

For spheroids containing a single cell type, MCF-7 green fluorescent protein (GFP) cells were harvested and diluted to concentrations of 6.25×10^5, 2.5×10^5, 1.25×10^5, and 2.5×10^4 cells/mL. When dispensed into the HDPs in 40 μL volumes, final cell concentrations equalled 25 000, 10 000, 5000 and 1000 cells per spheroid. For co-cultured spheroids, MCF-7 GFP and fibroblast cells were harvested and diluted to concentrations of 2.5×10^4, 1.25×10^4 and 6.25×10^3 cells/mL. The cells were combined together in a single final volume to create spheroids containing 2000, 1000 or 500 cells per spheroid and equal numbers of each cell type.

In order to reduce the dispense speed and allow the cell suspension to dispense in a droplet rather than a stream of liquid, a 5 μL cassette was placed into the MultiFlo FX's primary or secondary peristaltic pump position, while a 10 μL cassette type was selected in the dispenser utilities interface in the same position. This deviation caused half the volume normally expected to move through the cassette tubing. A 'peri-pump dispense step' was then selected. In order to dispense 40 μL using the specialised method, 80 μL was entered as the dispense volume. Further dispense settings included selecting a 'low' flow rate, in addition to positioning co-ordinates of Z:249, X:10 and Y:0. The X co-ordinate shifts the dispense tips to the right of centre, while the Z co-ordinate further ensures that drops consistently form beneath each well. In this manner, 40 μL of cell suspension was slowly dispensed as a droplet without falling through the well bottom, and with optimal contact between the well sides and the droplet.

The spheroids were imaged using the Cytation 3, which can focus through the clear tray below the hanging drops.

Images of a well containing 5000 MCF-7 GFP cells were captured after one, two, three and four days of incubation (Figure 3.13). Fluorescence imaging using the GFP channel (see Figure 3.13a–d), as well as bright-field imaging (see Figure 3.13e–h), was carried out to fully visualise cell configuration on each day. Bright-field and GFP images were then overlaid to create the final composite images (see Figure 3.13i–l).

Spheroids were also created containing a co-culture of MCF-7 and human dermal fibroblast cells. The important role that fibroblast cells play in tumour progression has been previously shown (Lin & Chang, 2008). Therefore, forming 3D cell structures which contain co-cultured cancer cells, in addition to fibroblasts, further promotes the creation of an *in vitro* cellular model which more closely resembles a true *in vivo* tumour environment. Spheroids containing 2000, 1000 or 500 total cells, with equal numbers of cancer and fibroblast cells, were created and imaged using fluorescence GFP or red fluorescent protein (RFP), as well as bright-field imaging channels (Figure 3.14).

Troubleshooting

Spheroid formation basics

There are cell-intrinsic differences between cell lines, which influence spheroid formation. As an example, three hepatoma cell lines (Hep3B, HepG2 and PLC/PRF/5) showed very different spheroid morphology after culture for 24 hours. Hep3B cells formed a tight format where cell borders were no longer visible, HepG2 cells formed a loose aggregate that had not yet formed a spheroid, and finally PLC/PRF/5 cells fell into the bottom of a drop, not aggregating or touching one another (Lin *et al.*, 2006).

To understand the intrinsic differences, the function of proteins that were likely to play a role in cellular adhesion was examined (Lin *et al.*, 2006). Integrins are a group of proteins used to connect cells with ECM, and are made up of two

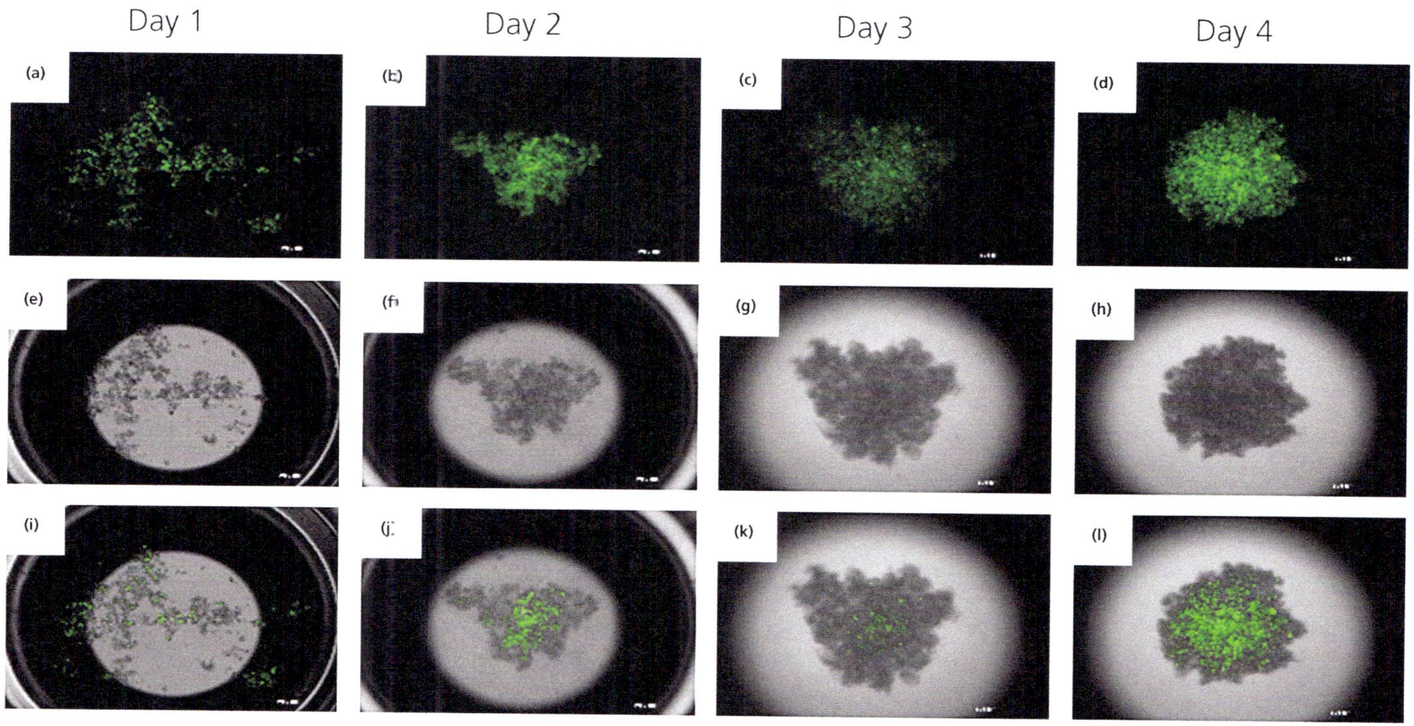

Figure 3.13 5000 cell spheroid formation tracking. MCF-7 spheroid formation imaged over a four-day period using GFP fluorescent channel (a–d) or bright-field imaging (e–h). Gen5 software was used to create final overlaid composite images (i–l). Source: BioTek Instruments, Inc.

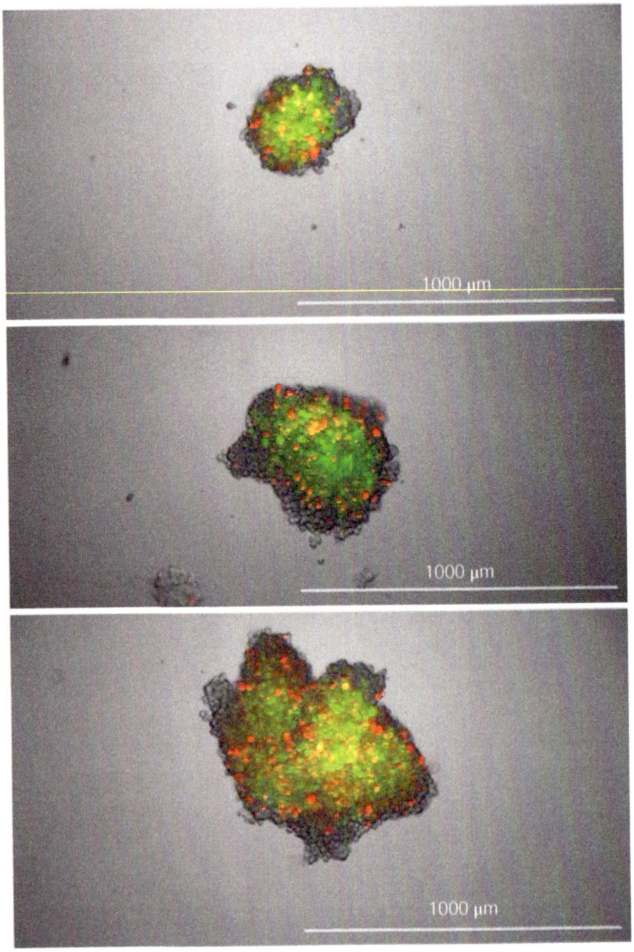

Figure 3.14 **Co-cultured cell spheroid formation tracking.** Overlaid fluorescence and bright-field images of MCF-7/human dermal fibroblast spheroids containing 500 (*top*), 1000 (*middle*) or 2000 (*bottom*) cells. GFP expression from MCF-7 cells and RFP expression from fibroblast cells were measured. Source: BioTek Instruments, Inc.

subtypes, α and β, with lots of combinations. β1-integrin is crucial for adherence to ECM. All three cell lines expressed high and approximately equal amounts of β1-integrin. Therefore, β1-integrin expression was not affecting spheroid formation.

Next, E-cadherin expression was studied. E-cadherins are another family of proteins important in cell aggregation; after the cells touch, E-cadherins help form connections between them. E-cadherin expression is common in epithelial cells. Hep3B cells expressed 8.5-fold higher and HepG2 cells fourfold higher E-cadherin compared to the PLC/PRF/5 cells. The difference in E-cadherin expression explained the tight versus loose aggregates, but not the initial spheroid formation.

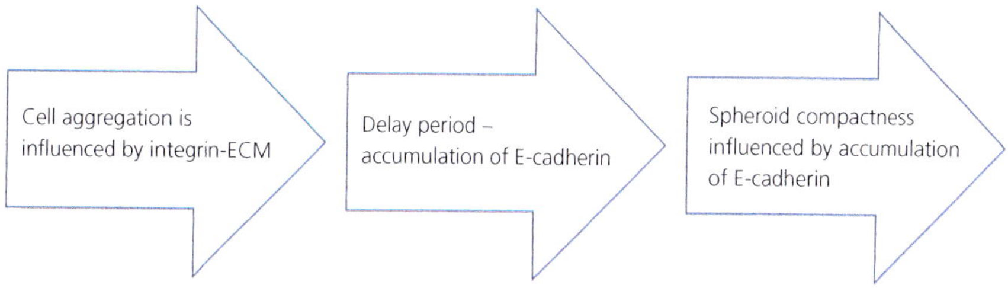

Figure 3.15 Spheroid formation model. A spheroid formation model developed by Lin and Chang suggests cell aggregation into spheroids is initially influenced by integrin and ECM, and spheroid compactness is influenced by accumulation of E-cadherin. Source: Lin 2008. Reproduced with permission of John Wiley & Sons.

Extracellular matrix secretion was studied next. Hep3B cells secreted the most ECM, HepG2 cells a fair amount and PLC/PRF/5 cells only a minimal amount. Therefore the conclusion was that ECM secretion played a role in spheroid formation.

Finally, E-cadherin expression in HepG2 cells grown either as spheroids or in single cell suspension culture was evaluated. HepG2 cells increased E-cadherin expression over 48 hours compared to the cells grown in suspension. The results allowed development of a spheroid formation model (Lin & Chang, 2008).

When the HepG2 cells were placed in suspension culture integrin-binding ECM drew the cells into a loose aggregate. Then there was a delay period as the cells accumulated E-cadherin expression which influences compaction, explaining why added ECM promotes spheroid formation (Figure 3.15).

Cell lines that establish relatively weak intercellular interactions, forming aggregates with less tightly associated cells, are not the best *in vitro* models since they lack diffusion barriers, as well as a stratified cellular composition with proliferating cells in the rim and quiescent and necrotic cells in the core. With the addition of exogenous ECM, these aggregates are able to form compact spheroids, which resemble more closely the morphology and diffusion properties of xenografts derived from these tumour cell lines.

The cell-specific matrix requirements – media-soluble ECM, individual matrix components (such as laminin, collagen or fibronectin) or methylcellulose – and final concentrations should be determined experimentally for each cell type and downstream application.

Example cell lines that form compact spheroids with the addition of ECM include MDA-MB-231 human breast adenocarcinoma cells, HepG2 human hepatocellular carcinoma cells and SK-OV-3 human ovarian adenocarcinoma cells.

Spheroid size

Spheroid size critically affects results. Large spheroids have hypoxic cores and can be more challenging for imaging and analysis but the hypoxic core may be necessary, depending on the model system.

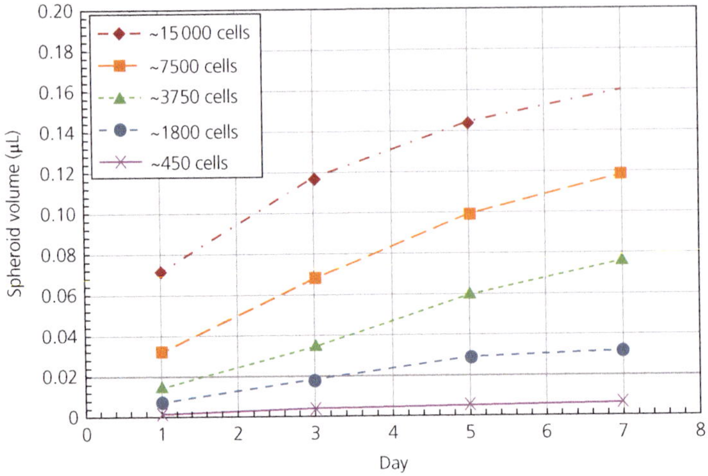

Figure 3.16 Determining spheroid size. Spheroid size is determined by plating cell number, and needs to be determined experimentally as it is also dependent on the size of the cell and the compactness of the spheroid. Source: Tung 2011. Reproduced with permission of the Royal Society of Chemistry.

Spheroid size is determined by plating cell number as illustrated by A431-H9 cells grown over a seven-day period in the 384-well Perfecta3D HDP (Figure 3.16) (Tung *et al.*, 2011). Spheroid size needs to be determined experimentally as it is also dependent on the size of the cell and the compactness of the spheroid.

Optimising assays for 3D

If there is a need to use assays that are not preoptimised for 3D cell culture, a main consideration is the thickness differences between 2D and 3D cultures since thickness plays a role in reagent penetration. With spheroids grown in the Perfecta3D HDPs, cultures can easily be exposed to different incubation times, temperature and even reagent concentrations. Samples can be fixed and sectioned to determine the penetration depth. Complete lysis can be done but it should be kept in mind that lysis is an endpoint whereas other analysis can allow for multiple read-outs from each spheroid.

Spheroid compatible assays include viability, proliferation, toxicity, invasion and migration and stem cell differentiation. An example is illustrated in Figure 3.17.

Results will be different in 3D

Because cell-cell interaction in 3D is more physiologically relevant, results attained in 3D cell culture will be different from those attained in 2D cultures, as illustrated by the example below. DU145 prostate cancer cells were treated with carboplatin, a platinum-based antineoplastic agent that interacts with DNA to interfere with DNA repair, then analysed with the Promega CellTiter Glo®

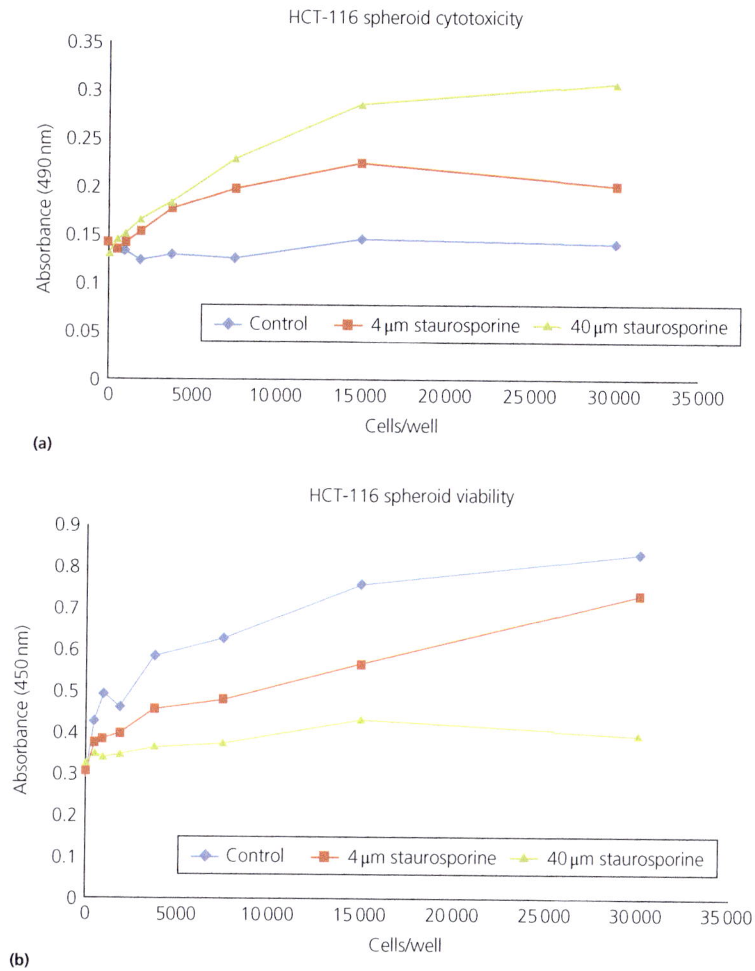

Figure 3.17 HCT-116 spheroid cytotoxicity and viability. (a) Cytotoxicity graph from human colorectal carcinoma cells (HCT-116) seeded into the HDP at 30 000, 15 000, 7500, 3750, 1875, 938 and 469 cells per 50 µL hanging drop of McCoy's 5A medium with 10% fetal bovine serum and cultured for five days at 37 °C with 5% CO_2 to generate uniform spheroids. For compound treatment, 4 µM staurosporine, 40 µM staurosporine or DMSO alone (control) were added for 48 hours. A 10 µL media sample was removed from each HDP well and evaluated using the Cayman Perfecta3D cytoxicity assay kit. (b) Graph of viability results from HCT-116 spheroids in (a) measured using the Cayman Perfecta3D viability kit, nicely illustrating reciprocal results compared to the cytotoxicity assay. Source: the Cayman Chemical Company.

luminescent assay to investigate cell viability and metabolic activity on the BioTek Synergy™ Neo microplate reader (Figure 3.18).

Typically, in 2D cell culture, a linear decrease in cells would correspond with an increased dosage of drug. However, in 3D cell culture there appears to be an increase in viability at low doses. Most likely, low doses of carboplatin kill the

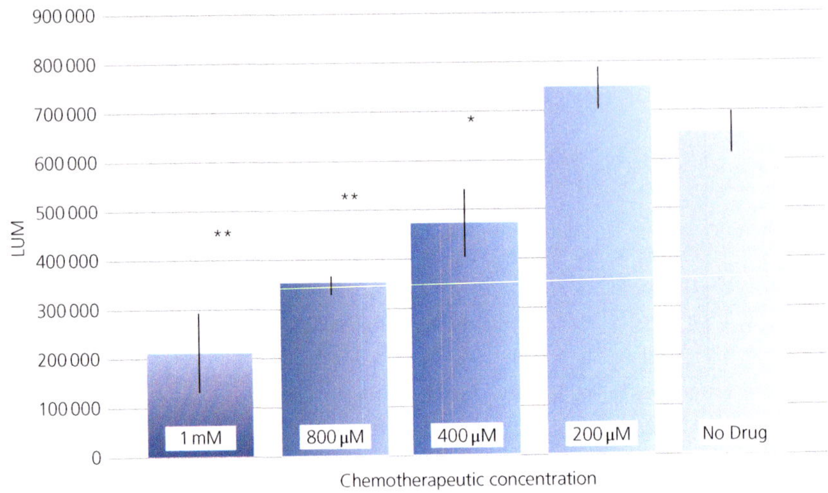

Figure 3.18 DU145 prostate cancer cells treated with carboplatin. DU145 prostate cancer cells were treated with carboplatin, a platinum-based antineoplastic agent that interacts with DNA to interfere with DNA repair, then analysed with the Promega CellTiter Glo luminescent assay to investigate cell viability and metabolic activity on the BioTek Synergy Neo microplate reader.

proliferating cells around the outside of the spheroid, which allows more cells access to the nutrients and oxygen and initiates proliferation at low levels. As dosage was increased, the drug was able to penetrate through the layer of dead cells to reach and destroy cells on the inside of the spheroid.

Conclusion

3D cell culture methods were once expensive, messy, laborious and difficult to adapt to existing procedures. Today, researchers can choose from an array of 3D cell culture tools, ranging from simple to complex, that provide an environment that mimics the structure, cell-cell and cell-matrix interactions of the *in vivo* environment. There is no 'one size fits all' solution for 3D applications. In addition, some 3D tools can be used in conjunction with each other.

One of the most exciting prospects among recent advances in 3D cell culture is standardised mass production of 3D spheroids for general and HTS applications. The 96- and 384-well Perfecta3D HDPs from 3D Biomatrix, used by scientists worldwide in industry and academia for cancer, stem cell and toxicology research, are designed to enable consistent formation of spheroids using conventional liquid handling, plate readers and high-content imaging tools. This scaffold-free platform is inexpensive and simple to use, and generates spheroids with consistent sizes and shapes for controllable and reliable testing. Adjusting the seeding density allows production of varying spheroid sizes, and co-cultures with other cell types (i.e. endothelial, stromal, epithelial cells) extend the predictive cytotoxicity capabilities of this 3D cell culture system.

The HDPs for spheroid growth are simple to implement and highly beneficial for scientists who are interested in a 3D *in vitro* environment for long- or short-term cultures or co-cultures that do not contain added biomaterials.

Acknowledgements

3D Biomatrix would like to acknowledge the many individuals at Cayman Chemical Company, BioTek Instruments, Inc., TTP Labtech Ltd. and Genomics Institute of the Novartis Research Foundation who contributed to application development.

References

Del Duca, D., Werbowetski, T. & del Maestro, R. F. (2004) Spheroid preparation from hanging drops: characterization of a model of brain tumor invasion. *Journal of Neuro-oncology*, **67**, 295–303.

Friedrich, J., Ebner, R. & Kunz-Schughart, L. A. (2007) Experimental anti-tumor therapy in 3-D: spheroids – old hat or new challenge? *International Journal of Radiation Biology*, **83**, 849–871.

Friedrich, J., Seidel, C., Ebner, R. & Kunz-Schughart, L. A. (2009) Spheroid-based drug screen: considerations and practical approach. *Nature Protocols*, **4**, 309–324.

Horman, S. R., To, J., Orth, A. P. *et al.* (2013a) 3D high-content analysis of spheroids: new methodology shown to improve screening. *Genetic Engineering & Biotechnology News*, **33**.

Horman, S. R., To, J., Orth, A. P. *et al.* (2013b) High-content analysis of three-dimensional tumor spheroids: investigating signaling pathways using small hairpin RNA. *Nature Methods*, **10**.

Hsiao, A. Y., Tung, Y. C., Qu X. *et al.* (2012) 384 hanging drop arrays give excellent Z-factors and allow versatile formation of co-culture spheroids. *Biotechnology and Bioengineering*, **109**, 1293–1304.

Hwanga, Y. S., Chung, B. G., Ortmanna, D. *et al.* (2009) Microwell-mediated control of embryoid body size regulates embryonic stem cell fate via differential expression of WNT5a and WNT11. *Proceedings of the National Academy of Sciences USA*, **106**, 16978–16983.

Junttila, M. R. & de Sauvage, F. J. (2013) Influence of tumour micro-environment heterogeneity on therapeutic response. *Nature*, **501**, 346–354.

Kubbies, I. A. M (2006) Rapid generation of single-tumor spheroids for high-throughput cell function and toxicity analysis. *Journal of Biomolecular Screening*, **11**, 922–932.

Kunz-Schughart, L. A., Freyer, J. P., Hofstaedter, F. & Ebner, R. (2004) The use of 3-D cultures for high-throughput screening: the multicellular spheroid model. *Journal of Biomolecular Screening*, **9**, 273–285.

Lin, R. Z. & Chang, H. Y. (2008) Recent advances in three-dimensional multicellular spheroid culture for biomedical research. *Biotechnology Journal*, **3**, 1172–1184.

Lin, R. Z., Chou, L. F., Chien, C. C. M. & Chang, H. Y. (2006) Dynamic analysis of hepatoma spheroid formation: roles of E-cadherin and β1-integrin. *Cell Tissue Research*, **324**, 411–422.

Sutherland, R. M. (1988) Cell and environment interactions in tumor microregions: the multicell spheroid model. *Science*, **240**, 177–184.

Tung, Y. C., Hsiao, A. Y., Allen, S. G. *et al.* (2011) High-throughput 3D spheroid culture and drug testing using a 384 hanging drop array. *Analyst*, **136**, 473–478.

CHAPTER 4

NanoCulture Plate: A scaffold-based high-throughput three-dimensional cell culture system suitable for live imaging and co-culture

Manabu Itoh[1], Kazuya Arai[1], Hiromi Miura[2] and M. Mamunur Rahman[2]

[1] *SCIVAX Life Sciences, Inc., Kawasaki, Kanagawa, Japan*

[2] *SCIVAX USA, Inc., Woburn, Massachusetts, USA*

Introduction

Three-dimensional (3D) cell culture has acquired substantial attention in recent years, due to its potential to deliver higher quantity cell culture information that is more representative of real tissue morphology and the cancer microenvironment, as well as being predictive of drug responses *in vivo* (Chiba *et al.*, 2014; Karlsson *et al.*, 2012). The transition from cell culture on the flat surface of a conventional two-dimensional (2D) structure to an environment, matrix or scaffold with 3D architecture is providing much needed support for emerging applications in cancer research, toxicology research, regenerative research and stem cell research (Arai *et al.*, 2013; Ivascu & Kubbies, 2006; Nakamura *et al.*, 2011a; Ribeiro *et al.*, 2012).

3D culture enables the formation of multicellular clusters (spheroids) that are driven by cell-cell and cell-matrix contacts, rather than cell-plastic interactions. Basically, there are two ways to accomplish this: (i) growing the cells in a scaffold and (ii) scaffold free, forcing cells to grow in suspension (Rimann & Graf-Hausner, 2012). There are different kinds of gels (animal derived or synthetic) and plates (low attachment) available for 3D cell culture. Though it is infrequently reported, it is a known fact that there are differences in cell growth, survival rate and drug sensitivity between scaffold-based and suspension-based 3D cell culture technologies. For example, human liver cancer cell line HepG2 hardly proliferates in suspension-based 3D cell culture while maintaining growth in scaffold-based 3D cell culture (Figure 4.1).

It is argued that scaffold-based 3D culture systems are superior to suspension-based 3D culture systems. However, their utilisation depends on the purpose of

---

Technology Platforms for 3D Cell Culture: A User's Guide, First Edition. Edited by Stefan Przyborski.

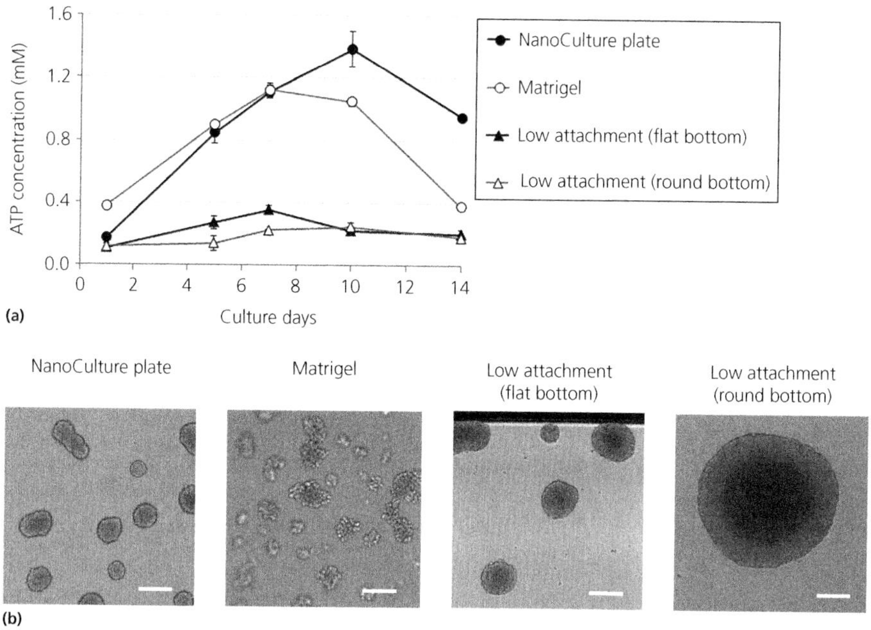

Figure 4.1 3D cell culture of HepG2 cells. (a) Proliferation curves of HepG2 cells under scaffold- or suspension-type 3D culture conditions. (b) HepG2 cells spheroid morphology on different types of 3D culture system. Scale bar, 100 µm.

the experiment. Suspension culture can be chosen when the interaction between the cells and the scaffold needs to be minimised. On the other hand, if the *in vivo* microenvironment and the interactions between the surrounding cells and extra-cellular matrix need to be reproduced, then a scaffold should be chosen.

Various publications have reported using soft agar, collagen, synthetic peptides and Matrigel® (Bentires *et al.*, 2006; Folkman & Hochberg, 1973; Zamora *et al.*, 1980) to act as the scaffolding material where cells can grow and proliferate. These methods require troublesome preparation, and as the culture medium gelates, it can cause various problems when developing an assay method for drug screening, for example affecting drug dispersion, difficulty in harvesting, difficulty in imaging and focusing as the spheroids lie scattered towards the z-axis. Moreover, gels containing biological substances vary from batch to batch, and thus cannot be used by pharmaceutical companies for drug screening, where reproducibility is vital. On the other hand, cells' growth rate is often slow in suspension 3D cell culture systems because the cells are not attached to any substrate. Furthermore, cells tend to aggregate by gravity or convection, causing uncontrollable spontaneous fusion that leads to variation in sphere sizes. The formation of very large spheres may cause unwanted cell death and/or spontaneous differentiation due to low nutrient and oxygen availability (Bauwens *et al.*, 2008; Serra *et al.*, 2012).

Under such circumstances, we aimed to develop a natural or biological substance-free, novel scaffold 3D cell culture system. It is well known that hydrogels form microsize network structures (Schuetz *et al.*, 2013). Using this knowledge, we applied a microsize pattern (grid) onto the culture surface of a scaffold. By altering several physical properties of culture surface and pattern size, we were able to confirm and optimise cell spheroid formation in several combinations. Further, we were able to demonstrate that the continuous patterned profile, smaller than the cells, acts as an optimised structure for culturing cancer cell lines. Also, it is known that the precision of the pattern strongly affects the ability to form spheroids.

The NanoCulture Plate (NCP) was released onto the market in 2006. It is now sold in more than 10 countries and used by over 300 research groups. The plate is easy to handle and ready to use. Cells are simply seeded using conventional 2D culture techniques and will form spheroids as they migrate along the micropatterned plate. Precise engineering of the patterns results in nearly zero well-to-well and lot-to-lot variability, a feature of the NCP that cannot be achieved by either gel- or matrix-based cell culture. This product is not only suitable for high-throughput screening, it is also compatible with various types of imaging apparatus and co-culture study.

Description of technology

Micropatterned surface technology

The NCP is a SBS format micropatterned plate. It has a black casing and clear flat bottom that is suitable for bright-field imaging or fluorescent measurement (Figure 4.2). The microsize pattern is applied on the culture surface plate, forming the cell's scaffold. The pattern types and sizes can be optimised for spheroid formation of cancer cell lines. At present, the commercialised product comes in a honeycomb (hexagonal) pattern or a square pattern grid (Figure 4.3b,c). Since the NCP has been developed for drug screening purposes, the lot-to-lot or well-to-well variation of the spheroid formation is minimised by maintaining a continuous profile pattern on the scaffold, applied evenly and precisely inside the plate and between the plates.

Spheroid formation process on the NCP

Figure 4.4a clearly shows how spheroids are formed when grown with the NCP, where the micropattern on the plate acts as the cell's scaffold. After seeding into the NCP, cells become attached to the well bottom film surface in single cell dispersion. However, in contrast to conventional monolayer culture plates, this adhesion is not very strong. So cells maintain a spherical shape and are weakly attached to the culture surface using pseudopodia. Then, cells start actively migrating on the well bottom film surface, and start forming spheroids by interacting with surrounding cells and the extracellular matrix. Together with cell

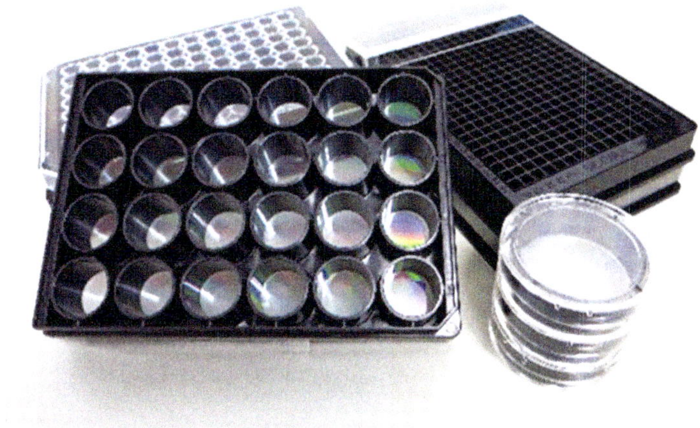

Figure 4.2 **Various types of NanoCulture Plates and dishes.**

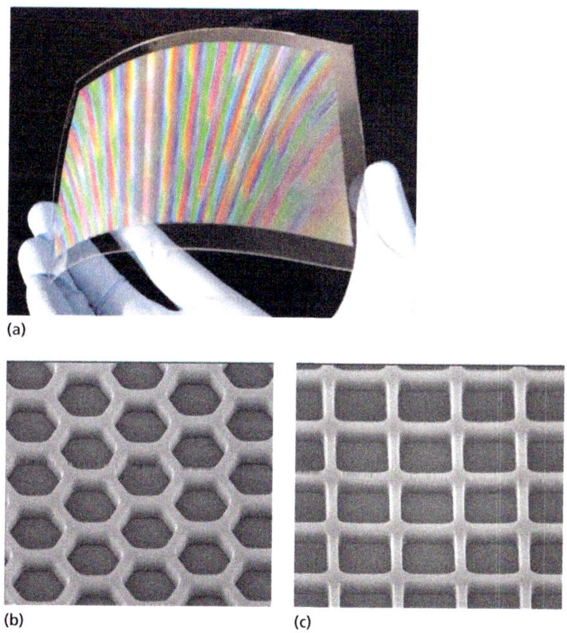

Figure 4.3 **NanoCulture Plate formats.** (a) Micropattern moulded film. (b) SEM image of honeycomb grid on NCP film. (c) Square grid on NCP film.

proliferation inside the spheroid and further migration, spheroids fuse amongst each other, growing in size. Figure 4.4b shows actual time-lapse images of HeLa cells after seeding. Some cell types form spheroids covered with extracellular matrix by day 6 (Figure 4.4c). Also, it is possible to observe that the cells have minimum attachment to the culture surface using pseudopodia (Figure 4.4d).

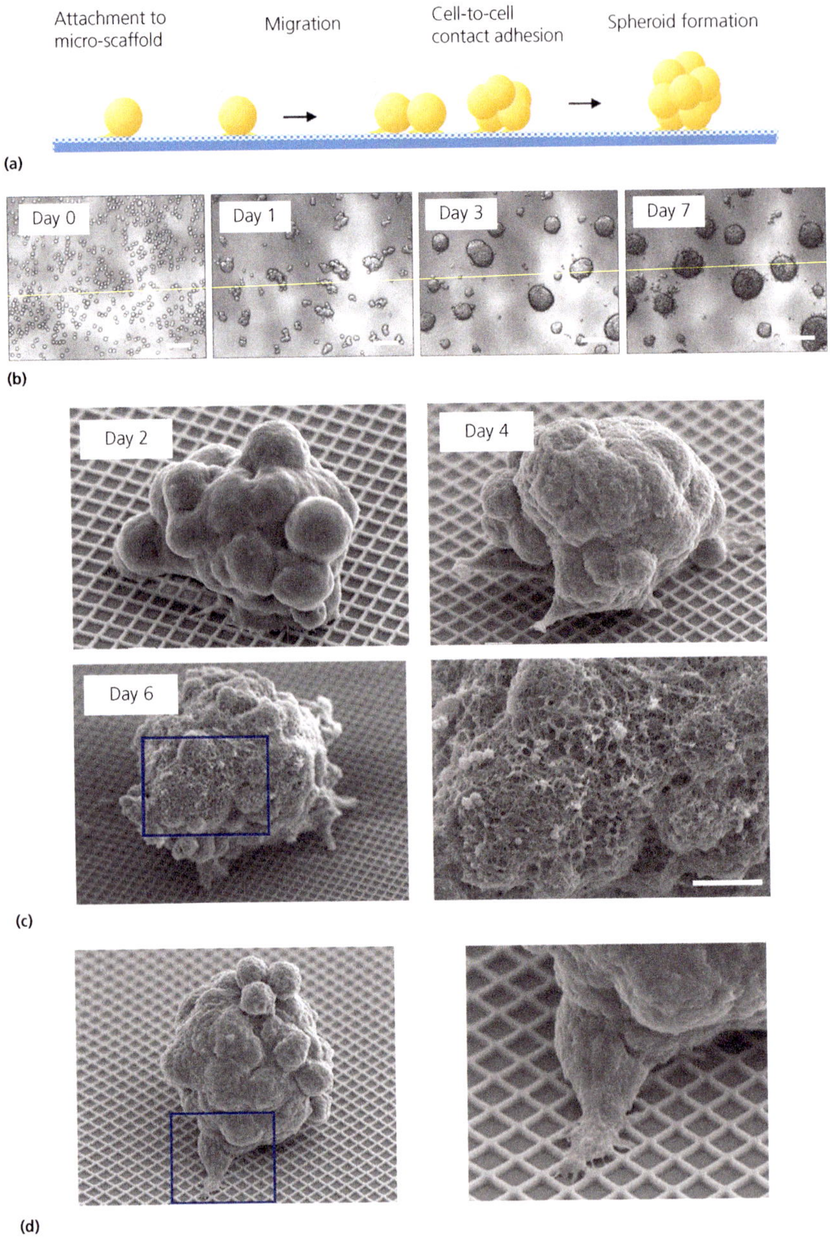

Figure 4.4 **Spheroid formation with the NCP.** (a) Scheme diagram of spheroid formation technology with the NCP. (b) Bright-field microscopic images of spheroid formation after seeding HeLa cells on a 96-well NCP (at day 0, 1, 3 and 7). Scale bar, 100 μm. (c) Spheroids covered with ECM by day 6. Scale bar, 10 μm. (d) Cells grip to the culture surface using pseudopodia.

Cell adhesiveness

The NCP has two commercial grades designed for cell adhesion strength. Low-binding grades have been optimised to be suitable for most cancer cell lines. High-binding grades are suitable for cell lines that cannot attach to the culture surface of low-binding plates, such as adipocytes, low-adhesion liver cells or stem cells. These two types of plates have the same material but different cell adhesiveness, which is obtained by modifying their hydrophobicity by physical treatment.

Product line-up

The NCP has a 35 mm dish for larger cell volume, together with the various SBS format multiwell plates (Tables 4.1 and 4.2).

Table 4.1 NanoCulture Plate (NCP) product list.

NCP format	Adhesiveness	Micropattern
384-well microplate	Low	Square Honeycomb
96-well microplate	Low	Square Honeycomb
24-well microplate	Low	Square Honeycomb
35 mm dish	Low	Square Honeycomb
384-well microplate	High	Square Honeycomb
96-well microplate	High	Square Honeycomb
24-well microplate	High	Square Honeycomb
35 mm dish	High	Square Honeycomb

Table 4.2 Dimensions of the NanoCulture Plate (NCP) (mm).

Parameters		NCP		
		384-well	96-well	24-well
A	Base length	127.8	127.76	127.76
B	Base width	85.5	85.48	85.48
C	Overall height	14.4	14.35	20
D	Well top diameter	3.65	6.69	16.2
D′	Well bottom diameter	3.2	6.58	15.7
E	Well depth	11.5	10.9	17.5
F	Flange height	7	6.35	12
G	A1 location (x)	12.15	11.24	15.13
H	A1 location (y)	9	14.38	13.49
I	Well-to-well spacing	4.5	9	19.5

Example of application

Culture conditions

Culture medium

No special medium is required for 3D cell culture with the NCP. The monolayer-based medium recommended by the cell vendors, such as ATCC, can be used. Also, if there is insufficient spheroid formation, extracellular matrix extracts can be added to enhance spheroid formation. Addition of 0.5 ~ 1% Matrigel is effective in some cases.

Recommended cell seeding number

The recommended cell seeding numbers are indicated in Table 4.3. The cell viability of many cell lines has been confirmed in one-week culture periods without changing the medium. Experimental conditions, including cell types, culture medium types, media change and culture period, must be considered depending on the purpose of the experiment. For example, increasing seeding cell density results in quicker spheroid formation, whilst reducing it can slow down spheroid formation. Furthermore, spheroid size can limit cell viability because the core can become hypoxic/malnourished, resulting in necrosis when the spheroid size becomes larger than 200–300 μm.

Standard procedure for spheroid culture of cancer cell lines with the NCP 96-well

Cell culture protocol using the 96-well NCP is as follows.

Procedure

1 Add medium to well (50 μL/well).
2 Centrifuge the plate at 1000× g for five minutes to remove bubbles (or pipetting).
3 Leave to stand for 15–30 minutes on the bench to remove microbubbles.
4 Seed the cells (10 000 cells/50 μL/well).
5 Leave to stand for 10–15 minutes on the bench until cell attachment occurs.
6 Incubate at 37 °C, 5% CO_2 or at ATCC's recommended incubation condition.
7 Measure with microscope or cell imager. Primary spheroid formation can be observed by day 3 after cell seeding. Mature spheroid size occurs by day 7.

Table 4.3 Recommended cell seeding number.

Format	Recommended seeding no.	Well area (mm^2)
384-well microplate	3000 cells/well	10
96-well microplate	10 000 cells/well	34
24-well microplate	50 000 cells/well	193
35 mm dish	200 000 cells/well	754

Example data

Figure 4.5 shows microscopic images of different cancer cell lines at day 7. Multiple spheroids formed inside the well similar to other gel-based 3D cell culture technologies. These spheroids show stable particle size distribution. Morphology differences between the cell types can be noticed. Cells with strong cell-cell adhesion forming compact spheroids, such as BxPC-3 and Capan-2 (Figures 4.5 and 4.6a), maintain epithelial cell-like characteristics, with upregulated E-cadherin expression (Figure 4.6b). Conversely, cells with weak cell-cell adhesion forming loose cluster-like spheroids, such as PANC-1 and MIAPaCa-2 (see Figures 4.5 and 4.6a), have downregulation of E-cadherin and show mesenchymal cell-like characteristics (Figure 4.6b). MCF-7 cells form hollow or

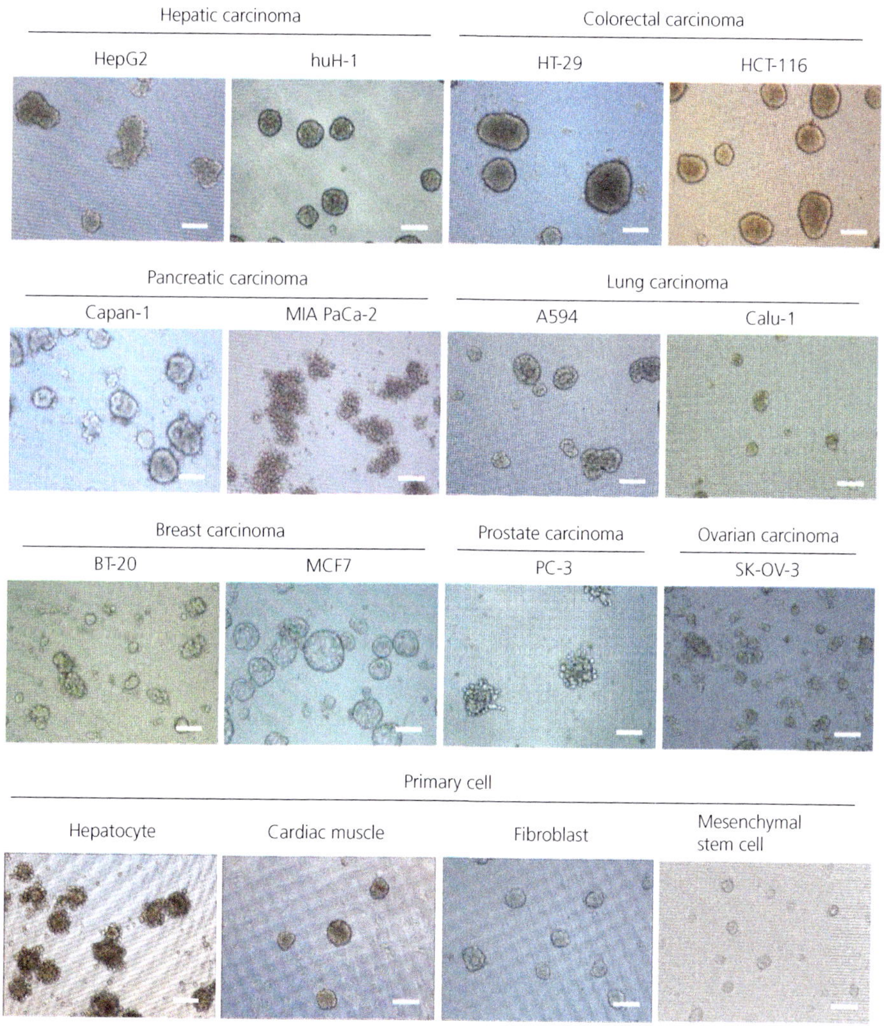

Figure 4.5 **Spheroid images of different types of cell lines on the NCP**. Scale bar, 100 μm.

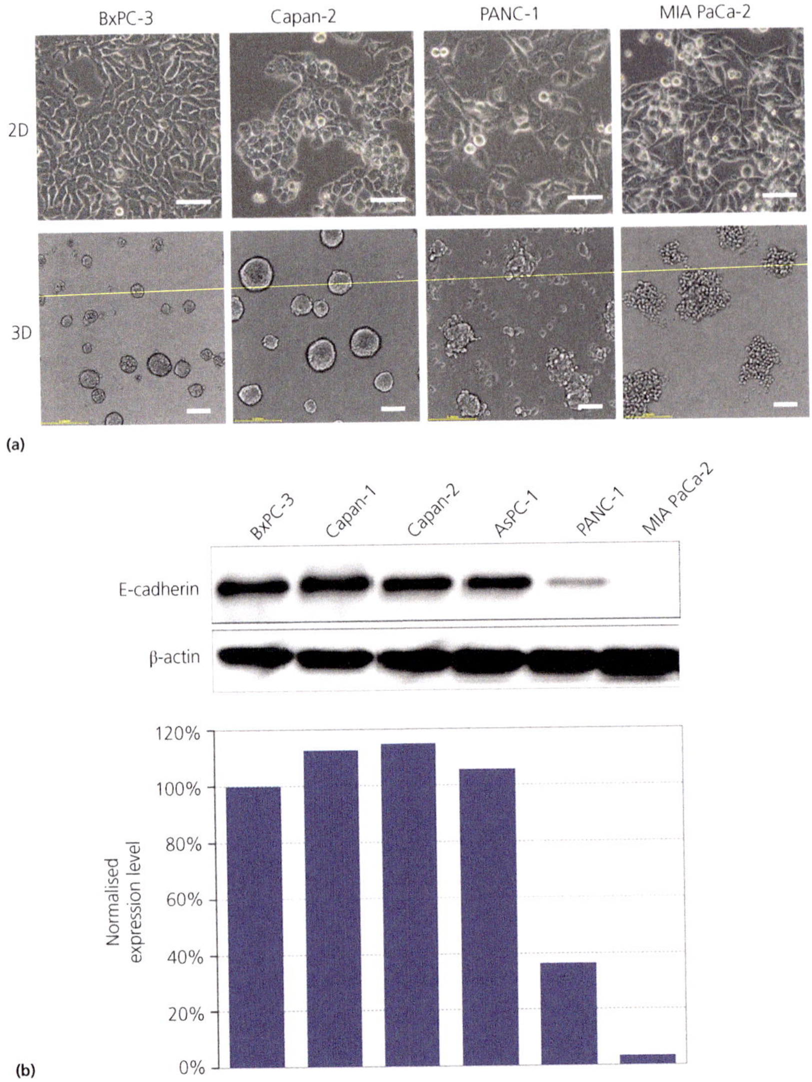

Figure 4.6 Spheroid morphology of different cell lines. (a) Cell morphology comparison between cells, with strong or weak cell-to-cell adhesion, and on both 2D and 3D cell culture systems. Scale bar, 100 μm. (b) Correlation between spheroid morphologies, tight or loose cell spheroids and E-cadherin expression levels.

tubular spheroids (see Figure 4.5). Comparison of data between E-cadherin expression level and NCP spheroid morphology is shown in Figure 4.6.

Methods of measurement

Other than bright-field and fluorescent imaging, the NCP can be used with various cell imaging systems; these include the In Cell Analyzer (GE Healthcare, Little Chalfont, UK) (Figure 4.7), Opera/Operetta (PerkinElmer, Waltham, USA),

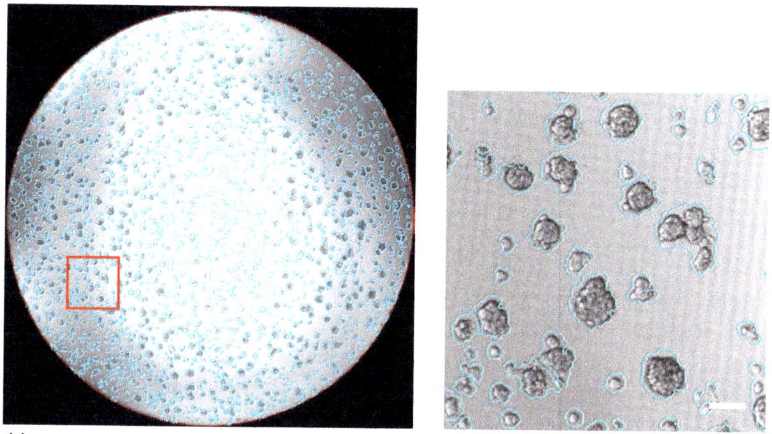

(a) (b)

Figure 4.7 **Measuring spheroid size.** (a) Well image of NCP 96-well measured by In Cell Analyzer 2200. (b) Highly magnified images of cell spheroids in selected area of the well – bright-field channel, objective lens×2, NCP-LS96, medium volume 100 μL. Scale bar, 100 μm.

Celigo® (Nexcelom Bioscience, St Lawrence, USA), ImageXpress® (Molecular Devices, Toronto, Canada), and Cell³iMager (Dainippon, Osaka, Japan) to obtain whole-well images for analysis. In addition, the NCP is suitable for live imaging, using time-lapse imaging equipment, e.g. IncuCyte Zoom® (ESSEN Biosciences, Ann Arbor, USA) or BioStation CT (Nikon, Kingston upon Thames, UK) (Karlsson *et al.*, 2012; Sakamoto *et al.*, 2015). As for viability measurement, though various methods are available, systems like MTT assay (WST-8 assay) are not recommended because the reagent does not penetrate into the spheroid core. It is therefore necessary to disperse the spheroids prior to measurement with those methods. In order to measure viability directly from the spheroid, an adenosine triphosphate (ATP) assay by using CellTiter-Glo® (CTG; Promega, Madison, USA) is recommended. A CTG assay uses surfactants to release the ATP inside cells into the medium, and then measures viability by bioluminescence without dispersing the spheroids. The NCP is also compatible with photoluminescence measurement, fluorometry and spectrometry, and it is capable of undergoing an ATP assay directly on the plate.

Spheroid list

Spheroid formation in the two types of microstructure well-bottom surface is mostly similar in all cell types, but there are some cases where there is better spheroid morphology in one of the patterns. We have confirmed spheroid formation for more than 100 cell types, mostly cancer cells, with the NCP. A list of recommended cancer cell line culture conditions is given in the appendix to this chapter. Spheroid culture of fibroblasts, hepatocytes, myocardial cells, embryonic stem cells, mesenchymal stem cells (MSCs) and neuronal precursor cells was performed with the NCP.

Long-term culture

The NCP is capable of long-term spheroid culture. For the 96-well NCP, we recommend 10 000 cells/well seeding density for one-week culture conditions. However, long-term culture is possible by lowering cell seeding density together with media change.

Procedure

1. Add medium to well (100 µL/well).
2. Centrifuge the plate at 1000× g for five minutes to remove bubbles.
3. Leave it standing for 15–30 minutes on the bench to remove microbubbles.
4. Seed cells (1000 cells/100 µL/well).
5. Leave it standing for 10–15 minutes on the bench until cell attachment occurs.
6. Incubate at 37 °C, 5% CO_2.
7. Change 50% of the medium every 2–3 days.
8. Measure ATP concentration with CellTiter-Glo luminescent cell viability assay kit at days 0, 1, 3, 7, 14, 21 and 35.
   - Remove supernatant (100 µL/well).
   - Add CTG reagents (100 µL/well).
   - Shake for 10 minutes.
   - Leave it standing for 20 minutes.
   - Measure using luminescence reader.

Example data

Figure 4.8 shows the results of the spheroid culture of HepG2 in DMEM containing 0 ~ 10% FBS. ATP concentration reached a peak by one week and then declined. This is due to the hypoxic environment that arises inside the spheroid as it grows in size, which ultimately can result in cell death. By seeding low-density cells, 1000 cells/well, the spheroid size growth time can be extended to maintain spheroid growth for 2–3 weeks (see Figure 4.8).

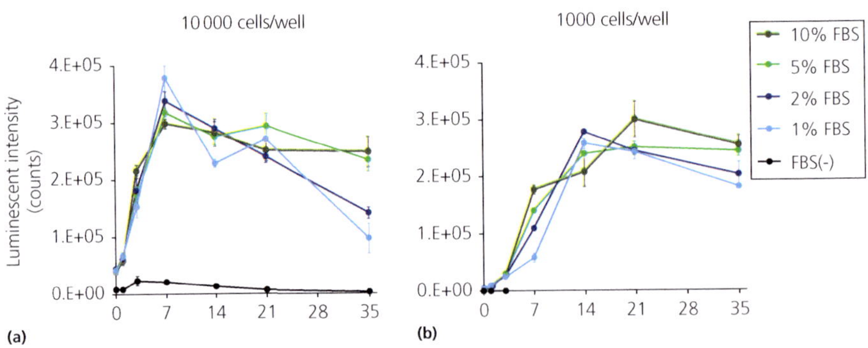

Figure 4.8 Proliferation curves of HepG2. HepG2 cells were cultured with various FBS concentrations for 35 days after seeding at density of (a) 10 000 cells/well and (b) 1000 cells/well on 96-well NCP.

Co-culture method

As cells actively aggregate in the NCP, it is possible to obtain co-culture images, depending on the cell's characteristics.

Procedure

Prewetting 96-well plate

1 Add medium to the plate (100 µL).
2 Centrifuge the plate at 1000× g for five minutes to remove bubbles (or pipetting).
3 Leave it standing for 15–30 minutes on the bench to remove microbubbles.

Preparation of cells

1 Stain tumour cell line, BxPC-3, with PKH-26 (red fluorescent dye).
2 Stain cancer-associated fibroblasts (CAFs) (pancreatic and lung CAFs) with PKH-67 (green fluorescent dye).
3 Mix BxPC-3 and CAFs equally.

Culture and measurement

1 Seed the cell mixture (1000 cells/100 µL/well).
2 Leave it standing for 10–15 minutes on the bench until cell attachment occurs.
3 Incubate at 37 °C, 5% CO_2.
4 Observe using bright-field and fluorescent images.

Example data

BxPC-3 is a pancreatic cancer cell line, which was co-cultured with lung cancer-derived CAFs. Each cell type formed spheroids and did not mix together. By day 3, lung CAFs spheroids were decreasing (Figure 4.9). Meanwhile, by co-culturing BxPC-3 with pancreatic cancer-derived CAFs, both cell types mixed well, and red and green merged and were seen as yellow spheroids (see Figure 4.9). Pancreatic CAFs spheroids remained healthy. Such phenomena may be related to the cell-to-cell affinity.

Antitumour screening

Standard procedure with NCP 384-well plate

The following protocol is a standard anticancer drug screening method.

Procedure

1 Add medium to well (20 µL/well).
2 Centrifuge the plate at 1000× g for five minutes to remove bubbles.
3 Leave it standing for 15–30 minutes on the bench to remove microbubbles.
4 Seed the cells (3000 cells/25 µL/well).
5 Leave it standing for 10–15 minutes on the bench until cell attachment occurs.
6 Incubate for three days at 37 °C, 5% CO_2.

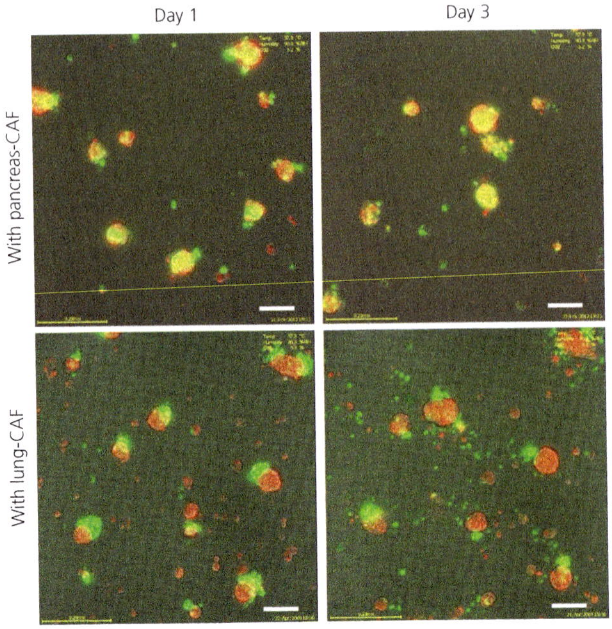

Figure 4.9 Co-culture of tumour cells and CAF. Pancreatic tumour cells BxPC-3 (stained with red fluorescent dye PKH-26) co-cultured with pancreas CAF (*upper panel*) and lung CAF (*lower panel*), both stained with green fluorescent dye PKH-67. At day 3, lung CAF spheroids were decreasing whereas pancreatic CAF spheroids remained healthy. Scale bar, 100 μm.

7 Add drugs to the cell (5 μL/well).

8 Incubate for four days at 37 °C, 5% CO_2.

9 Observe spheroids.

10 Measure ATP concentration with CellTiter-Glo luminescent cell viability assay kit.
- Add CTG reagents (50 μL/well).
- Shake for 10 minutes.
- Stand for 20 minutes.
- Measure luminescence.

Example data

We evaluated the suppression of cell growth between cell monolayers and 3D cell culture with lung cancer cell line A549, using a 1300 compound library ($LOPAC^{1280}$) (Figure 4.10a). While there were many compounds affecting the suppression of cell growth in the monolayer cultures, only four compounds showed stronger suppression of cell growth in the 3D cell cultures. This is known as multicellular resistance, whereby cells obtain drug resistance by forming spheroids; most of the inhibitors showed a reduced effect against multicellular resistance. *In vivo* solid tumours are also known to have multicellular resistance, and thus NCP-based 3D cell culture drug screening can be valuable in developing clinically effective anticancer drugs (Kerbel *et al.*, 1996).

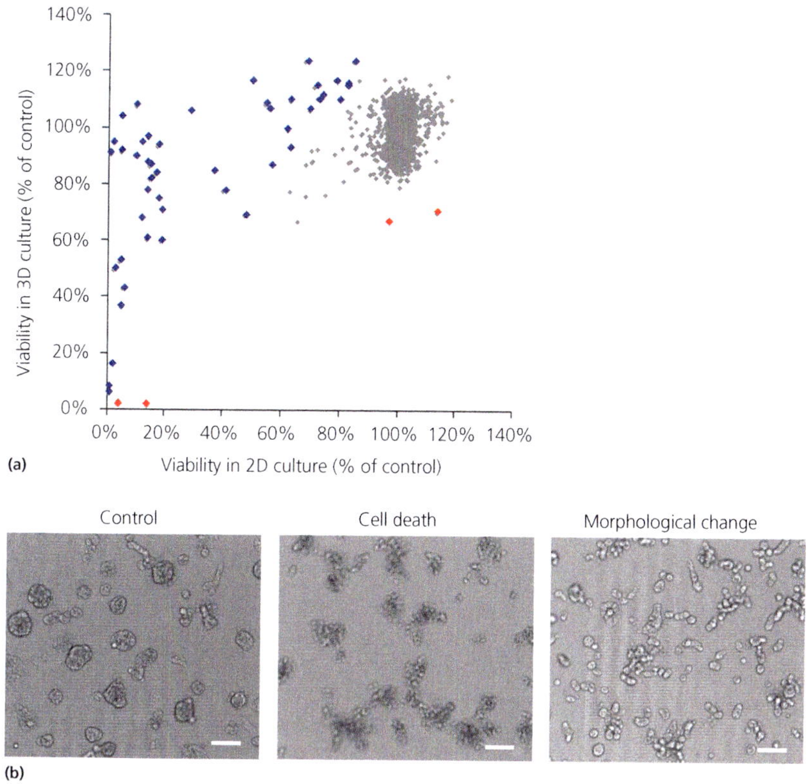

Figure 4.10 Anticancer drug screening with lung cancer cell line A549.
(a) Antiproliferation activities of LOPAC^{1280} under monolayer or NCP 3D cell culture conditions. 40 blue dots represent selective compounds for 2D culture with only four red dots for 3D culture. 3D cultured spheroids represent a multicellular resistance-like *in vivo* solid tumour and provide a better way to identify lead compounds. (b) The NCP can obtain spheroid morphology data, i.e. cell death, cell migration or cell-to-cell adhesion, by using imaging apparatus as a secondary drug screening system. Scale bar, 100 μm.

Moreover, the NCP can not only produce cell growth suppression data using an ATP assay, but it can also obtain spheroid morphology data through different imaging systems. In this way, it is possible to define the dead cells visually, and even if it does not show the growth suppression effect, it enables the discovery of target compounds that can affect cell migration or cell-cell adhesion (Figure 4.10b).

EMT inhibitor screening

Background

Epithelial-mesenchymal transition (EMT) is a procedure in which epithelial cells lose cell polarity or cell-cell adhesion with surrounding cells, and acquire a mesenchymal-like phenotype by obtaining capabilities for migration and invasion (Tsai & Yang, 2013). As this phenomenon is related to cancer invasion and metastasis, inhibiting this transition opens a new avenue for cancer therapy (Park *et al.*, 2011). Here, we introduce a cell-based assay method for the screening of EMT inhibition compounds. We treated a lung cancer cell line with transforming growth factor (TGF)-β2 to stimulate the spheroid to undergo EMT. While the extent of EMT can be observed by the morphological change under bright-field observation, we developed a measurement method using a hypoxia probe, in order to quantify more accurate data. Normally, epithelial cells have strong cell-cell adhesion but when stimulated, spheroids can transition into mesenchymal-like cells that have loose cell-cell adhesion, becoming less hypoxic. In this way, this assay can monitor this phenomenon by using a hypoxia-detecting probe reagent.

Procedure

1. Add 10% FBS contained in DMEM to well (20 μL/well).
2. Centrifuge at 1000× g for five minutes to remove bubbles.
3. Leave it standing for 15–30 minutes on the bench to remove microbubbles.
4. Seed A549 cells suspended in 10% FBS contained in DMEM (3000 cells/25 μL/well).
5. Leave it standing for 10–15 minutes on the bench until cell attachment occurs.
6. Incubate for three days at 37 °C, 5% CO_2.
7. Add mixture of 50 ng/mL of TGF-β2 and 10 μM of drugs (5 μL/well). (Final concentrations of TGF-β2 and drugs are 5 ng/mL and 1 μM, respectively.)
8. Incubate for three days at 37 °C, 5% CO_2.
9. Add hypoxia probe LOX-1 (final concentration 2 μM/well).
10. Incubate for one day at 37 °C, 5% CO_2.
11. Observe spheroids and measure hypoxic signals by imaging cytometer.
12. Measure ATP concentration with CellTiter-Glo luminescent cell viability assay kit.
    - Add CTG reagents (50 μL/well).
    - Shake for 10 minutes.
    - Leave it standing for 20 minutes.
    - Measure luminescence.

Example data

Figure 4.11a shows a diagram of the EMT method. We studied the morphological changes of A549 after stimulating cell spheroids with TGF-β2. Figure 4.11b shows the control without EMT induction, cell morphology after EMT induction, and the inhibitory effect after treatment with a TGF-β inhibitor. The red colour is a signal from the hypoxia probe and the more intense the fluorescence, the more hypoxic the environment is, thus indicating stronger cell-cell adhesion. The gene expression of E-cadherin, an epithelial marker lost during EMT, was also studied. By adding a TGF-β inhibitor, the downregulation of E-cadherin

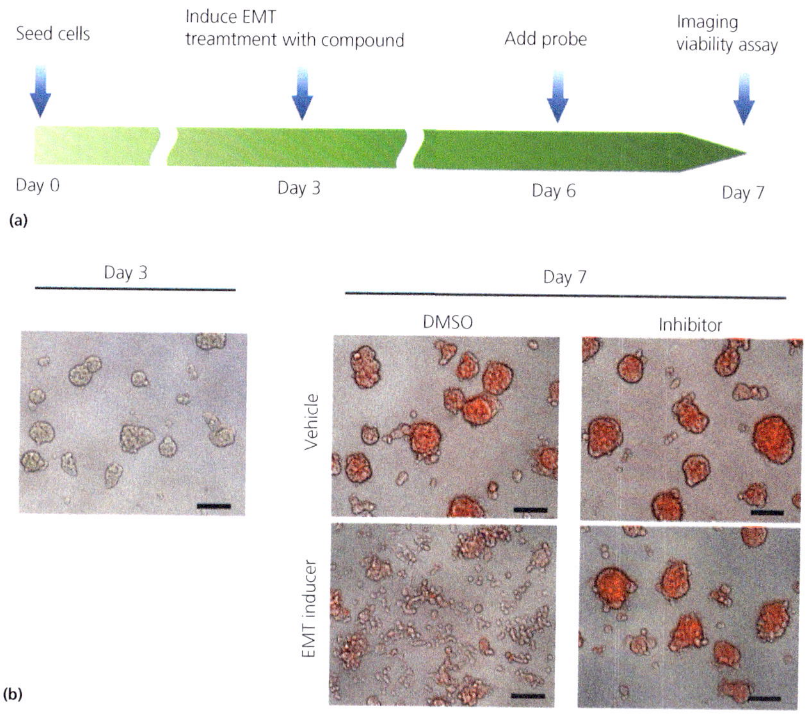

Figure 4.11 EMT spheroid assay. (a) Schematic diagram of EMT inhibition assay. (b) Detection of EMT induction and inhibition by hypoxia probe LOX-1 with cell spheroid morphology alteration. Red colour indicates more hypoxic, stronger cell-to-cell adhesion. Scale bar, 100 μm.

can be stopped (Figure 4.12a). Moreover, EMT also induces expression of mesenchymal markers N-cadherin, vimentin and zeb-1, which in turn were downregulated by the drug compound. The changing phase of molecular markers is proving that EMT can be reproduced and studied on the NCP. By screening compounds for an EMT inhibiting effect, using a 1330 compound library (LOPAC^{1280}), we were able to find several effective compounds with a different point of action, other than the known TGF-β inhibitor (Figure 4.12b).

Toxicology and regenerative research

APAP-induced hepatotoxicity model

In maintaining normal physiology and homeostasis in the body, the liver plays many critical roles. Among these, the metabolism of chemicals, toxins and drugs in hepatocytes is especially important. A breakdown in homeostasis is often associated with hepatotoxicity. Therefore, a simple and rapid human *in vitro* liver model is needed to examine the toxicity and adverse effects of chemicals in the field of pathophysiology and toxicology. We strongly believe that the NCP can fill this gap.

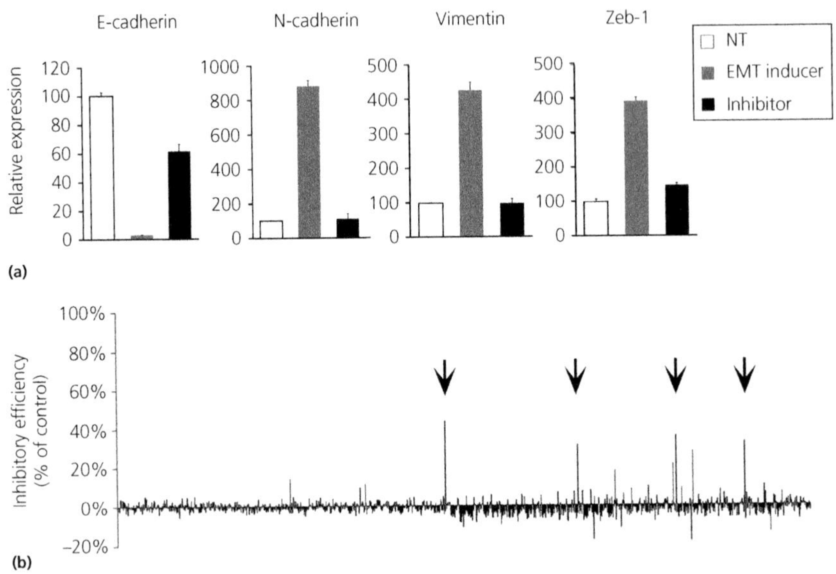

Figure 4.12 Gene expression in EMT. (a) Gene expression data after treating with EMT inducer TGF-β and drug inhibitor. Mesenchymal markers (N-cadherin, vimentin, zeb-1) were upregulated while epithelial marker E-cadherin was decreased. The changing phase of molecular markers proves that EMT is working on the NCP. (b) In addition to the known TGF-β inhibitor, several effective compounds have been found by screening a 1300 compound library (LOPAC[1280]) for EMT inhibiting effects.

Reports have demonstrated that the NCP 3D culture system can induce hepatocyte-specific functions (Nakamura *et al.*, 2011b). Higher expression of mRNA and protein levels of cytochrome including CYP1A1, 1A2, 2B6, 2D6 and 3A4 were observed when HepG2 cells were cultured on the NCP, compared to a 2D culture system. The cause of such upregulation is considered to be an increase in transcription factors like C/EBPa or HNF4a, in the 3D cell culture condition (Nakamura *et al.*, 2011b). Also, Aritomi *et al.* (2014) have shown upregulation of CYP2E1 expression in HepG2 cells cultured on the NCP.

Liver injury induced by overdose of acetaminophen is initiated by the reactive metabolite of acetaminophen, metabolised by CYP2E1 into iminoquinone (NAPQI) that covalently bonds to proteins or genes. Subsequently, glutathione works to detoxify NAPQI and in turn, the depletion of glutathione causes toxicity. N-acetylcysteine, a precursor of glutathione, is the only clinically proven therapeutic drug for liver damage by acetaminophen. This toxic response by the reactive metabolites of the drug, while occurring *in vivo*, was difficult to reproduce with a cell-based assay using cultured cells. Aritomi *et al.* reproduced the metabolism of acetaminophen into a mercapturic acid conjugate and, even further, the toxic response by acetaminophen rescued by N-acetylcysteine by using the 3D cell cultured HepG2 on the NCP.

Stem cells and tissue regeneration

Cell-cell interactions are crucial for preserving the stemness and differentiation of stem cells. The tissue microenvironment plays a critical role in guiding the differentiation of stem cells during normal regenerative processes (Djouad *et al.*, 2007; Engler *et al.* 2006). In 3D culture systems, cell shape, morphogenesis, motility, gene expression and signalling are often closer to real organs and tissues *in vivo* than ordinary 2D cultures. Essentially, 3D cell culture can mimic *in vivo* tissue development. There are several aspects to culturing stem cells such as maintenance and growth culture, embryoid body (EB) formation, and differentiation. Suspension-type 3D cell culture systems are being standardised for EB formation. Scaffold-based 3D cell culture technologies like NCP or hydrogel-based cultures, which provide an *in vivo*-like environment for cells, are expected to be effective for differentiation processes or enhancement and maintenance of cell function after differentiation. Miyagawa *et al.* (2011) showed differentiation of NCP-cultured bone marrow-derived MSCs into adipocytes. It is clear that the MSC differentiation is much more effective on the NCP, with enhanced expression of various differentiation markers, including peroxisome proliferator-activated receptor (PPAR)-γ, C/EBPa, adiponectin and aP2, compared to differentiation using a monolayer culture. These results indicated that the NCP could serve as an excellent scaffold for human MSCs during *ex vivo* differentiation. In addition, this system does not contain animal substances so it can be used for studying cellular properties and the development of tissue engineering applications for humans under xeno-free conditions.

Primary cancer cell culture kit for 3D cell culture

To carry out patient-specific anticancer drug screening and understand the tumour biology that leads to improving patient quality of life, it is necessary to harvest primary cancer cells from patient tumours. In general cell culture methods, primary cancer cells are difficult to culture because fibroblast cells increase more than cancer cells. To resolve the problem, a collagen gel embedding culture method has been developed. However, this method has some problems, including the complexity of cell culture preparation from small cancer tissues such as a biopsy sample, low efficiency of primary cancer cells and difficulty in performing multidrug sensitivity tests. Moreover, it does not reflect the *in vivo* microenvironment and it is more appropriate to harvest primary cancer cells from patient tumours and culture them in a 3D system to more closely mimic *in vivo* conditions. Therefore, we conducted primary tumour culture on the NCP, without isolating cancer cells from stroma cells. Finally, we have developed a 3D primary cancer cell culture kit by selecting the optimum culture medium and dispersion enzyme. The kit includes dissociation solution and tools, primary culture medium and the NCP. Researchers can evaluate drug sensitivity of patient-derived tumour cells on the NCP by luminescence ATP viability assay (Figure 4.13).

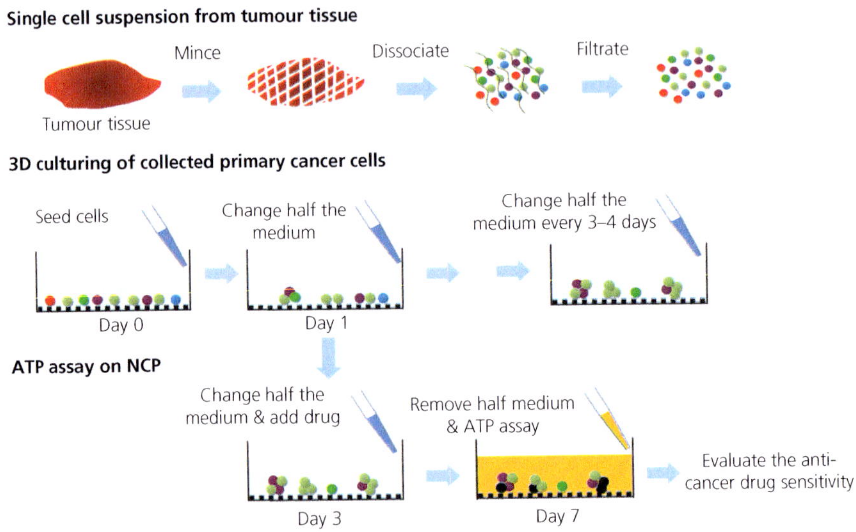

Figure 4.13 Procedure for 3D primary cancer cell culture and anticancer drug test. After harvesting a specimen (minimum 300 mg) under sterile conditions, the washed tissue was put on a chilled petri dish on ice and minced into 1 mm cubes with surgical scissors. The tissue was then dispersed using dispersion solution and seeded $1–2 \times 10^4$ cells/well/250 µL on a 96-well NCP after being filtered. Half the medium was changed according to protocol and the morphology of the spheroids was observed (*middle panel*). For drug sensitivity test and ATP assay, half the medium was removed and new drug added at day 3. Then an ATP assay was performed at day 7 (*lower panel*).

Troubleshooting

Cause	Solution
Uneven distribution of cells	• Apply prewetting process • Prevent movement until cells adhere to the well bottom film after seeding • In case of 96-well or 24-well microplate, put water or media into space between the wells
Forms monolayer Cells does not form spheroids	• Use heat-inactivated serum • Use cells with high E-cadherin expression • Add 0.5–1% Matrigel® to medium
Spheroids detach during the medium change	• Change 50% of the medium, as slowly as possible • Recommended automatic pipetting device: Biomek (Beckman Coulter, Brea, USA), EDR-384SX (BioTec, Tokyo, Japan) • For immunostaining, harvesting all spheroids and staining inside centrifuge tube is recommended
Grid inside NCP interferes when imaging spheroids	• Use fixed focus at several µm over the culture surface • Use 4× lens for IncuCyte Zoom

Summary

The NCP is a groundbreaking scaffold-based 3D cell culture system. It is the most simple, robust and reproducible 3D cell culture system available today. It is suitable for live imaging and is compatible with high-throughput screening. Cells seeded on the NCP actively start forming cell aggregates, as with gel-type culture. However, the NCP does not suffer from the drawbacks of conventional gel-type culture, such as lot-to-lot variation or a troublesome experiment procedure; on the contrary, it is quite easy to use. Using the NCP, specialised assays like co-culture or EMT inhibitor screening can be performed very easily. Lastly, this new 3D spheroid cell culture system provides a convenient and clinically relevant model for the identification and characterisation of cancer drugs for solid tumours.

Acknowledgements

We thank the members of GE Healthcare Japan for providing data on the In Cell Analyzer, Professor Namiki of Waseda University for providing SEM data, Dr Nakatsura and Ms Shimomura of the National Cancer Center for obtaining cancer cell CAFs co-culture data, Dr Okita of the National Research Institute for Child Health for providing the MSC differentiation data, and the University of Tokyo Open Innovation Center for Drug Discovery for providing the compound library.

References

Arai, K., Sakamoto, R., Kubota, D. & Kondo, T. (2013) Proteomic approach toward molecular backgrounds of drug resistance of osteosarcoma cells in spheroid culture system. *Proteomics*, **13**, 2351–2360.

Aritomi, K., Ishitsuka, Y., Tomishima, Y. *et al.* (2014) Evaluation of three-dimensional cultured HepG2 cells in a nano culture plate system: an in vitro human model of acetaminophen hepatotoxicity. *Journal of Pharmacological Sciences*, **124**, 218–229.

Bauwens, C. L., Peerani, R., Niebruegge, S. *et al.* (2008) Control of human embryonic stem cell colony and aggregate size heterogeneity influences differentiation trajectories. *Stem Cells*, **26**, 2300–2310.

Bentires-Alj, M., Gil, S. G., Chan, R. *et al.* (2006) A role for the scaffolding adapter GAB2 in breast cancer. *Nature Medicine*, **12**, 114–121.

Chiba, T., Suzuki, E., Yuki, K. *et al.* (2014) Disulfiram eradicates tumor-initiating hepatocellular carcinoma cells in ROS-p38 MAPK pathway-dependent and -independent manners. *Plos One*, **9**, e84807.

Djouad, F., Delorme, B., Maurice, M. *et al.* (2007) Microenviromental changes during differentiation of mesenchymal stem cells towards chondrocytes. *Arthritis Research and Therapy*, **9**, R33.

Engler, A. J., Sen, S., Sweeney, H. L. & Discher, D. E. (2006) Matrix elasticity directs stem cell lineage specification. *Cell*, **126**, 677–689.

Folkman, J. & Hochberg, M. (1973) Self-regulation of growth in three dimensions. *Journal of Experimental Medicine*, **138**, 745–753.

Ivascu, A. & Kubbies, M. (2006) Rapid generation of single tumor spheroids for high-throughput cell function and toxicity analysis. *Journal of Biomolecular Screening*, **11**, 922–932.

Karlsson, H., Fryknas, M., Larsson, R. & Nygren, P. (2012) Loss of cancer drug activity in colon cancer HCT-116 cells during spheroid formation in a new 3-D spheroid cell culture system. *Experimental Cell Research*, **318**, 1577–1585.

Kerbel, R. S., St Croix, B., Florenes, V. A. & Rak, J. (1996) Induction and reversal of cell adhesion-dependent multicellular drug resistance in solid breast tumors. *Human Cell*, **9**, 257–264.

Miyagawa, Y., Okita, H., Hiroyama, M. *et al.* (2011) A microfabricated scaffold induces the spheroid formation of human bone marrow-derived mesenchymal progenitor cells and promotes efficient adipogenic differentiation. *Tissue Engineering Part A*, **17**, 13–21.

Nakamura, K., Mizutani, R., Sanbe, A. *et al.* (2011a) Evaluation of drug toxicity with hepatocytes cultured in a micro-space cell cuture system. *Journal of Bioscience and Bioengineering*, **111**, 78–84.

Nakamura, K., Kato, N., Aizawa, K. *et al.* (2011b) Expression of albumin and cytochrome P450 enzymes in HepG2 cells cultured with a nanotechnology-based culture plate with microfabricated scaffold. *Journal of Toxicological Sciences*, **36**, 625–633.

Park, C. Y., Kim, D. K. & Sheen, Y. Y. (2011) EW-7203, a novel small molecule inhibitor of transforming growth factor-β (TGF-β) type I receptor/activin receptor-like kinase-5, blocks TGF-β1-mediated epithelial-to-mesenchymal transition in mammary epithelial cells. *Cancer Science*, **102**, 1889–1896.

Ribeiro, A., Vargo, S., Powell, E. M. & Leach, J. B. (2012) Substrate three-dimensionality induces elemental morphological transformation of sensory neurons on a physiologic timescale. *Tissue Engineering Part A*, **18**, 93–102.

Rimann, M. & Graf-Hausner, U. (2012) Synthetic 3D multicellular systems for drug development. *Current Opinion in Biotechnology*, **23**, 803–809.

Sakamoto, R., Rahman, M. M., Shimomura, M. *et al.* (2015) Time-lapse imaging assay using the BioStation CT: a sensitive drug-screening method for three-dimensional cell culture. *Cancer Science*, **106**, 757–765.

Schuetz, T., Richmond, N., Harmon, M. E. *et al.* (2013) The microstructure of collagen type I gel cross-linked with gold nanoparticles. *Colloids and Surfaces B: Biointerfaces*, **101**, 118–125.

Serra, M., Brito, C., Correia, C. & Alves, P. M. (2012) Process engineering of human pluripotent stem cells for clinical application. *Trends in Biotechnology*, **30**, 350–359.

Tsai, J. H. & Yang, J. (2013) Epithelial-mesenchymal plasticity in carcinoma metastasis. *Genes and Development*, **27**, 2192–2206.

Zamora, P. O., Danielson, K. G. & Hosick, H. L. (1980) Invasion of endothelial cell monolayers on collagen gels by cells from mammary tumor spheroids. *Cancer Research*, **40**, 4631–4639.

Appendix Spheroid formation confirmed with the NanoCulture Plate (NCP).

	Cell line	Culture condition		Morphology	
		NCP type	Medium	Spheroid type	Size
Lung cancer	A549	LS or LH	DMEM-10% FBS	Loose spheroid	Medium
	Calu-1	LS or LH	DMEM-10% FBS	Compact spheroid	Medium
	NCI-H1650	LS or LH	DMEM-10% FBS	Compact spheroid	Medium
	NCI-H1975	LS or LH	DMEM-10% FBS	Loose spheroid	Medium
	NCI-H1993	LS or LH	DMEM-10% FBS	Compact spheroid	Big
	NCI-H226	LS or LH	DMEM-10% FBS	Compact spheroid	Medium
	NCI-H23	LS or LH	DMEM-10% FBS	Loose spheroid	Medium
	NCI-H441	LS or LH	DMEM-10% FBS	Compact spheroid	Big
Breast cancer	BT-20	LS or LH	DMEM-10% FBS	Loose spheroid	Small
	BT-474	LS or LH	DMEM-10% FBS	Aggregate	Big
	BT-549	LS or LH	DMEM-10% FBS	Compact spheroid	Small
	HCC1806	LS or LH	DMEM-10% FBS	Compact spheroid	Medium
	HCC1954	LS or LH	DMEM-10% FBS	Compact spheroid	Medium
	Hs578T	LS or LH	DMEM-10% FBS	Loose spheroid	Small
	JIMT-1	LS or LH	DMEM-10% FBS	Loose spheroid	Small
	MCF7	LS or LH	DMEM-10% FBS with insulin	Lumen/acinus	Medium
	MDA-MB-231	LS or LH	DMEM-10% FBS	Aggregate	Medium
	MDA-MB-435S	LS or LH	DMEM-10% FBS with 0.5% Matrigel	Loose spheroid	Big
	SK-BR-3	LS or LH	DMEM-10% FBS	Aggregate	Medium
	T47-D	LS or LH	DMEM-10% FBS	Compact spheroid	Medium
	ZR-75-30	LS or LH	DMEM-10% FBS	Aggregate	Medium
Gastric cancer	MKN-45	LS or LH	DMEM-10% FBS	Aggregate	Medium
	MKN-74	LS or LH	DMEM-10% FBS	Compact spheroid	Big
Colon cancer	Colo-201	LS or LH	DMEM-10% FBS	Aggregate	Big
	Colo-205	LS or LH	DMEM-10% FBS with 0.5% Matrigel	Loose spheroid	Medium
	DLD-1	LS or LH	DMEM-10% FBS	Compact spheroid	Medium
	HCT-116	LS or LH	DMEM-10% FBS	Compact spheroid	Medium
	HT-29	LS or LH	DMEM-10% FBS	Compact spheroid	Big
	SW480	LS or LH	DMEM-10% FBS with 0.5% Matrigel	Loose spheroid	Medium
	SW620	LS or LH	DMEM-10% FBS	Loose spheroid	Medium
	WiDr	LS or LH	DMEM-10% FBS	Compact spheroid	Big
Pancreatic cancer	AsPC-1	LH	DMEM-10% FBS with 0.5% Matrigel	Loose spheroid	Medium
	BxPC-3	LS or LH	DMEM-10% FBS	Lumen/acinus	Small
	Capan-1	LS or LH	DMEM-10% FBS	Compact spheroid	Medium
	Capan-2	LS or LH	DMEM-10% FBS	Compact spheroid	Big
	MiaPaCa-2	LS or LH	DMEM-10% FBS	Aggregate	Medium
	PANC-1	LS or LH	DMEM-10% FBS with 0.5% Matrigel	Loose spheroid	Medium
	SUIT-2	LS or LH	DMEM-10% FBS	Loose spheroid	Small
	DU145	LS or LH	DMEM-10% FBS	Compact spheroid	Small

(Continued)

Appendix (Continued)

	Cell line	Culture condition		Morphology	
		NCP type	**Medium**	**Spheroid type**	**Size**
	PC-3	LS or LH	DMEM-10% FBS	Loose spheroid	Medium
Hepatic cancer	Hep3B	LH	DMEM-10% FBS	Compact spheroid	Big
	HepG2	LS or LH	DMEM-10% FBS	Compact spheroid	Medium
	HLE	LS or LH	DMEM-10% FBS	Aggregate	Medium
	HLF	LS or LH	DMEM-10% FBS	Loose spheroid	Medium
	huH-1	LS or LH	DMEM-10% FBS	Compact spheroid	Medium
	JHH-2	LS or LH	DMEM-10% FBS with 0.5% Matrigel	Loose spheroid	Small
	JHH-5	LS or LH	DMEM-10% FBS	Compact spheroid	Big
	JHH-6	LS or LH	DMEM-10% FBS	Loose spheroid	Small
	JHH-7	LS or LH	DMEM-10% FBS	Compact spheroid	Medium
	Li-7	LS or LH	DMEM-10% FBS	Compact spheroid	Small
	PLC/PRF/5	LS or LH	DMEM-10% FBS	Compact spheroid	Medium
Cholangio carcinoma	RBE	LS or LH	DMEM-10% FBS	Aggregate	Small
Kidney cancer	ACHN	LH	DMEM-10% FBS	Loose spheroid	Medium
Ovarian cancer	SK-OV-3	LS	DMEM-10% FBS	Aggregate	Small
Cervical cancer	HeLa	LS or LH	DMEM-10% FBS	Loose spheroid	Medium
Glioma	U-251	LS or LH	DMEM-10% FBS	Compact spheroid	Small
Osteosarcoma	HOS	LS or LH	DMEM-10% FBS	Loose spheroid	Medium
	MG-63	LS or LH	DMEM-10% FBS with 0.5% Matrigel	Compact spheroid	Medium
	143B	LS or LH	DMEM-10% FBS	Loose spheroid	Medium
	KHOS/NP	LS or LH	DMEM-10% FBS	Loose spheroid	Medium
	SJSA-1	LS or LH	DMEM-10% FBS	Compact spheroid	Medium
	Saos-2	LS or LH	DMEM-10% FBS with 0.5% Matrigel	Loose spheroid	Small
	NOS-10	LS or LH	DMEM-10% FBS	Compact spheroid	Big
	HuO9	LS or LH	DMEM-10% FBS	Loose spheroid	Medium
	MNNG-HOS	LS or LH	DMEM-10% FBS with 0.5% Matrigel	Compact spheroid	Big
	NOS-1	LS or LH	DMEM-10% FBS with 0.5% Matrigel	Loose spheroid	Medium
	HS-Os-1	LS or LH	DMEM-10% FBS	Compact spheroid	Medium

DMEM, Dulbecco's modified eagle medium; FBS, fetal bovine serum; LH, low binding NanoCulture plate – microhoneycomb pattern (NCP-LH); LS, low binding NanoCulture plate – microsquare pattern (NCP-LS).

CHAPTER 5

Micro-moulded non-adhesive hydrogels to form multicellular microtissues – the 3D Petri Dish®

Elizabeth Leary[1,2], Sean Curran[1,2], Michael Susienka[1,2], Kali L. Manning[1,2], Andrew M. Blakely[1,2,3] and Jeffrey R. Morgan[1,2]

[1] *Department of Molecular Pharmacology, Physiology and Biotechnology, Brown University, Providence, Rhode Island, USA*
[2] *Center for Biomedical Engineering, Brown University, Providence, Rhode Island, USA*
[3] *Rhode Island Hospital, Department of Surgery, Brown University, Providence, Rhode Island, USA*

Introduction

As this book demonstrates, three-dimensional (3D) cell culture models are growing in importance for study in the fields of developmental biology, cancer research, drug screening, toxicity testing, regenerative medicine and tissue engineering. Many of these approaches combine cells with a scaffold (natural or synthetic) to achieve 3D architecture and some have shown clinical promise as a substitute or replacement for diseased or injured tissues. For these scaffold-based approaches, the scaffold comprises a large proportion of the mass of these structures, and interaction of cells with the scaffold is a critical component. In many cases, cell-scaffold interactions are more prevalent than cell-cell interactions. However, in most natural tissues, such as the liver, the opposite occurs. The scaffold or extracellular matrix (ECM) comprises a relatively small proportion of the mass of the tissue, cell density is high and cell-cell interactions predominate.

In the absence of a scaffold for attachment, cells will aggregate and undergo a process known as self-assembly. In this scaffold free approach, mono-dispersed cells will spontaneously aggregate and form a 3D microtissue often called a spheroid. Self-assembly is thought to mimic natural processes that occur during embryogenesis, morphogenesis and organogenesis. In fact, many self-assembled spheroids mimic architectural and functional characteristics of their native tissue. Further supporting this idea is the observation that certain pairs of cell types will self-sort as they self-assemble. As a result of self-sorting, one cell type forms a central core in the spheroid, whereas the second cell type forms a coating around this core.

Technology Platforms for 3D Cell Culture: A User's Guide, First Edition. Edited by Stefan Przyborski.

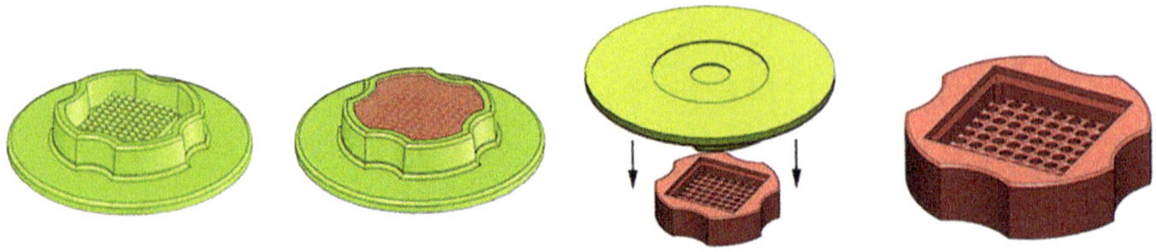

(a) (b) (c) (d)

Figure 5.1 Micromoulding of agarose, a non-adhesive hydrogel. A precision micromould containing a cavity with an array of small posts (a) is filled with molten agarose (b). After the agarose has solidified, the micromould is flexed to remove the agarose gel (c). The resulting agarose gel is a negative replica of the micromould and contains a seeding chamber and small recesses where microtissues form (d).

There are several methods to form scaffold-free spheroids, but this chapter focuses on the use of micromoulded agarose, a technology commercialised as the 3D Petri Dish® (www.microtissues.com) that is distributed by Sigma-Aldrich®. In this method, cells settle into small recesses micromoulded into a non-adhesive agarose gel and aggregate into multicellular microtissues. To form a micromoulded agarose gel, molten agarose (1–2%) is poured into a precision micromould containing custom-designed features that will form small recesses in the gel (Figure 5.1).

After the agarose has gelled, it is easily removed from the micromould by flexing the edges of the micromould. The agarose gel is then transferred to a multiwell plate where it is equilibrated with cell culture media. Moulded into the gel are two key structures: a large chamber for seeding cells and small recesses located below this chamber that collect cells as they settle. To form microtissues, a small volume of mono-dispersed cells (e.g. 75 μL or 190 μL) is seeded into the large seeding chamber (Figure 5.2)

The cells then settle by gravity into the small recesses below the seeding chamber. Since agarose is non-adhesive, cells are unable to attach to it. However, the concave bottoms of the agarose recesses ensure that the settled cells are in contact with one another. On this non-adhesive substrate, this cell-cell contact is enough to nucleate the aggregation and self-assembly process. Cells aggregate and self-assemble into a 3D multicellular microtissue at the bottom of each recess. Typically, self-assembly is complete within 24 hours but, depending on cell type, this may take longer. This non-adhesive hydrogel environment and the resulting microtissues have numerous distinct advantages.

- Cell-cell interactions dominate over cell-substrate (tissue culture plastic) interactions.
- Cell density is high and more comparable to native tissue and organs.
- Microtissues are surrounded with cell culture medium (hydrogel is not a barrier to mass transport).
- Cell-cell communication is maximised.
- Cells are able to move over each other and the ECM they produce.

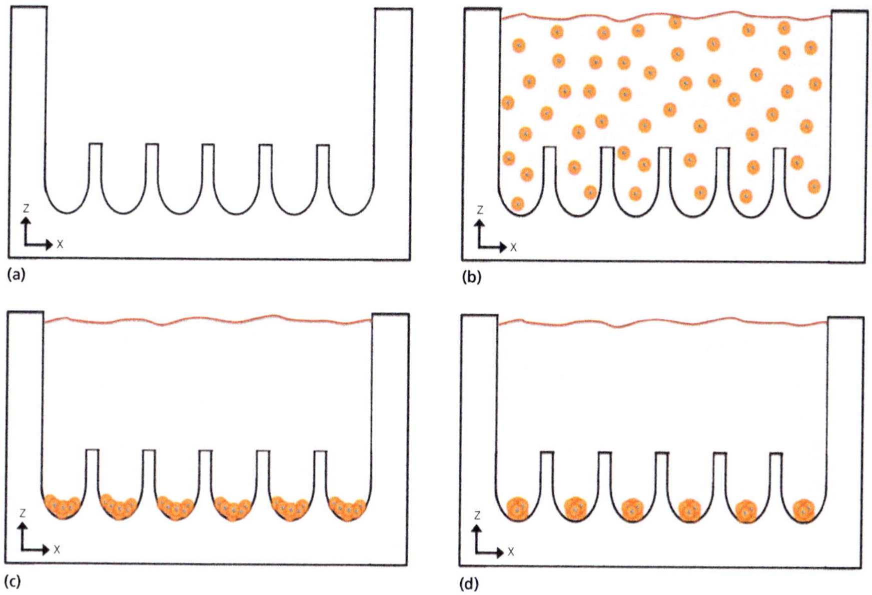

Figure 5.2 **Schematic of the steps in spheroid formation.** A micromoulded agarose gel that had been equilibrated with culture media (a) is seeded with mono-dispersed cells by pipetting into the large seeding chamber (b). After 30 minutes, cells settle by gravity to the bottom of the small recesses (microwells) that have concave bottoms (c). After 24 hours, cells have aggregated and self-assembled a spheroid in each of the small recesses (d).

- Cells are able to exert forces on each other and the ECM they produce.
- Cells are free to self-organise and undergo morphological changes according to biological principles (e.g. self-sort).

Description of technology

There are eight different micromould designs and each forms a unique micromoulded agarose gel (Figures 5.3 and 5.4) The eight micromoulds and their respective gels differ with regard to the size, number and shape of microtissues they produce as well as the multiwell plates in which they are to be used. Four micromoulds (12-series: 12-256, 12-81, 12-36TO, 12-60TR) form gels that fit in 12-well plates and four micromoulds (24-series: 24-96, 24-35, 24-24TR, 24-H) form gels that fit in 24-well plates.

Each micromould has a single cavity to hold molten agarose, either 330 μL or 500 μL, depending on the size of the micromould. After the agarose solidifies, it is removed from the cavity and inverted to reveal a negative replica of the key features within the micromould. There are two important general features: a large rectangular-shaped box and small features that protrude upward from the top surface of this box. These small features create small recesses in the agarose

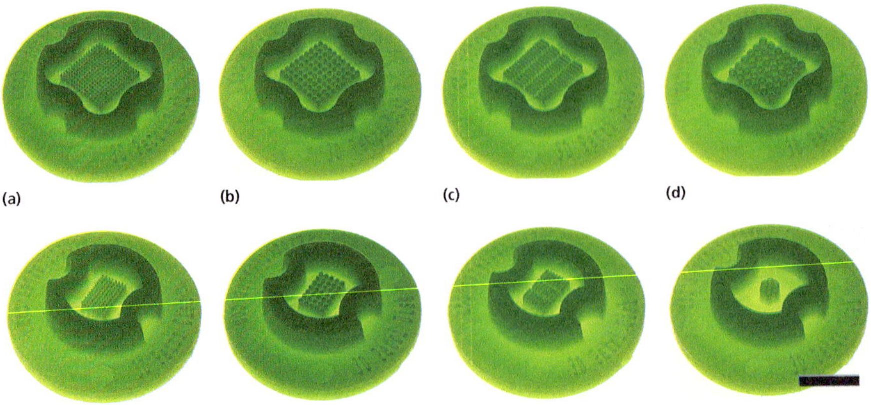

(a) (b) (c) (d)

(e) (f) (g) (h)

Figure 5.3 **Precision micromoulds.** Photographs of the eight micromoulds that are commercially available. Micromoulds that create gels for 12-well plates (a–d) or 24-well plates (e–h). Micromoulds that create gels for the formation of small spheroids (a, e), large spheroids (b, f), multicellular rods (c, g), toroids (d) and a honeycomb (h). Scale bar is 1 cm.

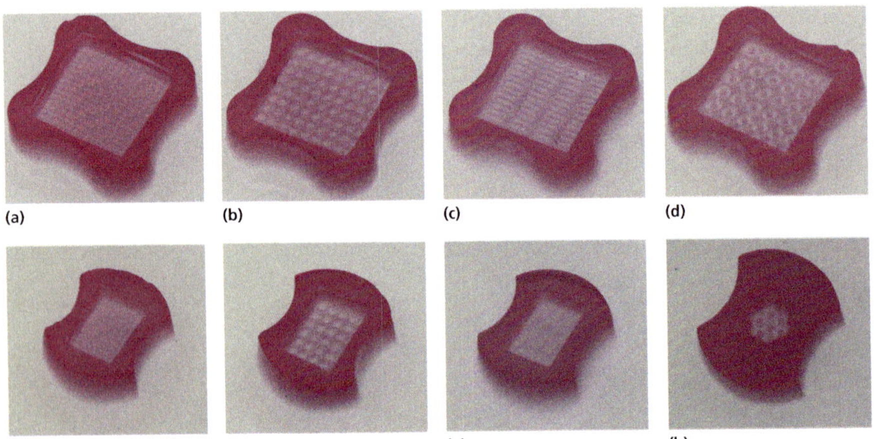

(a) (b) (c) (d)

(e) (f) (g) (h)

Figure 5.4 **Micromoulded agarose gels.** Photographs of the agarose gels cast in the eight commercially available micromoulds. Agarose gels for 12-well plates (a–d) or 24-well plates (e–h). Agarose gels for the formation of small spheroids (a, e), large spheroids (b, f), multicellular rods (c, g), toroids (d) and a honeycomb (h).

into which cells will settle and self-assemble 3D microtissues. For example, the 12-256 micromould has 256 small cylindrical posts protruding from the rectangular box that create 256 small recesses in the agarose where the cells will undergo self-assembly. The rectangular box of the micromould creates a seeding chamber in the agarose gel that is above the small recesses. This chamber can accommodate either 75 μL or 190 μL of cell suspension, depending on the

Table 5.1 Specifications of 3D Petri Dishes and the micromoulds used to cast them.

Catalogue no.	Features	Number of features	Array	Agarose volume	Cell seeding volume
12-256	Small spheroids	256	16×16	500 μL	190 μL
12-81	Larger spheroids	81	9×9	500 μL	190 μL
12-60TR	Troughs	60	4×15	500 μL	190 μL
12-36TO	Toroids	36	6×6	500 μL	190 μL
24-96	Small spheroids	96	8×12	330 μL	75 μL
24-35	Larger spheroids	35	5×7	330 μL	75 μL
24-24TR	Troughs	24	3×8	330 μL	70 μL
24-H	Honeycomb	1	1	330 μL	20 μL

Source: Reproduced with permission of Microtissues, Inc.

size of the micromould. For example, when using gels formed by the 12-256 micromould, a single pipetting step (filling the seeding chamber) results in the formation of 256 spheroids. Thus, it is easy to scale up the production of spheroids.

Four of the eight micromoulds form gels to produce spheroids and these are either small or large spheroids depending on the size and number of the recesses that are formed in the gel. For small spheroids, micromoulds 12-256 and 24-96 have 256 posts (16×16 array) and 96 posts (8×12 array), respectively. They form gels with 256 and 96 small recesses, respectively (Table 5.1).

The nominal sizes of the cylindrical recesses that form small spheroids are 400 μm in diameter and 800 μm deep whereas, for large spheroids, micromoulds 12-81 and 24-35 have 81 posts (9×9 array) and 35 posts (5×7 array) and they form gels with 81 and 35 large recesses, respectively. The nominal sizes of the cylindrical recesses that form large spheroids are 800 μm in diameter and 800 μm deep. All posts have rounded tops and thus create cylindrical recesses with concave bottoms in the agarose. This concave bottom ensures that all cells contact each other as they settle, thus resulting in a single pile of cells at the bottom of each recess. This single pile of cells will self-assemble into a single multicellular microtissue. Early prototypes with flat bottom recesses proved unsuccessful as they resulted in multiple non-uniform piles of cells that self-assembled into multiple spheroids of varying sizes within each recess (Napolitano *et al.*, 2007b).

There are two methods of controlling the size of spheroids: the number of cells seeded and micromould choice. One of the fundamental principles by which the gels work is that the cells settle by gravity from the seeding chamber into the small recesses below the chamber. Due to the concave bottom of each recess, all cells captured by a recess will collect and form one spheroid per recess. Thus, spheroid size can be controlled by changing the total number of cells seeded into the chamber. Spheroid size is directly proportional to the number of cells seeded, with more cells resulting in larger spheroids. An estimate of the size of a spheroid as a function of the number of cells seeded is provided (Tables 5.2 and 5.3).

Table 5.2 Estimates of cell seeding numbers for smaller spheroids in the 3D Petri Dish.

		No. 12-256 **12-well plate** **Small spheroids: 256**	**No. 24-96** **24-well plate** **Small spheroids: 96**
***Nominal spheroid diameter (μm)**	**~Cells/spheroid**	**Total cells seeded (cells/190 μL)**	**Total cells seeded (cells/75 μL)**
50	15	3840/190 μL	1440/75 μL
100	125	32 000/190 μL	12 000/75 μL
150	421	107 000/190 μL	40 000/75 μL
200	1000	256 000/190 μL	96 000/75 μL
250	1953	500 000/190 μL	187 000/75 μL
300	3375	860 000/190 μL	324 000/75 μL

* These are estimates only. Estimates of spheroid diameter are based on the following assumptions. (i) Diameter of input cells is 20 μm (4180 μm^3). (ii) Total spheroid volume equals cell volume times the number of input cells. (iii) The spheroid is a perfect sphere. (iv) Cells are perfectly distributed to all recesses. In practice, spheroid size and shape will vary depending on cell type. Source: Reproduced with permission of Microtissues, Inc.

Table 5.3 Estimates of cell seeding numbers for larger spheroids in the 3D Petri Dish.

		No. 12-81 **12-well plate** **Larger spheroids: 81**	**No. 24-35** **24-well plate** **Larger spheroids: 35**
***Nominal spheroid diameter (μm)**	**~Cells/spheroid**	**Total cells seeded (cells/190 μL)**	**Total cells seeded (cells/75 μL)**
200	1000	81 000/190 μL	35 000/75 μL
300	3375	273 000/190 μL	118 000/75 μL
400	8000	648 000/190 μL	280 000/75 μL
500	15 625	1.3×10^6/190 μL	547 000/75 μL
600	27 000	2.2×10^6/190 μL	945 000/75 μL
700	42 875	3.5×10^6/190 μL	1.5×10^6/75 μL

* These are estimates only. Estimates of spheroid diameter are based on the following assumptions. (i) Diameter of input cells is 20 μm (4180 μm^3). (ii) Total spheroid volume equals cell volume times the number of input cells. (iii) The spheroid is a perfect sphere. (iv) Cells are perfectly distributed to all recesses. In practice, spheroid size and shape will vary depending on cell type. Source: Reproduced with permission of Microtissues, Inc.

Spheroid size is also controlled by the choice of micromould. In addition to the total number of cells seeded, the number of cells in each recess is a function of the surface area of the mouth of each recess. Those recesses with larger mouths capture more settling cells and give rise to larger spheroids. For example, if gels formed from micromoulds 24-96 and 24-35 are both seeded with 100 000 cells, the spheroids formed by the 24-96 gels will have on average 1042 cells/spheroid, whereas the spheroids formed by 24-35 gels will have on average 2857 cells/spheroid with estimated diameters of 202 μm and 284 μm, respectively.

When choosing the size of a spheroid to make, there are limits to spheroid diameter that must be considered. Since nutrients and waste products enter and exit the spheroid by diffusion, there is a limit to the maximum diameter of a healthy spheroid. Large spheroids that are diffusion limited can develop necrotic cores where cells are dead, dying or undergoing apoptosis. A widely cited maximum diameter for a spheroid to avoid diffusion limitations is 200 μm (Curcio *et al.*, 2007). However, this maximum diameter can be cell type dependent. Some cell types, such as hepatocytes and cardiomyocytes, are in highly vascular organs *in vivo*, consume lots of oxygen and nutrients and may be healthier in small diameters. In contrast, cells such as chondrocytes from cartilage, a non-vascularised tissue, can tolerate larger diameters.

Another factor to remember is that not all spheroids are perfectly spherical (Dean *et al.*, 2007). Spheroids have *x*, *y* and *z* dimensions. When viewed with a conventional microscope such as an inverted microscope, the *x*, *y* dimensions are evident, and a spheroid can appear to be a perfect sphere. However, the *z* dimension may not necessarily equal the *x*, *y* dimensions. Some spheroids are oblate in shape and the *z* dimension is less than the *x*, *y* dimensions. Side-view microscopy is able to assess the *z* dimension (Dean *et al.*, 2007).

Four of the eight micromoulds (12-36TO, 12-60TR, 24-24TR, 24-H) form gels that produce microtissues much different in size and shape from a simple spheroid (see Figures 5.3 and 5.4 and Table 5.1). These gels direct cells to form toroids (12-36TO), rods (12-60TR, 24-24TR) and a honeycomb (24-H). These micromoulds contain more complex features. The 12-36TO micromould has features that create 36 toroid-shaped recesses in the gel (6 × 6 array). Each feature in the gel is a circular trough 400 μm wide, 800 μm deep with a concave bottom and an outer diameter of 1400 μm. At the centre is a circular agarose post with an outer diameter of 600 μm. When cells settle into these circular troughs, they form 36 multicellular toroids or donuts per gel and each toroid has about 20 000–25 000 cells depending on the number cells seeded (Table 5.4).

The 12-60TR and 24-24TR micromoulds create 60 troughs (4 × 15 array) and 24 troughs (3 × 8 array) respectively in the gel. These straight troughs with concave bottoms are 400 μm wide, 800 μm deep and 2200 μm in length. When cells settle into these troughs, they form multicellular rod-shaped microtissues, 60 or 24 at a time, depending on the choice of micromould. Each rod has about 20 000–25 000 cells, depending on the number of cells seeded. The 24-H micromould creates a

Table 5.4 Estimates of cell seeding numbers for making microtissues with complex shapes in the 3D Petri Dish.

Catalogue no.	Feature: no.	Seeding volume	~Cells/feature	*Total cells seeded
12-60TR	Troughs: 60	190 µL	20 000	1.2×10^6/190 µL
24-24TR	Troughs: 24	70 µL	20 000	480 000/75 µL
12-36TO	Toroids: 36	190 µL	20 000	720 000/190 µL
24-H	Honeycomb: 1	20 µL	140 000	140 000/20 µL

* These cell seeding densities are suggestions for initial experiments. Results can vary with cell type, so a range of cell seeding densities should be tested. Source: Reproduced with permission of Microtissues, Inc.

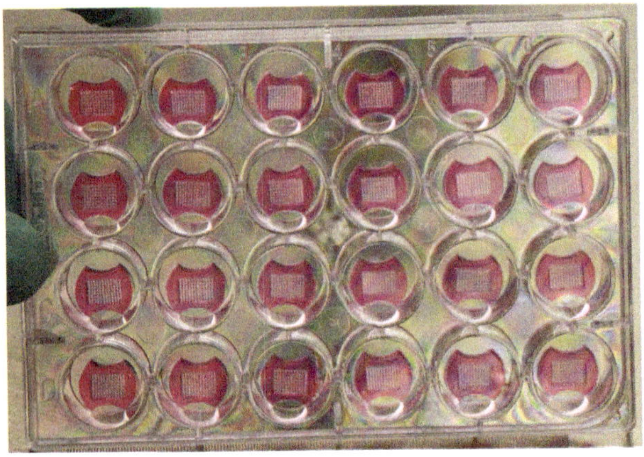

Figure 5.5 Multiwell plate filled with agarose gels. Photograph of a 24-well plate with each well containing a micromoulded agarose gel. Gels (2% agarose) cast in micromoulds (#24-96) were equilibrated in culture media. The agarose gel in each well can form 96 small spheroids.

single honeycomb-shaped trough in the gel. This honeycomb is a single continuous trough 400 µm wide, 800 µm deep with a concave bottom. The maximum distance across this honeycomb recess is 3.4 mm. Within this honeycomb pattern are seven agarose posts 600 µm in diameter: one central post surrounded by six equidistantly spaced posts. When cells settle into this honeycomb-shaped trough, they form one multicellular honeycomb per gel and each honeycomb has about 140 000–250 000 cells, depending on the number of cells seeded.

All micromoulds are autoclaved before use to ensure sterility. The micromoulds are made of an FDA-approved polymer and are flexible, which aids in the removal of the agarose gels. Once removed from the micromould, the agarose gels are transferred to either a 12-well or 24-well plate for equilibration with culture media prior to seeding with cells (Figure 5.5). When not in use, the precision micromoulds are stored in a dust-free container.

Applications

Spheroids can be formed from many cell types and over a range of sizes

A growing list of cell types have been used to form spheroids in the micromoulded agarose gels. Cell types include primary cells such as mesenchymal stem cells, embryoid bodies, human fibroblasts, rat cardiomyocytes, human endothelial cells, human chondrocytes, rat neural cells and zebrafish progenitor cells (Birenboim *et al.*, 2013; Dean *et al.*, 2007; Desroches *et al.*, 2012; Gwyther *et al.*,2011; Hoffecker & Iwata, 2014; Klopper *et al.*, 2010;Napolitano *et al.*, 2007b). Likewise, many cell lines and tumour cell lines from a variety of tissues and organs have been used to form spheroids, including MCF-7 cells, HepG2 cells, M28 cells, REN cells, KGN cells, TCL cells, RG2 cells and HeLa cells (Bao *et al.*, 2011; Barbone *et al.*, 2008; Ho *et al.*, 2012; Krotz *et al.*, 2010; Napolitano *et al.*, 2007a; Robins *et al.*, 2011; Yang *et al.*, 2009). As mentioned previously, spheroid size is controlled by the choice of micromould or the number of cells seeded (Figure 5.6).

The recesses in agarose gels produced by micromoulds for small and large spheroids have maximum diameters of 400 μm and 800 μm respectively and so

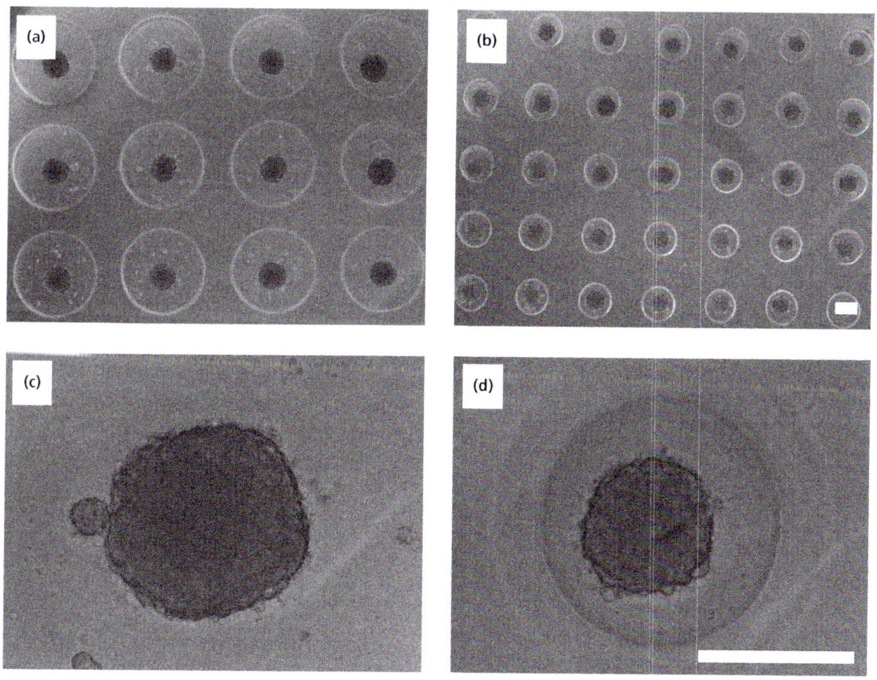

Figure 5.6 Formation of spheroids of varying sizes. Mono-dispersed cells of the human ovarian granulosa cell line KGN were seeded onto agarose gels with recesses designed for either large spheroids (a, c) or small spheroids (b, d). After 24 hours of self-assembly, spheroids of varying sizes were formed and imaged. Scale bar is 200 μm.

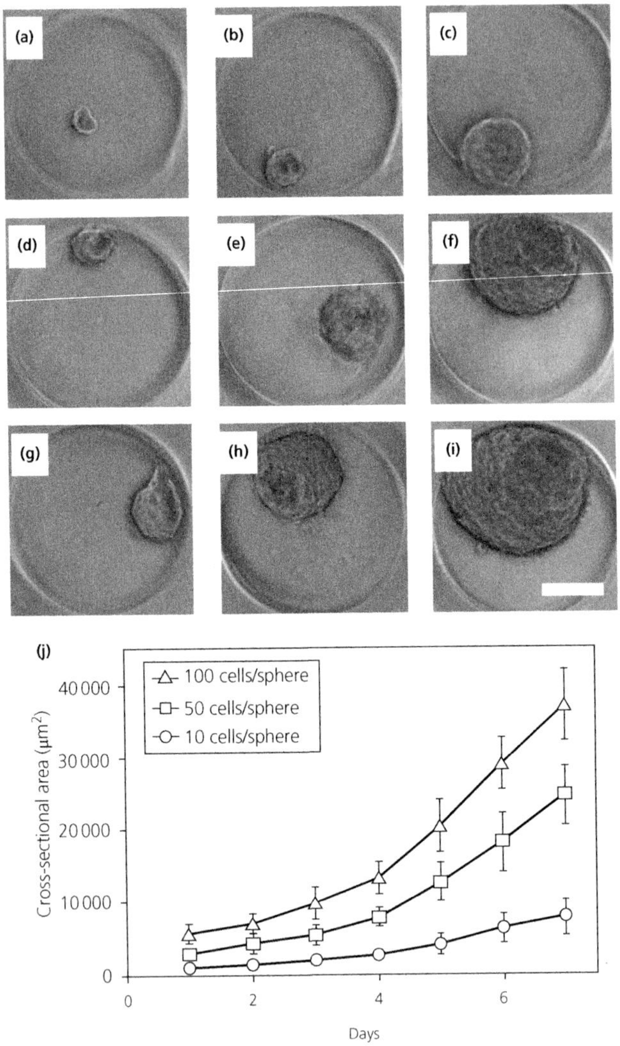

Figure 5.7 Initial cell seeding density controls the rate of spheroid growth. Mono-dispersed cells of the human breast cancer cell line MCF-7 were seeded onto agarose gels at a density of 10 (a–c), 50 (d–f) or 100 cells/recess (g–i). Spheroids were imaged over the course of seven days at day 1 (a, d, g), day 4 (b, e, h) and day 7 (c, f, i). The area of the spheroids (average with standard deviations, n = 20) is plotted over the course of seven days (j). Scale bar is 100 μm.

these provide the upper limits of spheroid size. The typical range of sizes for small spheroids varies from ~15 cells/spheroid (diameter ~50 μm) up to ~3375 cells/spheroid (diameter ~300 μm). The typical range of sizes for large spheroids varies from ~1000 cells/spheroid (diameter 200 μm) up to ~42 875 cells/spheroid (diameter ~700 μm). Spheroids can readily grow in the recesses of the agarose gels (Figure 5.7)

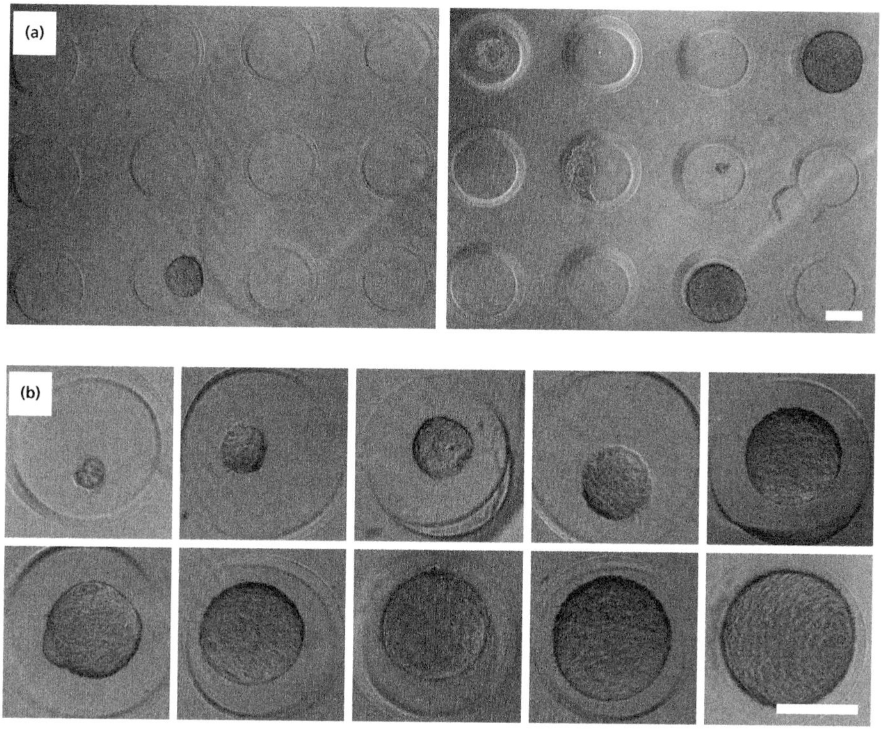

Figure 5.8 Single cells can grow into spheroids. Mono-dispersed cells of the human breast cancer cell line MCF-7 were seeded onto agarose gels at clonal density (<1 cell/recess) and allowed to grow. After 14 days, clonal spheroids were counted and imaged. Representative images of clonal spheroids at low magnification (a) and high magnification (b) show heterogeneity in spheroid size. The efficiency of clone formation was greater in 3D versus 2D, 62.3 ± 1.6 % versus 39 ± 3.6%. Scale bar is 200 μm.

Spheroid growth is evident and can be quantified from images showing the increase in spheroid diameter over time, as well as through the use of standard metabolic indicators such as the cell proliferation reagent WST-1 (4-[3- (4-iodophenyl)-2-(4-nitrophenyl)-2H-5-tetrazolio]-1,3-benzene disulfonate) (Napolitano *et al.*, 2007a). Spheroid growth is cell type dependent. Some cell types, such as the tumour line RG2, grow very rapidly, whereas spheroids of normal human fibroblasts show minimal growth over the course of two weeks (Napolitano *et al.*, 2007b).

Spheroids can be grown from single cells

Depending on cell type, a spheroid can be grown from a single cell (Figure 5.8). In this clonal assay, very few cells are seeded per gel so that only one cell seeds a single recess. For example, gels formed by the 12-256 micromould have 256 recesses. Seeding this gel with 100 cells will result in a majority of empty recesses and many recesses having a single cell. Over the course of two weeks, these single cells will divide and grow into spheroids. The percentage of single cells that gives rise to a spheroid is the efficiency of clonal spheroid formation. The efficiency of clonal spheroid formation was determined for MCF-7 cells, a breast cancer cell line, and compared directly to the efficiency of clonal formation in conventional 2D cell culture. MCF-7 cells were seeded onto either agarose gels or tissue culture plates, allowed to grow for two weeks and then the number of spheroids or clones counted. The efficiency of clone formation was greater in 3D versus 2D, with 62.3±1.6% of the cells forming spheroids in 3D as opposed to only 39±3.6% cells forming clones in 2D. Clones in 2D and 3D were heterogeneous in size. Spheroids ranged in size with diameters of ~50 μm up to ~300 μm. The stable culture environment provided by the agarose gel and the ability to replenish cell culture media facilitate the long-term growth needed to assess clonal growth. Additionally, the ordered array of recesses, whose bottoms are all on the same optical plane, facilitates the repeated imaging of spheroids over time.

Spheroid clonal growth has applications in cancer stem cell research and the assessment of drugs to halt clonal growth (Reya *et al.*, 2001). This single cell assay with micromoulded agarose gel has been used to quantify anoikis, a form of programmed cell death that occurs for some anchorage-dependent cells when they cannot attach to a substrate (Rago *et al.*, 2008).

Some pairs of cells will self-sort during self-assembly

Prior studies have established that certain types of cells will not only self-assemble into a mixed spheroid but will also self-sort, with one cell type forming the inner core and the other cell type forming the outer coating of the spheroid (Figure 5.9).

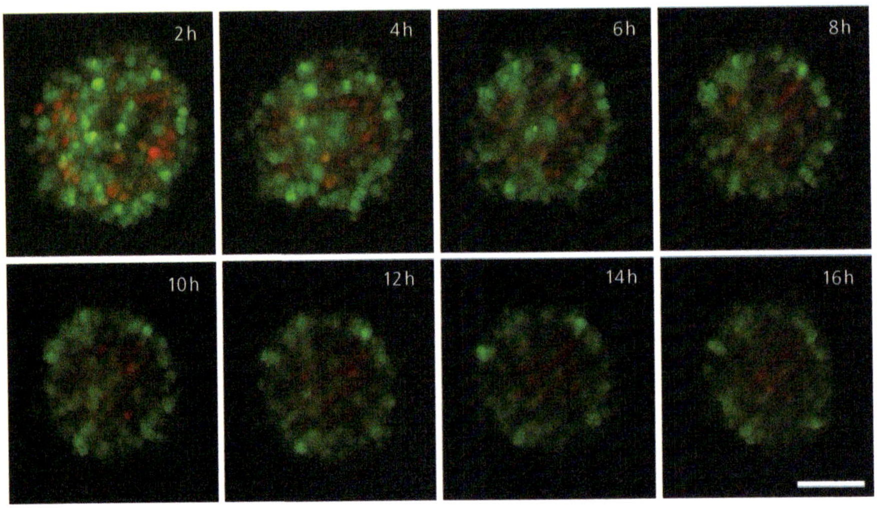

Figure 5.9 Some pairs of cell types will self-sort. CellTracker stained H35 hepatocyte cell line (green) and normal human fibroblasts (red) were mixed in a 1:1 ratio, seeded onto gels and allowed to assemble overnight. Shown are combined red and green fluorescent images over time. The initial random mix of cells self-assembled and self-sorted over time. Scale bar is 100 μm.

This sorting behaviour is highly dependent on the specific pairs of cell types used. Some pairs will self-sort, while others will be randomly mixed in the spheroid, and some pairs will form two separate and distinct spheroids (Achilli *et al.*, 2012; Hoffecker & Iwata, 2014). A recent paper that quantified the kinetics of sorting showed that it occurred coincident in time with self-assembly and was the same when the ratio of cell types was varied over a sixfold range (Achilli *et al.*, 2012). Sorting is thought to be due to differences in cell surface adhesion (Martz *et al.*, 1974), but recent data has also suggested that differences in cytoskeletal contraction can contribute to sorting (Dean & Morgan, 2008; Hoffecker & Iwata, 2014).

Sorting is assayed by labelling one cell type with a fluorescent dye, such as CellTracker™ Green, and the other cell type with a different fluorescent dye, such as CellTracker Red. The cell types are mixed at a desired ratio and the mixture is seeded onto the agarose gel. The cell mixture settles into the recesses and then self-assembles into mixed spheroids. Fluorescent images of the mixed spheroid reveal the relative positions of each cell type in the final spheroid. Spheroids with three different cell types have been produced using the same method (Napolitano *et al.*, 2007a). Since both normal tissues and organs and cancerous tumours are composed of a mixture of different cell types, mixed spheroids may serve to better replicate the complexities of both the healthy and diseased *in vivo* environment.

Spheroids can be used to study drug uptake and diffusion in 3D

Tissues and organs *in vivo* often have multiple cell layers that present complex barriers to the uptake and transport of drugs. In contrast to 2D cell monolayers, 3D spheroids have multiple cell layers and thus create a barrier to drug uptake and transport (Figure 5.10).

Uptake of fluorescent drugs or dyes (reporter drugs) is easy to quantify by fluorescent microscopy. Incubation of spheroids with calcein-AM or Hoechst 33342 results in fluorescent spheroids and this fluorescence increases with time. Calcein-AM and Hoechst 33342 can both diffuse across the cell membrane. Calcein-AM is converted to fluorescent calcein by cellular esterases and Hoechst 33342 becomes fluorescent when it binds to nuclear DNA. A series of time-lapse fluorescent images can be used to quantify the uptake of these dyes.

A recent paper focusing on the actions of a drug efflux pump and gap junction intercellular communication used spheroids to quantify the uptake and inward diffusion of calcein-AM and its derivative calcein (Achilli *et al.*, 2014). Spheroids in the agarose gels were incubated with calcein-AM (non-fluorescent) that can diffuse through the lipid bilayer. Once inside the cell, calcein-AM is converted by esterases into fluorescent calcein. Calcein can no longer diffuse across the cell membrane, but can diffuse into neighbouring cells and deeper into a spheroid via gap junctions. Calcein-AM is a substrate for P-glycoprotein (Pgp), a well-known member of the ATP-dependent family of drug efflux

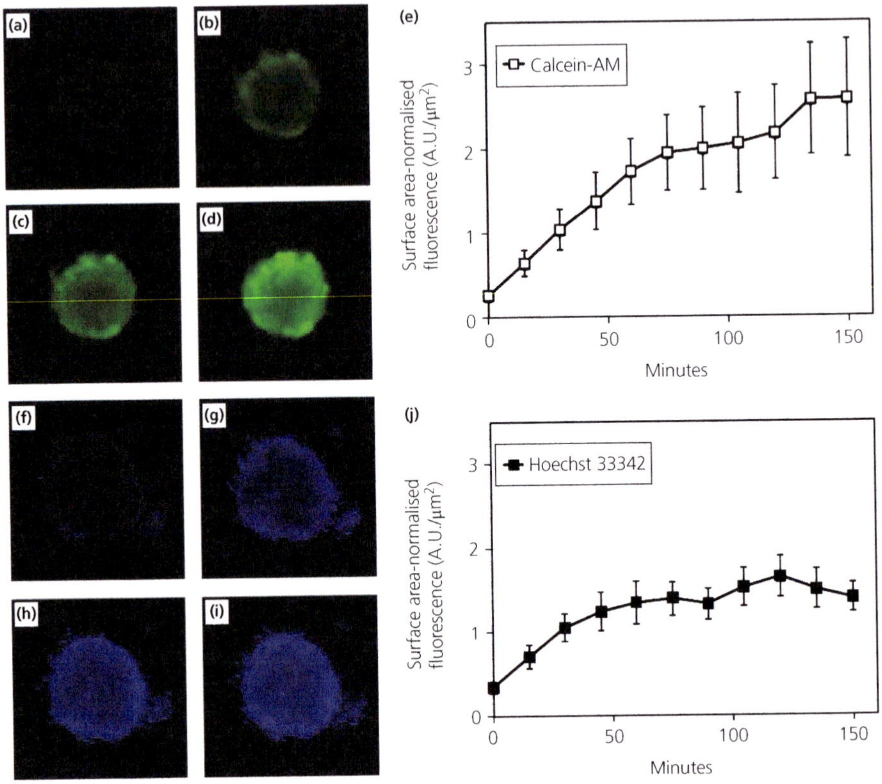

Figure 5.10 Fluorescence uptake and quantification in spheroids. Spheroids self-assembled from the human ovarian granulosa cell line KGN were incubated in the gel with either calcein-AM (a–e) or Hoechst 33342 (f–j). Images of fluorescent spheroids were taken every 15 minutes for the course of 150 minutes, with representative images shown at 0 (a, f), 45 (b, g), 105 (c, h) and 150 minutes (d, i). The total fluorescence of the spheroids (average with standard deviations, n = 20) with respect to time is plotted for both calcein-AM (e) and Hoechst 33342 (j).

transporters, which means Pgp prevents calcein-AM from entering the spheroid. Thus, drug uptake and diffusion within a spheroid were quantified by measuring the fluorescent calcein in the entire spheroid (Pgp activity) and the proportion of fluorescent calcein that enters the core of the spheroid (gap junction activity). The role of the efflux pump ABCG2 has also been investigated using its fluorescent substrate, Hoechst 33342 (Curran *et al.*, 2015a,b). Spheroids have also been used to measure the uptake and tissue penetration of nanoparticles (Achilli *et al.* 2014; Ho *et al.* 2012).

Spheroids can be used as building blocks to form larger tissues

In contrast to cells grown in a scaffold or gel, spheroids will fuse with one another when brought into close contact (Figure 5.11) This phenomenon has been known for many years and was first used as a confrontational assay with

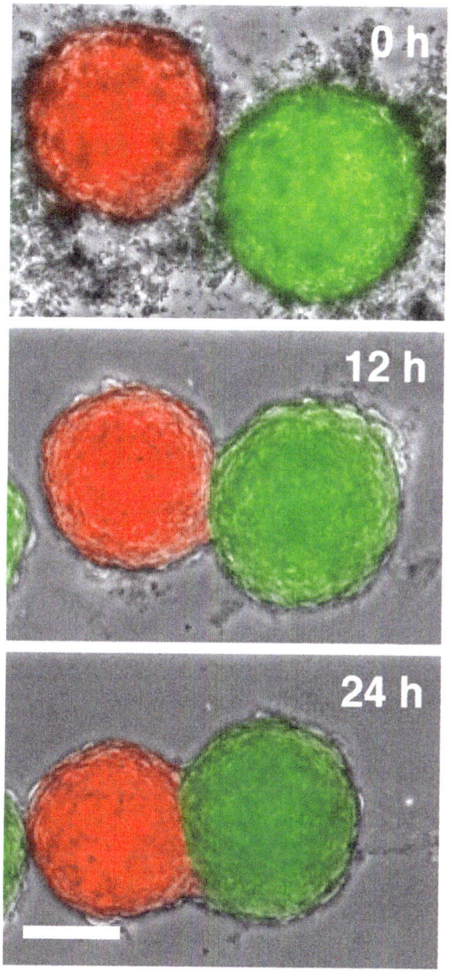

Figure 5.11 **Spheroids can fuse.** Primary human chondrocytes stained with CellTracker were self-assembled into spheroids (~1500 cells/spheroid) for one or two days (red and green, respectively). Spheroids were harvested from the gels and pipetted onto a second gel to bring them into close proximity. Three images (phase contrast, red, green fluorescence) were taken at each time point (once per hour for 24 hours) and the combined images for each time point are shown. Scale bar is 100 μm.

multicellular tumour spheroids to investigate tumour invasiveness (Hattermann *et al.*, 2011; Kunz-Schughart *et al.*, 1998; Yamada & Cukierman, 2007).

More recently, spheroid fusion has been investigated in the context of tissue engineering and the use of spheroids as potential building blocks to form larger tissue constructs. Micromoulded gels are amenable to the scale-up of spheroid production needed for tissue engineering applications (Elbert, 2011; Mironov *et al.*, 2011; Tan *et al.*, 2014). For example, gels made from the 12-256 micro-mould form 256 spheroids with one pipetting step. A single 12-well plate of these gels would produce 3072 spheroids.

Spheroids are easily harvested from the agarose gels (discussed in Protocol 2, below) and can be used to study fusion by transferring spheroids to a secondary agarose gel to promote more controlled spheroid-spheroid interactions (Rago *et al.*, 2009). Spheroid fusion generally occurs over 24–48 hours and is thought to be driven by cell-cell and cell-ECM interactions between the two spheroids. The kinetics of spheroid fusion has been quantified by measuring the change in the long axis of two or more spheroids or the angle at the interface of two spheroids undergoing fusion. The rate and extent of fusion were dependent on the length of time spheroids were precultured or matured prior to fusion. The longer the maturation time, the slower the rate and the lower the extent of fusion.

Spheroid building blocks were used to create toroids and honeycombs. Harvested spheroids were seeded onto agarose gels with either toroid-shaped or honeycomb-shaped recesses. The spheroids underwent fusion and the mould design directed them to form either a toroid or a honeycomb microtissue (Rago *et al.*, 2009). Fusion is not limited to a spheroid – other shapes such as toroids will undergo fusion (Livoti & Morgan, 2010). Multicellular toroids were harvested from their gels and brought into contact either edge to edge (*x*, *y* plane) or stacked on top of each other (*z* plane). Depending on their point of contact, toroids fused to form either a double-lumen structure (figure-of-eight) or a tubular structure, also within 24–48 hours.

Multicellular contraction can be quantified using rod-shaped microtissues

When cells are seeded in agarose gels with trough-shaped recesses, the cells form a long multicellular rod structure with a length (*x*-dimension) that is significantly larger than its width (*y*-dimension) (Figure 5.12). Over time, this rod undergoes contraction as self-assembly proceeds. Self-assembly is a function of many cellular processes including cell surface adhesion and cytoskeletal contraction. The rate of contraction can be quantified by measuring the long axis of the rod over time. Cell types differ in their rate and extent of rod contraction. Rods of normal human fibroblasts contract very rapidly at a near exponential rate and ultimately form a structure that approximates a sphere (Dean & Morgan, 2008; Dean *et al.*, 2007). Whereas rods of a hepatocyte cell line contracted more slowly at a linear rate and still retained a rod structure even after four days. In cell mixtures, small numbers of fibroblasts mixed with hepatocytes were able to drive the rate and extent of rod contraction.

The rod contraction assay has been used to investigate the role of cytoskeletal contraction in self-assembly (Dean & Morgan, 2008). Addition of Y-27632 (a well-known inhibitor of cytoskeletal-mediated contraction) inhibited rod contraction in a dose-dependent and reversible manner. Increasing doses of Y-27632 slowed contraction and maximal inhibition was determined to be 50%, suggesting that Y-27632 cannot completely inhibit self-assembly presumably because surface adhesion molecules also drive self-assembly.

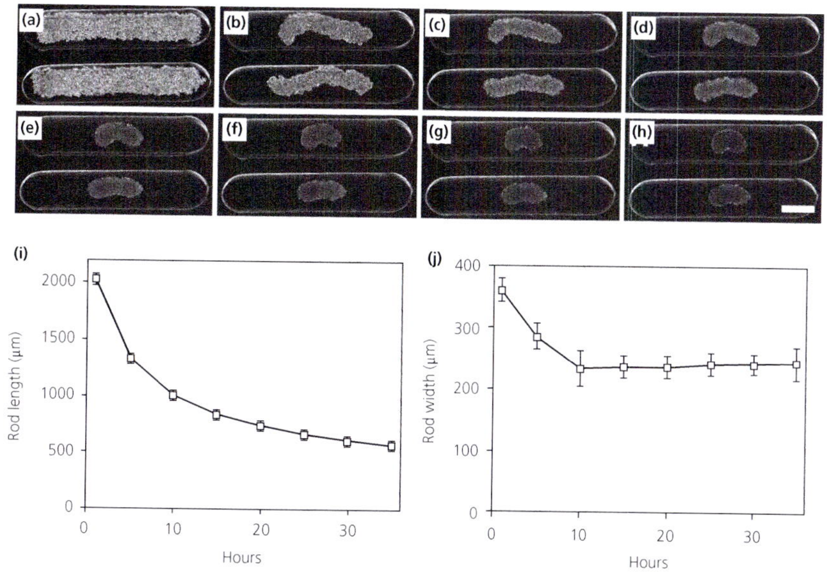

Figure 5.12 Cells **can self-assemble rod-shaped microtissues that contract over time.** Mono-dispersed KGN cells were seeded onto gels with trough-shaped recesses where they formed rod-shaped microtissues that contracted over time. Images were taken at 1 (a), 5 (b), 10 (c), 15 (d), 20 (e), 25 (f), 30 (g) and 35 hours (h) after cell seeding. Rod contraction was determined by measuring the decrease in length (i) and width (j) throughout the course of 35 hours. Shown are average values with standard deviations (n = 13) Scale bar is 400 µm.

Likewise, rod contraction has been used to investigate the contribution of specific surface adhesion molecules and surface receptors. The rate of contraction in rods of cells expressing connexon 43 (Cx43, gap junctions) was slowed by the addition of neutralising antibodies and carbenoxolone, a drug that targets gap junctions (Bao *et al.*, 2011). Rods of C6 glioma cells transfected to overexpress Pannexin-1 (Panx1), a surface protein related to gap junctions but with tumour-suppressive properties, contracted faster than rods of control C6 cells (Bao *et al.*, 2012). Treatment with the drugs carbenoxolone and probenecid, that block Panx1, reversed the effect.

Cells can self-assemble microtissues with complex shapes

In addition to spheroids, cells can self-assemble complex shapes such as a toroid or a honeycomb (Figure 5.13).

Prior to the use of micromoulded hydrogels, the self-assembly of cells was thought to be limited to the spheroid, the geometry that minimises the surface free energy. Use of micromoulded hydrogels with toroid- and honeycomb-shaped recesses demonstrated for the first time that cells could self-assemble complex shaped structures and that these structures are stable when removed from the gel (Dean *et al.*, 2007; Livoti & Morgan, 2010;Napolitano *et al.*, 2007a). A toroid (~25 000 cells) and a honeycomb (~250 000) have significantly more

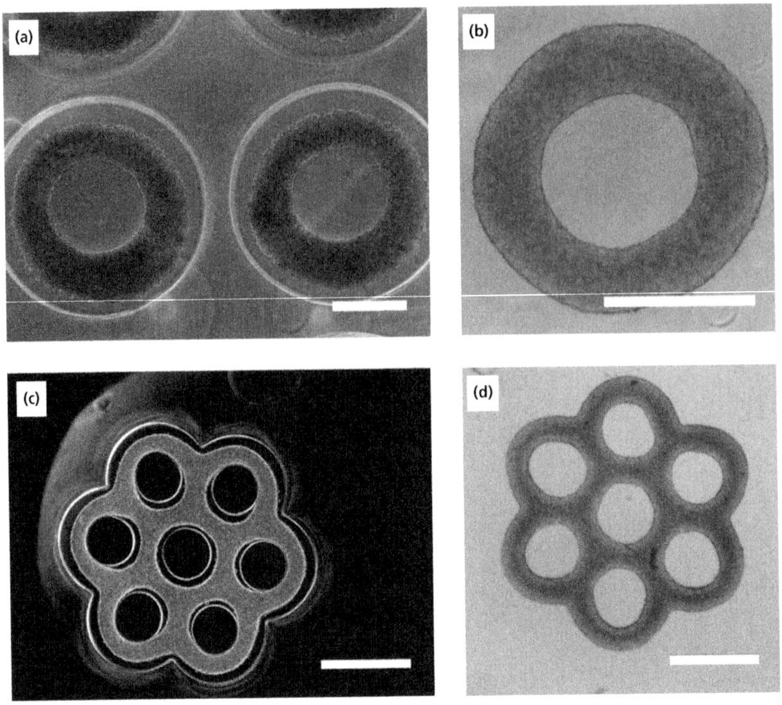

Figure 5.13 Cells can self-assemble microtissues with complex shapes. Cells were seeded onto gels with either toroid- or honeycomb-shaped features. (a) Phase contrast image of H35 toroids (40 000 cells/toroid) in the gel after 24 hours of self-assembly. (b) Image of a self-assembled toroid of KGN cells (35 000 cells/toroid) after harvesting from the gel. (c) Image of a self-assembled honeycomb of MCF-7 cells (250 000 cells/honeycomb) in the gel after 24 hours of self-assembly and (d) outside of the gel. Scale bars are 500 μm (a, b) and 1000 μm (c, d).

cells than a spheroid (~1000). These larger structures are able to survive because the agarose posts in the gel create lumens that facilitate diffusion for nutrient supply and waste removal. Through utilising structures containing lumens, it is possible to make very large microtissues. In a gel made from a custom-designed micromould, cells have self-assembled a honeycomb ~2 cm across with over 6 million cells (Tejavibulya *et al.*, 2011).

Multicellular microtissues can be encapsulated

Microencapsulation and immunoisolation of allogenic and xenogenic cells is a promising method for the treatment of a variety of metabolic disorders. Many years have been spent optimising spherical microcapsules, yet microencapsulation has not achieved its full clinical potential. The limitations of spherical microcapsules include difficulty in retrieving implanted capsules, and relatively few cells per capsule leading to minimal cell-cell interactions that are beneficial for cell function and inefficient cell packing which causes an implant of capsules to have a relatively large volume due to the large proportion of alginate (Mikos *et al.*, 1994; Wilson & Chaikof, 2008). As an alternative to spherical microcapsules, self-assembled microtissues have been encapsulated in alginate (Rago & Morgan, 2009). Mono-dispersed HepG2 cells were seeded onto a micromoulded agarose gel and self-assembled spheroids within 24 hours. This array of densely packed spheroids was encapsulated *in situ* using alginate. When separated from the agarose gel, the encapsulated array had HepG2 microtissues in close proximity to its surface that was further modified by a simple dipping process (e.g. poly-L-lysine). Microtissue size, viability and albumin secretion were all controllable by the number of cells seeded onto the micromoulded agarose gel, and microtissue shape (toroids, honeycombs) and spacing were controllable by micromould choice.

This approach to encapsulation and the use of self-assembled/self-packing 3D microtissues offers new design possibilities that may help to address certain limitations of conventional microcapsules.

- The self-packing that occurs during the self-assembly maximises cell density and cell function while minimising volume of the implant.
- All microtissues are positioned as an ordered array on a single *z* plane close to the surface, enabling uniform mass transport to all microtissues of the implant.
- The encapsulated microtissue array is a sheet, thus its surgical placement and removal are relatively easy.

Custom-designed micromoulds

In addition to the commercially available micromoulds, custom-designed micromoulds have been used for numerous applications. A toroid-on-cone has been used to quantify the forces of self-assembly (Youssef *et al.*, 2011, 2012). Gels produced by this micromould are similar to the toroid micromoulds described above except that the agarose post at the centre of the toroid is in the shape of a cone as opposed to a cylinder. When seeded with cells, a toroid is formed and the forces of self-assembly cause the toroid to move up the cone. The rate of movement of the toroid up the cone and toroid mass were used to calculate the power exerted to overcome gravity. The power of a fibroblast toroid was greater than 10 times that of a hepatocyte toroid. Both could be slowed by the addition of a drug targeting the cytoskeleton, demonstrating that cellular contraction plays an important role in self-assembly. The toroid-on-cone assay was also used to quantify the power of toroids composed of a mixture of fibroblasts and hepatocytes and this demonstrated the importance of heterotypic cell interactions (Youssef *et al.*, 2012).

Another custom micromould, in the shape of a loop-ended dogbone, has been used to investigate the differences between constrained and unconstrained microtissues and the morphological changes that occur to cells and tissues when the cells exert forces during self-assembly (Dean *et al.*, 2009; Svoronos *et al.*, 2014). Of all the cell types tested to date, normal human fibroblasts exert the greatest force during self-assembly and these forces can cause complex-shaped microtissues to fail. The loop-ended dogbone micromould is two toroids joined by a connecting rod. Cells contract around the agarose posts as they self-assemble and exert forces that cause the connecting rod to thin and the cells in that area to elongate, ultimately leading to tissue rupture. The timing and position of tissue rupture are a function of the length and width of the connecting rod. Moreover, rupture can be delayed or stopped by the addition of drugs such as blebbistatin or Y-27632 that target cytoskeletal contraction. The morphological changes that occur to the dogbone-shaped microtissue during self-assembly have been simulated using a finite element model (Wang *et al.*, 2013).

Protocols

Protocol 1: casting, equilibrating and seeding the 3D Petri Dish (©Microtissues, Inc., used by permission)

1 Rinse micromoulds for casting 3D Petri Dishes in dH_2O.
2 Place micromoulds for casting 3D Petri Dishes in a container suitable for autoclave sterilisation.
3 Measure 1 g of high-quality pure agarose powder and place into a dry 100 mL autoclave-safe glass bottle. Be sure bottle and agarose powder are dry with no liquid or moisture. Screw on the lid so that it is secure but still loose.
4 Autoclave micromoulds for casting 3D Petri Dishes and bottle with agarose powder for 30 minutes on dry cycle.
5 In a biosafety cabinet using aseptic technique, add 50 mL of sterile saline (0.9% (w/v) NaCl) to the bottle containing sterilised agarose powder. Screw on the lid so that it is secure but still loose. Swirl the bottle to mix the agarose powder.
6 Use a microwave oven to boil and completely dissolve the agarose powder. Stop the microwave every 10 seconds and swirl the bottle to help dissolve the agarose. Be sure there are no small bits of translucent undissolved agarose.
CAUTION: Molten agarose is very hot, can boil over when the bottle is swirled and can cause severe skin burns. Always use appropriate personal protection (gloves, oven mitt, eye protection and lab coat) when working with molten agarose.
7 Allow the molten agarose to cool to about 60–70 °C. In a biosafety cabinet using aseptic technique, pipette 500 µL of molten agarose into a **12-series** micromould or 330 µL into a **24-series** micromould. Avoid creating bubbles while mixing or pipetting agarose.
8 Remove any small bubbles that may be trapped in the small features of the micromould by pipetting or gentle scraping before the agarose gels.
9 After the agarose has gelled, carefully flex the micromould to remove the 3D Petri Dish. Do not overflex the micromould as it may damage the 3D Petri Dish.

Some practice may be necessary to remove the 3D Petri Dish from the micromould. Transfer the 3D Petri Dish to a 12-well or 24-well tissue culture plate.

10 To equilibrate the 3D Petri Dish, add cell culture medium (2.5 mL/well for 12-well plate and 1.0 mL/well for 24-well plate). Incubate for 30 minutes or longer. Remove culture medium and replace with fresh medium. Repeat once more to equilibrate the 3D Petri Dish with culture medium.

11 Trypsinise cells, count and prepare the desired cell seeding number in the specified volume for the 3D Petri Dish of choice (see Tables 5.1–5.4).

12 Remove culture medium surrounding the outside of the 3D Petri Dish and tilt the tissue culture plate so that the medium from the cell seeding chamber of the 3D Petri Dish can also be carefully removed.

13 Carefully seed cell suspension dropwise into the cell seeding chamber.

14 Allow ~10 minutes for cells to settle into the features of the 3D Petri Dish. Then add additional medium to the outside of the 3D Petri Dish (2.5 mL/well for 12-well plate and 1.0 ml/well for 24-well plate).

15 Place the tissue culture plate in the cell culture incubator and exchange medium surrounding the 3D Petri Dish as needed.

16 When not in use, store micromoulds in a dust- and fibre-free environment. Periodically examine the micromoulds under a dissecting microscope to be sure that small features are free of dust and small fibres.

CAUTION: For Research Use Only, Not for Clinical, Diagnostic or Therapeutic Use.

Protocol 2: harvesting multicellular microtissues from the 3D Petri Dish (©Microtissues, Inc., used by permission)

The multicellular microtissues do not adhere to the gel, so they are easy to harvest from the 3D Petri Dish. Use a micropipette to carefully squirt culture medium into the seeding chamber to dislodge the microtissues from the microwells. The microtissues are then harvested by drawing them up into the micropipette. The harvested microtissues can be washed and processed for subsequent applications (RT-PCR, Western blots, histology, immunostaining). This method is easy, fast and works very well for spheroids. For some applications, the following centrifuge method may be useful.

1 Remove cell culture medium surrounding the 3D Petri Dish containing the multicellular microtissues.

2 Place fresh cell culture medium into the well of a new 12- or 24-well plate, depending on the size of the 3D Petri Dish used (1.0 mL for 12 series or 0.5 mL for 24 series).

3 Carefully remove the 3D Petri Dish with a sterile spatula or sterile tweezers. Invert the 3D Petri Dish into the well of the new plates. Be careful not to create air bubbles between the upside-down 3D Petri Dish and the bottom of the new culture plate. Cover the plate and place the six-well plate in a swinging bucket rotor in a centrifuge and spin for five minutes at 500 rpm.

4 After centrifugation, view the inverted 3D Petri Dish under the microscope to ensure that the multicellular microtissues have been released.
5 Remove the empty 3D Petri Dish with a sterile spatula or sterile tweezers.
6 The harvested multicellular microtissues remaining in the cell culture medium can be washed and further processed depending on the application (immunostaining, Western blot, RT-PCR, etc.).
CAUTION: For Research Use Only, Not for Clinical, Diagnostic or Therapeutic Use.

Troubleshooting

Category	Issue	Solutions
Hydrogel formation	Agarose solidifies too quickly	Reheat the agarose, but always beware of boiling off too much water If still viscous, make fresh agarose
	Bottom of hydrogel is not flat	If convex, scrape molten agarose off the top with sterile spatula If slanted or concave, increase volume of agarose Use glass coverslip to flatten top of gel before it solidifies
Spheroid formation	Spheroids not forming in microwells	Remove air bubbles from recesses Ensure seeding chamber and recesses are fully intact Allow longer time for self-assembly (48 hours) After seeding, do not tip or slosh hydrogels After seeding, do not pipette or aspirate directly into the seeding chamber Freshly isolated cells may need to be cultured in 2D prior to seeding in 3D
	Multiple small spheroids form per microwell	Increase the number of cells seeded Use appropriate micromould design, switch to smallest recess diameter
	Spheroids migrate/climb out of microwell	Lower the number of cells seeded
	Spheroids are not of uniform size	Make sure bottom of hydrogel is flat Ensure hydrogels are placed on level surface during self-assembly After seeding, place on rotating shaker for 5 minutes to disperse cells evenly among recesses
	Spheroids are forming unexpected shapes	Equilibrate gels in serum-free media to prevent protein deposition on gel Make fresh agarose

(Continued)

(Continued)

Category	Issue	Solutions
Additional issues	Hydrogels shift during time-lapse imaging	Secure hydrogel in place by adding small volume of agarose around outside of hydrogel
	Correct concentration of spheroid treatment conditions (e.g. drug or growth factor)	In your calculations be sure to account for volume of hydrogel because compound will diffuse into and equilibrate with the hydrogel
	High background signal when fluorescent dye is added to spheroids in hydrogel	Fluorescent dye can become trapped inside hydrogel, leading to high background; perform more wash steps or transfer spheroids to new hydrogel
	Spheroids are growing too large	Different cell types have different growth properties in 2D versus 3D; try seeding hydrogels with lower cell densities

Conclusion

This chapter has provided an in-depth description of the technology, applications and protocols of the 3D Petri Dish, a commercially available system for the production of multicellular spheroids and microtissues with large and complex shapes. The technology uses precision micromoulds to form moulded agarose gels with arrays of small recesses. When these gels are seeded with cells, the cells settle into the recesses, aggregate and self-assemble into scaffold-free multicellular microtissues. This process occurs overnight, allowing for the creation of hundreds of microtissues in a single gel. Within these multicellular structures lies a cell density comparable to native tissues and organs, maximising interactions, communication and biomechanics compared to 2D monolayers and 3D aggregates with scaffolds. This environment promotes improved differentiation, function and a phenotype that more closely approximates the *in vivo* environment

Most adherent cell types, including primary cells and stem cells, as well as cell lines and cancer cell lines will self-assemble and form microtissues in the 3D Petri Dish. Self-assembly is typically complete within about 24–48 hours, although the rate of self-assembly varies with cell type. By seeding a single gel with two or more cell types, mixed microtissues can be created. Certain mixtures of cell types may self-sort as they self-assemble, creating an architecture with one cell type forming the core and the other cell type forming an outer coating. Other mixtures of cell types may result in two separate but conjoined spheroids. Mixed microtissues are another step closer to the complexity of the *in vivo* environment. Once assembled, microtissues formed in the 3D Petri Dish are easily harvested for histology, immunostaining, Western blots and RT-PCR analyses.

Microtissue size is controlled by micromould choice and the number of cells seeded. Large recesses can give rise to larger microtissues and so can seeding more cells onto the gel. Microtissue shape is dictated by micromould design, which, in addition to making spheroids, can direct cells to self-assemble into rods, toroids or honeycombs. Prior to the development of this technology, it was thought that cells would only self-assemble a spheroid, the geometry that minimises surface free energy. Micromould designs can also direct the self-assembly of large microtissues with complex structures, up to 2 cm wide and containing tens of millions of cells.

The useful applications of this technology continue to increase in number as it is adopted by a wider variety of subject areas, including toxicity testing, phenotypic drug discovery, studies of drug uptake/diffusion, the scale-up of spheroid production, clonal growth of spheroids, quantification of cell adhesion and microtissue contraction, drug inhibition of self-assembly, encapsulation of microtissues and use of spheroids and honeycomb-shaped microtissues in tissue engineering.

Acknowledgements

This work was funded in part by the NSF CMMI 1312392, seed funds from the Johnson & Johnson Corporate Office of Science & Technology, the Brown Technology Ventures Office and the NIH T32 grant GM065085-09. The authors would also like to acknowledge the generous support of Donna McGraw Weiss '89 and Jason Weiss. JRM has an equity interest in Microtissues, Inc. This relationship has been reviewed and managed by Brown University in accordance with its conflict of interest policies.

References

Achilli, T. M., McCalla, S., Tripathi, A. & Morgan, J. R. (2012) Quantification of the kinetics and extent of self-sorting in three dimensional spheroids. *Tissue Engineering Part C Methods*, **18**, 302–309.

Achilli, T. M., Mccalla, S., Meyer, J., Tripathi, A. & Morgan, J. R. (2014) Multilayer spheroids to quantify drug uptake and diffusion in 3D. *Molecular Pharmacology*, **11**, 2071–2081.

Bao, B., Jiang, J., Yanase, T., Nishi, Y. & Morgan, J. R. (2011) Connexon-Mediated cell adhesion drives microtissue self-assembly. *Faseb Journal*, **25**, 255–264.

Bao, B. A., Lai, C. P., Naus, C. C. & Morgan, J. R. (2012) Pannexin1 Drives multicellular aggregate compaction via a signaling cascade that remodels the actin cytoskeleton. *Journal of Biological Chemistry*, **287**, 8407–8416.

Barbone, D., Yang, T. M., Morgan, J. R., Gaudino, G. & Broaddus, V. C. (2008) Mammalian target of rapamycin contributes to the acquired apoptotic resistance of human mesothelioma multicellular spheroids. *Journal of Biological Chemistry*, **283**, 13021–13030.

Birenboim, R., Markus, A. & Goldstein, R. S. (2013) Simple generation of neurons from human embryonic stem cells using agarose multiwell dishes. *Journal of Neuroscience Methods*, **214**, 9–14.

Curcio, E., Salerno, S., Barbieri, G., de Bartolo, L., Drioli, E. & Bader, A. (2007) Mass transfer and metabolic reactions in hepatocyte spheroids cultured in rotating wall gas-permeable membrane system. *Biomaterials*, **28**, 5487–5497.

Curran, S., Achilli, T. M., Leary, E., Wilks, B. *et al.* (2015a) A 3D spheroid system to evaluate inhibitors of the Abcg2 transporter in drug uptake and penetration. *Technology*, **3**, 54–63.

Curran, S., Vantangoli, M. M., Boekelheide, K. & Morgan, J. R. (2015) Architecture of chimeric spheroids controls drug transport. *Cancer Microenvironment*, **8**, 101–109.

Dean, D. M. & Morgan, J. R. (2008) Cytoskeletal-mediated tension modulates the directed self-assembly of microtissues. *Tissue Engineering Part A*, **14**, 1989–1997.

Dean, D. M., Napolitano, A. P., Youssef, J. & Morgan, J. R. (2007) Rods, tori, and honeycombs: the directed self-assembly of microtissues with prescribed microscale geometries. *Faseb Journal*, **21**, 4005–4012.

Dean, D. M., Rago, A. P. & Morgan, J. R. (2009) Fibroblast elongation and dendritic extensions in constrained versus unconstrained microtissues. *Cell Motility Cytoskeleton*, **66**, 129–141.

Desroches, B. R., Zhang, P., Choi, B. R. *et al.* (2012) Functional scaffold-free 3-D cardiac microtissues: a novel model for the investigation of heart cells. *American Journal of Physiology: Heath and Circulatory Physiology*, **302**, H2031–2042.

Elbert, D. L. (2011) Bottom-up tissue engineering. *Currend Opinion in Biotechnology*, **22**, 674–680.

Gwyther, T. A., Hu, J. Z., Christakis, A. G. *et al.* (2011) Engineered vascular tissue fabricated from aggregated smooth muscle cells. *Cells Tissues Organs*, **194**, 13–24.

Hattermann, K., Held-Feindt, J. & Mentlein, R. (2011) Spheroid confrontation assay: a simple method to monitor the three-dimensional migration of different cell types in vitro. *Annals of Anatomy*, **193**, 181–184.

Ho, D. N., Kohler, N., Sigdel, A. *et al.* (2012) Penetration of endothelial cell coated multicellular tumour spheroids by iron oxide nanoparticles. *Theranostics*, **2**, 66–75.

Hoffecker, I. T. & Iwata, H. (2014) Manipulation of cell sorting within mesenchymal stromal cell-islet cell multicellular spheroids. *Tissue Engineering Part A*, **20**, 1643–1653.

Klopper, A. V., Krens, G., Grill, S. W. & Heisenberg, C. P. (2010) Finite-size corrections to scaling behavior in sorted cell aggregates. *European Physical Journal E Soft Matter*, **33**, 99–103.

Krotz, S. P., Robins, J. C., Ferruccio, T. M. *et al.* (2010) In vitro maturation of oocytes via the pre-fabricated self-assembled artificial human ovary. *Journal of Assisted Reproduction and Genetics*, **27**, 743–50.

Kunz-Schughart, L. A., Kreutz, M. & Knuechel, R. (1998) Multicellular spheroids: a three-dimensional in vitro culture system to study tumour biology. *International Journal of Experimental Pathology*, **79**, 1–23.

Livoti, C. M. & Morgan, J. R. (2010) Self-assembly and tissue fusion of toroid-shaped minimal building units. *Tissue Engineering Part A*, **16**, 2051–2061.

Martz, E., Phillips, H. M. & Steinberg, M. S. (1974) Contact inhibitions of overlapping and differential cell adhesion: a sufficient model for the control of certain cell culture morphologies. *Journal of Cell Science*, **16**, 401–419.

Mikos, A. G., Papadaki, M. G., Kouvroukoglou, S., Ishaug, S. L. & Thomson, R. C. (1994) Mini-review: islet transplantation to create a bioartificial pancreas. *Biotechnology and Bioengineering*, **43**, 673–677.

Mironov, V., Kasyanov, V. & Markwald, R. R. (2011) Organ printing: from bioprinter to organ biofabrication line. *Current Opinion in Biotechnology*, **22**, 667–673.

Napolitano, A., Dean, D., Man, A. *et al.* (2007a) Scaffold-free three-dimensional cell culture utilizing micromoulded nonadhesive hydrogels. *Biotechniques*, **43**, 494–500.

Napolitano, A. P., Chai, P., Dean, D. M. & Morgan, J. R. (2007b) Dynamics of the self-assembly of complex cellular aggregates on micromoulded nonadhesive hydrogels. *Tissue Engineering*, **13**, 2087–2094.

Rago A. P. & Morgan, C. P. Jr (2009) Encapsulated arrays of self-assembled microtissues: an alternative to spherical microcapsules. *Tissue Engineering Part A*, **15**, 387–395.

Rago, A. P., Napolitano, A. P., Dean, D. M., Chai, P. R. & Morgan, J. R. (2008) Miniaturization of an anoikis assay using non-adhesive micromoulded hydrogels. *Cytotechnology*, **56**, 81–90.

Rago, A. P., Dean, D. M. & Morgan, J. R. (2009) Controlling cell position in complex heterotypic 3D microtissues by tissue fusion. *Biotechnology and Bioengineering*, **102**, 1231–1241.

Reya, T., Morrison, S. J., Clarke, M. F. & Weissman, I. L. (2001) Stem cells, cancer, and cancer stem cells. *Nature*, **414**, 105–111.

Robins, J. C., Morgan, J. R., Krueger, P. & Carson, S. A. (2011) Bioengineering anembryonic human trophoblast vesicles. *Reproductive Science*, **18**, 128–135.

Svoronos, A. A., Tejavibulya, N., Schell, J. Y., Shenoy, V. B. & Morgan, J. R. (2014) Micro-mould design controls the 3D morphological evolution of self-assembling multicellular microtissues. *Tissue Engineering Part A*, **20**, 1134–1144.

Tan, Y., Richards, D. J., Trusk, T. C. *et al.* (2014) 3D Printing facilitated scaffold-free tissue unit fabrication. *Biofabrication*, **6**, 024111.

Tejavibulya, N., Youssef, J., Bao, B., Ferruccio, T. M. & Morgan, J. R. (2011) Directed self-assembly of large scaffold-free multi-cellular honeycomb structures. *Biofabrication*, **3**, 034110.

Wang, H., Svoronos, A. A., Boudou, T. *et al.* (2013) Necking and failure of constrained 3D microtissues induced by cellular tension. *Proceedings of the National Academy of Sciences USA*, **110**, 20923–20928.

Wilson, J. T. & Chaikof, E. L. (2008) Challenges and emerging technologies in the immunoisolation of cells and tissues. *Advanced Drug Delivery Reviews*, **60**, 124–145.

Yamada, K. M. & Cukierman, E. (2007) Modeling tissue morphogenesis and cancer in 3D. *Cell*, **130**, 601–610.

Yang, T. M., Barbone, D., Fennell, D. A. & Broaddus, V. C. (2009) Bcl-2 family proteins contribute to apoptotic resistance in lung cancer multicellular spheroids. *American Journal of Respiratory Cell and Molecular Biology*, **41**, 14–23.

Youssef, J., Nurse, A. K., Freund, L. B. & Morgan, J. R. (2011) Quantification of the forces driving self-assembly of three-dimensional microtissues. *Proceedings of the National Academy of Sciences USA*, **108**, 6993–6998.

Youssef, J., Chen, P., Shenoy, V. B. & Morgan, J. R. (2012) Mechanotransduction is enhanced by the synergistic action of heterotypic cell interactions and Tgf-Beta1. *Faseb Journal*, **26**, 2522–2530.

CHAPTER 6

Organotypic microtissues on an air-liquid interface

Lars E. Sundstrom[1], Igor Charvet[2] and Luc Stoppini[2]
[1] *Elizabeth Blackwell Institute for Health Research, School of Clinical Sciences, University of Bristol, Bristol, UK*
[2] *Hepia/HES-SO Campus Biotech, Geneva, Switzerland*

Introduction

Since the very early days of tissue culture, a major driver has been to develop systems that mimic the functions of organs outside the body. The pioneering work of Ross Harrison at the beginning of the last century demonstrated that tissues could be maintained *in vitro* by cultivating frog neural tube explants in small drops of medium, hanging upside down on a coverslip (Harrison, 1907). In this single seminal experiment, tissue culture, the hanging drop method and air-liquid interfaces were first described together. Carrel and his colleagues (Carrel, 1912) extended these findings, demonstrating that organs from warm-blooded animals could retain function outside the body by isolating beating chick hearts and keeping them alive and beating *ex vivo* for extended periods of time (in one case over a 30-year period) (Landecker, 2007).

Since these early days, however, difficulty in keeping explanted tissues from differentiated organs viable for sufficient periods of time and the relative complexity in generating them have limited their use in research. Instead, monolayer culture of single cell types has become the standard method in tissue culture. 2D culture systems are generally easy to produce, are relatively stable and easy to manipulate under experimental conditions and for this reason have become a standard tools for drug discovery research. However, the translational validity of such 2D culture systems, often using transformed cell lines which are also often from non-human origin, is increasingly becoming questioned (Sundstrom, 2007; Sundstrom *et al.*, 2005). The development of new *in vitro* systems that provide better information on how substances will behave in humans from both an efficacy and a toxicity perspective is imperative to improve productivity in drug development.

In our view, a key missing link is test systems that mimic human organ function *in vitro*, which can be produced in sufficient quantity and quality to be useful

Technology Platforms for 3D Cell Culture: A User's Guide, First Edition. Edited by Stefan Przyborski.

for drug testing. 3D tissue culture systems are uniquely positioned to provide this missing link and fill a clear gap between simple cellular systems and human biology (Sundstrom, 2007). A significant advantage of many 3D systems is the ability to use mixed cell types and generate tissue-like structures (microtissues) in which cell-cell interactions and 'organotypic' functions are present. Recently, the ability to use human stem cell-based material as a starting point for 3D culture has created an additional advantage by providing a new source of human cells that can be replicated and differentiated under laboratory conditions.

The advantages of 3D over 2D culture systems are becoming increasingly accepted but so far uptake of 3D culture methods has been slower than anticipated (Editorial, 2003), mainly as a result of the complexity involved in generating 3D cultures compared to conventional 2D tissue culture methods. However, developing robust, reliable methods to grow cultures in three dimensions has been at the forefront of tissue culture research for the past few decades. Recent advances now make it possible to produce these at a cost and scale that should allow these technologies to become a mainstream tool in research and in the development of better *in vitro* systems for drug testing (Pampaloni, 2007).

A major advance has been the development of reaggregation approaches to tissues that can form the basis of a 3D multicellular culture system with 'organotypic' properties (Layer *et al.*, 2002). A popular variant of this method has been the generation of spheroid cultures, in which cells are aggregated to form spherical 3D cultures (Honegger, 1989). Spheroids have the advantage of being relatively easy to produce from a variety of materials but have limitations in terms of viability and ease of manipulation and often require specialised equipment such as rotating vessel incubators or spinner flasks to maintain. The fact that the cultures are floating in many cases adds an additional manipulation for imaging or physiological measurements. There are also limitations in that spheroids above a certain diameter may suffer from hypoxia developing at the core of the culture, as oxygen penetration is often insufficient.

Classic methods of growing 2D cultures in submerged systems are generally difficult to extrapolate directly to 3D organotypic systems. At cell densities that relate to tissue-like structures (around 10^8 cells/mL), oxygen penetration of the tissue is usually insufficient to maintain long-term survival in classic submerged systems without the use of bioreactors to maintain homeostasis (Portner, 2007). Bioreactor-based systems developed so far are relatively complex and while of great use for large-scale cell culture, they do not lend themselves easily to drug-testing applications, as scalability, cost and throughput are significant issues.

Methods of maintaining or generating microtissues at the air-liquid interface are therefore becoming increasingly popular as tissues survive comparatively longer at the air-liquid interface than in other formats such as submerged cultures since oxygen can diffuse directly into the culture.

We have developed a method of generating stable organotypic microtissues (micro-organoids) by allowing these to form on a semi-porous membrane where

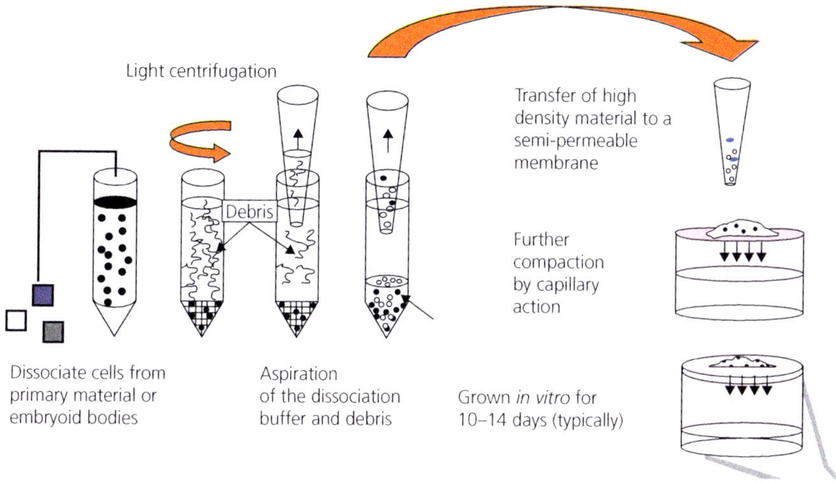

Figure 6.1 Method of generating stable organotypic microtissues. Primary tissues or embryoid bodies from human or animal origin are dissociated and recompacted by light centrifugation, placed on a semi-permeable membrane at the air-liquid interface and grown *in vitro* for 10–14 days before use. Source: Biggs 2011. Reproduced with permission of Sage Publications.

the tissue is on the opposite side to the culture medium separated by a semi-permeable membrane. Cultures are fed and maintained by diffusion of nutrients through the membrane and the culture is exposed on the other side of the membrane to the air through a thin layer of liquid at the air-liquid interface (see Figure 6.1). A key advantage of this method is that cultures are relatively thin (around 100–150 µM) which allows penetration of oxygen into the tissue and long-term survival, which is essential to tissues that suffer acutely from hypoxia such as brain or heart. There is also no need for scaffolds or complex culture vessels, so scalable systems can be developed such as higher throughput formats (e.g. 96-well systems). The cultures are also stable and static, which offers advantages for imaging and electrophysiological read-outs.

The method is similar to that originally described for generating organotypic brain slice cultures outlined by Stoppini (Stoppini *et al.*, 1991), but it involves disaggregation of the starting material and then reaggregation of the cells onto a semi-permeable membrane. An advantage of this method is that a variety of different materials can be used to generate the cultures from primary tissues, embryoid bodies or other stem cell-derived animal or human material, thus offering a standardised method for multiple tissues and comparison between animal and human tissues. We have recently compared our method of generating reconstituted tissues on an air-liquid interface with other 3D methods for applications relating to drug development (Sundstrom, 2012). The fact that the tissues are thinner than conventional spheroids offers advantages such as oxygen diffusion into the tissue, so avoiding hypoxia in the centre of the culture. Moreover, there is no need for specialised equipment or scaffolds to generate these cultures. The fact that the cultures are static and not free-floating has significant advantages for imaging applications and electrophysiological approaches. The ability to use both primary and stem cell-based material using a single approach means it is possible to compare different starting materials and compare results on the same platform. Since the material is first dissociated, it is also possible to control what is put back into the culture, for example by altering cell stoichiometry or introducing reporter systems into the culture mix.

Description of technology

General method (Figure 6.1)

The starting material can either be primary tissues or material differentiated from stem cells such as embryoid bodies. It is first dissociated enzymatically or mechanically and resuspended in an appropriate buffer, e.g. Earl balanced salt solution. Following a light centrifugation step to pellet the cells (e.g. 1660g for five minutes), these can be resuspended in an appropriate culture medium (e.g. Dulbecco's modified eagle medium) at an appropriate cell density. It is worth experimenting to find the optimal cell density and the best semi-permeable membranes to use for the tissue in question, but typically, a density of around 50 000 cells per microliter onto a Millipore® Millicell®-CM membrane is a good starting point. The cell suspension is then transferred by pipette directly onto the membrane and the membrane with the cultures on it is then placed in culture medium for 7–14 days, changing the medium twice a week. Culture conditions are similar to those described for organotypic slice culture (Stoppini *et al.*, 1991). For a standard 6-well plate with inserts, we typically use 1 mL of culture medium per well with 20% fetal bovine serum and grow up to six cultures per insert. Media should be changed twice a week.

After 7–14 days cultures should become translucent and can be used for experiments usually after 10–14 days. Length in culture varies depending on the starting material. Rodent primary material, for instance, will lose viability after around 21 days (Bailey, 2011). However, cultures derived from fetal or stem cell material may survive for several months, and in some cases we have maintained tissue functional for over a year.

Introducing reporter systems (Figure 6.2)

Introducing reporter systems into the cellular composition can be a major advantage in developing new screening tools, as can be seen in a recent example where LN18 human glioblastoma cell lines were first transfected to express fluorescent proteins and then reintroduced into dissociated primary rat brain material before being reaggregated. In this example of a chimeric culture, the labelled human tumour cells in rat primary brain tissue, which start out dispersed, reaggregate to form microtumour-like focal regions within the brain tissue over

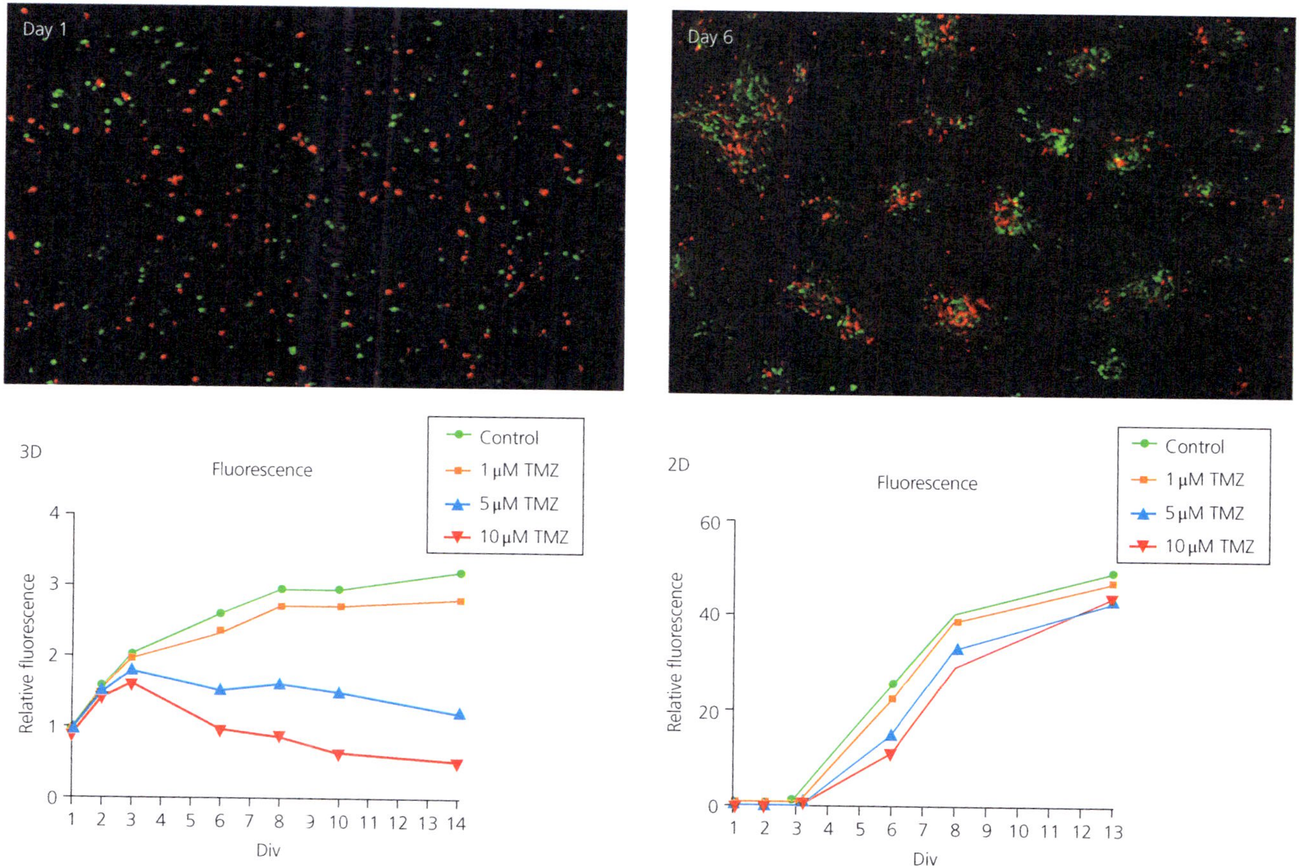

Figure 6.2 Human LN18 glioma cells reintroduced to disaggregated rat brain tissue labelled with fluorescent reporters. This demonstrates aggregation into microtumours after six days *in vitro* which is dose dependently inhibited by temozolomide. Addition of temozolomide to standard monolayer culture of LN18 glioma cell lines has no effect on these cultures. Source: Biggs 2011. Reproduced with permission of Sage Publications.

a seven-day period. Interestingly, this phenomenon appears to be highly orthotopic, as for example human ovarian tumour cell lines do not reaggregate in this way in rat brain tissue. Measuring the rate and extent of this aggregation in the presence or absence of antitumour compounds provides a useful assay for determining the effectiveness of compounds in preventing cell movement and aggregation. For example, temozolomide and taxol, which are both used clinically, show strong dose-dependent effects on cell migration and aggregation but show no effects on standard 2D tumour cell lines (see Biggs *et al.*, 2011 for a full description).

Self-directed reorganisation

Self-induced reorganisation of primary material to form tissue-like structure seems to be a feature of this culture method. For example, disaggregating P0 rat brain and then reaggregating it on the membrane typically results in a layered structure within 7–10 days, with astrocytes migrating to face the membrane, neuronal cell bodies above the astrocytes in the centre of the culture and neuronal projections processes mostly above the cell bodies (Biggs *et al.*, 2011). While the organisation of the resulting cultures does not recapitulate the original anatomical structure, rather remarkably homotopic relationships of cells to one another do seem to reform so that relative connectivity between excitatory and inhibitory cells, for example, is maintained, resulting in cultures that mimic the original tissues. These cultures form functional synapses, which are spontaneously active and respond in similar ways to a number of pharmacological agents and, apart from the lack of anatomical structure, are to a large extent indistinguishable from the parent tissue (Bailey *et al.*, 2011).

The ability to use human material as a starting point has obvious advantages for drug testing and the method we describe has been used to generate a number of important microtissues for efficacy and drug safety or toxicity testing, as reviewed by Sundstrom *et al.* (2012). For example, functional human pancreatic microtissues can be generated from human adult primary islets that recapitulate islet beta cell function, releasing insulin in response to glucose. These 'pancreatic' microtissues can be used to study the effects of insulin secretagogues and other compounds affecting insulin release. Human 'cardiac microtissues' can also be generated from cardiac stem cells (embryonic or induced pluripotent stem cell derived) that form spontaneously beating microtissues whose activity can be relatively easily recorded by using microelectrode array systems (Figure 6.3). A significant advantage of cardiac microtissues derived in this way is the fact that electrophysiological signals are far larger than those recorded in 2D systems, providing a much better signal-to-noise ratio. This is particularly advantageous as fewer cultures are needed to see drug effects, which can sometimes be relatively subtle. We have also shown that it is possible to generate multielectrode array systems etched directly onto a semi-permeable membrane (Stoppini, 1997), enabling long-term and repeated recordings from excitable

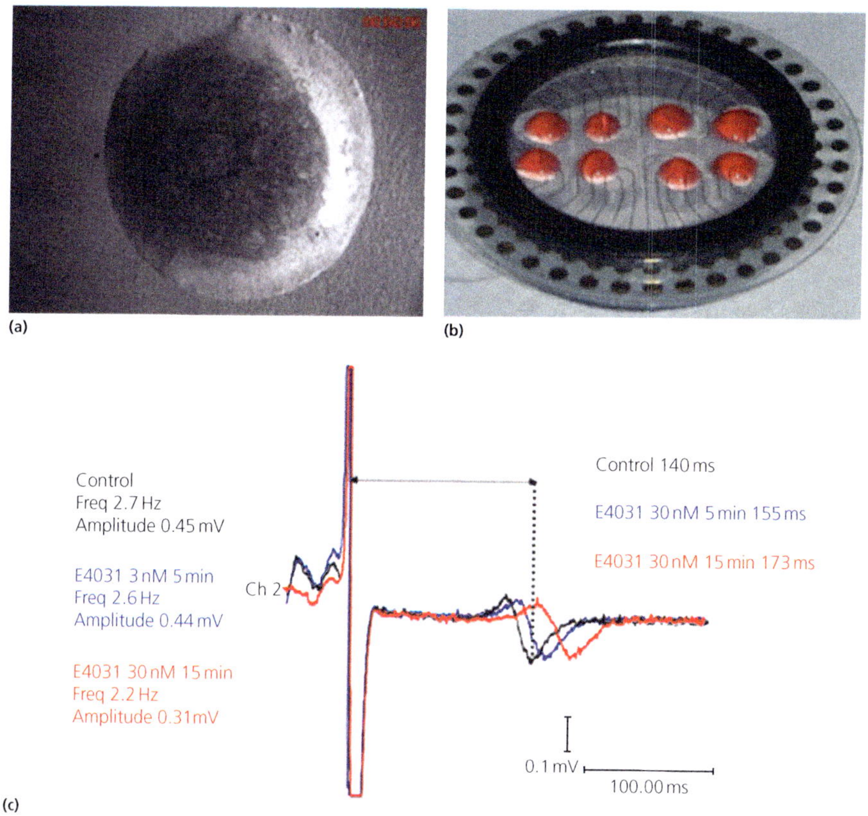

Figure 6.3 Cardiac microtissues formed from reaggregating human primary tissues. Example of a cardiac microtissue (a). These cardiac microtissues beat spontaneously, generating large field potentials that resemble remarkably well the human electrocardiogram and which can easily be recorded on planar electrode arrays (b). The reference compound E4031 (930 nM) demonstrates a classic shift of the t-component of the complex (c).

tissues. Recording activity from spontaneously beating cardiac tissues derived from human embryonic stem cell reveals a response pattern that is strikingly reminiscent of an electrocardiogram (with easily distinguishable PQRT components), and QT prolongation with HERG channel blockers is also evident, opening up possibilities for cardiac safety testing.

Similarly, Rowe *et al.* (2013) have shown that compared to conventional extracellular matrix sandwich techniques, human adult and fetal primary hepatocytes grown by this method on an air-liquid interface survive better, retain better function, have enhanced metabolic function and develop more anatomical microstructures such as biliary canaliculi. This suggests that this method should also provide an enhancement over existing liver toxicity methods using HepG2 or human primary material for toxicity testing and transformation of compounds that are metabolised in the liver.

3D neural OrganDOT™ tissues for neurotoxicology studies

Experimental design

In this example application, we will present how 3D stable and long-term *in vitro* neural cultures may be used as a model for *in vitro* neurotoxicity testing. In this model, human stem cells (H9) were differentiated to generate neuroprogenitor cells. Functional 3D neural networks were examined after up to two months in culture. To characterise the models, we measured changes in gene and protein expression of neural markers by real-time polymerase chain reaction (RT-PCR) and Western blotting, as well as morphological analyses. Extracellular recordings of electrophysiological activities were used to verify the functionality of the cultures. 3D cultures expressed markers of both mature neurons and astrocytes, and intricate neurite networks were observed in those cultures. To assess the predictive capacity of our *in vitro* test system to verify acute neurotoxic effects toxic compounds, we studied two environmentally important and well-known neurotoxic compounds: trimethyltin chloride (TMTC) and methyl mercury chloride (MeHg). The results showed a clear decrease in gene expression of neuronal markers, β3 tubulin and synapsin-1 and MAP2, as well as an increased level of the glial marker GFAP. This is indicative of neuronal failure and astrogliosis. Further development of these human *in vitro* models could enable screening of toxicological profiles of chemicals or of new drug candidates.

Materials and methods

Embryonic human stem cell cultures

Undifferentiated human embryonic stem cell line 9 from the University of Wisconsin (HUES9) was cultivated on BD Matrigel™ hESC-qualified matrix in mTeSR™ 2 medium supplemented with ROCK inhibitor (Y-27632, Cellagentech) to a final concentration of 5 μM.

- mTeSR medium is a defined, serum-free medium formulation which has been developed to standardise feeder-independent hES cell culture and maintain and expand hES cells in an undifferentiated state when used with BD Matrigel hESC-qualified matrix as a substrate. Details about culture of H.1 can be found in the manual *Maintenance of Human Embryonic Stem Cells in mTeSR™* (StemCell Technologies, 29106).
- The ROCK inhibitor Y-27632 improves the survival of freeze-thawed hES cells (Watanabe *et al.*, 2007).
- HUES9 should be passaged every 4–5 days, before they become subconfluent. In these conditions, H.9 cells can be cultured for up to 20 passages.
- HUES9 were passaged using Accutase® (Millipore) at a dilution of 1:10.

Neural induction

HUES9 at about 20% confluence were disaggregated into single cells using Accutase for 10 minutes and pre-plated on a CELLstart™ (Life Technologies)-coated surface in mTeSR medium supplemented with ROCK inhibitor 5 μM. The next day, the ROCK inhibitor was withdrawn and spontaneous neural differentiation was initiated in medium composed of DMEM/F12-N2 (50%) Neurobasal-B27 (50%), including 10 nM transforming growth factor (TGF)-β inhibitor (SB431542, Tocris), 10 nM dorsomorphin (Tocris) and 500 ng/mL of Noggin (R&D Systems). At day 8, we verified the formation of neural tube-like rosettes, i.e. the signature of neural precursor cells (NPC) in formation (Wilson & Stice, 2006). Induction was maintained upon days 15–20 and the obtained NPC were allowed to expand in Knock out DMEM/F1-neural StemPro® (Life Technologies) supplemented with fibroblast growth factor (FGF)8b (20 ng/mL) and epidermal growth factor (EGF) (20 ng/mL) until they were nearly confluent.

Neuronal differentiation in 3D

To generate 3D neural microtissues (N-OrganDOTs), 3 μL drops of NPCs in suspension (150 000 cells/μL) were deposed onto precut patches of a hydrophilic polytetrafluoroethylene (PTFE) membrane (6 mm diameter, 0.4 μm; BioCell-Interface) which were placed on top of a Millicell-CM (0.4 μm) culture plate insert (Millipore). Inserts were placed in a 6-well plate, each insert bearing four N-OrganDOTs maintained at the air-liquid interface. Differentiation medium was composed of DMEM/F12-N2 (33%), Neurobasal-B27 (33%), Knock out DMEM/F12-Neural StemPro (33%), supplemented with 1% FBS (Invitrogen™). Cultures were maintained *in vitro* for two months to form 3D neural structures.

Electrophysiological recordings from two-months cultured N-OrganDOTs

After two months in culture on Millicel membrane, the N-OrganDOTs were controlled by electrophysiological recordings using a multielectrode array system (BioCell-Interface). The BioCell multielectrode array consists of a network of 40 golden microelectrodes (30 μm thick) embedded on a perforated polyimide support so that perfusion solutions from the lower chamber can reach the tissue placed onto the array. The console unit comprises a thermoregulated electronic module that houses the disposable cartridge that includes the multielectrode array, a series of signal preamplifiers and conditioners as well as multiplexed channels. An analogue/digital interface (National Instrument Corporation) was installed inside a computer and digitises signals from the network of electrodes in contact with the tissue. Online or offline analyses are carried out using dedicated software (Hakkoum *et al.*, 2006).

Patches of PTFE membranes bearing N-OrganDOTs were laid upside down onto microelectrode arrays (MEA) designed onto porous membranes (Biocell-Interface cartridges). Stimuli from 1000–4000 mV were applied to the N-OrganDOTs in order to generate evoked field potentials. N-OrganDOTs generating significantly strong evoked field potentials were then selected for the toxicological tests.

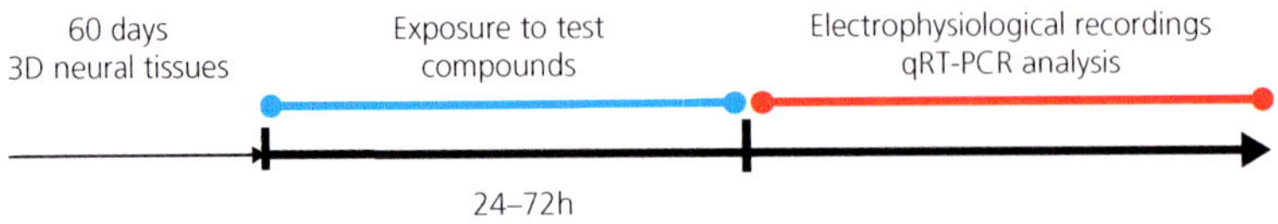

Figure 6.4 3D neural tissues timeline. 3D neural tissues were grown for up to 60 days to mature before their exposure to toxic compounds for between 24 and 72 hours. Electrophysiological recordings were performed to assess functionality before analysis of the different neural gene expressions by qRT-PCR.

Toxic molecule exposure

N-OrganDOTs with positive electrophysiological measurements were treated for 48 hours with the two neurotoxic molecules TMTC and MeHg diluted in differentiation medium. Non-toxic D-mannitol was also applied to concentrations from 0.01 mM to 100 mM (Figure 6.4).

Real-time quantitative PCR analysis of tubulin β3, synaptophysin and GFAP expression

RNA isolation

After toxic compound treatment, two-month-old N-OrganDOTs were picked up from PTFE patch, rinsed with medium and processed for RNA isolation (or stored in RNA later until isolation of RNA). N-OrganDOTs were first homogenised in RLT lysis buffer (Qiagen) and total RNA was extracted using an RNeasy Mini Kit, in combination with a QIAshredder and including a DNA-ase treatment (Qiagen). The RNA was quantified spectrophotometrically at 260 nm (ND-1000, Nanodrop Technologies) and stored at –70 °C.

Primer design

The oligonucleotides used as primers are listed below.

Human tubulin β3	
Name	Hs__F
Sequence	GCAACTACGTGGGCGACT
Name	Hs_TUBB3_R
Sequence	CGAGGCACGTACTTGTGAGA
Human synaptophysin	
Name	Hs_Syp_ex4-5-184F
Sequence	AGGGCCCATGCTGGACTT
Name	Hs_Syp_ex4-5-254R
Sequence	CCCATGCCGATGAGCTAACT

Human glial fibrillary acidic protein (GFAP)

Name	Hs_GFAP_ex4-73F
Sequence	CTCACCGCAGCCCTGAAA
Name	Hs_GFAP_ex4-123R
Sequence	GGACGCCATTGCCTCATACT

Human neuronal nuclear protein (NeuN)

Name	Hs_NeuN_ex5-6-28F
Sequence	CAGTCTACGGGCCTGAATTCTATG
Name	Hs_NeuN_ex5-6-87R
Sequence	CTGTGCCGGTGGTGGG

Human synapsin-1

Name	Hs_Syn1_ex2-2F
Sequence	GCAAAATACTTCAAAGGGAAAAAGA
Name	Hs_Syn1_ex2-54R
Sequence	TCTACTTTAATGTCAATTTCTCCATGGA

Quantitative real-time PCR

First strand cDNA synthesis was performed with PrimeScript reverse transcriptase enzyme kit (Takara Bio Inc.). A quantity of 0.5 µg total RNA was added to the mixture containing both oligo(dT) and random hexamer primers, following the suppliers instructions (www.takara-bio.com/). SYBR green assays were designed using the program Primer Express v 2.0 (Applied Biosystems) with default parameters.

Amplicon sequences were aligned against the human genome by BLAST to ensure that they were specific for the gene being tested. Oligonucleotides were obtained from Invitrogen. The efficiency of each design was tested with serial dilutions of cDNA. Two stably expressed housekeeping genes were used as internal standards for the normalisation of expression signals. Here, the obtained values were normalised to the mean of expression of the two housekeeping genes EEF1 (elongation factor 1 complex) and GusB (β-glucuronidase). PCR reactions (10 µL volume) contained diluted cDNA, 2×Power SYBR® Green Master Mix (Applied Biosystems), 300 nM of forward and reverse primers. PCR was performed on a SDS 7900 HT instrument (Applied Biosystems) with the following parameters: 50 °C for two minutes, 95 °C for 10 minutes and 45 cycles of 95 °C 15 seconds/60 °C one minute. Each reaction was performed in three replicates on a 384-well plate.

Data analysis

Expression levels were determined as the number of cycles needed for the amplification to reach a threshold (C_T) fixed in the exponential phase of PCR reaction. The $\mathbf{C}_T$ were transformed into quantities using PCR efficiencies in order to use

GeNorm software. Raw C_T values obtained with SDS 2.2 (Applied Biosystems) were imported in Excel and normalisation factor and fold changes were calculated using the GeNorm method (Vandesompele *et al.*, 2002). C_T values from the ABI PRISM 7000 Sequence Detection System (Applied Biosystems) were analysed. A two-sample *F*-test was performed in order to compare two population variances. A *P*-value superior to 0.05 indicated that no difference of variation of expression could be deduced.

Immunocytochemistry

N-OrganDOTs samples were fixed in 4% paraformaldehyde for 20 minutes at 4 °C and embedded in Paraplast paraffin wax (Sigma) using standard laboratory procedures. Ten micron wax sections were cut and dried overnight at 37 °C. The following day, the paraffin sections were rehydrated and permeabilised with 0.3% Triton X-100. Afterwards, they were incubated with blocking solution (consisting of 5% goat serum and 1% BSA in PBS) at room temperature for 30 minutes. Staining was performed with different primary antibodies, incubated overnight at 4 °C, followed by secondary antibody incubation for two hours at room temperature and counterstained with DAPI nuclear stain (Sigma) for 15 minutes. The primary antibodies used were β tubulin III (dilution 1:200; Chemicon), Map2 (dilution 1:200; Millipore), Neurofilament 68kD (dilution 1:200; Abcam) and GFAP (dilution 1:400; Abcam). The appropriate secondary antibodies used for their visualisations were anti-mouse IgG-Cy3 and anti-rabbit IgG -FITC or anti-rabbit IgG-FITC (Chemicon).

Paraffin sections were incubated overnight at 4 °C with primary antibodies of rabbit/mouse monoclonal Tub3 (dilution 1:200; Chemicon), Neurofilament 68kD (dilution 1:200; Abcam) or GFAP (dilution 1:400; Abcam). The following day, the paraffin sections were washed in PBS and then incubated with goat anti-rabbit/mouse secondary antibody (dilution 1:400; Invitrogen) in darkness for two hours at room temperature.

Results

Immunohistochemistry

Transversal immunofluorescence staining of 3D N-OrganDOTs showed a polarisation of the neural parenchyma with the formation of a glia limitans-type of barrier close to the PTFE porous membrane (Figure 6.5e). Neurons were found embedded within a dense network of neurites (Figure 6.5d).

Electrophysiology

In order to assess the functionality of the neural networks, 3D neural tissues were placed in perfused chambers onto multiple electrode arrays to record spontaneous electrophysiological activities as well as to evoke field potentials induced by electrical stimulation. Cultures showing electrophysiological activities were selected to be included in the toxicological studies.

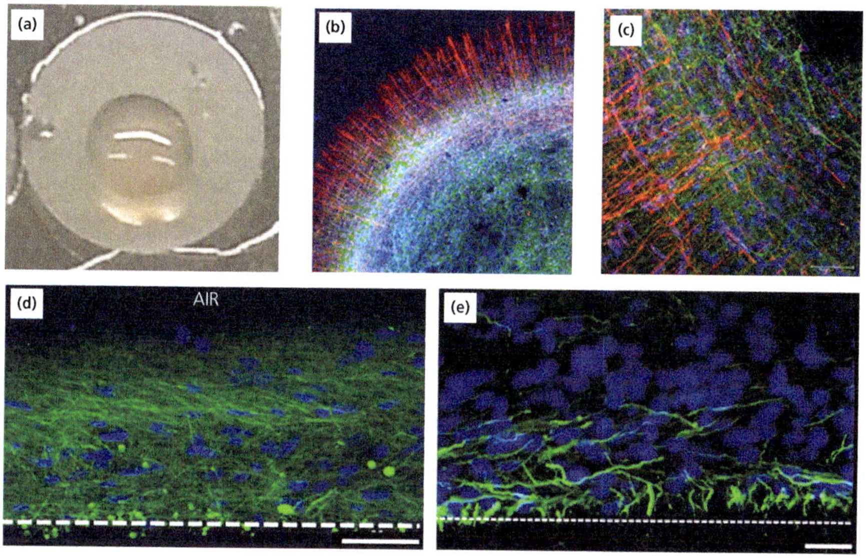

Figure 6.5 **3D air-liquid neural cultures derived from human embryonic stem cells.** (a) Macrophotography of 3D neural neural tissue grown onto porous membrane at the air-liquid interface after two months in culture. (b) Immunofluorescent staining of the same 3D neural tissue where astrocytes can be visualised in red, MAP2 in green (×100). (c) Higher magnification (×200). Transverse sections of 3D neural cultures show the abundant presence of neurites within the entire section (d). Astrocytes stained by GFAP could be found mostly located close to the synthetic PTFE membrane of the culture inserts represented by the white dotted line (e). All immunostaining was counterstained by DAPI to visualise the cell nuclei in blue.

Evoked population spikes

To evoke field potentials (EFPs), 3D neuronal tissues were stimulated by bipolar stimulation (Figure 6.6a). Despite the MEA having 40 electrodes, the software only allows simultaneous recording using eight recording channels, so the eight most suitable electrodes to evoke and record a population spike (two electrodes for bipolar stimulation and six electrodes for recording) were selected using the BioCell software. During experiments, population spike amplitudes were continuously evoked by stimulation every five seconds. The stimulating pulse amplitudes ranged between 1000 and 4000 mV, with a duration of 100 μS. We could observe that the amplitude of the second EFP is slightly decreased (Figure 6.6A). This type of pairing of two pulses in rapid succession led to an inhibition known as paired pulse inhibition (PPI) and was generally due to the recurrent inhibition of the subsequent response initiated in the neural network by the second stimulus delivered shortly after the first one.

Spontaneous activity

Spikes were monitored by eight independent electrodes. Cultures start to show spike activity after four weeks and bursts were found in older cultures (6–8 weeks).

Gene expression analysis (Figures 6.6 and 6.7)

Two-month-old N-OrganDOTs in culture were treated for 48 hours by vehicle control or by 1 nM, 10 nM, 100 nM and 1000 nM TMTC as well as MeHg 0.125, 1 and 5 μM (a number of 3 N-OrganDOTs in culture were used for each drug concentration and control).

We could observe dose–response toxicity with a decrease of the expression of β3 tubulin, synaptophysin, synapsin-1 and MAP-2, whereas we found an increase of GFAP expression as it was already shown in other *in vivo* and *in vitro* studies (Monnet-Tschudi *et al.*, 2007). On the other hand, the level expression of β3 tubulin remained stable when tissues were treated with D-mannitol even at high concentrations (data not shown).

Discussion

Compared to other currently available *in vitro* neuronal models, 3D microtissues generated by our N-OrganDOT culture method represent a good compromise between histological complexity and throughput. They are less complex than

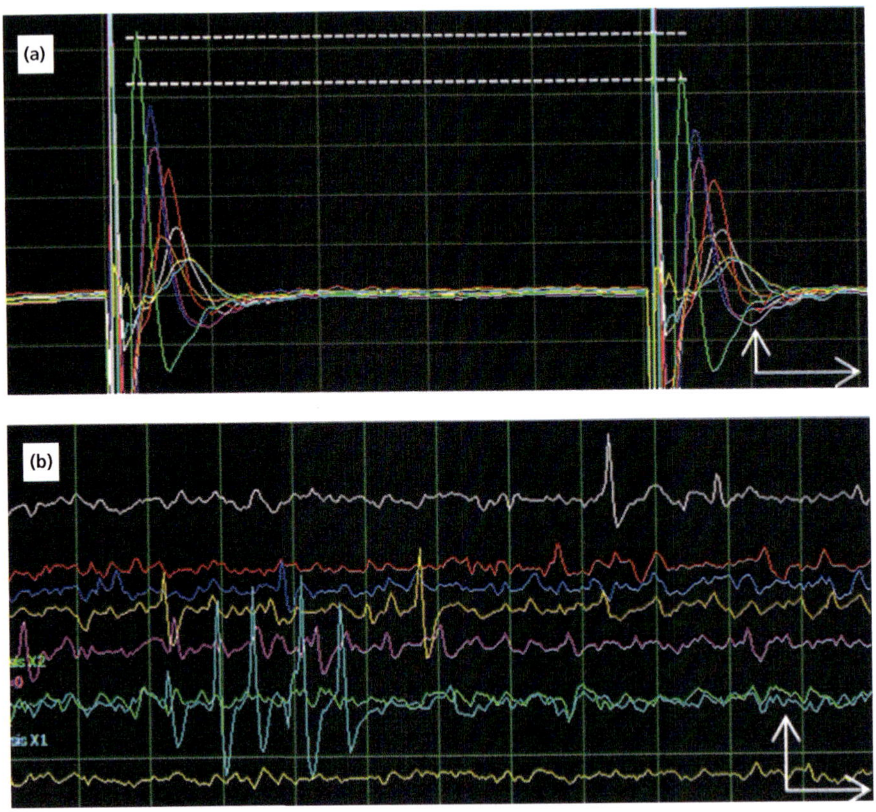

Figure 6.6 Spontaneous activity. (a) Evoked synaptic activity recorded by eight selected channels in response to paired pulse stimulation. The diminution of the amplitude of the second EFP is showed by the dotted lines. (b) Spontaneous activity recordings show individual spikes as well as the presence of bursts. (Scale: A 100 μV and 100 msec; B 10 μV and 10 msec).

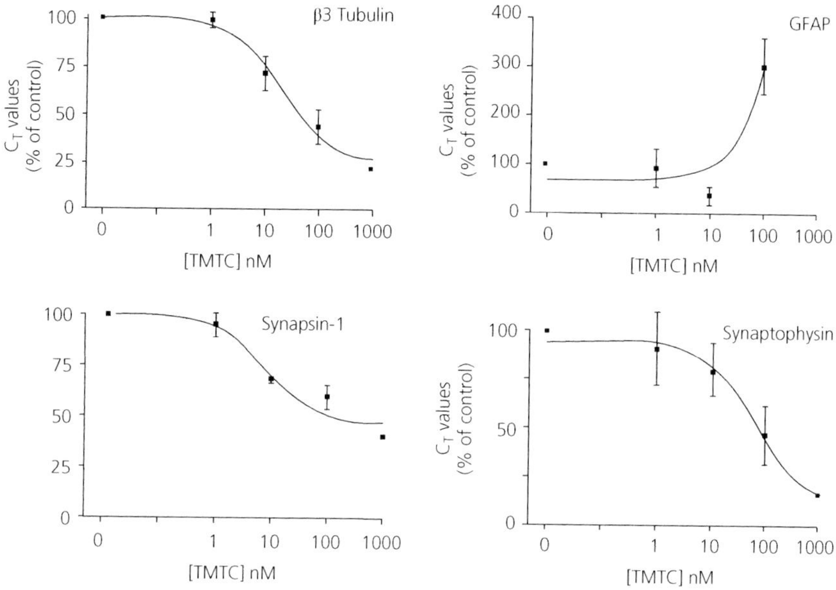

Figure 6.7 Gene expression analysis. Mean ± SD of normalised C_T values (using cycle time threshold) from three independent assays are expressed as % relative expression compared to untreated controls and are fitted in sigmoid dose–response curves. Expressions of β3 tubulin, synapsin-1 and synaptophysin are clearly decreased dose dependently while GFAP expression increases dramatically at the higher concentration of trimethyltin chloride [TMTC].

brain explants, which totally maintain the brain cytoarchitecture, but their 3D structure allows the formation of cell-cell interactions very similar to that of organotypic brain slices. This is important because the interactions between different cell types such as microglia, astrocytes, oligodendrocytes and neurons play an important role in neurotoxicity (Eskes *et al.*, 2002; Zurich *et al.*, 2004). On the morphological level, our cultures showed that 3D N-OrganDOT cell cultures grow well on hydrophilic PTFE membranes and develop into dense neuronal networks (see Figure 6.5d, e). Electrophysiological activity recording was used to assess the functionality of the 3D N-OrganDOT cultures. Paired pulse stimulation revealed a classic PPI of the second field potential, indicating the presence of short-term synaptic plasticity via a functional inhibitory network. Thus, despite the lack of cytoarchitecture, 3D N-OrganDOT cell cultures displayed essential electrophysiological properties of neurons *in situ* (excitability, synaptic transmission) and also featured more complex neural networks (short-term plasticity).

We studied two environmentally important and well-known neurotoxic compounds to determine the predictive capacity of our *in vitro* test system to assess acute neurotoxic effects. Trimethyltin chloride is an antifouling compound, which is known to affect neuronal activity by blockade of glutamatergic and GABAergic receptor channels (Kruger, 2005). Methyl mercury chloride is a standard compound intensively studied in the field of neurotoxicity.

It was shown to inhibit voltage-activated calcium and potassium channels in rat hippocampal neurons (Szucs *et al.*, 1997) and dorsal root ganglion neurons (Leonhardt *et al.*, 1996). Our results confirm the acute toxicological action of trimethyltin chloride and methyl mercury (see Figures 6.7 and 6.8) on neurons as well as an increase of expression of GFAP *in vitro* as reported by Jebbett *et al.* (2013).

Therefore, follow-up of the gene expression of neural markers in 3D human OrganDOT cultures seems to be relevant for the prediction of neurotoxic effects and may lead to the reduction or replacement of more costly and elaborate animal-based studies.

Conclusions and perspectives

We have shown that the same tissue culture method (OrganDOT) can be adapted to generate 3D microtissues at the air-liquid interface that exhibit organotypic characteristics and physiological functions from a variety of human and animal tissues. The microtissues remain viable for at least several weeks and in some cases several months, they are stable, easily produced and consistent from batch to batch. In the case of excitable tissues such as brain and heart, microtissues form aggregates that show spontaneous activity and evoked potentials that are

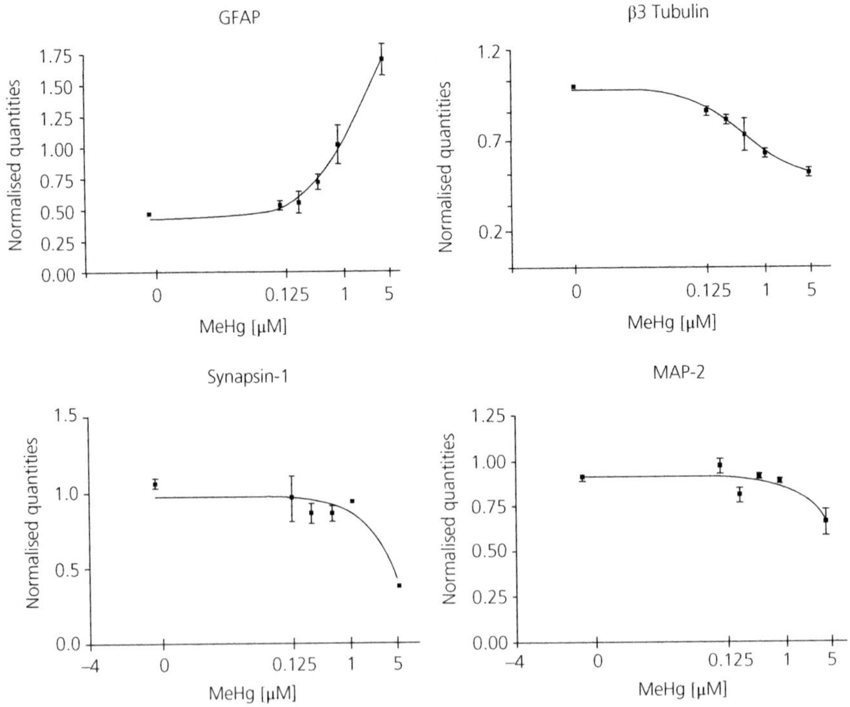

Figure 6.8 Gene expression analysis. Mean ± SD of normalised C_T values (using cycle time threshold) from three independent assays are fitted in sigmoid dose–response curves. Expressions of β3 tubulin, synapsin-1 and MAP-2 are decreased dose dependently while GFAP expression increases with the concentration of the toxic compound methyl mercury (MeHg).

comparable to explanted *in vivo* tissues and in many cases to the organs of origin. These have potential for the development of efficacy models in drug development and for development of models of pathology as well as applications in drug safety and toxicity that are as close as possible to the original tissue sample.

The fact that the same culture method can be used with human and animal primary material and stem cells offers great versatility for comparisons between species and tissue sources at a translational level. The air-liquid interface seems to offer greater stability and long-term viability compared to some other 3D culture methods, and the fact that the cellular stoichiometry can be manipulated at the point of culture is another advantage, as reporter and read-out systems can be built into the cultures at the outset.

There are, however, trade-offs to be made in the design and development of OrganDOT systems. For example, human primary material is unarguably the most authentic starting material but obtaining a suitable, stable and uniform supply is often a rate-limiting factor. Survival of adult differentiated tissues is often poorer than less differentiated tissues such neonatal animal primary tissues or embryoid bodies made from stem cells. In many cases, animal primary tissues are good surrogates and their supply can be less difficult (e.g. human brain tissue supply is often very limited) but the validity of such systems needs to be compared against human tissues wherever possible.

The use of embryoid bodies and stem cell material as a source of starting material offers great potential, as it is an almost infinitely renewable source of starting material that can be generated relatively uniformly under laboratory conditions. However, the degree of differentiation *in vitro* can be a limiting factor for some tissues (e.g. hepatocytes) and also cost can be high since the method uses between 150000 and 500000 cells per culture. Nevertheless, the current method lends itself well to the development of medium-throughput assays that have much greater fidelity with human physiology than 2D culture-based systems. It is likely that these will fill the gap between classic 2D high-throughput and high-content screening systems and *in vivo* models, as these also address the issue of the lack of throughput of *in vivo* systems compared to current *in vitro* methods.

Troubleshooting

Induction of NPC

Embryonic stem cells are usually hard to convert to a pure and stable population of NPCs (i.e. without contaminant flat cells). To prevent induction of flat cells, it is essential to use both Noggin and Dorsomorphin, and SB431542 simultaneously (usually only Noggin or Dorsomorphin and SB431542 are used), and to maintain this combination at least for 10 days. If these two points are not respected, flat cells will appear in NPC cultures (usually after the third or fourth passage), and will rapidly become predominant.

Culture of NPC

No viable NPC can appear after thawing stock; the reason is that cells have not been handled gently. NPC are fragile and need to be treated with great care: do not vortex and then dislodge the cells with gentle banging of the flasks, instead centrifuge the cells at low speed. To make the stock aliquots, slow freeze the cells and finally fast thaw the aliquots to start the cultures. If the NPCs are passaged more than 10 times, the cells will grow slowly. Use NPCs under passage 10 and do not let them overgrow. If the cells are plated too sparsely, it may induce a spontaneous differentiation. Seed cells at a minimal density around 6×10^6 cells/cm^2.

Differentiation of NPC in neurons and glial cells

The NPC can fail to differentiate if they have been passaged too many times, or if the cell density is too high (endogenous basic FGF (bFGF) is produced and prevents differentiation).

Neurosphere culture

For the formation of neurospheres in 24-well plates in the presence of bFGF and EGF, the suspension of NPCs must not exceed 50 000 cells per well. Beyond this density, the neurospheres will not be calibrated and secondary spheres will be generated. Once the neurospheres are formed in 24-well plates, they can be transferred to Erlenmeyer flasks for long-term culture with medium without growth factors and with FBS 1% to induce neuronal and glial differentiation. During this operation, there is a risk that the neurospheres will aggregate due to growth factors remaining in the differentiation medium. So it is worth starting the differentiation in 24-well plates during the first 3–4 days, and only after this time transferring the neurospheres to Erlenmeyer flasks.

Acknowledgements

Supported by the HES-SO, TSB, CTI/KTI FP7 (ESNATS) and the Swiss Centre for Applied Human Toxicology (SCAHT).

References

Bailey, J. L., O'connor, V., Hannah, M. *et al.* (2011) In vitro CNS tissue analogues formed by self-organisation of reaggregated post-natal brain tissue. *Journal of Neurochemistry*, **117**, 1020–1032.

Biggs, T., Foreman, J., Sundstrom, L., Regenass, U. & Lehembre, F. (2011) Antitumor compound testing in glioblastoma organotypic brain cultures. *Journal of Biomolecular Screening*, **16**, 805–817.

Carrel, A. (1912) On the permanent life of tissues outside of the organism. *Journal of Experimental Medicine*, **15**, 516–528.

Editorial (2003) Goodbye, flat biology? Seductive higher dimensions could consign glass dishes to history. *Nature*, **424**, 861.

Eskes, C., Honegger, P., Juillerat-Jeanneret, L. & Monnet-Tschudi, F. (2002) Microglial reaction induced by noncytotoxic methylmercury treatment leads to neuroprotection via interactions with astrocytes and IL-6 release. *Glia*, **37**, 43–52.

Hakkoum, D., Muller, D. & Stoppini, L. (2006) Electrophysiological monitoring of hippocampal slice cultures using MEA on porous membrane, in *Advances in Network Electrophysiology: Using Multi-Electrode Arrays* (eds M. Baudry & M. Taketani), Springer, New York, pp.112–126.

Harrison, R. (1907) Observations on the living developing nerve fiber. *Anatomical Record*, **1**, 116–128.

Honegger, P. (1989) Aggregate cell cultures. *ALTEX*, **6**, 29–36.

Jebbett, N. J., Hamilton, J. W., Rand, M. D.& Eckenstein, F. (2013) Low level methylmercury enhances CNTF-evoked STAT3 signaling and glial differentiation in cultured cortical progenitor cells. *Neurotoxicology*, **38**, 91–100.

Kruger, K., Diepgrond, V., Ahnefeld, M. *et al.* (2005) Blockade of glutamatergic and GABAergic receptor channels by trimethyltin chloride. *British Journal of Pharmacology*, **144**, 283–292.

Landecker, H. (2007) *Culturing Life: How Cells Became Technologies*. Harvard University Press, Cambridge, Massachusetts.

Layer, P. G., Robitzki, A., Rothermel, A. & Willbold, E. (2002) Of layers and spheres: the reaggregate approach in tissue engineering. *Trends in Neurosciences*, **25**, 131–134.

Leonhardt, R., Haas, H. & Busselberg, D. (1996) Methyl mercury reduces voltage-activated currents of rat dorsal root ganglion neurons. *Naunyn Schmiedebergs Archives of Pharmacology*, **354**, 532–538.

Marx, U. (2007) How drug development of the 21st century could benefit from human micro-organoid in-vitro technologies, in *Drug Testing in-Vitro, Breakthroughs and Trends in Cell Culture Technology* (eds U. Marx & V. Sandig), Wiley-VCH, Weinheim, Germany.

Monnet-Tschudi, F., Zurich, M. G. & Honegger, P. (2007) Neurotoxicant-induced inflammatory response in three-dimensional brain cell cultures. *Human Experimental Toxicology*, **26**, 339–346.

Pampaloni, F., Reynaud, E. G. & Stelzer, E. H. (2007) The third dimension bridges the gap between cell culture and live tissue. *Nature Reviews Molecular Cell Biology*, **8**, 839–845.

Portner, R. & Giese, C. (2007) An overview of bioreactor design and process control for reproducible three-dimensionsal tissue culture, in *Drug Testing in-Vitro, Breakthroughs and Trends in Cell Culture Technology* (eds U. Marx & V. Sandig), Wiley-VCH, Weinheim, Germany.

Rowe, C., Gerrard, D. T., Jenkins, R. *et al.* (2013) Proteome-wide analyses of human hepatocytes during differentiation and dedifferentiation. *Hepatology*, **58**, 799–809.

Stoppini, L., Buchs, P. A. & Muller, D. (1991) A simple method for organotypic cultures of nervous tissue. *Journal of Neuroscience Methods*, **37**, 173–182.

Stoppini, L., Duport, S. & Corrèges, P. (1997) A new extracellular multirecording system for electrophysiological studies: application to hippocampal organotypic cultures. *Journal of Neuroscience Methods*, **72**, 23–33.

Sundstrom, L. E. (2007) Thinking inside the box. To cope with an increasing disease burden, drug discovery needs biologically relevant and predictive testing systems. *EMBO Reports*, **8**, S40–S43.

Sundstrom, L. E., Morrison 3rd, B., Bradley, M. & Pringle, A. (2005) Organotypic cultures as tools for functional screening in the CNS. *Drug Discovery Today*, **10**, 993–1000.

Sundstrom, L. E., Biggs, T., Laskowski, A. & Stoppini, L. (2012) OrganDots – an organotypic 3D tissue culture platform for drug development. *Expert Opinion on Drug Delivery*, **7**, 525–534.

Szucs, A., Angiello, C., Salanki, J. & Carpenter, D. O. (1997) Effects of inorganic mercury and methylmercury on the ionic currents of cultured rat hippocampal neurons. *Cellular and Molecular Neurobiology*, **17**, 273–288.

Vandesompele, J., de Preter, K., Pattyn, F. *et al.* (2002) Accurate normalisation of real-time quantitative RT-PCR data by geometric averaging of multiple internal control genes. *Genome Biology*, **3**, research0034.1–research0034.11.

Watanabe, K., Ueno, M., Kamiya, D. *et al.* (2007) A ROCK inhibitor permits survival of dissociated human embryonic stem cells. *Nature Biotechnology*, **25**, 681–686.
Wilson, P. G. & Stice, S. S. (2006) Development and differentiation of neural rosettes derived from human embryonic stem cells. *Stem Cell Reviews*, **2**, 67–77.
Zurich, M. G., Honegger, P., Schilter, B. *et al.* (2004) Involvement of glial cells in the neurotoxicity of parathion and chlorpyrifos. *Toxicology and Applied Pharmacology*, **201**, 97–104.

PART II
Hydrogels

CHAPTER 7

Materials and assay systems used for three-dimensional cell culture

Suparna Sanyal and Marshall Kosovsky
Corning Life Sciences, Tewksbury, Massachusetts, USA

Introduction

Mammalian cell culture has served as an invaluable tool for cell biology research. Monolayers of adherent cells grown on flat and rigid two-dimensional (2D) substrates, such as polystyrene or glass, have evolved as a standard approach for conventional cell culture. 2D cell culture studies have played a pivotal role in furthering our understanding of developmental biology, tissue morphogenesis, disease mechanisms, drug discovery, large-scale protein production, tissue engineering and regenerative medicine. However, there are inadequacies and limitations associated with 2D culture systems, especially regarding their inability to emulate *in vivo* conditions and provide physiological relevance.

In the body, nearly all cells in tissues reside in an extracellular matrix (ECM) consisting of a complex three-dimensional (3D) architecture, and interact with neighbouring cells through biochemical and mechanical cues (Lodish *et al.*, 2002). Cell-cell and cell-ECM interactions establish a 3D communication network that maintains the specificity and homeostasis of the tissue (Kleinman *et al.*, 2003). Key events in the life cycle of a cell are regulated by organising principles that are determined by the surrounding cellular microenvironment (Bissell *et al.*, 2003). The inability of cells to achieve *in vivo*-like structural organisation and connectivity in 2D cell culture assays can limit or diminish properties such as cellular morphology, viability, proliferation, differentiation, gene and protein expression, response to stimuli, drug metabolism and general cell functionality.

These limitations of 2D culture have contributed to poor predictive power of preclinical cell-based drug and toxicity screening assays. More than 90% of drugs that pass through *in vitro* preclinical studies fail to meet the desired efficacy or safety margins required in subsequent clinical trials (van der Worp *et al.*, 2010). The failure rate is even higher for cancer drugs (Hutchinson & Kirk, 2011; Westhouse, 2010) as 2D culture systems are often inadequate to effectively

Technology Platforms for 3D Cell Culture: A User's Guide, First Edition. Edited by Stefan Przyborski.

model tumour biology (Aggarwal *et al.*, 2009; Hait, 2010). Furthermore, there is a strong dependence on concomitant use of animal models for bioavailability and toxicological studies during preclinical drug development when 2D cell culture models are utilised. The high rates of drug candidate failure suggest that animal models may not be suitable and/or representative for safety assessment of therapeutics intended for human use (Li, 2004; Polson & Fuji, 2012; van der Worp *et al.*, 2010).

To overcome some of these shortfalls, numerous 3D cell culture models have been developed over the past two decades to address the diverse requirements of different cell types and applications. Emerging evidence strongly suggests that 3D cell cultures that establish physiological cell-cell and cell-ECM interactions can mimic the specificity of native tissue with greater physiological relevance than conventional 2D cultures (Lee *et al.*, 2008; Haycock, 2011; Pamapaloni *et al.*, 2007). According to a 3D cell culture trends report (HTStec, 2010), natural ECM-based hydrogels (e.g. Corning® Matrigel® matrix and collagen) and 3D cell aggregates known as spheroids are among the most widely used models for 3D culture *in vitro*. These 3D models have been used to demonstrate physiologically relevant properties of diverse cell types, and have been extensively applied to stem cell differentiation, tumourigenesis and drug discovery research. Spheroids and/or natural hydrogels are advantageous since they can be manipulated into more complex *in vivo*-like co-culture models that incorporate multiple cell types. Furthermore, devices such as permeable supports (e.g. Corning® Transwell® inserts) can be incorporated into 3D culture models with a hydrogel substrate(s) to facilitate the analysis of interactions between different cell types, soluble factors and the culture microenvironment.

This chapter will focus on general principles, advantages and caveats of *in vitro* 3D cell culture systems, with particular attention paid to the following three areas: cell biology and tissue modelling, drug discovery and drug toxicity. These overviews will highlight the matrices most commonly used for *in vitro* 3D cell culture such as ECM-based hydrogels.

2D versus 3D cell culture models

Fundamental differences in the microenvironment of 2D and 3D cell culture systems influence various cellular behaviours including cell attachment, growth, morphology and polarity, gene and protein expression, response to stimuli, cellular metabolism and overall function. One of the primary differences observed when comparing cells in 2D and 3D cultures is the dissimilarity in cell morphology. Cells adopt 2D or 3D confirmation primarily based on the orientation of integrin-mediated adhesions to the ECM. In the case of 2D cultures, cell attachment occurs on one side of the cell (the side that contacts the 2D surface), whereas 3D cell culture promotes attachment around the entire surface of the cell (Baker & Chen, 2012). In general, cell attachment and spreading on restraint-free 2D substrates occur within minutes. In contrast, cell attachment and spreading in 3D culture systems are preceded by proteolytic degradation of their physical environment, which can occur over the course of hours, or even days in some instances (Khetan & Burdick, 2010).

It has been suggested that the degree of cell spreading can impact cell proliferation, apoptosis and differentiation (Chen *et al.*, 1997; McBeath *et al.*, 2004; Singhvi *et al.*, 1994; Thomas *et al.*, 2002). Many cells, when isolated from tissues and placed onto planar cell culture surfaces, become progressively flatter, divide aberrantly and lose their differentiated phenotype (Petersen *et al.*, 1992; von der Mark *et al.*, 1977). Interestingly, some of these cell types can regain their physiological form and function when embedded in a 3D culture environment. For instance, encapsulation of dedifferentiated chondrocytes in 3D cultures restores their physiological phenotype, including cell shape and the expression of cartilaginous markers (Benya & Shaffer, 1982). Similarly, mammary epithelial cells embedded in a 3D environment halt uncontrolled division, assemble into acinar-like structures and establish a *de novo* basement membrane (Emerman & Pitelka, 1997; Lee *et al.*, 1984; Petersen *et al.*, 1992).

Cellular morphology can also affect the functionality of the cell (Brock *et al.*, 2003; Kilian *et al.*, 2010; Mahmud *et al.*, 2009; Théry *et al.*, 2006, 2007). For example, glandular epithelial cell organisation, signalling and secretion more closely resemble the properties observed *in vivo* when cultured in a 3D environment, in contrast to the behaviour that occurs on 2D surfaces (Debnath & Brugge, 2005; Nelson & Bissell, 2006). The morphologies of fibroblasts, including cytoskeletal organisation and the types of cell adhesions, are also more similar to their *in vivo* behaviour when the fibroblasts are grown in a 3D matrix compared to 2D (Yamada & Cukierman, 2007). Moreover, a 3D environment has been found to be optimal for supporting fibroblast intracellular signalling characteristics (Cukierman *et al.*, 2001; Grinnell, 2003; Walpita & Hay, 2002).

Gene expression and mRNA splicing patterns can also vary considerably depending on whether cells are cultured under 2D versus 3D conditions (Birgersdotter *et al.*, 2005; Li *et al.*, 2006). For example, melanoma cells exhibit distinct gene expression patterns when cultured on 2D substrates, in contrast to the properties observed with melanoma cell spheroids cultured in a 3D environment. Genes upregulated in the 3D spheroids are also found to be upregulated in tumours *in vivo* (Ghosh *et al.*, 2005). When mammary epithelial cells are cultured on 2D substrates, these cells exhibit a dramatic upregulation of $\alpha 1$ integrin mRNA expression. In contrast, when these cells are cultured under 3D conditions using a reconstituted basement membrane such as Corning Matrigel matrix, the cells exhibit mRNA expression levels comparable with those observed in native breast tissue (Baker & Chen, 2012; Delcommenne & Streuli, 1995).

The differentiation niches of primary cells and stem cells are inherently 3D, and their biochemistry and topology have been found to dramatically affect the differentiation process (Fuchs *et al.*, 2004). For example, primary hepatocytes cultured as a monolayer dedifferentiate and die within a few days (Gómez-Lechón

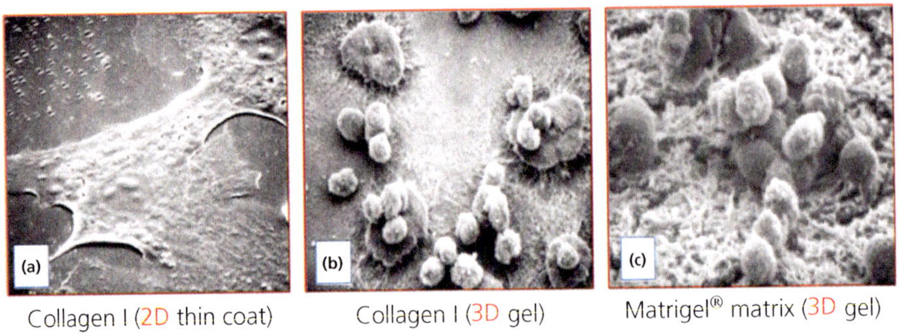

Figure 7.1 Morphology of primary rat hepatocytes in 3D cell culture. Primary rat hepatocytes exhibit differentiated morphology on 3D growth substrates composed of collagen I (b) and Matrigel matrix (c), yet exhibit flattened, dedifferentiated morphology on 2D collagen I (a).

et al., 1998). Dedifferentiated hepatocytes lose the ability to synthesise drug-metabolising enzymes that are essential for toxicity assays in pharmaceutical research (Gómez-Lechón *et al.*, 1998). The deficits in hepatocyte function observed on 2D surfaces can be overcome by embedding primary hepatic cells within 3D matrices such as collagen I (Berthiaume *et al.*, 1996), Corning Matrigel matrix (Berthiaume *et al.*, 1996), synthetic peptide scaffolds (Semino *et al.*, 2003), or by maintaining them under conditions of perfusion flow (Powers *et al.*, 2002). Hepatocytes cultured on 3D matrices of collagen I or Matrigel matrix form clusters and exhibit a rounded morphology that is indicative of the differentiated state (Figure 7.1b,c), yet were found to adopt a flattened, dedifferentiated morphology on a 2D collagen I substrate (Figure 7.1a). Under these conditions, hepatocytes cultured on 3D substrates exhibited optimal expression of the liver-enriched transcription factor C/EBP and elevated levels of cytochrome P450 enzymes, and were found to exhibit markedly reduced activities on the 2D surface (data not shown). Collectively, these observations demonstrate that 3D cell culture systems are capable of supporting varying degrees of cell complexity and functionality that are observed *in vivo*, which is dependent on the cell type and culture conditions. In contrast, 2D culture models usually support limited cell differentiation and *in vivo*-like functionality.

Description of technology

3D cell culture techniques

A critical need for *in vitro* 3D cell culture systems is to effectively mimic specific aspects of *in vivo* cell behaviour to provide relevant models for studying tissue development and morphogenesis, cellular differentiation, genotypic and/or phenotypic response to compounds in drug and toxicity screening assays. Some basic 3D models have involved the culture of cellular aggregates in suspension without the use of matrix-based substrata. However, the majority of more complex 3D cell culture models utilise either hydrogel-based matrices or solid scaffolds. A vast array of materials and fabrication techniques has been employed to develop scaffolds with varying physical and biological characteristics to address the requirements of different cell types. Of the current materials in use, naturally derived ECM-based hydrogels are most commonly used for *in vitro* applications of 3D cell culture.

3D spheroid cultures

Cellular spheroids are simple 3D models that can be generated from a wide range of cell types, which form spheroids because of the tendency of adherent cells to aggregate. Common examples of spheroids include embryoid bodies, mammospheres, tumour spheroids, hepatospheres and neurospheres. Adherent cells have a natural tendency to aggregate and form spheroids under circumstances that impede adhesion to cell culture substrates. Common matrix-free methods employed for generating spheroids include the use of attachment-resistant cell culture surfaces such as the Corning 96- and 384-well spheroid microplate with ultra-low attachment (ULA) surface, or by maintaining the cells as suspension cultures in media (e.g. hanging drop technology, rotary cultures and bioreactors). Several cell types also form spheroids in 3D hydrogels and, to a limited extent, in some solid scaffolds depending on the structural and physical properties of the material. The overall size of spheroids is limited to a few hundred micrometres, beyond which necrosis ensues within the core of the spheroids (Mueller-Klieser, 1986).

Spheroids naturally mimic various aspects of solid tissues and are equipped with inherent gradients for efficient diffusion of oxygen and nutrients as well as the removal of metabolic wastes. These cellular aggregates can emulate avascular, solid tumour behaviour more effectively than standard 2D environments because spheroids, much like tumours, usually contain a heterogeneous population of surface-exposed and deeply buried cells, proliferating and non proliferating cells, and well-oxygenated and hypoxic cells (Frieboes *et al.*, 2006). Additionally, differentiation of pluripotent stem cells (PSCs) typically involves formation of spherical structures called embryoid bodies, an important step for subsequent cell differentiation studies *in vitro*.

Spheroids thus represent an especially good physiological 3D model for studying solid tumourigenesis and stem cell differentiation. In addition, spheroids can be readily analysed by imaging using light, fluorescence and confocal microscopy, which is an advantage over more complex 3D cell culture models. Furthermore, it is relatively simple to mass-produce uniformly sized 3D spheroids, making them highly amenable for many *in vitro* high-throughput and toxicity screening applications.

3D cultures using hydrogels and extracellular matrices

Hydrogels are composed of networks of crosslinked polymer chains or complex protein molecules of natural or synthetic origin. Due to their significant water content, hydrogels possess biophysical characteristics very similar to natural tissue, and serve as highly effective matrices for 3D cell culture. Hydrogels can be used as stand-alone 3D matrices or combined with other technologies, such as solid scaffolds, permeable supports, cellular microarrays and microfluidics devices.

Hydrogels can be employed in 3D culture systems in a variety of ways. They can be used as a coating reagent for various cell culture surfaces, including solid scaffolds. Alternatively, cells can be encapsulated in or sandwiched between these matrices. The morphology, growth and functionality of cells within the hydrogel matrices depend on the presentation of biophysical and biochemical cues, as well as physical properties such as permeability and matrix stiffness.

Natural hydrogels and extracellular matrices

Naturally derived hydrogels for cell culture are typically formed of proteins and ECM components such as collagen, laminin, fibrin, hyaluronic acid, chitosan or Corning Matrigel matrix. Derived from natural sources, these gels are inherently biocompatible and bioactive (Dawson *et al.*, 2008). They also promote many cellular functions due to the presence of various endogenous factors, which can be advantageous for supporting viability, proliferation, function and development of many cell types (Tibbitt & Anseth, 2009).

The ECM provides a complex, nanoscale architecture of structural proteins such as collagen, laminin and fibronectin to create the mechanical properties inherent in the cellular microenvironment (Daley *et al.*, 2008; Hynes, 2009). Cells sense these mechanics through their cell surface integrins, and bind to specific adhesion motifs present on the ECM proteins. The ECM is vital for sequestering soluble biomolecules and growth factors, and releasing these signalling molecules with spatial-temporal control to guide processes such as cell migration, matrix degradation and deposition (Daley *et al.*, 2008; Hynes, 2009). ECM remodelling is imperative for achieving tissue homeostasis and is particularly pronounced during development and diseases. Thus, to truly mimic the ECM *in vivo*, it is necessary to develop 3D culture models that exhibit the mechanical and chemical properties of the ECM, not only at the initial stage of cell seeding but also in a dynamic and tunable manner as the cells grow and develop.

Corning Matrigel matrix is an ECM-based natural hydrogel that has been used extensively for 3D cell cultures *in vitro* and *in vivo*. This reconstituted basement membrane is extracted from Engelbreth-Holm-Swarm (EHS) mouse tumours and contains all of the common ECM molecules present within basement membrane (i.e. laminin, collagen IV, heparin sulfate proteoglycan and nidogen/entactin) (Kleinman & Martin, 2005). The ECM components of Matrigel activate cancer cell signalling pathways that regulate angiogenesis (Languino

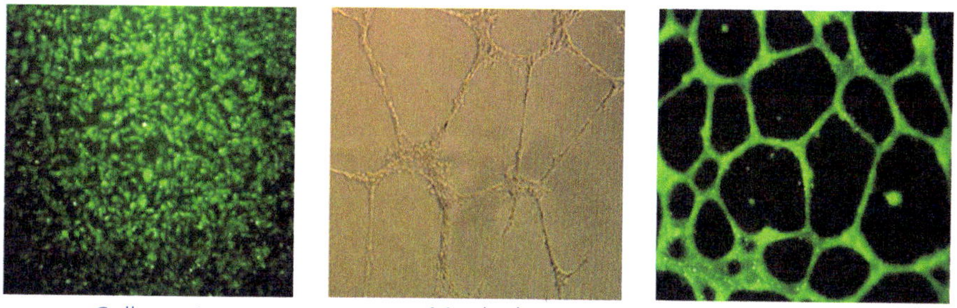

Figure 7.2 Human microvascular endothelial cells cultured in Matrigel matrix. Human microvascular endothelial cells form tubules on a 3D substrate composed of Matrigel matrix.

et al., 1989; Zhou *et al.*, 2004), tumour cell motility (Carpenter *et al.*, 2009) and sensitivity to antitumour drugs (Miyamoto *et al.*, 2004). Since Matrigel mimics the *in vivo* basement membrane, it is often used for studying cancer cells that resemble cells that reside in epithelial tissues (Gurski *et al.*, 2010). Dr Mina Bissell's pioneering research on mammary epithelial cells and breast cancer has utilised Matrigel matrix, or an equivalent, which demonstrated the enabling power of 3D cell culture for creating *in vivo*-like model systems (Barcellos-Hoff, 1989) and the importance of integrin signalling in cancer (Bissell *et al.*, 2003; Weaver, 1997).

Endothelial cells line the closed internal body cavities (e.g. blood vessels, bile ducts) and play an integral role during angiogenesis. The primary stages of angiogenesis include endothelial cell migration and invasion through the vascular basement membrane, and the subsequent process of tubulogenesis (tube formation). To study endothelial cell tubulogenesis, many laboratories utilise 3D cell culture environments composed of Corning Matrigel matrix to promote endothelial cell differentiation and the formation of capillary-like structures *in vitro* (Figure 7.2).

Collagen type I is another natural hydrogel that is commonly used for 3D cell culture. Collagen I can be isolated from various biological sources, including bovine skin, rat tail tendon and human placenta. This ECM molecule can also be electrospun into membranes (Hartman *et al.*, 2009; Sisson *et al.*, 2009) and can support 3D cell growth and differentiation. Additionally, collagen I interacts with integrin receptors to modulate gene expression (Kiefer & Farach-Carson, 2001). The target genes modulate the production of matrix metalloproteinases (MMPs), the enzymes that degrade ECM components during tumour metastasis (Ellerbroek *et al.*, 2001), as well as genes that affect cell sensitivity to drugs (Kim *et al.*, 2009) and cell migratory function (Hall *et al.*, 2008; Menke *et al.*, 2001).

Since natural hydrogels are isolated from animal-derived sources, these materials may not be suitable for cell therapy or regenerative medicine studies that rely on highly defined cell culture systems. To circumvent some of the issues associated with animal-derived biomaterials, matrices have been developed in organisms that are animal free or derived from recombinant nucleic acid technology. Hyaluronic acid (hyaluronan or HA) is an increasingly popular biologically derived matrix (Chen *et al.*, 2005; David *et al.*, 2008; Gurski *et al.*, 2009). Most commercial-grade HA is of bacterial origin and characterised by high purity and homogeneous quality. These gels may be modified by the addition of ECM components to improve cell attachment and growth properties.

Synthetic hydrogels

Synthetic hydrogels are a good choice for 3D cell culture applications when naturally derived biological matrices are unsuitable. Synthetic hydrogels are composed of purely non-natural molecules (reviewed in DeVolder & Kong, 2012) such as polyethylene glycol (PEG) (Sawhney *et al.*, 1993), polyvinyl alcohol (Martens & Anseth, 2000),and poly-2-hydroxyethylmethacrylate-78. They are biologically inert, but provide structural support for various cell types. PEG hydrogels have been shown to maintain the viability of encapsulated cells while allowing for ECM deposition as the hydrogel degrades (Bryant & Anseth, 2002), thereby demonstrating that synthetic gels can function as 3D cell culture platforms in the absence of integrin-binding ligands.

It is also possible to modify inert synthetic hydrogels with appropriate biological components. An example of a synthetic hydrogel that is tunable is Corning PuraMatrix™ peptide hydrogel, a self-assembling, synthetic oligopeptide that exhibits nanometre scale fibres. Nanometre-sized fibres and pores are essential to ensure a true 3D environment for the cell (Gelain *et al.*, 2006; Horii *et al.*, 2007). An additional advantage of PuraMatrix peptide hydrogel is the ability to customise the material with specific peptide sequences to improve cell attachment, cell homing and other behaviours (Zhang, 2004). To achieve optimal cell growth and differentiation using PuraMatrix peptide hydrogel, it is necessary to supplement the hydrogel with appropriate bioactive molecules (e.g. growth factors, ECM proteins and/or other molecules). Functionalisation with appropriate biological factors can convert inert synthetic matrices such as PuraMatrix peptide hydrogel into useful *in vitro* models for supporting differentiation of stem and progenitor cells. As this field advances, there will be a need for matrices with combined properties of natural and synthetic hydrogels.

Applications of 3D cultures *in vitro*

Cell biology and tissue models

Stem cell differentiation and organoid generation

Remarkable advances in the culture and differentiation of stem cells and committed progenitor cells have been realised with the use of 3D cell culture systems. Stem cells, particularly PSCs, have great potential for generating pure populations of any cell type in the body. Pure populations of progenitor or terminally differentiated cells, especially those that are difficult to isolate from tissues, could be invaluable for enabling drug discovery, cell therapy and tissue regeneration.

Major breakthroughs have been recently achieved in the area of stem cell differentiation using 3D culture systems that recapitulate *in vivo* development and temporal control of signalling pathways (Table 7.1). Notably, in a sophisticated study by Koehler and colleagues, murine embryonic stem cells (mESCs) were successfully differentiated to inner ear sensory epithelia when cultured as floating cell aggregates in media containing Matrigel matrix (Koehler *et al.*, 2013). From the sensory epithelia, hair cells with structural and functional properties of native mechanosensitive hair cells of the inner ear were shown to spontaneously arise (Koehler *et al.*, 2013). This novel approach has enabled elucidation of complex mechanisms underlying inner ear development, and may be useful for *in vitro* disease modelling, drug discovery or cell therapy.

Recent studies have enabled the derivation of renal lineage from stem cells. For example, mESCs have been shown to differentiate into complex renal tubular cells when grown as a 3D culture using Matrigel matrix (Morizane *et al.*, 2013). In a study by Xia *et al.* (2013), hPSCs were differentiated into renal progenitor-like cells. Further maturation of these cells into ureteric bud structures was accomplished by establishing a 3D culture system in which differentiated human cells assembled and integrated alongside murine cells to form chimeric ureteric buds (Xia *et al.*, 2013). Both of these studies identified key steps in modelling kidney development, and collectively highlight the importance of 3D systems for supporting the differentiation of cells associated with highly organised organs.

Another major advancement in stem cell differentiation is the formation of self-organising 3D mini 'organs' known as organoids. Organoids can recapitulate histological details and provide functional representation of multiple cell types that are present within the native organs. Organoids representative of intestinal (McCracken *et al.*, 2011; Spence *et al.*, 2011) retinal (Eiraku *et al.*, 2011), pancreatic (Greggio *et al.*, 2013), mammary (Gray *et al.*, 2010), colonic (Wang *et al.*, 2013) and cerebral tissues (Bershteyn & Kriegstein, 2013) have been developed recently using 3D models composed of Matrigel matrix (see Table 7.1). By manipulating the medium composition, researchers have generated either hollow spheres, which are mainly composed of pancreatic progenitors, or complex organoids that spontaneously undergo pancreatic morphogenesis and differentiation. *In vitro* culture of organoids is a major step toward elucidating the principles of organ development and the mechanisms responsible for genetic diseases.

Cancer and tumour cell biology

Perhaps the most notable application of Matrigel matrix for 3D cell culture is in the field of cancer biology. Since Matrigel matrix is a reconstituted basement membrane isolated from mouse tumours, this physiological material

Table 7.1 Tissue and disease models.

Application	Cell type	3D model	Culture substrate and matrix	Reference
STEM CELL DIFFERENTIATION	Human PSCs	Differentiation to ureteric bud committed renal progenitor cells	Corning Matrigel matrix	(Xia *et al.* 2013)
	Human PSCs	Differentiation to cardiac microtissues	Aligned collagen I	(Thavandiran *et al.* 2013)
	Human PSCs	Differentiation to endoderm	Collagen I	(Turovets *et al.* 2012)
	Human ESCs	Differentiation to hepatocytes	Collagen I scaffold; 2D collagen I-coated dish	(Baharvand *et al.* 2006)
	Human ESCs	Differentiation to smooth muscle, neurons and hepatocyte-like cells	Polymer scaffold with ECM coating	(Carlson *et al.* 2012)
	Human ESCs	Differentiation to neural precursors	Hyaluronic acid matrix	(Brannvall *et al.* 2007)
	Mouse ESCs	Differentiation to renal tubular cells	Corning Matrigel matrix; ultra-low attachment plates	(Morizane *et al.* 2013)
	Mouse ESCs	Differentiation to inner ear sensory epithelia	Corning Matrigel matrix	(Koehler *et al.* 2013)
	Mouse ESCs	Differentiation to neurons	Collagen I scaffold	(Li *et al.* 2012)
	Mouse ESCs	Differentiation to neurons and astrocytes	Fibrin scaffold	(Willerth *et al.* 2006)
	Mouse ESCs	Differentiation to chondrocytes	PEG scaffold	(Hwang *et al.* 2006)
	Human NPCs	Differentiation to neuronal cells	Corning PuraMatrix (RADA-16) scaffold	(Liedmann *et al.* 2012)
	Human NPCs	Differentiation to neuronal cells	Corning PuraMatrix ± functionalised matrix (laminin I)	(Ortinau *et al.* 2010)
	Mouse primary neuronal cells	Differentiation to osteogenic lineage cells in 2D and 3D systems	PLLA and polystyrene polymer scaffolds coated with collagen I	(Lai *et al.* 2012)
	Human MSCs	Differentiation to osteocytes	Corning PuraMatrix peptide hydrogel	(Chen *et al.* 2013)
	Canine AdMSCs	Differentiation to osteogenic lineage cells in 2D, 3D and *in vivo* (implantation in dog) systems	Corning Matrigel matrix	(Kang *et al.* 2012)
	Human dental pulp stem cells	Differentiation to osteocytes	Corning Matrigel matrix; collagen sponge	(Riccio *et al.* 2010)

ORGANOID GENERATION	Human PSCs	Intestinal organoids	Corning Matrigel matrix	(Spence *et al.* 2011; McCracken *et al.* 2011)
	Mouse ESCs	Retinal optic cup structure	Corning Matrigel matrix	(Eiraku *et al.* 2011)
	Mouse embryonic pancreatic progenitors	Pancreatic organoid	Corning Matrigel matrix	(Greggio *et al.* 2013)
	Mouse epithelia	Mammary organoid	Corning Matrigel matrix	(Ewald *et al.* 2008)
	Mouse primary colon tumour cells	Primary mouse colon tumour organoids	Corning Matrigel matrix	(Xue & Shah, 2013)
	Mouse ESCs	Functional thyroid follicular cells	Corning Matrigel matrix	(Antonica *et al.* 2012)
	Human ESCs	Cerebral organoid model of human brain development and microcephaly	Corning Matrigel matrix	(Lancaster *et al.* 2013)

AdMSC, adipose-derived mesencymal stem cell; ECM, extracellular matrix; ESC, embryonic stem cell; MSC, mesenchymal stem cell; NPC, neural precursor cell; PEG, polyethylene glycol; PLLA, poly-L-lactide; PSC, pluripotent stem cell.

appears to be well equipped with microenvironment characteristics that are important for mimicking or supporting key steps involved in the progression of tumour growth, vascularisation and metastasis. Tumour cells typically form spheroids when grown in 3D suspension cultures or within hydrogels. Cells within the spheroids are heterogeneous and organised in a manner that enables gradient formation and diffusion dynamics that correspond more closely to tumours formed *in vivo* (Gorlach *et al.*, 1994). The inner core of the spheroids exhibits a hollow lumen resembling the necrotic areas of *in vivo* cancers (Gorlach *et al.*, 1994). These areas are hypoxic and are usually observed at a distance from nutrient and oxygen supplies. In addition, the proliferation of tumour cells cultured in 3D is typically slower than that observed in monolayer cultures (Feder-Mengus *et al.*, 2008; Gorlach *et al.*, 1994).

Tumour cells of epithelial origin that are cultured in 3D have been shown to change shape and lose polarity, a feature typically associated with tumour progression *in vivo* (Yamada & Cukierman, 2007). Other parameters, such as proliferation, gene expression and drug sensitivity of tumour cells cultured in 3D environments, are also more representative of *in vivo* tumours compared to those cultured on 2D surfaces (Gurski *et al.*, 2010; Kim, 2005) (Table 7.2). For example, Cichon, *et al.* (2012) demonstrated that lung cancer cells adopt a flattened morphology when grown on 2D tissue culture plastic, but can form smooth, round spheroids, grape-like structures or invasive, branching structures when grown in 3D Matrigel matrix. Moreover, gene expression analysis revealed differences that distinguish lung cell lines that grow as smooth versus branched structures in this 3D model. This type of information may be useful for understanding differential prognostic outcomes of adenocarcinoma patients (Cichon *et al.*, 2012). In a study by Härmä *et al.* (2010), primary and non-transformed prostate epithelial cells as well as prostate cancer epithelial (PrC) cell lines formed well-differentiated round spheroids, while most PrC lines formed poorly differentiated spheroids or aggressively invading structures in 3D Matrigel matrix cultures. Additionally, specific signalling pathways were activated in the invasive PrC aggregates, suggesting potential targets for blocking invasion of tumour cells in 3D models. Other studies have investigated the impact of 3D microenvironments on phenotype, gene expression and epidermal growth factor receptor (EGFR) inhibition of common colorectal cancer cell lines (Luca *et al.*, 2013). Culture of PrC cell lines in 3D Matrigel matrix resulted in a drastically altered phenotype and significant changes in gene expression patterns compared to 2D culture, which was consistent with sensitivity to anti-EGFR targeted therapy *in vivo* (Luca *et al.*, 2013). The ability to model phenotypic and genotypic changes in cancer cells that are present in live tumours (especially tumours that are malignant or metastatic) is critical for developing more effective and predictive screening assays for chemotherapeutics.

Table 7.2 3D Cancer and tumour cell models.

Application	Cell type	3D model	Culture substrate and matrix	Comments/benefits of 3D model	Reference
CANCER AND TUMOUR CELL BIOLOGY	Normal and breast cancer cell lines	A model to evaluate effects of BMP4 on breast cancer cells in a 3D system	Corning Matrigel matrix; PEG	*In vivo*-like morphology, polarity observed in Matrigel matrix but not PEG gel	(Ampuja *et al.* 2013)
	Lung cancer cell lines	Compared cell morphology and gene expression in 2D and 3D systems	Corning Matrigel matrix	Gene expression profile and tumour cell morphology varied for different cancer cell lines in 3D but not 2D model	(Cichon *et al.* 2012)
	Primary mouse endometrial epithelial cells	Evaluated effects of growth factors, drugs and RNAi-induced gene alterations on cell proliferation, polarisation and glandular formation	Corning Matrigel matrix	Single-lumen, polarised glandular structures resembling endometrial carcinogenic tissue *in vivo*. Model suitable for studying effects of drugs, growth factors and gene alterations	(Eritja *et al.* 2010)
	Prostate cancer cell line (panel)	Compared tumour morphogenesis, gene expression and metabolism in 2D and 3D models	Corning Matrigel matrix	Some cancer cell lines demonstrated differences in signalling pathway activation and poorly differentiated spheroids and invasive structures that could be blocked in 3D but not 2D systems	(Härmä *et al.* 2010)
	Breast adinocarcinoma fibroblast-like, epithelial-like and human fibrosarcoma cell lines	Developed an XTT proliferation assay in 2D and 3D systems for HTP drug screening	Corning Matrigel matrix; collagen I	Cell proliferation kinetics, antiproliferative and cytotoxic effects of drugs were established in a more physiologically relevant 3D model compared to 2D systems	(Huyck *et al.* 2012)
	Mouse breast cancer cell line + stromal fibroblasts	Co-culture model to study the effects of stromal cells on breast cancer progression	Corning Matrigel matrix	Fibroblasts conferred chemoresistance and protected cancer cells from cytotoxic compounds in 3D cultures	(Li & Lu, 2011)
	Colorectal cancer cells	Investigated impact of 3D microenvironment on phenotype, gene expression and EGFR inhibition of common colorectal cancer cell lines	Corning Matrigel matrix	Significant changes in cell phenotype and gene expression were observed in 3D but not 2D model; data from 3D studies correlated with sensitivity to anti-EGFR targeted therapy *in vivo*	(Luca *et al.* 2013)
	Human intestinal cancer cell line (Caco-2)	Studied the effects of CLCA1 on cancer proliferation and differentiation	Corning Matrigel matrix	Found that CLCA1 may contribute to spontaneous differentiation and reduce proliferation of Caco-2	(Yang *et al.* 2013)

EGFR, epidermal growth factor receptor; HTP, high throughput; PEG, polyethylene glycol.

Cancer co-culture models

Most *in vitro* cell culture models involve growing a single cell type to enable investigation of its growth and behaviour in isolation. However, communication between different cell types is critical for maintaining tissue homeostasis *in vivo*. Simulating these interactions between two or more pertinent cell types using co-culture systems can improve the overall biological relevance of *in vitro* cell culture models.

Three-dimensional co-culture models have been very useful for studying complex interactions between cancer cells and other cell types, and elucidating their contribution to tumour growth, vascularisation and metastasis (Table 7.3). For example, stromal cells have been shown to induce chemoresistance and protect tumour cells from the toxic effects of anticancer drugs (Li & Lu, 2011). Endothelial cells and the associated vasculature provide a blood supply that enables tumour growth and survival, which is also responsible for carrying therapeutic compounds to the cancer cells (Carmeliet & Jain, 2000). In a recent coculture study, it was shown that endothelial cells can sensitise tumour spheroids to the toxic effects of chemotherapeutic drugs *in vitro*, and can induce angiogenesis and metastasis following implantation *in vivo* (Upreti *et al.*, 2011).

Tissue co-culture models

Permeable supports (e.g. Transwell inserts) offer unique advantages for co-culture studies and are thus widely incorporated into these systems (see Table 7.3). The pore-containing membranes of permeable supports are available with small pore sizes (e.g. 0.4 micron) that allow the passage of bioactive factors or soluble cues across the membrane, yet serve to physically separate cells located on opposite sides of the membrane.

In two studies, permeable supports were used to investigate the role of endothelial cells in the growth of breast epithelial cells (Ingthorsson *et al.*, 2010) and human lung epithelial cells (Franzdóttir *et al.*, 2010) embedded within 3D Matrigel matrix. In both instances, the endothelial cells were found to induce growth and morphological changes of the cancer epithelial cells, which could be achieved, at least in part, through secretion of soluble cues. In a different study, fibroblasts and endothelial cells were co-cultured with hepatocyte spheroids and found to significantly enhance hepatocyte function (Otsuka *et al.*, 2013). Using permeable supports, it was further determined that endothelial cells required direct cell-cell contact to enhance the function of hepatocytes, while fibroblasts were capable of this behaviour via the action of soluble factors (paracrine regulation) or in response to direct contact with the hepatocytes.

In a more complex permeable support co-culture model, four cell types (alveolar, macrophage-like, mast and endothelial cells) were grown together to generate a model that mimics the alveolar barrier to study the potential toxic effects of particles on the lung (Klein *et al.*, 2013). In this study, the membrane

Table 7.3 3D Co-culture models.

Application	Cell type	3D model	Culture substrate and matrix	Reference
CANCER CO-CULTURE MODELS	Mouse mammary tumour cells + endothelial cells	Co-culture of tumour spheroid to study impact of endothelial cells on tumour growth, vascularisation and metastasis *in vitro* and *in vivo*	Hanging drop; implantation in nude mice	(Upreti *et al.* 2011)
	Primary breast epithelial cells + endothelial cells	Co-culture model to study the role of endothelial cells in growth of normal and cancerous breast epithelial cells in 3D cultures	Transwell filters (0.4 μm); Corning Matrigel matrix	(Ingthorsson *et al.* 2010)
	Mesenchymal stem cells (MSCs) + ovarian cancer cells (OCC)	Co-culture model to study the influence of MSCs on OCC migration and invasion in an amniochorionic membrane model	Transwell filters; ultra-low attachment plates; Corning Matrigel matrix; amniotic membrane scaffold	(Touboul *et al.* 2013)
	Mouse breast cancer cell line + stromal fibroblasts	Co-culture model to study the effects of stromal cells on breast cancer progression	Corning Matrigel matrix	(Li & Lu, 2011)
	Keratinocytes + melanocytes + dermal stem cells + melanoma cells	A 3D skin reconstruct model to study melanoma progression in human skin	Bovine collagen I	(Li *et al.* 2011)
	Human melanoma cells + keratinocytes	Organotypic skin melanoma spheroid model for *in vitro* drug testing	Collagen I	(Vörsmann *et al.* 2013)
MULTICELLULAR CO-CULTURE MODELS	Human primary brain endothelial cells + human primary pericytes + human primary astrocytes	Blood–brain barrier model	Transwell inserts (0.4 μm)	(Urich *et al.* 2013)
	Hepatocytes + mouse fibroblasts or bovine endothelial cells	Micropatterned co-culture of hepatocyte spheroids	Micropatterned PEG hydrogel; gelatin	(Otsuka *et al.* 2013)
	Human bronchial epithelial cells + endothelial cells	A model to recapitulate lung morphogenesis	Transwell inserts (0.4 μm); Corning Matrigel matrix	(Franzdóttir *et al.* 2010)

(*Continued*)

Table 7.3 (Continued)

Application	Cell type	3D model	Culture substrate and matrix	Reference
	Primary rat hepatocytes + rat liver endothelial cells	A physiologically relevant co-culture model of hepatic sinusoids	Collagen I	(Kim & Rajagopalan, 2010)
	Alveolar cell line + macrophage-like cells + mast cells + endothelial cells	Tetraculture model mimicking alveolar barrier to study toxic effects of particles on the lungs; cells cultured at air-liquid interface (ALI)	Transwell inserts (0.4 μm)	(Klein *et al*. 2013)
	Mesenchymal stem cells + alveolar epithelial cells	Novel collagen drop cell migration assay for wound repair model	Transwell inserts (3 μm); collagen I	(Akram *et al*. 2013)
	Human retinal progenitor cells + endothelial cells	Hypoxia-induced retinal neovascularisation with and without endothelial cells	Transwell inserts (0.4 μm); Corning Matrigel matrix	(Kumar *et al*. 2011)
	Human mammary epithelial cells + human fibroblasts + adipocytes	A physiologically relevant triculture system to model human breast tissue	Corning Matrigel matrix + collagen I; porous silk protein scaffold	(Wang *et al*. 2010)
	Human bronchial epithelial cells	Differentiation of cells to glandular acini on ALI	Transwell collagen IV-coated; Corning Matrigel matrix	(Wu *et al*. 2011)
	Polarised airway epithelial cells	Infection of polarised epithelial cells from normal and cystic fibrosis patients. Cells were grown at ALI	Transwell inserts (0.4 μm)	(Mitchell *et al*. 2011)

PEG, polyethylene glycol.

played a central role by providing support for the complete tetraculture system, and enabled 3D organisation of cells that closely resembled the *in vivo* histology of the alveolar barrier. Endothelial cells were seeded on the basolateral side of the microporous membrane, and the other three cell types were seeded in the apical compartment and cultivated at the air-liquid interface (ALI) (Klein *et al.*, 2013).

In addition to cell permeation responses, permeable supports are also used to study cell migration and invasion. In an interesting 3D co-culture model, mesenchymal stem cells within an amnionic membrane scaffold were found to enhance the migration and invasion of ovarian cancer cells via the action of MSC-secreted IL-6 (Touboul *et al.*, 2013). Taken together, these results suggest that permeable support co-culture models may provide advantages and improvements over the current methods used for drug screening and tissue regeneration.

Drug discovery

The cost of drug development is becoming unsustainable due to the high attrition rate of drug candidates in late-stage human clinical trials or in some cases post marketing (Kola, 2008; Li, 2004). The predominant reasons for drug failure are lack of efficacy, poor pharmacokinetics,and adverse or toxic side effects that are not identified during early preclinical studies involving cell-based assays and animal testing (Bissell *et al.*, 2002; Singh, 2006; Westhouse, 2010). These issues highlight the need for improved preclinical models. Such models will need to better recapitulate pathobiological processes underlying diseases of specific tissues and organs in humans, and also more accurately predict physiological responses to therapeutic compounds and toxicity screens.

Physiologically relevant 3D cell culture models are ideally suited to bridge the gap between conventional 2D preclinical models and *in vivo* clinical studies in humans. Great advances have been achieved over the past few decades in this field, yet there are numerous practical issues that limit the widespread use of 3D cell culture models for drug discovery. For example, many novel 3D models have not been validated against established *in vivo* responses. Additionally, the outcome of 3D assays can be highly variable due to a lack of standardised protocols, the use of heterogeneous cell populations and limited permeability or poor diffusion dynamics. Also, 3D culture models are costly and not easily amenable to automation or scale-up for high-throughput screening. In spite of these limitations, new technologies and improved 3D *in vitro* assays are being developed to overcome these difficult challenges.

Matrigel matrix is commonly used for cancer research and drug development. This material has served as the long-standing gold standard matrix for 3D cell culture and the development of tumour and angiogenesis models. The most common *in vitro* 3D applications for this material are tumour cell invasion and angiogenesis (i.e. endothelial cell tube formation) assays in cancer drug development. One challenge limiting the use of Matrigel matrix in drug discovery is the need to dispense small volumes of this viscous gel reproducibly for high-throughput 3D assays. However, a recent study demonstrated the successful use of Matrigel for a high-throughput 3D screening assay. To establish the assay system, a liquid handling system was used to prepare 384-well plates with a diluted preparation of Matrigel matrix (1:3 dilution) (Hongisto *et al.*, 2013). When 63 prominent cancer drug candidates were analysed using this Matrigel-based system, the sensitivity and gene expression profile for breast cancer cells (JIMT1) were found to correlate with tumour xenografts (Hongisto *et al.*, 2013). In the same study, cancer cells grown in polyHEMA-induced anchorage-independent 3D culture failed to show similar drug sensitivity or gene expression profiles, and more closely resembled the behaviour observed in a 2D system. This study exemplifies the type of assay system that may ultimately replace 2D preclinical models.

Toxicity screening

State-of-the-art drug metabolism and toxicity screening assays often use difficult-to-source and expensive primary cells such as hepatocytes, cardiomyocytes and neurons in 2D cell culture models. Long-term toxicity tests are usually performed in animal models. As discussed earlier, these models can be highly inaccurate and misleading in predicting the *in vivo* human responses to drugs (Bissell *et al.*, 2002; Li, 2004; Polson & Fuji, 2012). Several 3D cell culture models have the potential to improve the predictive outcome of preclinical toxicity studies.

Microfabrication and microfluidics technologies have been used to generate several organ-on-chip models (Huh *et al.*, 2011). These are essentially miniaturised 3D models of organs with *in vivo*-like functionality, which may prove to be very useful for toxicity testing (Table 7.4). For example, Toh *et al.* (2009) developed a 3D hepatocyte chip (3D HepaTox chip) for *in vitro* toxicity testing. In this model, the hepatocytes retain their biosynthetic and metabolic functionality, and multiple drugs can be tested simultaneously using this system (Toh *et al.*, 2009). The IC_{50} values of drugs tested in this model closely correlated with reported *in vivo* LD_{50} values. Lee *et al.* (2008) created another 3D microarray for high-throughput screening of drug candidates and their cytochrome p450-generated metabolites. This microarray chip consists of cells encapsulated in collagen or alginate gels arrayed on a functionalised glass slide and a complementary human P450-containing microarray, and is also suitable for screening multiple compounds simultaneously (Lee *et al.*, 2008).

Culturing cells at an ALI using permeable supports has enabled the development of many *in vitro* 3D models for drug discovery (see Table 7.4). One of the well-known applications of ALI cultures is 3D modelling of lung epithelium. These cellular models mimic native tissues and demonstrate apical-basal polarity and cilia on the apical end, and are capable of secreting mucus (Huang *et al.*, 2011).

Table 7.4 Drug discovery and toxicity models.

Application	Cell type	3D model	Culture substrate and matrix	Comments/benefits of 3D model	Reference
DRUG DISCOVERY AND TOXICITY	Human epithelial-like breast cancer cell line (JIMT1)	High-throughput 3D screening assay	Corning Matrigel matrix; polyHEMA	Using a liquid handling system and diluted Matrigel matrix (1:3), drug sensitivity and gene expression of 63 drugs were compared simultaneously. Drug response in Matrigel matrix closely matched tumour xenografts while that on polyHEMA anchorage-independent 3D culture resembled profiles closer to 2D cell culture	(Hongisto *et al.* 2013)
	Rat primary hepatocytes	A microfluidic 3D hepatocyte chip for drug toxicity testing	3D matrix made with positively charged methylated collagen + negatively charged HEMA-MMA-MAA terpolymer; microfluidic perfusion enabled	High-throughput model for testing multiple drugs simultaneously in perfused hepatocytes that retain synthetic and metabolic function	(Toh *et al.* 2009)
	Human alveolar epithelial cells	A human lung-on-a-chip microdevice for nanotoxicology testing	PDMS microscaffold coated with fibronectin and collagen for supporting cells	Mimicked organ-level lung function for studying effects of nanotoxic particles on lung; also able to model impact of mechanical strain on lung function and inflammation	(Huh *et al.* 2011)
	Human breast cancer cell line (MCF7) and human hepatoma cell line (Hep3B)	Microarray for HTP toxicology assay screening of drugs and their P450-generated metabolites	Cells encapsulated in collagen I or alginate gel arrayed on a functionalised glass slide with complementary P450-containing microarray	Enabled testing of multiple compounds and their metabolites generated by CP1A2, CYP2D6, CYP3A4 and a mixture of the three P450s	(Lee *et al.* 2008)

(*Continued*)

Table 7.4 (Continued)

Application	Cell type	3D model	Culture substrate and matrix	Comments/benefits of 3D model	Reference
	EGFP-expressing colon cancer and mouse ESCs	High-throughput 3D proliferation and cytotoxic assay	Microbioreactors were made from multiwell plates. PET fibrous scaffold discs were fitted in the microwells within the bioreactors. Scaffolds were coated with fibronectin	Significantly increased fluorescence signals and high signal-to-noise ratio (10-fold) in this 3D system compared to 2D cultures. Could be used with conventional fluorescence microplate readers	(Zhang & Yang, 2011)
	Human primary nasal or bronchial epithelial cells	MucilAir - human airway epithelium model for toxicity testing	Cells cultured using air-liquid interface method on Transwell inserts (0.4 μm)	Mimics physiological characteristics of lung epithelia (e.g. morphology, apicobasal polarity, cilia formation, and mucus secretion). Validated model	(Huang *et al.* 2011)
	Alveolar cell line + macrophage-like cells + mast cells + endothelial cells	Tetraculture model mimicking alveolar barrier to study toxic effects of particles on the lungs; cells cultured at air-liquid interface	Transwell inserts (0.4 μm)	In this study, the Transwell membrane played a central role by providing support for the complete tetraculture system and enabled 3D organisation of cells that closely resembled *in vivo* histology of the alveolar barrier	(Klein *et al.* 2013)

EGFP, enhanced green fluorescent protein; ESC, embryonic stem cell; HTP, high throughput; PDMS, poly-dimethyl-siloxane; PET, polyethylene terephthalate.

ALI cultures made from freshly isolated nasal and bronchial biopsies are commercially available as MucilAir™ (Epithelix) and are useful for studying inhalation-based toxic responses to nanoparticles, gases, smoke and viral infections. Using cells from diseased donors for MucilAir enables the modelling of pulmonary disorders such as asthma, allergies, viral infections and cystic fibrosis. The ALI cultures can be maintained in a homeostatic state for long periods of time (3–4 weeks), which is typically feasible in animal models only and not in 2D cell culture models *in vitro* (Huang *et al.*, 2011). Another commercially available ALI culture is EpiDerm™ (MatTek), a model for human epidermis derived by differentiating epidermal keratinocytes on chemically modified collagen-coated cell culture inserts. This *in vitro* skin model reproduces morphological, structural and barrier function properties of normal human skin, and is used for studying dermal toxicity.

Another potential method for improving drug discovery is through production of PSC-derived differentiated cells. PSCs can provide an unlimited source of scarce or difficult-to-isolate cells for evaluating disease mechanisms, drug screening and toxicity testing. In particular, differentiated cardiomyocytes, hepatocytes and neuronal cells are highly desirable for toxicity screening. One critical issue that hinders the use of PSCs is that they are susceptible to apoptosis upon enzymatic dissociation (Watanabe *et al.*, 2007), which limits large-scale production of these cells. Faulkner-Jones *et al.* (2013) recently demonstrated for the first time that ESCs can be bio-printed into microdroplets that can be cultured in hanging drops to generate uniform spheroids while maintaining viability and functionality. These findings suggest there is potential for high-throughput screening assays using directed differentiation of PSCs. Using hPSC-derived cells could circumvent the problems of species-specific differences in drug response that are often encountered when using animal models to predict toxicity in humans.

Conclusion

In vitro 3D cell culture models have become increasingly sophisticated and widely used to study cell growth, tissue morphogenesis, stem cell differentiation, disease modelling and drug discovery. It is also evident that different 3D models with varying characteristics are required to meet the needs of specific cell types or applications. To date, many complex 3D models of tumourigenesis, stem cell differentiation and organoid formation have made use of natural ECM-based hydrogels, such as Matrigel matrix or collagen. Combining newer technologies, such as microfluidics or bio-printing, with biologically relevant materials, such as Matrigel matrix, may help to provide novel systems and models for cell biology research.

References

Aggarwal, B. B., Danda, D., Gupta, S. & Gehlot, P. (2009) Models for prevention and treatment of cancer: problems vs promises. *Biochemical Pharmacology*, **78**, 1083–1094.

Akram, K. M., Samad, S., Spiteri, M. A. & Forsyth, N. R. (2013) Mesenchymal stem cells promote alveolar epithelial cell wound repair in vitro through distinct migratory and paracrine mechanisms. *Respiratory Research*, **14**, 9.

Ampuja, M., Jokimäki, R., Juuti-Uusitalo, K. *et al.* (2013) BMP4 inhibits the proliferation of breast cancer cells and induces an MMP-dependent migratory phenotype in MDA-MB-231 cells in 3D environment. *BMC Cancer*, **13**, 429.

Antonica, F., Kasprzyk, D. F., Opitz, R. *et al.* (2012) Generation of functional thyroid from embryonic stem cells. *Nature*, **49**, 66–71.

Baharvand, H., Hashemi, S., Kazemi Ashtiani, S. &. Farrokhi, A. (2006) Differentiation of human embryonic stem cells into hepatocytes in 2D and 3D culture systems in vitro. *International Journal of Development and Biology*, **50**, 645–652.

Baker, B. M. & Chen, C. S. (2012) Deconstructing the third dimension – how 3D culture microenvironments alter cellular cues. *Journal of Cell Science*, **125**, 3015–3024.

Barcellos-Hoff, M. H., Aggeler, J., Ram, T. G. & Bissell, M. J. (1989) Functional differentiation and alveolar morphogenesis of primary mammary cultures on reconstituted basement membrane. *Development*, **105**, 223–235.

Benya, P. D. & Shaffer, J. D. (1982) Dedifferentiated chondrocytes reexpress the differentiated collagen phenotype when cultured in agarose gels. *Cell*, **30**, 215–224.

Bershteyn, M. & Kriegstein, A. R. (2013) Cerebral organoids in a dish: progress and prospects. *Cell*, **155**, 19–20.

Berthiaume, F., Moghe, P. V., Toner, M. & Yarmush, M. L. (1996) Effect of extracellular matrix topology on cell structure, function and physiological responsiveness: hepatocytes cultured in a sandwich configuration. *FASEB Journal*, **10**, 1471–1484.

Birgersdotter, A., Sandberg, R. & Emberg, I. (2005) Gene expression perturbation *in vitro* – growing case for three-dimensional (3D) culture systems. *Seminars in Cancer Biology*, **15**, 405–412.

Bissell, M. J., Radisky, D. C., Rizki, A., Weaver, V. M. & Petersen, O. W. (2002) The organizing principle: microenvironmental influences in the normal and malignant breast. *Differentiation: Research in Biological Diversity*, **70**, 537–546.

Bissell, M. J., Rizki, A. & Saira Mian, I. (2003) Tissue architecture: the ultimate regulator of breast epithelial function. *Current Opinion in Cell Biology*, **15**, 753–762.

Brannvall, K., Bergmann, K., Wallenquist, U. *et al.* (2007) Enhanced neuronal differentiation in a three-dimensional collagen-hyaluronan matrix. *Journal of Neuroscience Research*, **85**, 2138–2146.

Brock, A., Chang, E., Ho, C. C. *et al.* (2003) Geometric determinants of directional cell motility revealed using microcontact printing. *Langmuir*, **19**, 1611–1617.

Bryant, S. J. & Anseth, K. S. (2002) Hydrogel properties influence ECM production by chondrocytes photoencapsulated in poly(ethylene glycol) hydrogels. *Journal of Biomedical Materials Research*, **59**, 63–72.

Carlson, A. L. Florek, C. A., Kim, J. J. *et al.* (2012) Microfibrous substrate geometry as a critical trigger for organization, self-renewal, and differentiation of human embryonic stem cells within synthetic 3-dimensional microenvironments. *FASEB Journal*, **26**, 3240–3251.

Carmeliet, P. & Jain, R. K. (2000) Angiogenesis in cancer and other diseases. *Nature*, **407**, 249–257.

Carpenter, P. M., Dao, A. V., Arain, Z. S. *et al.* (2009) Motility induction in breast carcinoma by mammary epithelial laminin 332 (laminin 5). *Molecular Cancer Research*, **7**, 462–475.

Chen, C., Chan, M. C., Hsieh, R. K. *et al.* (2005) Activation of CD44 facilitates DNA repair in T-cell lymphoma but has differential effects on apoptosis induced by chemotherapeutic agents and ionizing radiation. *Leukemia & Lymphoma*, **46**, 1785–1795.

Chen, C.S., Mrksich, M., Huang, S. *et al.* (1997) Geometric control of cell life and death. *Science*, **276**, 1425–1428.

Chen, J., Shi, Z. D., Ji, X. *et al.* (2013) Enhanced osteogenesis of human mesenchymal stem cells by periodic heat shock in self-assembling peptide hydrogel. *Tissue Engineering Part A*, **19**, 716–728.

Chirila, T. V., Constable I. J., Crawford, G. J. *et al.* (1993) Poly(2-hydroxyethyl methacrylate) sponges as implant materials: in vivo and in vitro evaluation of cellular invasion. *Biomaterials*, **14**, 26–38.

Cichon, M. A., Gainullin, V. G., Zhang, Y. & Radisky, D. C. (2012) Growth of lung cancer cells in three-dimensional microenvironments reveals key features of tumor malignancy. *Integrative Biology*, **4**, 440–448.

Cukierman, E., Pankov, R., Stevens, D. R. & Yamada, K. M. (2001) Taking cell-matrix adhesions to the third dimension, *Science*, **294**, 1708–1712.

Daley, W. P., Peters, S. B. & Larsen, M. (2008) Extracellular matrix dynamics in development and regenerative medicine. *Journal of Cell Science*, **121**, 255–264.

David, L., Dulong, V., Le Cerf, D. *et al.* (2008) Hyaluronan hydrogel: an appropriate three-dimensional model for evaluation of anticancer drug sensitivity. *Acta Biomaterialia*, **4**, 256–263.

Dawson E., Mapili, G., Erickson, K. *et al* (2008) Biomaterials for stem cell differentiation. *Advanced Drug Delivery Reviews*, **60**, 215–228.

Debnath, J. & Brugge, J. S. (2005) Modelling glandular epithelial cancers in three-dimensional cultures *Nature Reviews Cancer*, **5**, 675–688.

Delcommenne, M. & Streuli, C. H. (1995) Control of integrin expression by extracellular matrix. *Journal of Biological Chemistry*, **270**, 26794–26801.

Devolder, R. & Kong, H. J. (2012) Hydrogels for in vivo-like three-dimensional cellular studies. *Wiley Interdisciplinary Reviews: Systems Biology and Medicine*, **4**, 351–365.

Eiraku, M., Takata, N., Ishibashi, H. *et al.* (2011) Self-organizing optic-cup morphogenesis in three-dimensional culture. *Nature*, **472**, 51–56.

Ellerbroek, S. M., Wu, Y.I., Overall, C. M. & Stack, M. S. (2001) Functional interplay between type i collagen and cell surface matrix metalloproteinase activity. *Journal of Biological Chemistry*, **276**, 24833–24842.

Emerman, J. T. & Pitelka, D. R. (1977) Maintenance and induction of morphological differentiation in dissociated mammary epithelium on floating collagen membranes. *In Vitro*, **13**, 316–328.

Eritja, N., Llobet Navas, D., Domingo, M. *et al* (2010) A novel three-dimensional culture system of polarized epithelial cells to study endometrial carcinogenesis. *American Journal of Pathology*, **176**, 2722–2731.

Ewald, A. J., Brenot, A., Duong, M. *et al.* (2008) Collective epithelial migration and cell rearrangements drive mammary branching morphogenesis. *Developmental Cell*, **14**, 570–581.

Faulkner-Jones, A., Greenhough, S., King, J. A. *et al.* (2013) Development of a valve-based cell printer for the formation of human embryonic stem cell spheroid aggregates. *Biofabrication*, **5**, 015013.

Feder-Mengus, C., Ghosh, S., Reschner, A. *et al.* (2008) New dimensions in tumor immunology: what does 3D culture reveal? *Trends in Molecular Medicine*, **14**, 333–340.

Franzdóttir, S. R., Axelsson, I. T., Arason, A. J. *et al.* (2010) Airway branching morphogenesis in three- dimensional culture. *Respiratory Research*, **11**, 162–171.

Frieboes, H. B., Zheng, X., Sun, C.H. *et al.* (2006) An integrated computational/experimental model of tumor invasion. *Cancer Research*, **66**, 1597–1604.

Fuchs, E., Tumbar, T. & Guasch, G. (2004) Socializing with the neighbors: stem cells and their niche. *Cell*, **116**, 769.

Gelain, F., Bottai, D., Vescovi, A. & Zhang, S. (2006) Designer self-assembling peptide nanofiber scaffolds for adult mouse neural stem cell 3-dimensional cultures. *PLoS One*, **1**, e119.

Ghosh, S., Spagnoli, G. C., Martin, I. *et al.* (2005) Three-dimensional culture of melanoma cells profoundly affects gene expression profile: a high density oligonucleotide array study. *Journal of Cellular Physiology*, **204**, 522–531.

Gómez-Lechón, M. J., Jover, R., Donato, T. *et al.* (1998) Long-term expression of differentiated functions in hepatocytes cultured in three-dimensional collagen matrix. *Journal of Cellular Physiology*, **177**, 553–562.

Gorlach, A., Herter, P., Hentschel, H. *et al.* (1994) Effects of nIFN beta and rIFN gamma on growth and morphology of two human melanoma cell lines: comparison between two- and three-dimensional culture. *International Journal of Cancer*, **56**, 249–254.

Gray, R. S., Cheung, K. J. & Ewald, A. J. (2010) Cellular mechanisms regulating epithelial morphogenesis and cancer invasion. *Current Opinion in Cell Biology*, **22**, 640–650.

Greggio, C., de Franceschi, F., Figueiredo-Larsen, M. *et al.* (2013) Artificial three-dimensional niches deconstruct pancreas development in vitro. *Development*, **140**, 4452–4462.

Grinnell, F. (2003) Fibroblast biology in three-dimensional collagen matrices. *Trends in Cell Biology*, **13**, 264–269.

Gurski, L. A., Jha, A. K., Zhang, C. *et al.* (2009) Hyaluronic acid-based hydrogels as 3D matrices for in vitro evaluation of chemotherapeutic drugs using poorly adherent prostate cancer cells. *Biomaterials*, **30**, 6076–6085.

Gurski, L. A., Petrelli, N. J., Jia, X. *et al.* (2010) 3D matrices for anti-cancer drug testing and development. *Oncology Issues*, **25**, 20–25.

Hait, W. N. (2010) Anticancer drug development: the grand challenges. *Nature Reviews Drug Discovery*, **9**, 253–254.

Hall, C. L., Dubyk, C. W., Riesenberger, T. A. *et al.* (2008) Type I collagen receptor ($\alpha 2\beta 1$) signaling promotes prostate cancer invasion through RhoC GTPase. *Neoplasia*, **10**, 797–803.

Härmä, V., Virtanen, J., Rami, M. *et al.* (2010) A comprehensive panel of three-dimensional models for studies of prostate cancer growth, invasion and drug responses. *PLoS One*, **5**, e10431.

Hartman, O., Zhang, C., Adams, E. L. *et al.* (2009) Microfabricated electrospun collagen membranes for 3-D cancer models and drug screening applications. *Biomacromolecules*, **10**, 2019–2032.

Haycock, J. W. (2011) 3D cell culture: a review of current approaches and techniques. *Methods in Molecular Biology*, **695**, 1–15.

Hongisto, V., Jernström, H., Fey, V. *et al.* (2013) High-throughput 3D screening reveals differences in drug sensitivities between culture models of JIMT1 breast cancer cells. *PLoS One*, **8**, e77232.

Horii, A., Wang, X., Gelain, F. & Zhang S. (2007) Biological designer self-assembling peptide nanofiber scaffolds significantly enhance osteoblast proliferation, differentiation and 3-D migration. *PLoS One*, **2**(2), e190.

HTSTec (2010) *3D Cell Culture Trends 2010 Report*. HTStec Limited, Cambridge, UK.

Huang, S., Wiszniewski, L. & Constant, S. (2011) The use of in vitro 3D cell models in drug development for respiratory diseases, in *Drug Discovery and Development – Present and Future* (ed. I. Kapetanovi), InTech, Rijeka, Croatia.

Huh, D., Hamilton, G. & Ingber, D. E. (2011) From three-dimensional cell culture to organs-on-chips. *Trends in Cell Biology*, **21**, 745–754.

Hutchinson, L. & Kirk, R. (2011) High drug attrition rates – where are we going wrong? *Nature Reviews Clinical Oncology*, **8**, 189–190.

Huyck, L., Ampe, C. & van Troys, M. (2012) The XTT cell proliferation assay applied to cell layers embedded in three-dimensional matrix. *Assay Drug Development Technologies*, **10**, 382–392.

Hwang, N. S., Kim, M. S., Sampattavanich, S. *et al.* (2006) Effects of three-dimensional culture and growth factors on the chondrogenic differentiation of murine embryonic stem cells. *Stem Cells*, **24**, 284–291.

Hynes, O. (2009) The extracellular matrix: not just pretty fibrils. *Science*, **326**, 1216–1219.

Ingthorsson, S., Sigurdsson, V., Fridriksdottir, A. J. R. *et al.* (2010) Endothelial cells stimulate growth of normal and cancerous breast epithelial cells in 3D culture. *BMC Research Notes*, **3**, 184.

Kang, B. J., Ryu, H. H., Park, S. S. *et al.* (2012) Effect of Matrigel on the osteogenic potential of canine adipose tissue-derived mesenchymal stem cells. *Journal of Veterinary Medicine and Science*, **74**, 827–836.

Khetan, S. & Burdick, J. A. (2010) Patterning network structure to spatially control cellular remodeling and stem cell fate within 3-dimensional hydrogels. *Biomaterials*, **31**, 8228–8234.

Kiefer, J. A. & Farach-Carson, M. C. (2001) Type I collagen-mediated proliferation of PC3 prostate carcinoma cell line: implications for enhanced growth in the bone microenvironment. *Matrix Biology*, **20**, 429–437.

Kilian, K. A., Bugarija, B., Lahn, B. T. & Mrksich, M. (2010) Geometric cues for directing the differentiation of mesenchymal stem cells. *Proceedings of the National Academy of Sciences USA*, **107**, 4872–4877.

Kim, J. B. (2005) Three-dimensional tissue culture models in cancer biology. *Seminars in Cancer Biology*, **15**, 365–377.

Kim, Y. & Rajagopalan, P. (2010) 3D hepatic cultures simultaneously maintain primary hepatocyte and liver sinusoidal endothelial cell phenotypes. *PLoS One*, **5**, e15456.

Kim, Y. J., Bae, H. I., Kwon, O. K. & Choi, M. S. (2009) Three-dimensional gastric cancer cell culture using nanofiber scaffold for chemosensitivity test. *International Journal of Biology and Macromolecules*, **45**, 65–71.

Klein, S. G., Serchi, T., Hoffman, L. *et al.* (2013) An improved 3D tetraculture system mimicking the cellular organisation at the alveolar barrier to study the potential toxic effects of particles on the lung. *Particle and Fibre Toxicology*, **10**, 31–47.

Kleinman, H. K. & Martin G. R. (2005) Matrigel: basement membrane matrix with biological activity. *Seminars in Cancer Biology*, **15**, 378–386.

Kleinman, H. K., Philp, D. & Hoffman, M. P. (2003) Role of the extracellular matrix in morphogenesis. *Current Opinion in Biotechnology*, **14**, 526–532.

Koehler, K. R., Mikosz, A. M., Molosh, A. L. *et al.* (2013) Generation of inner ear sensory epithelia from pluripotent stem cells in 3D culture. *Nature*, **500**, 217–221.

Kola, I. (2008) The state of innovation in drug development. *Clinical Pharmacology & Therapeutics*, **83**, 227–230.

Kumar, R., Harris-Hooker, S., Ritesh, K. & Sanford, G. (2011) Co-culture of retinal and endothelial cells results in the modulation of genes critical to retinal neovascularization. *Vascular Cell*, **3**, 27–41.

Lai, Y., Cheng, K. & Kisaalita, W. (2012) Three dimensional neuronal cell cultures more accurately model voltage gated calcium channel functionality in freshly dissected nerve tissue. *PLoS One*, **7**, e45074.

Lancaster, M. A., Renner, M., Martin, C. A. *et al.* (2013) Cerebral organoids model human brain development and microcephaly. *Nature*, **501**, 373–379.

Languino, L. R., Gehlsen, K. R., Wayner, E. *et al.* (1989) Endothelial cells use alpha 2 beta 1 integrin as a laminin receptor. *Journal of Cell Biology*, **109**, 2455–2462.

Lee, E. Y., Parry, G. & Bissell, M. J. (1984) Modulation of secreted proteins of mouse mammary epithelial cells by the collagenous substrata. *Journal of Cell Biology*, **98**, 146–155.

Lee, J., Cuddihy, M. J., Kotov, N. A. (2008) Three-dimensional cell culture matrices: state of the art. *Tissue Engineering*, **14**, 61–86.

Lee, M. Y., Kumar, R. A., Sukumaran, S. M. *et al.* (2008) Three-dimensional cellular microarray for high-throughput toxicology assays. *Proceedings of the National Academy of Sciences USA*, **105**, 59–63.

Li, A. P. (2004) Accurate prediction of human drug toxicity: a major challenge in drug development. *Chemico-Biological Interactions*, **150**, 3–7.

Li, C., Heidt, D. G., Dalerba, P. *et al.* (2006) Identification of pancreatic cancer stem cells. *Cancer Research*, **66**, 1990–1999.

Li, L. & Lu, Y. (2011) Optimizing a 3D culture system to study the interaction between epithelial breast cancer and its surrounding fibroblasts. *Journal of Cancer*, **2**, 458–466.

Li, L., Fukunaga-Kalabis, M. & Herlyn, M. (2011) The three-dimensional human skin reconstruct model: a tool to study normal skin and melanoma progression. *Journal of Visualised Experiments*, **54**, e2937.

Li, X., Katsanevakis, E. & Liu, X. (2012) Engineering neural stem cell fates with hydrogel design for central nervous system regeneration. *Progress in Polymer Science*, **37**, 1105–1129.

Liedmann, A., Frech, S., Morgan, P. J. *et al.* (2012) Differentiation of human neural progenitor cells in functionalized hydrogel matrices. *Bioresearch Open Access*, **1**, 16–24.

Lodish, H., Berk, A., Zipursky, S. L., Matsudaira, P., Baltimore, D. & Darnell, J. (2002) *Molecular Cell Biology*. W.H. Freeman, New York.

Luca, A. C., Mersch, S., Deenen, R. *et al.* (2013) Impact of the 3D microenvironment on phenotype, gene expression, and EGFR inhibition of colorectal cancer cell Lines. *PLoS One*, **8**, e59689.

Mahmud, G., Campbell, C. J., Bishop, K. J. M. *et al.* (2009) Directing cell motions on micropatterned ratchets. *Nature Physics*, **5**, 606–612.

Martens, P. & Anseth, K. S. (2000) Characterization of hydrogels formed from acrylate modified poly(vinyl alcohol) macromers. *Polymer*, **41**, 7715–7722.

McBeath, R., Pirone, D. M., Nelson, C. M. *et al.* (2004) Cell shape, cytoskeletal tension, and RhoA regulate stem cell lineage commitment. *Developmental Cell*, **6**, 483–495.

McCracken, K. W., Howell, J. C., Wells, J. M. *et al.* (2011) Generating human intestinal tissue from pluripotent stem cells in vitro. *Nature Protocols*, **6**, 1920–1928.

Menke, A., Philippi, C., Vogelmann, R. *et al.* (2001) Down-regulation of E-cadherin gene expression by collagen type I and type III in pancreatic cancer cell lines. *Cancer Research*, **61**, 3508–3517.

Mitchell, G., Grondin, G., Bilodeau, G. *et al.* (2011) Infection of polarized airway epithelial cells by normal and small-colony variant strains of Staphylococcus aureus is increased in cells with abnormal cystic fibrosis transmembrane conductance regulator function and is influenced by NF-κB. *Infection and Immunity*, **79**, 3542–3551.

Miyamoto, H., Murakami, T., Tsuchida, K. *et al.* (2004) Tumor-stroma interaction of human pancreatic cancer: acquired resistance to anticancer drugs and proliferation regulation is dependent on extracellular matrix proteins. *Pancreas*, **28**, 38–44.

Morizane, R., Monkawa, T., Fujii, S. *et al.* (2013) Kidney specific protein-positive cells derived from embryonic stem cells reproduce tubular structures in vitro and differentiate into renal tubular cells. *PLoS One*, **8**, e64843.

Mueller-Klieser, M., Freyer, J. P. & Sutherland, R. M. (1986) Influence of glucose and oxygen supply conditions on the oxygenation of multicellular spheroids. *British Journal of Cancer*, **53**, 345–353.

Nelson, C. M. & Bissell, M. J. (2006) Of extracellular matrix, scaffolds, and signaling: tissue architecture regulates development, homeostasis, and cancer. *Annual Review of Cell and Developmental Biology*, **22**, 287–309.

Ortinau, S., Schmich, J., Block, S. *et al.* (2010) Effect of 3D-scaffold formation on differentiation and survival in human neural progenitor cells. *BioMedical Engineering OnLine*, **9**, 70–87.

Otsuka, H., Sasaki, K., Okimura, S. *et al.* (2013) Micropatterned co-culture of hepatocyte spheroids layered on non-parenchymal cells to understand heterotypic cellular interactions. *Science and Technology of Advanced Materials*, **14**, 065003.

Pampaloni, F., Reynaud, E. G. & Stelzer, E. H. K. (2007) The third dimension bridges the gap between cell culture and live tissue. *Nature Reviews Molecular Cell Biology*, **8**, 839–845.

Petersen, O. W., Rønnov-Jessen, L., Howlett, A. R. & Bissell, M. J. (1992) Interaction with basement membrane serves to rapidly distinguish growth and differentiation pattern of normal and malignant human breast epithelial cells. *Proceedings of the National Academy of Sciences USA*, **89**, 9064–9068.

Polson, A. G. & Fuji, R. N. (2012) The successes and limitations of preclinical studies in predicting the pharmacodynamics and safety of cell-surface-targeted biological agents in patients. *British Journal of Pharmacology*, **166**, 1600–1602.

Powers, M. J., Janigian, D. M., Wack, K. E. *et al.* (2002) Functional behavior of primary rat liver cells in a three-dimensional perfused microarray bioreactor. *Tissue Engineering*, **8**, 499–513.

Riccio, M., Resca, E., Maraldi, T. *et al.* (2010) Human dental pulp stem cells produce mineralized matrix in 2D and 3D cultures. *European Journal of Histochemistry*, **54**, e46.

Sawhney, A. S., Chandrashekhar, P. P. & Hubbell, J. A. (1993) Bioerodible hydrogels based on photopolymerized poly(ethylene glycol)-co-poly(alpha-hydroxy acid) diacrylate macromers. *Macromolecules*, **26**, 581–587.

Semino, C. E., Merok, J., Crane, G. C. *et al.* (2003) Functional differentiation of hepatocyte-like spheroid structures from putative liver progenitor cells in three-dimensional peptide scaffolds. *Differentiation*, **71**, 262–270.

Singh, S. S. (2006) Preclinical pharmacokinetics: an approach towards safer and efficacious drugs. *Current Drug Metabolism*, **7**, 165–182.

Singhvi, R., Kumar, A., Lopez, G. P. *et al.* (1994) Engineering cell and shape function. *Science*, **264**, 696–698.

Sisson, K., Zhang, C., Farach-Carson, M. C. *et al.* (2009) Evaluation of cross-linking methods for electrospun gelatin on cell growth and viability. *Biomacromolecules*, **10**, 1675–1680.

Spence, J. R., Mayhew, C. N., Rankin, S. A. *et al.* (2011) Directed differentiation of human pluripotent stem cells into intestinal tissue *in vitro*. *Nature*, **470**, 105–109.

Thavandiran, N., Dubois, N., Mikryukov, A. *et al.* (2013) Design and formulation of functional pluripotent stem cell-derived cardiac microtissues. *Proceedings of the National Academy of Sciences USA*, **110**, E4698–4707.

Théry, M., Racine, V., Piel, M. *et al.* (2006) Anisotropy of cell adhesive microenvironment governs cell internal organization and orientation of polarity. *Proceedings of the National Academy of Sciences USA*, **103**,19771–19776.

Théry, M., Jiménez-Dalmaroni, A., Racine, V. *et al.* (2007) Experimental and theoretical study of mitotic spindle orientation. *Nature*, **447**, 493–496.

Thomas, C. H., Collier, J. H., Sfeir, C. S. & Healy, K. E. (2002) Engineering gene expression and protein synthesis by modulation of nuclear shape. *Proceedings of the National Academy of Sciences USA*, **99**, 1972–1977.

Tibbitt, M. W. & Anseth, K. S. (2009) Hydrogels as extracellular matrix mimics for 3D cell culture. *Biotechnology and Bioengineering*, **103**, 655.

Toh, Y. C., Lim, T. C., Tai, D. *et al.* (2009) A microfluidic 3D hepatocyte chip for drug toxicity testing. *Lab on a Chip*, **9**, 2026–2035.

Touboul, C., Lis, R., Al Farsi, H. *et al.* (2013) Mesenchymal stem cells enhance ovarian cancer cell infiltration through IL6 secretion in an amniochorionic membrane based 3D model. *Journal of Translational Medicine*, **11**, 28.

Turovets, N., Fair, J., West, R. *et al.* (2012) Derivation of high-purity definitive endoderm from human parthenogenetic stem cells using an in vitro analog of the primitive streak. *Cell Transplantation*, **21**, 217–234.

Upreti, M., Jamshidi-Parsian, A., Koonce, N. A. *et al.* (2011) Tumor-endothelial cell three-dimensional spheroids: new aspects to enhance radiation and drug therapeutics. *Translational Oncology*, **4**, 365–376.

Urich, E., Patsch, C., Aigner, S. *et al.* (2013) Multicellular self-assembled spheroidal model of the blood brain barrier. *Scientific Reports*, **3**, 1500.

Van der Worp, H. B., Howells, D. W., Sena, E. S. *et al.* (2010) Can animal models of disease reliably inform human studies? *PLoS Medicine*, **7**, e1000245.

Von der Mark, K., Gauss, V., von der Mark, H. & Müller, P. (1977) Relationship between cell shape and type of collagen synthesised as chondrocytes lose their cartilage phenotype in culture. *Nature*, **267**, 531–532.

Vörsmann, H., Groeber, F., Walles, H. *et al.* (2013) Development of a human three-dimensional organotypic skin-melanoma spheroid model for in vitro drug testing. *Cell Death & Disease*, **4**, e719.

Walpita, D. & Hay, E. (2002) Studying actin-dependent processes in tissue culture. *Nature Reviews Molecular Cell Biology*, **3**, 137–141.

Wang, S., Ahmad, A. A., Shah, P. K. *et al.* (2013) Capture and 3D culture of colonic crypts and colonoids in a microarray platform. *Lab on a Chip*, **13**, 4625–4634.

Wang, X. Sun, L., Maffini, M. V. *et al.* (2010) A complex 3D human tissue culture system based on mammary stromal cells and silk scaffolds for modeling breast morphogenesis and function. *Biomaterials*, **31**, 3920–3929.

Watanabe, K., Ueno, M., Kamiya, D. *et al.* (2007) A ROCK inhibitor permits survival of dissociated human embryonic stem cells. *Nature Biotechnology*, **25**, 681–686.

Weaver, V. M., Petersen, O. W., Wang, F. *et al.* (1997) Reversion of the malignant phenotype of human breast cells in three-dimensional culture and *in vivo* by integrin blocking antibodies. *Journal of Cell Biology*, **137**, 231–245.

Westhouse, R.A. (2010) Safety assessment considerations and strategies for targeted small molecule cancer therapeutics in drug discovery. *Toxicologic Pathology*, **38**,165–168.

Willerth, S. M., Arendas, K. J., Gottlieb, D. I. *et al.* (2006) Optimization of fibrin scaffolds for differentiation of murine embryonic stem cells into neural lineage cells. *Biomaterials*, **27**, 5990–6003.

Wu, X., Peters-Hall, J. R., Bose, S. *et al.* (2011) Human bronchial epithelial cells differentiate to 3D glandular acini on basement membrane matrix. *American Journal of Respiratory Cell and Molecular Biology*, **44**, 914–921.

Xia, Y., Nivet, E., Sancho-Martinez, I. *et al.* (2013) Directed differentiation of human pluripotent cells to ureteric bud kidney progenitor-like cells. *Nature Cell Biology*, **15**, 1507–1515.

Xue, X. & Shah, Y. M. (2013) In vitro organoid culture of primary mouse colon tumors. *Journal of Visualised Experiments*, **75**, e50210.

Yamada, K. M. & Cukierman, E. (2007) Modeling tissue morphogenesis and cancer in 3D. *Cell*, **130**, 601–610.

Yang, B., Cao, L., Liu, B. *et al.* (2013) The transition from proliferation to differentiation in colorectal cancer is regulated by the calcium activated chloride channel A1. *PLoS One*, **8**, e60861.

Zhang, S. (2004) Beyond the Petri dish. *Nature Biotechnology*, **22**, 151–152.

Zhang, X. & Yang, S. T. (2011) High-throughput 3-D cell-based proliferation and cytotoxicity assays for drug screening and bioprocess development. *Journal of Biotechnology*, **151**, 186–193.

Zhou, Z., Wang, J., Cao, R. *et al.* (2004) Impaired angiogenesis, delayed wound healing and retarded tumor growth in perlecan heparan sulfate-deficient mice. *Cancer Research*, **64**, 4699–4702.

CHAPTER 8

HyStem®, a customisable hyaluronan-based hydrogel matrix for 3D cell culture

T. I. Zarembinski[1], B. J. Engel[2], N. J. Doty[1], P. E. Constantinou[2], M. V. Onorato[1], I. E. Erickson[1], E. L. S. Fong[3], M. Martinez[2], R. L. Milton[2], B. P. Danysh[2], N. A. Delk[4], D. A. Harrington[2], M. C. Farach-Carson[2,3] and D. D. Carson[2,5]

[1] *BioTime, Inc., Alameda, California, USA*

[2] *Department of BioSciences, Rice University, Houston, Texas, USA*

[3] *Department of Bioengineering, Rice University, Houston, Texas, USA*

[4] *Department of Biological Sciences, University of Texas at Dallas, Richardson, Texas, USA*

[5] *Department of Genetics, M. D. Anderson Cancer Center, Houston, Texas, USA*

Introduction

Cells in living tissues are surrounded by an extracellular matrix (ECM), a soft, water-swollen hydrogel network composed of glycosaminoglycans, proteoglycans, insoluble and soluble proteins, ions and small molecules (Lutolf & Hubbell, 2005). Importantly, this matrix provides not only structure but also cellular signalling crucial for cellular function such as proliferation, differentiation and migration (Lee & Spicer, 2000). Despite this reality, cells have been ripped from their native habitat for research purposes; they are traditionally studied in the absence of an ECM on a hard plastic surface (Tibbitt & Anseth, 2009).

One approach to recapitulating native cellular function *in vitro* is to encapsulate the cells with a manufactured hydrogel that mimics the ECM. While a number of naturally occurring biopolymers may provide the water-swollen properties of the ECM (Schmidt *et al.*, 2008), few fulfil the additional criteria required to be a suitable ECM mimic (Prestwich, 2008). Put simply, besides its water-rich, tissue-like character (Tibbitt & Anseth, 2009), the ideal hydrogel is composed of a biopolymer that is already present in the native ECM. This is in contrast to a synthetic polymer that may require significant engineering to add the signalling motifs back (Lutolf & Hubbell, 2005). The ideal hydrogel should also have few chemically defined components to enable deconvolution of extracellular signals provided to the cells and to provide batch-to-batch consistency. Additionally, the hydrogel must allow a researcher to tailor its formulation not only to a particular

Technology Platforms for 3D Cell Culture: A User's Guide, First Edition. Edited by Stefan Przyborski.

cell type (not all ECMs are equivalent in composition) but also to adjust qualities such as matrix stiffness or ligand identity to interrogate cellular signalling pathways. Finally, the matrix needs to be manufactured using rigorous quality systems to ensure reproducibility and needs to be both easy to use and amenable to downstream microscopic and immunohistochemical analyses required for high-content analysis and live cell imaging.

The HyStem® platform meets these criteria. At its heart, HyStem is a matrix based on hyaluronic acid (HA), a major component of the extracellular matrix. HA is what gives HyStem its cytocompatible and biocompatible character that can be exploited for the culture of cells *in vitro* and for subsequent implantation *in vivo* (Prestwich, 2007). HyStem's biggest advantage is that the final hydrogel can be assembled based on user-defined design criteria for mechanical properties, biological content and diffusion characteristics. This is because its thiol-based chemistry and *in situ* gelation offer the user a high degree of flexibility. An analogy to illustrate the breadth of options for HyStem customisation is that of a biomaterial LEGO® set: the user can engineer the final composition as well as spatiotemporal control of gelation into the final hydrogel by choosing the appropriate HA backbone, crosslinker, covalently linked backbone decorations, non-covalently linked additives and/or cells. Furthermore, HyStem has functionality in multiple phases, liquid, gel and solid, each uniquely important for specific applications.

HyStem as a biomaterial LEGO set

What HyStem is and how it works

The basic HyStem hydrogel platform is a modular system consisting of three components (Table 8.1). Glycosil® (thiolated HA) provides the basic space-filling volume for the hydrogel and its *in vivo*-like characteristics. Gelin®-S (thiolated porcine gelatin) is crucial for supporting cellular attachment for a wide variety of anchorage-dependent cells (Shu *et al.*, 2003). Extralink® (polyethylene glycol diacrylate, PEGDA) is the thiol reactive crosslinker. When all three components are mixed together, Extralink covalently links the thiolated species together via the Michael addition reaction (Figure 8.1) (Zheng Shu *et al.*, 2004).

The elegance of the Michael addition reaction is that it occurs at an acceptable rate at physiological temperatures and pH (from 5 to19 minutes

Table 8.1 The three basic components of HyStem-C.

Component	Base material	Reactive group	Purpose in hydrogel	Molecular weight (kDa)	Effect of increased concentration
Glycosil	Hyaluronic acid	Thiol	Basic scaffold	250	Stiffer hydrogel
Gelin-S	Porcine gelatin	Thiol	Cellular attachment	Wide range	No effect
Extralink	Polyethylene glycol	Acrylic acid	Crosslinker	3.4	Stiffer hydrogel

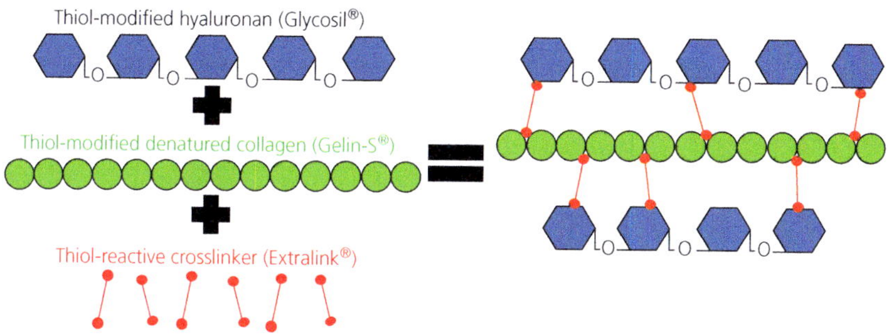

Figure 8.1 **HyStem-C crosslinking of Glycosil, Gelin-S and Extralink via Michael addition reaction.**

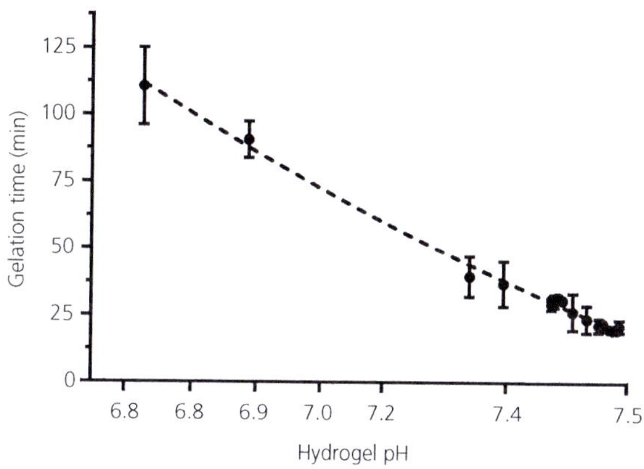

Figure 8.2 **Effect of pH on HyStem gelation time.**

depending on the ratio of the components), is non-exothermic and produces few by-products. Since the rate of the Michael addition reaction is dependent on the concentration of the reactive thiolate form (pKa for the thiols of Glycosil is 8.87), its gelation speed is tightly coupled to pH (Zheng Shu *et al.*, 2004). In fact, gelation time and pH are exponentially related; gelation time is reduced by half with every 0.3 pH unit decrease (Figure 8.2). Furthermore, the gelation kinetics are biphasic: while the gelation mostly occurs within 20 minutes of starting the reaction, final crosslinking continues at a slower rate for up to four days and is thought to be due to continued reaction between free thiols on the Glycosil molecules (Hanjaya-Putra *et al.*, 2010; Young & Engler, 2011).

Composition flexibility

A variety of additional biopolymers can be incorporated into the HyStem hydrogel platform since each can either be covalently linked to the Glycosil thiols or physically trapped within the HyStem matrix if it is large enough (a complete list of options is shown in Table 8.2).

Table 8.2 Additional menu items for HyStem customisation.

Component	Description	Key use	Reference
Carbohydrate backbone			
Heprasil®	Mixture of thiolated HA and thiolated heparin	Slow release of growth factors incorporated into the hydrogel	Cai *et al.* 2005; Peattie *et al.* 2008
Backbone decorations			
Thiolated proteins	Engineered protein fragments with exposed cysteine	Indirect covalent linkage (via Extralink)	Ghosh *et al.* 2006
Protein coupled by another crosslinker	Use of heterobifunctional crosslinker to couple backbone thiols to protein amine groups	Indirect covalent linkage to any biopolymer with a free amine group	Chopra *et al.* 2014
Maleimide-peptide	Maleimide reactive group placed on N-terminus of peptide	Direct covalent linkage of cell attachment peptides	Zarembinski *et al.* 2014
Crosslinkers			
GSSG	Oxidised glutathione	Faster gelation without increases in gel stiffness (thiol-disulfide exchange)	Zarembinski *et al.* 2014
PEGnor/I2959	Polyethylene glycol norbornene mixed with Irgacure 2959	Fastest gelation (radical-mediated photo-polymerisation)	Hynes *et al.* 2014
PEGSSDA	PEGDA with three internal disulfide bonds	Recovery of cells after 3D cell culture	Zhang *et al.* 2008

HA, hyaluronic acid; PEGDA, polyethylene glycol diacrylate; PEGSSDA, polyethylene glycol diacrylate with internal disulfide bonds.

For slow growth factor release studies, Heprasil® is the common choice (component of the HyStem-HP kit; Table 8.3). Heprasil contains both thiolated HA and small amounts of thiolated heparin, a negatively charged glycosaminoglycan which binds to and slowly releases small basic growth factors (Cai *et al.*, 2005; Peattie *et al.*, 2008). Since the hydrogel's average pore size slows the diffusion of proteins of approximately 66 kDa or greater (Cai *et al.*, 2005), any large biopolymer (e.g. laminins, fibronectin or antibodies) or cell can be incorporated into the hydrogel prior to gelation. These large biopolymers will eventually be released as the hydrogel is degraded. This strategy has been used for *in vivo* applications for inhibition of specific signal transduction pathway proteins (Clarkson *et al.*, 2011; Overman *et al.*, 2012).

Additional backbone decorations may be added that further replicate the cellular microenvironment. One common addition is a covalently linked protein or protein fragment for specific cellular attachment requirements. Indeed, covalent linkage of proteins is favoured particularly for cells that require a tethered

Table 8.3 The three HyStem kits.

Kit	Glycosil	Gelin-S	Extralink	Heprasil	Purpose
HyStem	x		x		3D cell culture of non-adherent cells
HyStem-C	x	x	x		3D cell culture of adherent cells
HyStem-HP		x	x	x	Slow growth factor release

cellular attachment substrate for appropriate sensing of their environmental compliance (Chopra *et al.*, 2012; Engler *et al.*, 2006). Covalent linkage of proteins and/or protein fragments may be accomplished by indirect or direct methods. One indirect approach is to pre-react exposed reduced cysteines on the protein fragment with one end of the PEGDA crosslinker; this functionalised protein then may be crosslinked into the Glycosil network during the standard gelation reaction (Ghosh *et al.*, 2006). As another indirect approach, a heterobifunctional polyethylene glycol may be used, which selectively reacts on one end with common functional groups on proteins and on the other end with Glycosil thiols. For example, maleimide-dPEG8-N-hydroxysuccinimide ester (M-dPEG-NHS) has been used to crosslink Glycosil thiols to fibronectin (Chopra *et al.*, 2014). The direct way to link biomolecules to thiols employs a thiol-reactive moiety present on the additive. Maleimides react rapidly with thiols via the Michael addition reaction, and may be readily placed on the N-terminus of peptides such as RGDS for efficient coupling of the sequence to the HyStem backbone to provide specific cellular attachment sequences (Zarembinski *et al.*, 2014). In theory, this approach can be used to covalently link any synthesised biopolymer (e.g. maleimide-modified RNAs or DNAs) to the hydrogel backbone.

While including additional components, biopolymers and/or decorations to the core HyStem product is often desired, HyStem can also be utilised to minimise cellular attachment (Shu *et al.*, 2003). Using only the Glycosil and Extralink components, the resulting biomaterial reduces cell interactions or adhesions that may be undesired for certain *in vitro* and *in vivo* applications. Liu *et al.* (2008) describe the use of HyStem for the reduction of peritendinous adhesions in rabbit flexor tendons after surgical repair. This application of HyStem is understood to reduce adhesions by providing a lubricating barrier that limits the ability of cells to attach, thereby reducing the fibrous entanglements that reduce range of motion after tendon repair.

Spatiotemporal control

Every user's unique hydrogel application will include specific requirements (e.g. gelation time and gel stiffness), which are crucial to the successful implementation of the resulting hydrogel. HyStem gelation time typically falls between 20 and 30 minutes, but it can be modulated to occur as fast as 30 seconds or

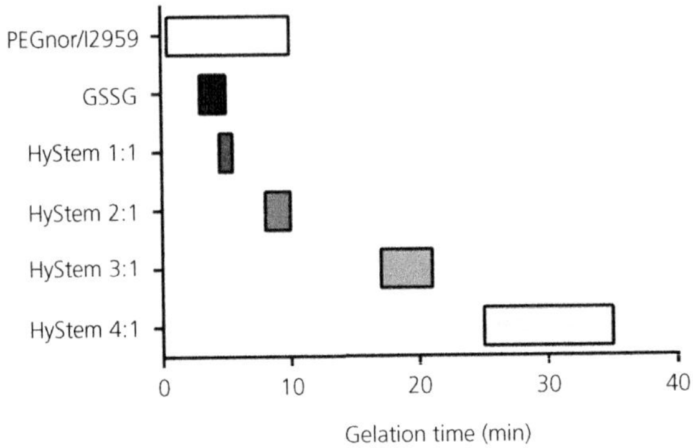

Figure 8.3 Typical HyStem gelation time ranges. Crosslinkers shown here are polyethylene glycol norbornene/Irgacure 2959 (PEGnor/I2959) and oxidised glutathione (GSSG). All ratios correspond to the ratio of Glycosil:Extralink.

as slow as 4–6 hours by changing the ratios of the components (Zheng Shu *et al.*, 2004) or the pH or through the use of alternative crosslinkers (Figure 8.3) (see Table 8.2). However, as crosslink density and PEG content are increased, the ability to recover cells after 3D culture progressively decreases. To compensate for this, a PEGSSDA crosslinker was developed which contains three internal disulfide bonds, which break upon addition of reducing agent (N-acetylcysteine or reduced glutathione) in one hour to enable cell recovery (Zhang *et al.*, 2008) (see Table 8.2).

To address diverse needs for hydrogel gelation, there are two additional chemical entities using non-Michael addition chemistry for crosslinking: oxidised glutathione (GSSG) and polyethylene glycol norbornene/Irgacure 2959 (PEGnor/I2959) (see Table 8.2). Both give rise to unique gelation properties. GSSG is not a crosslinker that is incorporated into the final matrix; instead, it catalyses disulfide formation between the thiolated molecules via the thiol-disulfide exchange reaction (Zarembinski *et al.*, 2014). The advantage of this crosslinker is that it enables substantially faster gelation times (3–5 minutes) without the concomitant increased gel stiffness normally obtained using increased Glycosil/Extralink concentration strategies (Zheng Shu *et al.*, 2004). GSSG also has a long history of safe use in the human body since it is a component of a commonly used ophthalmic irrigant, BSS Plus (Zarembinski *et al.*, 2014) as well as a naturally occurring component of the body. The PEGnor/I2959 system uses ultraviolet light to catalyse radical-based thiol-ene step growth photopolymerisation (Fairbanks *et al.*, 2009). The main advantage is rapid gelation (as little as 0.5 min), giving comparable spatiotemporal control of the PEG-based photopolymerised gels while providing a more *in vivo*-like and degradable HA backbone. Its main use is in bioprinting which requires rapid gelation after dispensing of nanoliter and picoliter quantities (Hynes *et al.*, 2014).

HyStem phases

HyStem before polymerisation – a liquid

Additional control can be achieved with different forms of the hydrogel. The initial form is as a liquid prior to gelation. This form is best for pipetting into wells for 3D cell culture (see below and Engel *et al.*, 2015), coating wells with thin layers of product (Covey *et al.*, 2012; Zarembinski *et al.*, 2014), delivery of cargo into animal models through a small gauge cannula (Compte *et al.*, 2009; Overman *et al.*, 2012; Zhong *et al.*, 2010) or used in conjunction with plasticware that enables additional functionality (e.g. hanging drop plates or bioreactors utilising dynamic perfusion). This is also the stage at which the composition and spatiotemporal control can be engineered into the final hydrogel as described previously (see Table 8.2). The stiffness of the final hydrogel can also be regulated at this point by changing the Glycosil and/or Extralink concentrations (range of shear elastic modulus: 11–3500 Pa) (Vanderhooft *et al.*, 2009).

HyStem as a water-swollen network – a hydrogel

The second form is the resulting hydrogel. HyStem hydrogels may be easily formed by aliquoting into any of several common cell culture substrates, such as 8-well chamber slides (e.g. the LabTek series), 96-well or 384-well plates, or other similarly sized containers. There are two additional shapes that have been shown to be useful: hydrogel discs for 3D cell culture and microspheres (microcarriers) for drug or cell delivery.

Free-floating hydrogel discs have multiple advantages. Cell culture media can penetrate the hydrogel from all sides, reducing diffusion effects (Fong *et al.*, 2014). Additionally, these free-floating gels eliminate potential cell interactions with a hard surface. The discs are prepared by casting into silicon moulds with cavities of desired diameter using commercially available histology slides as a backing. After polymerisation of the hydrogel-containing cells, the discs can be released from the moulds manually and subsequently cultured free floating in the well of a larger multiwell plate (Figure 8.4). See appendix to this chapter for more details in its preparation. Microspheres can be prepared by the water-in-oil techniques to prepare spheres of a variety of sizes. Smaller microspheres (5 micrometres) can be used for smart delivery of growth factors or therapeutics (Heris *et al.*, 2012) and larger microspheres (100–150 micrometres) for seamless cell culture/cell delivery (Figure 8.5).

HyStem as a film or sponge – a solid

The final forms are the solids: sponge and film. Both forms can be picked up with forceps and are generally sturdier than the hydrogel. Sponges are prepared by lyophilisation of the hydrogel. The sponge provides a massive 2D+ surface for cell culture (Liu *et al.*, 2004). Its handleability also allows its use for implantation in animal models, for either delivery of cells (Liu *et al.*, 2004) or small molecule cargo incorporated into the sponge prior to lyophilisation (Liu *et al.*, 2006).

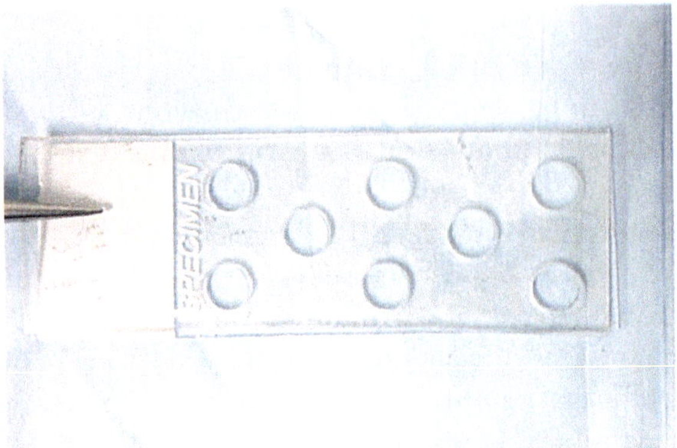

(a)

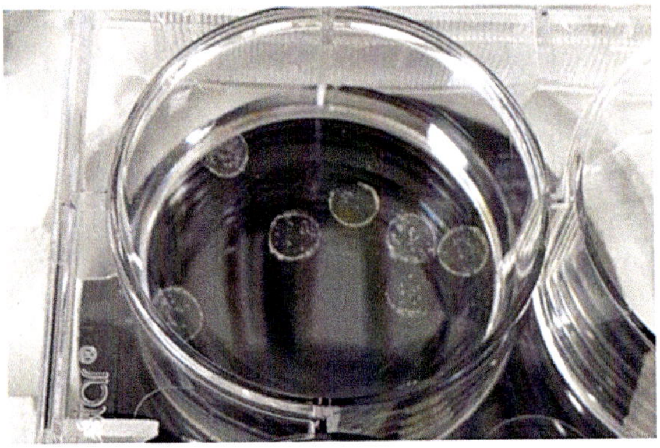

(b)

Figure 8.4 Reproducible preparation of hydrogel discs. (a) Cured silicon sheet with 6 mm circles cut out and layered on a typical histology slide to produce a hydrogel mould. (b) Free-floating hydrogel pucks in a 6-well plate.

Films are formed by dehydrating a hydrogel in a cell culture hood with overnight dry air flow (Liu *et al.*, 2005; Shu *et al.*, 2003). Films provide a form that permits a stable surface for future cell culture use (Shu *et al.*, 2003) or as implantable sheets for slow growth factor release (Peattie *et al.*, 2004).

Description of technology

The HyStem components have been bundled into three distinct kits commonly requested by basic and translational researchers. See Table 8.3 for a guide. The hydrogel is supplied as three individual vials with lyophilised components within each. Each component is then reconstituted in sterile water and finally combined. The standard HyStem-C hydrogel formulation gels in approximately 20 minutes and has a shear elastic modulus (G′) of 83 Pa (Vanderhooft *et al.*, 2009). See Figure 8.1 for schematic of gelation.

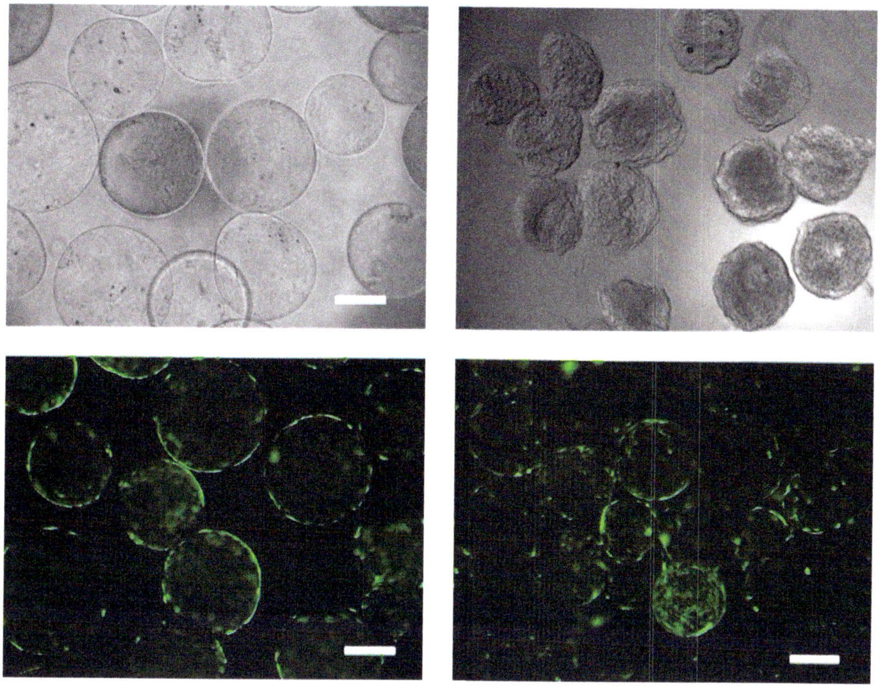

Figure 8.5 **Microcarriers formed by water-in-oil emulsion using HyStem-C hydrogel components.** Left column shows traditional hydrogel microcarriers while the right column highlights the less congruent form of microcarriers dried through lyophilisation. The bottom row demonstrates human mesenchymal stem cell attachment and viability (calcein-AM) when seeded and cultured on HyStem-C microcarriers. Scale bar = 250 μm.

Applications

An emerging trend in drug discovery is the use of medium-throughput, high-content analysis methods in conjunction with 3D cell culture as a secondary screen to assess positive hits in a high-throughput cancer drug discovery campaign (Godugu *et al.*, 2014; Hongisto *et al.*, 2013). A key challenge in using 3D cell culture as part of an integrated drug discovery pipeline is the ability to adapt existing machinery for its preparation. Tasks include:

- reproducible dispensing of the liquid hydrogel/cell mixture into multiwell plates (and in some cases reproducible preparation of multilayer constructs in each well)
- efficient microscopic techniques to find and analyse the resulting cell-based clusters formed in many different focal planes
- efficient immunostaining of the cellular mass for high-content analysis
- extraction of proteins and RNAs for subsequent Western or polymerase chain reaction (PCR) analysis.

Recently, we described adaptation of the HyStem system to existing high-throughput robotics and demonstrated feasibility of the system for high-throughput, high-content drug screening applications (Engel *et. al.* 2015). Here we provide our solutions to these challenges.

Hydrogel dispensing

There are two steps required for automated preparation of a 3D culture: mixing in the reagents of interest (e.g. cells for 3D cell culture; beads or fluorophores for tracking of different hydrogel layers during optimisation) followed by dispensing (Figure 8.6) (see below). Dispensing of an *in situ* crosslinking hydrogel presents unique challenges since dispensing must be completed prior to gelation. Depending on the application and the throughput needed, one of two approaches can be selected for automated dispensing of the HyStem hydrogel: premixing of the crosslinker with the other hydrogel components prior to dispensing or mixing of the crosslinker into the other components after dispensing the individual components (in-well mixing; see Figure 8.6). Premixing is advantageous in that it simplifies the dispensing protocol and ensures complete mixing of HA with the crosslinker. However, the number of plates that can be dispensed with one solution of HA and crosslinker is limited by the gelation time and the speed of the robotics system. The advantages of in-well mixing are that an unlimited number of plates can be dispensed without concern for gelation until the crosslinker is added. The difficulty of in-well mixing is ensuring adequate homogeneity of HA and crosslinker. If using an aspiration-based system for dispensing, adequate mixing can be achieved by repeated aspiration and dispensing within the well after addition of crosslinker. If using a unidirectional dispensing

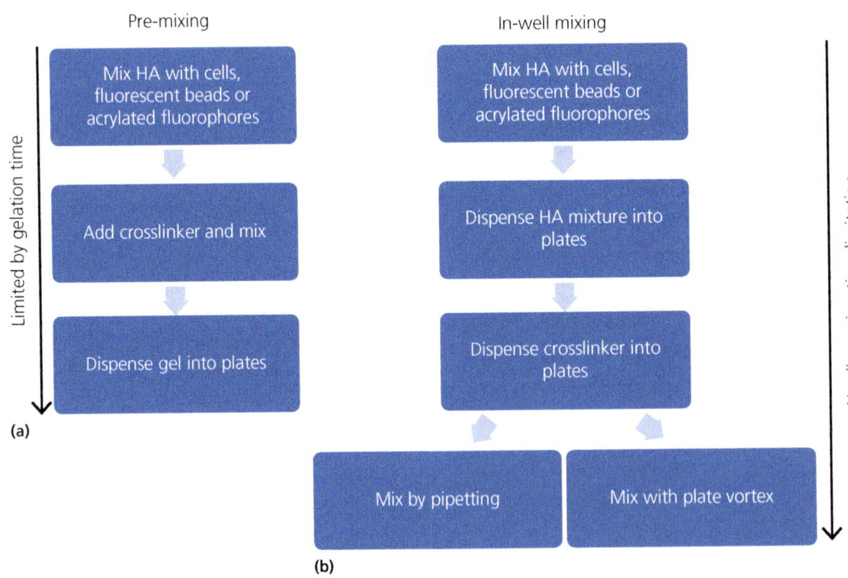

Figure 8.6 Workflow for automated dispensing of HyStem. Workflow shows one using premixing (a) or in-well mixing (b) of crosslinker.

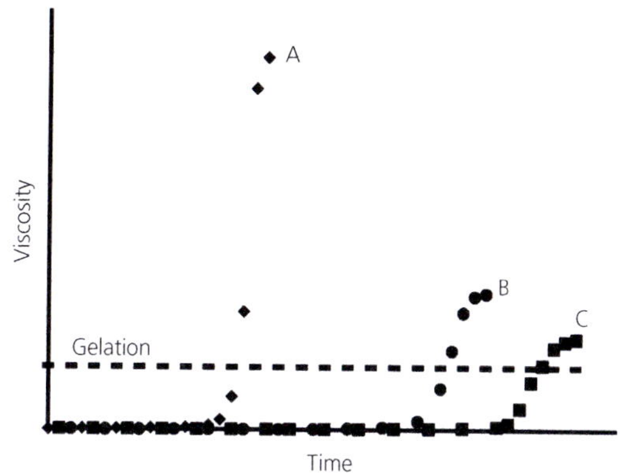

Figure 8.7 Viscometry measurements for three distinct HyStem formulations with differing gelation times. A Fast; B intermediate; C slow. Dotted line shows the viscosity level associated with visual gelation.

method (i.e. peristaltic pumps), mixing needs to be accomplished using a plate vortex mixer, which can prove difficult, especially with small-volume wells.

We prefer a premixing approach and we regularly use the benchtop Multidrop® Combi Reagent Dispenser (Thermo Scientific). The Multidrop can aliquot 0.5–50 microlitres (for the small cassette) of liquid hydrogel in an entire 96- or 384-well plate in 5–30 seconds, allowing the production of at least 20 384-well plates per hydrogel batch. This window of operation must account for sufficient time to allow for adequate washing of the tubing before gelation of the residual hydrogel occurs within the tubing. With HyStem, the viscosity of the hydrogel dramatically increases towards the end of its gelation window based on viscometry measurements (Figure 8.7) and sets the appropriate timing boundaries for plate set-up and apparatus cleaning. There are more complex methods of mixing, such as using tortuous path in-line mixing, which we have not yet evaluated. Any other systems would need to be validated to ensure adequate mixing and cleaning.

Characterisation of multilayer systems in high throughput

A common approach to making a more *in vivo*-like 3D cell culture is to incorporate more than one cell type where one cell type may provide signals to the other (Amann *et al.*, 2014; Chwalek *et al.*, 2014). Our approach has been to use sequential layers each containing a different cell type (Engel *et al.*, 2015). A key challenge in this approach is not only the accurate imaging and analysis of the individual cells but also accurately identifying the location of boundaries of the gel layers. In order to reliably image specific cell layers, the location of gel boundaries needs to be determined. This is especially important when differences in cellular morphology cannot be used reliably to distinguish cell layers.

There are two simple ways to correlate the volume of gel used with the thickness of a gel layer in any given well. Each relies on the use of fluorophores incorporated into or between the gel layers, which then can be accurately imaged.

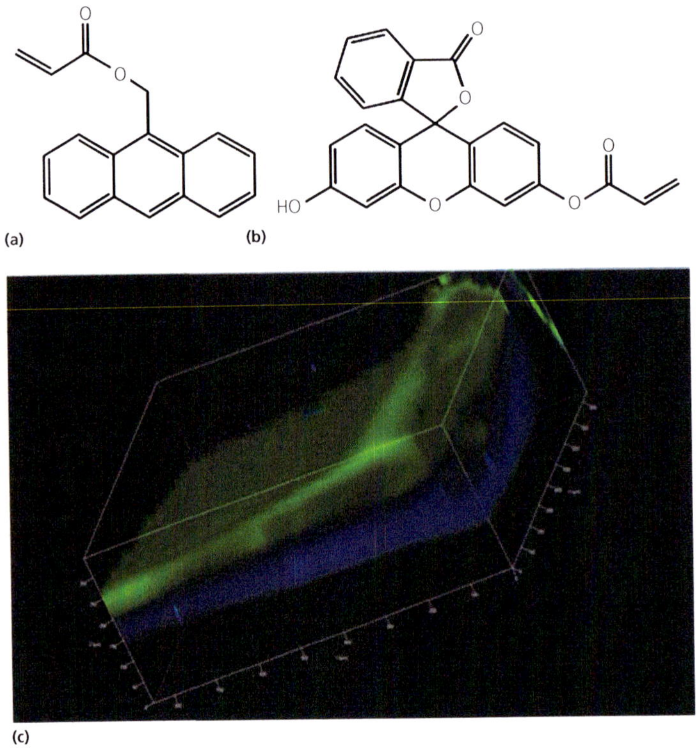

Figure 8.8 Acrylated fluorophore dyes commonly incorporated into HyStem gel layers. (a) Chemical structure of 9-anthracenylmethyl acrylate. (b) Chemical structure of fluorescein-o-acrylate. (c) 3D representation of a fluorescein-o-acrylate-containing HyStem gel layered on a 9-anthracenylmethyl acrylate-containing gel layer.

One method involves use of acrylated fluorophore dyes (e.g. 9-anthracenylmethyl acrylate, CAS 31645-34-8 (Figure 8.8a), and fluorescein-o-acrylate, CAS 193419-86-2 (Figure 8.8b)) at or below micromolar concentrations to avoid changes in gelation kinetics or stiffness. These acrylate-functional fluorophores can form covalent attachments to the thiolated-HA polymer backbone through the same Michael addition reaction chemistry used to fabricate the hydrogel. Alternating two fluorophores between sequential gel layers permits accurate imaging of all gel layers (Figure 8.8c) and measurement of intra- and interlayer distances by confocal microscopy.

By comparing the intensity of one fluorophore with the next while imaging through the gel stack, clear gel boundaries can be visualised (Figure 8.9, left); imaging depth is limited only by the working distance of the objective.

A second method to correlate the volume of gel used with the thickness of a gel layer in any given well utilises commercially available fluorophore-tagged microspheres. The optimal microsphere size is determined by the gel porosity,

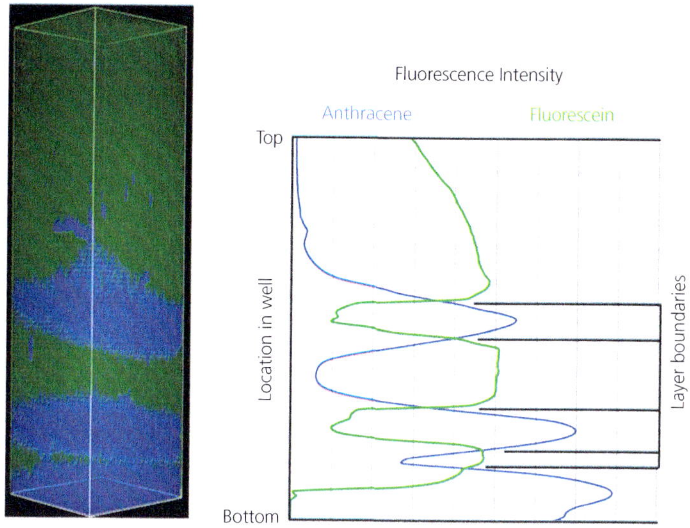

Figure 8.9 Fluorophore method. Confocal z-stack imaging of a typical acrylated fluorophore method for visualising (a) and calculating (b) gel layer thicknesses.

typical cell size and resolution limit of conventional confocal microscopes. For example, commercially available 6 μm beads are too large to pass through the hydrogel pores in a typical HyStem gel, and mimic the average size of a small mammalian cell. In general, beads between 1 and 10 μm in size are sufficiently greater than the Abbe limit to be well resolved, and still 1–2 orders of magnitude larger than the average pore size of an HA hydrogel. Placement of the fluorescent beads between gel layers provides a distinct 'bead boundary' that can be easily used to identify the location and measure the thickness of individual layers. Depending on how vigorous the dispensing method is, it may be necessary to mix the fluorescent beads with a small volume of uncrosslinked hydrogel before dispensing in order to ensure an even layer of beads. Imaging of the well with confocal microscopy will show regions of beads between layers with few beads in between, marking the boundaries of gel layers (Figure 8.10).

Instead of marking gel boundaries, another approach is to encapsulate fluorescent beads with different fluorophores uniformly in each gel layer (Figure 8.11). Encapsulated beads can be used to advantage in live imaging of cells migrating through the gel layers. The bead locations can be used as fixed reference points for measurement of cellular migration or traction force characteristics (Meseke & Forster, 2013). Finally, fluorescent beads and acrylated fluorescent dyes can be used simultaneously to provide two complementary methods for measuring gel layer thickness and determining microscopy parameters.

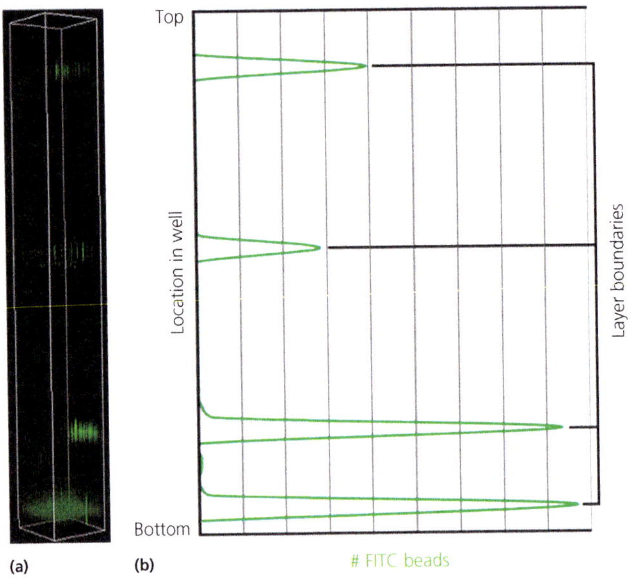

Figure 8.10 **Bead boundary method.** Confocal z-stack imaging of a typical bead boundary method for visualising (a) and calculating (b) gel layer thicknesses.

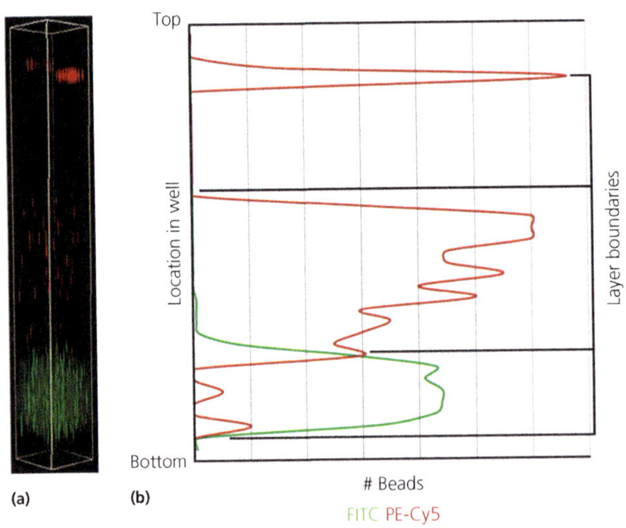

Figure 8.11 **Encapsulated beads method.** Confocal z-stack imaging of a typical encapsulated bead method for visualising (a) and calculating (b) gel layer thicknesses. The topmost PE-Cy5 beads are in a boundary layer, illustrating the utilisation of multiple imaging methods.

Automated image analysis considerations

Since a large number of images are collected for imaging multiple gel layers, automated image analysis is a requirement for analysing a multilayer gel system in high-throughput formats. The first step is to choose the appropriate software to process the images. There are a number of image analysis programs with a range of robustness and ease of use. For those with computer programming experience, MATLAB (The MathWorks, Inc.) can provide very flexible and custom methodology for image analysis. For those with limited computer programming experience, Pipeline Pilot (Accelrys, Inc.), the Broad Institute's CellProfiler (Carpenter *et al.*, 2006) and the National Institute of Health's ImageJ (Schneider *et al.*, 2012) provide more user-friendly programs for image analysis. Each program has pros and cons, but all are sufficient to perform the relatively simple batch image analysis necessary for measurement of gel thickness and analysis of drug testing data (see section on drug testing below).

There are two approaches to image analysis depending on whether an acrylated fluorescent dye or fluorescent bead approach was used (as described above). In the case of incorporation of acrylated fluorescent dyes into gel layers, images need to be analysed for the average image intensity of fluorescent signal for each corresponding dye. By plotting the fluorescent intensity of a given channel against the z-slice, we can observe clear gel boundaries (see Figure 8.9). Identifying the crossover point between gel layers will then define the gel boundaries. Measuring the number of consecutive slices between crossover points and multiplying by the distance between z-slices will provide an accurate measurement of the gel layer thickness. This can then be correlated with the z-location of the microscope to define imaging zones where one particular cell type can be observed.

In the case of fluorescent bead-based measurements, the major change in analysis is to measure the number of beads (objects) detected per image, rather than average image intensity. The number of beads per image can then be plotted against the z-slice. When using the bead boundary method, local maxima of bead counts define the gel boundaries (see Figure 8.10). If using fluorescent beads encapsulated into the gel layers, the analysis is similar to using acrylated fluorescent dyes with the exception that the threshold units are in number of beads per image (see Figure 8.11). The number of z-slices between bead count local maxima (or the number of slices with high bead counts for encapsulated beads) multiplied by the distance between z-slices will define the thickness of the gel layers and direct efforts to visualise a specific layer.

Drug testing and analysis

One outcome measure commonly used to identify promising compounds in 3D cell-based assays is live/dead staining. Typical live cell staining utilises calcein-AM, a compound that, when imported into the cell, is processed by cellular esterases to remove the acetomethoxy group (Poole *et al.*, 1993) which then allows fluorescence at 495/515 nm. This reaction requires cells to be metabolically active, so esterases can perform the modification, and intact membranes, so the fluorescent product is accumulated in the cell. Typical dead cell staining utilises ethidium homodimer-1 (EthD), which is membrane impermeant. After EthD enters a cell with compromised membranes, it binds specifically to DNA and produces fluorescence at 528/617 nm (Gaugain *et al.*, 1978). Addition of a membrane-permeable DNA dye, such as Hoechst 33342 (350/461 nm), to the live/dead assay allows for labelling of all cells present via staining of their nuclei.

A key challenge with live/dead staining with 3D culture is finding the correct focal plane for the analysis. In 3D culture, not all cells are in the same plane and many cell types form clusters, which makes it difficult to distinguish individual cells, especially true for calcein-AM staining of cell clusters. Performing automated image analysis adds another level of difficulty in detecting individual cells. This is in contrast to 2D culture: all cells are in a single plane and from a single image, a true ratio of dead to living cells (or total nuclei) can be determined by simple counting. To solve the focal plane problem, there are two approaches. Given sufficient resolution of imaging, it is possible to clearly distinguish individual nuclei stained with Hoechst 33342 or EthD. If resolution is not sufficient or calcein-AM is being used for cell clusters, measurement of total fluorescent area for each channel can be used as an approximation of individual cell counts (see appendix).

High-content analysis using immunostaining

High-content analysis relies on discerning subtle changes in cellular phenotype via immunostaining. 3D cultures in HyStem lend themselves to such analysis since they are optically transparent and can be fixed, permeabilised and stained without the need for sectioning. More complex assessment of cell phenotype, as conveyed through antibody staining, is possible for these hydrogels. As shown in Figure 8.12, we have cultured multicellular clusters of prostate cancer cell lines within HyStem hydrogels, and are able to fix, permeabilise and immunostain these specimens using slight modifications of standard methods. Because of the large size (>150 kDa) of most immunoglobulins, their diffusion rate is slowed through the hydrogel, and requires extended incubation times for immunostaining (see appendix). However, as shown in Figure 8.12, both intracellular and nuclear antigens can be routinely labelled by this method, in addition to structural markers (e.g. DAPI for nuclear staining and fluorescent phalloidin derivatives for actin filaments).

High-content analysis using RT-PCR and Western blotting

Additional assessment of cell phenotype using RT-PCR and Western blotting requires the recovery of cells from the hydrogel matrix. Use of the PEGSSDA crosslinker (see Table 8.2) allows for recovery of cells. The hydrogels can be dissociated by the addition of a reducing agent (N-acetylcysteine or reduced

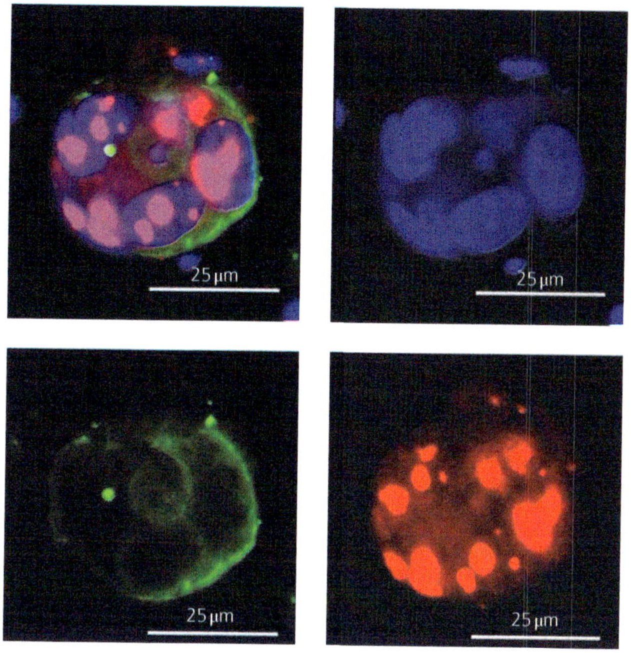

Figure 8.12 Immunostaining of prostate cancer cell clusters cultured in HyStem. DAPI (*blue*), cytokeratin (*green*) and Ki-67 (*red*) all stain well in HyStem cultured cells. Scale bar = 25 μm.

glutathione) followed by 37 °C incubation. Typically, one hour is sufficient but variation in size and stiffness of the hydrogel may require longer incubation times. Cells can then be spun down, washed and processed for RT-PCR and/or Western blot analysis. As an example of the PEGSSDA technology, endometrial, breast and prostate cancer cells were grown in HyStem. In the 3D environment where endometrial and breast cells were completely encapsulated in HyStem, levels of *MUC1* mRNA increased compared to cells that were cultured in 2D (Figure 8.13a). Cell lysates from a prostate cancer cell line grown in 3D HyStem culture were probed with β-actin antibody (Figure 8.13b).

Troubleshooting

General

Most technical issues around the basic hydrogel fall into two classes: reconstitution of the thiolated biopolymers (Glycosil and Gelin-S) and gelation time. Reconstitution of the product will become difficult after the expiry date has passed (one year from manufacturing date). Thiols are sensitive to oxidation by atmospheric oxygen. For this reason, manufacturing of the product requires purging each lyophilised vial with argon or nitrogen prior to closure, followed by

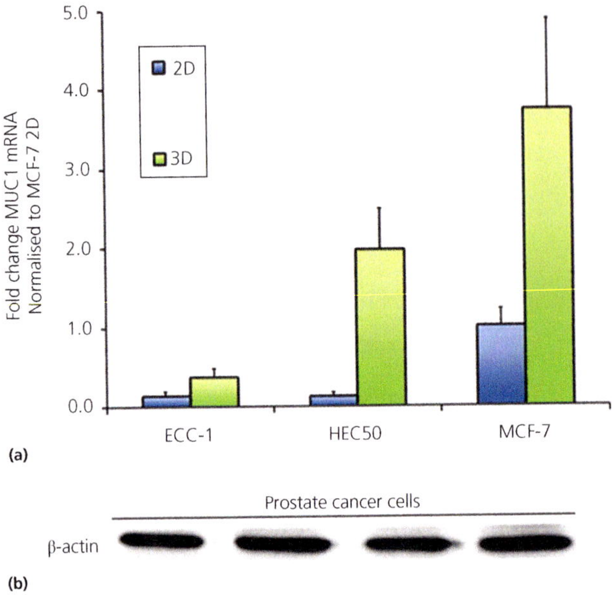

Figure 8.13 HyStem is amenable to cell recovery for RT-PCR and Western blot analysis. (a) Comparison of MUC1 mRNA levels of endometrial (ECC-1 and HEC50) or breast cancer (MCF-7) cell lines grown in 2D (*blue*) or 3D HyStem (*green*) cell culture. (b) Western blot of prostate cancer cells grown in 3D HyStem, probed with β-actin antibody.

storage at −20 °C. Nevertheless, storage at room temperature for more than a month or storage of previously reconstituted vials at −20 °C significantly shortens the product shelf-life. It is always recommended to use the entire contents of a vial after reconstitution as well as to always use the product within one year.

As mentioned previously, gelation time is exquisitely sensitive to pH (see Figure 8.2). It is for this reason that the final product is formulated with phosphate-buffered saline salts buffered at pH 7.4 prior to lyophilisation. Under certain circumstances, it is necessary to resuspend the product in a specific cell growth medium. This deviation presents several challenges: First, different media have different pH and may drastically affect gelation time. Second, the protein-rich sera added to minimal media may inhibit appropriate reconstitution of the product. It is recommended to reconstitute the individual components in the appropriate minimal medium first followed by the addition of the serum. Third, the salts in the media can contribute to an osmolality much higher than physiological conditions, which may be deleterious to the cells.

High-content screening

Difficulties that arise during the use of HyStem in high-content screening include changes in viscosity of the gel during the gelation process, variation and inaccuracy of dispensing due to tips/tubing containing residual hydrogel and meniscus effects in low-volume assays (e.g. 384-well plates). Since HA has higher viscosity than other aqueous solutions, the total volume dispensed per unit time is lower when using peristaltic pump-based dispensing systems. Careful calibration of individual systems to account for variation in dispensed volumes can compensate for differences. In the case of dispensing premixed HA-crosslinker, the viscosity will change over time. There is very little change in solution viscosity for the first 50% of a typical gelation process (see Figure 8.7) with a sharp increase near the time of gelation. As long as dispensing and clean-up are performed before this sharp increase, the effect of changing viscosity on volume dispensed is minimised.

If tubing or reusable tips are not carefully cleaned after each use with HyStem, they can be fouled with residual gel. This will result in inaccurate dispensing volumes and location. If tubing and reusable tips are compatible, residual gel can be dissolved using 0.5–1.5 N NaOH. The NaOH solution is run through the tubing/tips and incubated for 5–10 minutes. Flushing the system with water to remove dissolved gel and NaOH solution will restore normal operation.

When dispensing low-volume solutions into small wells, such as those of a 384-well plate, there is a significant meniscus effect. The meniscus will change the thickness of a gel layer depending on where within the well measurements are performed. For consistency, we recommend taking measurements and imaging in the centre of the well.

Conclusion

HyStem is a cytocompatible hydrogel matrix based on hyaluronic acid. Its thiol-based chemistry and Michael addition crosslinking chemistry allow for a wide range of user-mediated optimisations. In this chapter we have highlighted HyStem's wide range of protocols and adaptability to new workflows by developing easy-to-use protocols for assembling multicellular, multilayer 3D cell cultures in 384-well plates using equipment commonly used in drug development laboratories. Since HyStem has recently been manufactured pursuant to current Good Manufacturing Practices, we see it as being a unique scaffold to bridge *in vitro* experiments with *in vivo* and eventually human clinical investigations.

Acknowledgements

We thank Jennifer Bahr-Davidson and Tim Kang for careful review of the manuscript. We also thank Ramya Srinivas for excellent technical assistance.

Appendix

Preparation of hydrogel discs

We have cast thin slabs of commercially available silicons (e.g. Sylgard® 184, Dow Corning Corp.), and cut holes in these cured silicon sheets to create reusable moulds, which can be autoclaved repeatedly with no effect on mould size or shape. As shown in Figure 8.4, 6 mm circles were cut in a pattern with a laser cutter (X-660 automated CO_2 laser cutter, Universal Laser Systems) from a 1 mm thick cured silicon sheet. When trimmed to the size of a conventional 24 × 60 mm histology slide, the mould can be easily pressed onto the glass to form a sufficient seal that will allow the hydrogel precursors and media to be added without significant leakage. After the necessary crosslinking time, individual hydrogels can be freed from the silicone moulds by scoring around the edge of each well, and the resultant hydrogel pucks may be cultured free floating in a larger multiwell plate, as shown in Figure 8.4.

Covalent linkage of acrylated fluorophores

As a standard method, dissolve the acrylate-functional fluorophores in DMSO at 1 mg/mL and add 1 μL directly to 1 mL of standard HyStem solution (aqueous, 10 mg/mL in deionised water). Allow the dye and HyStem to react for 10 minutes prior to adding Extralink PEGDA. If multiple layers are needed, it is recommended to let the labelled hydrogel form completely, and to soak with multiple rinses of PBS prior to depositing a second layer with an alternative fluorophore. Although anthracene has maximum excitation and emission of ~335/420 nm, it can be imaged using a 405 nm laser and appropriate emission bandpass filters (e.g. for DAPI) on a standard confocal microscope. Fluorescein-functionalised hydrogels may be imaged using standard FITC settings (i.e. 488 nm laser, 518 nm maximum emission).

Immunostaining in 3D

Methods involving standard IgG antibodies are feasible for use with cells in hydrogels, particularly if the cells are well dispersed. Cell clusters may also be stained in an intact 3D hydrogel, but may require long incubation times, as listed below. Note that larger immunoglobulins (e.g. IgY) are in our experience too large to penetrate beyond the outer ~100 μm of a hydrogel, and may have excessive background staining. For all solutions listed below, use a sufficient volume to immerse the hydrogel, preferably 2–3× the gel volume whenever possible. All reagent concentrations should follow manufacturer recommendations, but users are advised to consider the volume of each gel when considering their calculations (e.g. 50 μL of antibody solution added to a 50 μL hydrogel should be prepared at a 2× concentration to account for the additional gel volume).

1. Fix hydrogel according to preferred method. For example, prepare a fresh 4% paraformaldehyde (PFA) solution in PBS, and submerge the hydrogel within this solution for 20 minutes at room temperature (RT).
2. Remove PFA solution and transfer to a labelled container for safe disposal. Wash 3 × 10 minutes with PBS.
3. Incubate in permeabilisation solution (50 μL of 0.2% Triton X-100 in PBS) for 10 minutes at RT with light shaking.

4 Remove permeabilisation solution. Wash 3 × 10 minutes with PBS.
5 Incubate in blocking solution (0.2% Triton X-100 + 3% of the desired blocking protein, such as bovine serum albumin, in PBS) for one hour at RT with light shaking.
6 Remove blocking solution. Add primary antibody, diluted in blocking solution. Incubate overnight at 4 °C, preferably with light shaking.
7 Remove primary antibody. Wash 3 × 10 minutes with PBS.
8 Add secondary antibody, diluted in blocking solution. Incubate for two hours at RT, preferably with light shaking.
9 Remove secondary antibody. Wash 3 × 10 minutes with PBS.
10 Add optional phalloidin-fluorophore counterstain. Incubate for 30 minutes.
11 Remove phalloidin solution. Wash 1 × 10 minutes with optional DAPI solution.
12 Remove DAPI solution.
13 Wash 2 × 10 minutes with PBS. Maintain gel in PBS solution during imaging.

Cellular extract preparation for Western blot analysis

1 In a microfuge tube, add 250 μL of lysis buffer per 50 μL gel.
2 Pipette until the mechanical force breaks down the gel.
   - Freeze/thaw cycles at -80 °C can help yield more protein.
   - The goal is to mechanically disrupt the gel enough to expose the cells to the lysis buffer.
3 Centrifuge the sample briefly in a microcentrifuge to separate the protein lysate from remaining gel.

Automated image analysis

Using an automated image analysis program, the total number and size of all clusters need to be measured and summed for each channel. The resultant data can then be analysed using one of two formulas. The formula for the 'Metabolic Survival Index' represents the ratio of live cell cluster area to the sum of the live cell and dead cell cluster area:

$$\textit{Metabolic Survival Index} = \frac{\textit{Total calcein} - \textit{AM channel area}}{\textit{Total calcein} - \textit{AM channel area} + \textit{total EtD channel area}} \times 100$$

The Metabolic Survival Index estimates the ratio of living cells to total cells using the calcein-AM signal and the EthD signal. This estimation assumes that the area of the signal for a single living cell is close to the signal from EthD for a single dead cell. As calcein-AM is a cytoplasmic marker and EthD stains only nuclei, this assumption is only accurate for cells with relatively limited cytoplasm. For example, cells that tend to form spheres in 3D culture would be more accurately represented by the Metabolic Survival Index than cells that have a more spread-out morphology. This index does, however, take into account cells which express esterases and can therefore be considered metabolically active.

An alternative formula, defined as the 'Nuclear Survival Index', is a ratio of the Hoechst 33342 cluster area to the sum of the Hoechst 33342 and EthD cluster areas:

$$\textit{Nuclear Survival Index} = \frac{\textit{Total Hoechst 33342 channel area}}{\textit{Total Hoechst 33342 channel area} + \textit{total EthD channel area}} \times 100$$

This index does not rely on comparisons of size of cellular cytoplasm and nuclei. Instead, this ratio directly compares the fluorescent area of all stained nuclei (Hoechst 33342) to that of nuclei within cells with compromised membranes (EthD). The differences in sizes between Hoechst 33342- and EthD-stained nuclei should be small. Therefore, error compared to a typical % living cell will be minimised. This index does not, however, take into account cells with intact membranes that are not metabolically active.

References

Amann, A., Zwierzina, M., Gamerith, G. *et al.* (2014) Development of an innovative 3D cell culture system to study tumour-stroma interactions in non-small cell lung cancer cells. *Plos One*, **9**, E92511.

Cai, S., Liu, Y., Zheng Shu, X. & Prestwich, G. D. (2005) Injectable glycosaminoglycan hydrogels for controlled release of human basic fibroblast growth factor. *Biomaterials*, **26**, 6054–6067.

Carpenter, A. E., Jones, T. R., Lamprecht, M. R. *et al.* (2006) Cellprofiler: image analysis software for identifying and quantifying cell phenotypes. *Genome Biology*, **7**, R100.

Chopra, A., Lin, V., Mccollough, A. *et al.* (2012) Reprogramming cardiomyocyte mechanosensing by crosstalk between integrins and hyaluronic acid receptors. *Journal of Biomechanics*, **45**, 824–831.

Chopra, A., Murray, M. E., Byfield, F. J. *et al.* (2014) Augmentation of integrin-mediated mechanotransduction by hyaluronic acid. *Biomaterials*, **35**, 71–82.

Chwalek, K., Tsurkan, M. V., Freudenberg, U. & Werner, C. (2014) Glycosaminoglycan-based hydrogels to modulate heterocellular communication in *in vitro* angiogenesis models. *Science Reports*, **4**, 4414.

Clarkson, A. N., Overman, J. J., Zhong, S., Mueller, R., Lynch, G. & Carmichael, S. T. (2011) Ampa receptor-induced local brain-derived neurotrophic factor signaling mediates motor recovery after stroke. *Journal of Neuroscience*, **31**, 3766–3775.

Compte, M., Cuesta, A. M., Sanchez-Martin, D. *et al.* (2009) Tumor immunotherapy using gene-modified human mesenchymal stem cells loaded into synthetic extracellular matrix scaffolds. *Stem Cells*, **27**, 753–760.

Covey, M. V., Streb, J. W., Spektor, R. & Ballas, N. (2012) Rest regulates the pool size of the different neural lineages by restricting the generation of neurons and oligodendrocytes from neural stem/progenitor cells. *Development*, **139**, 2878–2890.

Engel, B. J., Constantinou, P. E., Sablatura, L. K. *et al.* (2015) Multilayered, hyaluronic acid-based hydrogel formulations suitable for automated 3D high throughput drug screening of cancer-stromal cell cocultures. *Advanced Healthcare Materials*, **4**, 1664–1674.

Engler, A. J., Sen, S., Sweeney, H. L. & Discher, D. E. (2006) Matrix elasticity directs stem cell lineage specification. *Cell*, **126**, 677–689.

Fairbanks, B. D., Schwartz, M. P., Halevi, A. E., Nuttleman, C. R., Bowman, C. N. & Anseth, K. S. (2009). A versatile synthetic exracellular matrix mimic via thiol-norbornene photopolymerization. *Advanced Materials*, **21**, 5005–5010.

Fong, E. L., Martinez, M., Yang, J. C. *et al.* (2014) Hydrogel-based 3D model of patient-derived prostate xenograft tumors suitable for drug screening. *Molecular Pharmacology*, **11**, 2040–2050.

Gaugain, B., Barbet, J., Oberlin, R., Roques, B. P. & Le Pecq, J. B. (1978) DNA bifunctional intercalators. I. Synthesis and conformational properties of an ethidium homodimer and of an acridine ethidium heterodimer. *Biochemistry*, **17**, 5071–5078.

Ghosh, K., Ren, X. D., Shu, X. Z., Prestwich, G. D. & Clark, R. A. (2006) Fibronectin functional domains coupled to hyaluronan stimulate adult human dermal fibroblast responses critical for wound healing. *Tissue Engineering*, **12**, 601–613.

Godugu, C., Patel, A. R., Doddapaneni, R., Somagoni, J. & Singh, M. (2014) Approaches to improve the oral bioavailability and effects of novel anticancer drugs berberine and betulinic acid. *Plos One*, **9**, E89919.

Hanjaya-Putra, D., Yee, J., Ceci, D., Truitt, R., Yee, D. & Gerecht, S. (2010) Vascular endothelial growth factor and substrate mechanics regulate *in vitro* tubulogenesis of endothelial progenitor cells. *Journal of Cellular and Molecular Medicine*, **14**, 2436–2447.

Heris, H. K., Rahmat, M. & Mongeau, L. (2012) Characterization of a hierarchical network of hyaluronic acid/gelatin composite for use as a smart injectable biomaterial. *Macromolecular Bioscience*, **12**, 202–210.

Hongisto, V., Jernstrom, S., Fey, V. *et al.* (2013) High-throughput 3D screening reveals differences in drug sensitivities between culture models of Jimt1 breast cancer cells. *Plos One*, **8**, E77232.

Hynes, W. F., Doty, N. J., Zarembinski, T. I. *et al.* (2014) Micropatterning of 3d microenvironments for living biosensor applications. *Biosensors (Basel)*, **4**, 28–44.

Lee, J. Y. & Spicer, A. P. (2000) Hyaluronan: a multifunctional, megadalton, stealth molecule. *Current Opinion in Cell Biology*, **12**, 581–586.

Liu, Y., Shu, X. Z., Gray, S. D. & Prestwich, G. D. (2004) Disulfide-crosslinked hyaluronan-gelatin sponge: growth of fibrous tissue *in vivo*. *Journal of Biomedical Materials Research A*, **68**, 142–149.

Liu, Y., Zheng Shu, X. & Prestwich, G. D. (2005) Biocompatibility and stability of disulfide-crosslinked hyaluronan films. *Biomaterials*, **26**, 4737–4746.

Liu, Y., Ahmad, S., Shu, X. Z., Sanders, R. K., Kopesec, S. A. & Prestwich, G. D. (2006) Accelerated repair of cortical bone defects using a synthetic extracellular matrix to deliver human demineralized bone matrix. *Journal of Orthopaedic Research*, **24**, 1454–1462.

Liu, Y., Skardal, A., Shu, X. Z. & Prestwich, G. D. (2008) Prevention of peritendinous adhesions using a hyaluronan-derived hydrogel film following partial-thickness flexor tendon injury. *Journal of Orthopaedic Research*, **26**, 562–569.

Lutolf, M. & Hubbell, J. (2005) Synthetic biomaterials as instructive extracellular microenvironments for morphogenesis in tissue engineering. *Nature Biotechnology*, **23**, 47–55.

Meseke, M. & Forster, E. (2013) A 3d-Matrigel/microbead assay for the visualization of mechanical tractive forces at the neurite-substrate interface of cultured neurons. *Journal of Biomedical Materials Research A*, **101**, 1726–1733.

Overman, J. J., Clarkson, A. N., Wanner, I. B. *et al.* (2012) A role for ephrin-a5 in axonal sprouting, recovery, and activity-dependent plasticity after stroke. *Proceedings of the National Academy of Sciences USA*, **109**, E2230–2239.

Peattie, R. A., Nayate, A. P., Firpo, M. A., Shelby, J., Fisher, R. J. & Prestwich, G. D. (2004) Stimulation of *in vivo* angiogenesis by cytokine-loaded hyaluronic acid hydrogel implants. *Biomaterials*, **25**, 2789–2798.

Peattie, R. A., Pike, D. B., Yu, B. *et al.* (2008) Effect of gelatin on heparin regulation of cytokine release from hyaluronan-based hydrogels. *Drug Delivery*, **15**, 389–397.

Poole, C. A., Brookes, N. H. & Clover, G. M. (1993) Keratocyte networks visualised in the living cornea using vital dyes. *Journal of Cell Science*, **106**(Pt 2), 685–691.

Prestwich, G. D. (2007) Simplifying the extracellular matrix for 3-D cell culture and tissue engineering: a pragmatic approach. *Journal of Cellular Biochemistry*, **101**, 1370–1383.

Prestwich, G. D. (2008) Engineering a clinically-useful matrix for cell therapy. *Organogenesis*, **4**, 42–47.

Schmidt, J. J., Rowley, J., Kong, H. J. (2008) Hydrogels used for cell-based drug delivery. *Journal of Biomedical Materials Research A*, **87**, 1113–1122.

Schneider, C. A., Rasband, W. S. & Eliceiri, K. W. (2012) NIH Image to Imagej: 25 years of image analysis. *Nature Methods*, **9**, 671–675.

Shu, X. Z., Liu, Y., Palumbo, F. & Prestwich, G. D. (2003) Disulfide-crosslinked hyaluronan-gelatin hydrogel films: a covalent mimic of the extracellular matrix for *in vitro* cell growth. *Biomaterials*, **24**, 3825–3834.

Tibbitt, M. W. & Anseth, K. S. (2009) Hydrogels as extracellular matrix mimics for 3d cell culture. *Biotechnology and Bioengineering*, **104**, 655–663.

Vanderhooft, J. L., Alcoutlabi, M., Magda, J. J. & Prestwich, G. D. (2009) Rheological properties of cross-linked hyaluronan-gelatin hydrogels for tissue engineering. *Macromolecular Bioscience*, **9**, 20–28.

Young, J. L. & Engler, A. J. (2011) Hydrogels with time-dependent material properties enhance cardiomyocyte differentiation *in vitro*. *Biomaterials*, **32**, 1002–1009.

Zarembinski, T. I., Doty, N. J., Erickson, I. E., Srinivas, R., Wirostko, B. M. & Tew, W. P. (2014) Thiolated hyaluronan-based hydrogels crosslinked using oxidized glutathione: an injectable matrix designed for ophthalmic applications. *Acta Biomaterialia*, **10**, 94–103.

Zhang, J., Skardal, A. & Prestwich, G. D. (2008) Engineered extracellular matrices with cleavable crosslinkers for cell expansion and easy cell recovery. *Biomaterials*, **29**, 4521–4531.

Zheng Shu, X., Liu, Y., Palumbo, F. S., Luo, Y. & Prestwich, G. D. (2004) *In Situ* crosslinkable hyaluronan hydrogels for tissue engineering. *Biomaterials*, **25**, 1339–1348.

Zhong, J., Chan, A., Morad, L., Kornblum, H. I., Fan, G. & Carmichael, S. T. (2010) Hydrogel matrix to support stem cell survival after brain transplantation in stroke. *Neurorehabilitation and Neural Repair*, **24**, 636–644.

CHAPTER 9

3-D Life biomimetic hydrogels: A modular system for cell environment design

Brigitte M. Angres and Helmut Wurst
Cellendes GmbH, Reutlingen, Germany

Introduction

Biomimetic hydrogels for 3D cell culture

Among the many materials used for 3D cell culture, hydrogels are frequently favoured because their soft consistency resembles the consistency of the natural microenvironment of many cells. In addition, unlike solid carriers for 3D cell culture, hydrogels are usually highly transparent. They are easy to set up and it is possible to observe the cells during the cultivation period.

For many years, extracts of natural extracellular matrices (ECM) or purified collagen, a major constituent of the ECM of many tissues, have been used as hydrogels to mimic ECM *in vitro*. Despite their frequent use, these materials harbour some drawbacks when applied as matrices for 3D cell culture. Extracts of natural ECM, comprising ECM proteins, growth factors and other undefined components, cannot be controlled in their composition, and thus do not always represent the nature of the matrix by which a specific cell type is normally surrounded in its native tissue. When using these matrices for *in vitro* cultures, the natural cell phenotype and physiology might not be maintained. Another problem is the batch-to-batch variation of the product caused by the fact that this material is derived from naturally grown tissue and the difficulty in fully describing the types and concentrations of the many components of these hydrogels. Definition and reproducibility of a 3D cell culture material, however, are prerequisites for their use in a more regulated environment, for example in the preclinical drug development process.

When using 3D matrices mimicking the functions and properties of specific natural tissues to understand how cells react to particular signals in the extracellular environment, it is better to utilise a defined basic hydrogel matrix which, *a priori*, does not interact with cells but only provides a 3D scaffold in which cells

Technology Platforms for 3D Cell Culture: A User's Guide, First Edition. Edited by Stefan Przyborski.

are suspended. Such hydrogels are usually made of highly purified synthetic or biotechnologically produced polymers, which are modified with specific biochemical functionalities. With this approach, it is possible to build biomimetic hydrogels that support the specific phenotype of cells, and to understand how entire signalling cascades from the exterior environment cause cells to react in a specific way (Lutolf & Hubbell, 2005). Because of these possibilities, biomimetic hydrogels have become more and more attractive for the development of specific cell culture models used in basic research or drug development. However, to realise this potential, it is necessary to develop reagents and protocols to enable users who lack expertise in chemistry to utilise and translate this potential into useful experimental set-ups.

General properties of the *3-D Life* hydrogel system

The *3-D Life* hydrogel system described here addresses the needs for such defined biomimetic cell microenvironments. It consists of individual reagents that are combined by the user to form cell-inert hydrogels and can be modified to mimic various functions of the ECM. This approach results in a system with unparalleled flexibility, since many functions, such as cell adhesion, gel stiffness, cell degradability or user-controlled degradability, can be incorporated and independently varied from each other. The use of ultraviolet (UV) light is not required since *3-D Life* hydrogels are formed by a cell-compatible chemical reaction, which proceeds spontaneously upon mixing of two polymers with complementary reactive groups. Since *3-D Life* hydrogels are completely transparent, cells can be imaged by most light microscopic techniques. The low viscosity of the individual reagents before gel formation allows the use of pipetting robots and thus automation of the set-up of 3D cell cultures for higher throughput, as illustrated in Rimann *et al.* (2013). Gel formation occurs at room temperature and does not require any cooling or heating, in contrast to other hydrogels, such as Matrigel®. The hydrogels can be used in any shape and moulded in any container or injected into animals. To recover the cells grown in the gel, hydrogels can be degraded at any time without any adverse effects on cell integrity.

A cell-inert hydrogel backbone

The *3-D Life* hydrogel system consists at its core of highly hydrophilic and chemically defined polymers. These are maleimide-functionalised polyvinyl alcohol (Mal-PVA) or dextran (Mal-Dextran) and a thiol-functionalised polyethylene glycol (PEG-Link). When these polymers are combined, thiol groups spontaneously react with maleimide groups and form stable thioether bonds. The crosslinked polymers form a contiguous molecular network or, on a macroscopic level, a fully transparent hydrogel. If cells are present during this crosslinking reaction, they will be embedded inside the hydrogel.

Polyvinyl alcohol and dextran resist protein adsorption and cell attachment (Massia *et al.*, 2000; Nuttelman *et al.*, 2001) and are therefore ideal as a cell-inert backbone for the development of biomimetic hydrogels. They have been used many times for 3D cell culture (de Groot *et al.*, 2001; Draye *et al.*, 1998; Ferreira *et al.*, 2007; Lévesque & Shoichet, 2007; Liu & Chan-Park, 2010; Nuttelman *et al.*, 2001; Schmedlen *et al.*, 2002; Zajaczkowski *et al.*, 2003). Hydrogels based on PVA are stable and cannot be degraded. In contrast, hydrogels based on dextran can be degraded by the addition of the enzyme dextranase at any time during cultivation, permitting the recovery of intact cells from the hydrogel for further use.

The resistance of PEG to adsorption of proteins and cells is well known. Because of this property and its general compatibility with cells and tissues, it has been extensively used as a component of biomimetic hydrogels (DeVolder & Kong, 2012). Hydrogels made from Mal-PVA or Mal-Dextran and PEG-Link can create a cell-inert matrix with a high water content, which provides an ideal background for identifying and analysing effects caused by added bioactive compounds or for designing matrices with specific bioactive functionalities.

Matrix metalloproteinase-sensitive crosslinking

The crosslinking of Mal-PVA or Mal-Dextran with PEG-Link forms a cell-impenetrable network. Both polymers can alternatively be crosslinked with a PEG-based cell-degradable thiol polymer (CD-Link), which contains a matrix metalloproteinase (MMP)-cleavable peptide portion. Thus, MMP-producing cells can locally cleave the crosslinks and spread and migrate within the hydrogel as described in Seliktar *et al.* (2004).

Combining hydrogel components

The maleimide-functionalised polymers and thiol-functionalised crosslinkers can be mixed and matched to set up gels of defined properties. Either Mal-PVA or Mal-Dextran can be combined with PEG-Link or CD-Link to form gels with non-degradable, user-degradable and/or cell-degradable properties (Figure 9.1).

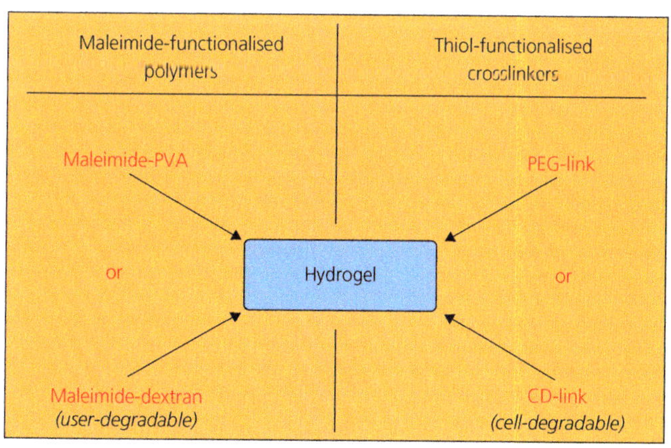

Figure 9.1 Combinatorial variations of *3-D Life* hydrogels.

Adding cell adhesion functions

Interactions between the cell surface and the extracellular matrix occur through the binding of cell surface receptors to specific amino acid sequences in extracellular proteins. Since extracellular proteins are often of high molecular weight, poorly soluble and difficult to obtain at sufficient amounts and purity, smaller portions of these proteins, in recombinant form or in the form of peptides, are often employed to analyse or induce the formation of cellular adhesions in hydrogels (Hern & Hubbell, 1998).

For example, the amino acid sequence Arg-Gly-Asp (RGD) interacts strongly with various integrins and is sufficient to induce the formation of a cell adhesion complex on the cell surface (Ruoslahti & Pierschbacher, 1987). If a peptide containing this motif is attached to a hydrogel matrix, cells respond to this signal (Woerly *et al.*, 1995). Peptides carrying a single thiol in the form of a cysteine residue can be attached to Mal-PVA or Mal-Dextran prior to crosslinking.

Applications

3-D Life hydrogels can be used in a wide variety of applications and it is impossible to list them all in this chapter. Instead, an overview of possibilities based on the gels' properties and a few selected examples illustrating these principles is provided.

In general, simple and complex 3D cell culture models can be built by embedding cells in hydrogels which mimic physical and chemical cues that support cell physiology or multicellular organisation. Cultures containing only single cell types include epithelial cells for the formation of epithelia, fibroblasts to represent a simple connective tissue culture or chondrocytes to study the formation of cartilage. Epithelia grown *in vitro* in 3D are important models for the investigation of normal and pathological tissue functions and research in regenerative medicine (Hynds & Giangreco, 2013; Mroue & Bissell, 2013). For example, in RGD-modified *3-D Life* hydrogels, renal epithelial cells become polarised and form epithelial monolayers surrounding a lumen (Figure 9.2), while the absence of the RGD motif causes them to form only aggregates of non-polarised cells. This receptor-ligand interaction is highly specific. Only a simple exchange in the order of two amino acids (RDG instead of RGD) results in a much lower efficiency at inducing cell polarisation (Figure 9.3). The example also illustrates how the hydrogel system can be used to investigate increased ligand densities and their biological effects in 3D culture. RGD is very efficient at inducing cyst formation in this 3D system. At a concentration of 5 μMol/L approximately half of the cellular aggregates form a cyst, and at a concentration of 50 μMol/L, nearly 100% of the cellular aggregates do so.

Co-cultures of multiple cell types are used to develop more complex tissue models of healthy or diseased states (see the tumour-stroma model below).

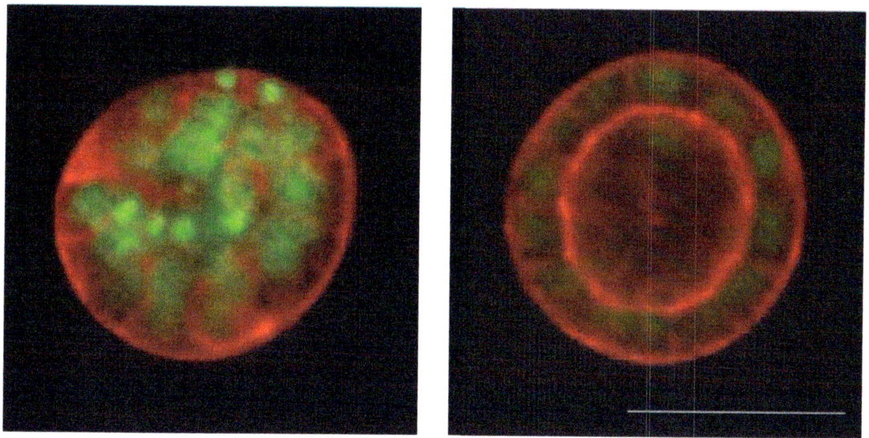

(a) (b)

Figure 9.2 Confocal laser scanning microscopy of MDCK cells in a *3-D Life* hydrogel. MDCK cells were cultured for nine days in *3-D Life* PVA-PEG hydrogel modified with 5 mMol/Lthioglycerol (a) or 5 mMol/L RGD peptide (b). Red: actin cytoskeleton; green: nuclei. Scale bar: 50 µm.

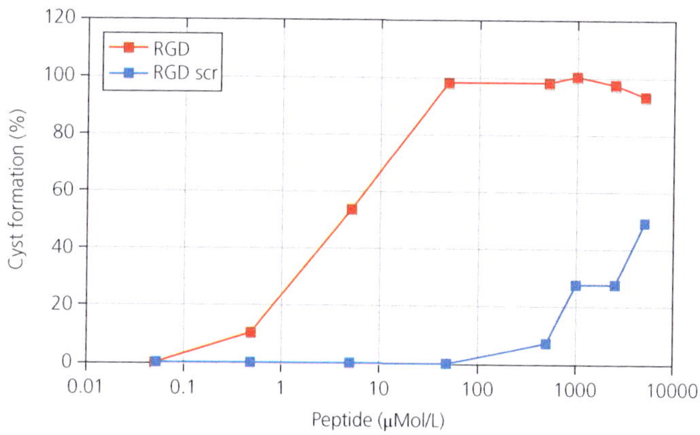

Figure 9.3 RGD peptide concentration versus cyst formation. Increasing amounts of RGD peptide or scrambled RGD peptide were covalently attached to PVA-PEG hydrogels. After nine days of cultivation, 100 aggregates in each culture were randomly selected and their phenotype (cyst or no cyst) was determined by visual inspection.

Such cultures would serve studies on migration, invasion or chemotaxis in cancer models, or angiogenesis and neuronal outgrowth, to name a few.

Besides growing cells inside gels, cells can also be grown on top of gels. Thus, tissue models consisting of a surface monolayer, such as epithelia, can be combined with an underlying stroma. The gels can also be used in analytical approaches, if bioactive modifications are omitted or selectively added. For example, peptides carrying cell adhesion motifs of different ECM proteins can be immobilised alone or in combination in the gel to analyse their specific effect on cell physiology and behaviour. Paracrine effects can be investigated by layering two gels harbouring two different cell populations on top of each other. The combinatorial use of PVA and dextran gels in this scenario provides the additional option to selectively recover and analyse cells from the dextran gel, because the degradation of the dextran gel by dextranase leaves the PVA gel intact.

Since *3-D Life* hydrogels are elastic and the stiffness can be adjusted, they can be used to investigate the influence of mechanical stimuli on the cellular phenotype. This can be performed in 3D cell cultures as well as in 2D cultures, as cells can also be cultured on top of the gels. Since the degradation of dextran gels by dextranase does not affect cells or ECM proteins, unmodified dextran gels can also be used as removable placeholders in the presence of cells and proteins. For example, dextran gels have been used to cover a centre spot of an ECM-coated surface to block the colonisation of cells subsequently seeded on the surface surrounding this spot. After removal of the gel with dextranase, migration of cells into the now cell-free space is quantified and used as a measure of the cells' migration activity (Oliver Peter, personal communication).

Description of technology

The basis of gel formation

The use of the *3-D Life* hydrogel system does not require a thorough understanding of the underlying chemistry. Essentially, it is a set of liquid reagents which, upon mixing, form a solid hydrogel. Nevertheless, a basic understanding of the underlying principles of gel formation is useful for developing novel 3D cell culture applications. Gel formation by covalent crosslinking of polymers with complementary reactive groups, such as thiol and maleimide groups, requires that one type of the polymers carries at least two and the other one at least three complementary reactive groups per polymer chain. At lower degrees of functionalisation, only linear or circular macromers, but no hydrogel networks, can be formed.

The PEG-based crosslinkers in the *3-D Life* system carry a thiol group at each end of the linear molecule. Mal-PVA and Mal-Dextran, on the other hand, carry over 20 maleimide groups along each polymer chain (Figure 9.4a). As a consequence, this high degree of functionalisation of the maleimide polymers makes the *3-D Life* system very robust at forming hydrogels. Even if a large portion of the maleimides on each polymer is used to attach peptides, which contain only a single thiol group (Figure 9.4b), the remaining maleimide groups still ensure that the polymer will be integrated in the hydrogel matrix.

PEG-Link and CD-Link

Maleimide polymers can be crosslinked with either the non-degradable PEG-Link or the cell-degradable CD-Link (see Figure 9.4b). PEG-Link is produced by modifying the two reactive terminal hydroxyls of PEG with a thiol group.

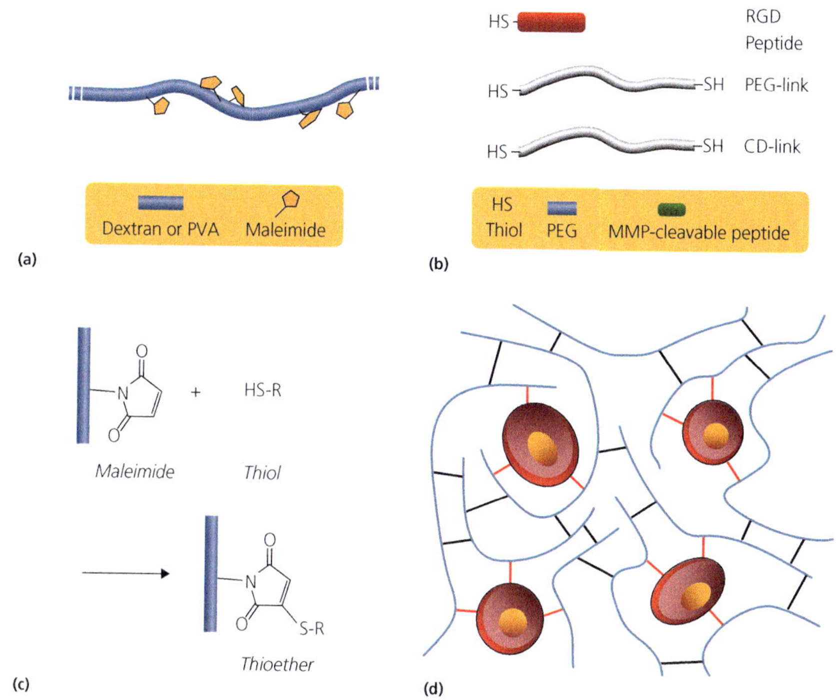

Figure 9.4 The basis of gel formation. (a) Maleimide-functionalised PVA or dextran. (b) Thiol-functionalised crosslinkers PEG-Link and CD-Link and RGD peptide. (c) Michael addition reaction for attachment of RGD peptide and polymer crosslinking (R: RGD Peptide, PEG-Link or CD-Link). (d) Schematic representation of gel structure and embedded cells.

CD-Link is a derivative of PEG-Link, in which a MMP-cleavable peptide has been positioned at one or both polymeric ends of the PEG and the terminal thiol group. This peptide contains the motif Pro-Leu-Gly-Leu-Trp-Ala (Stack & Gray, 1989), which is derived from a motif originally found at amino acid positions 773–778 in the polypeptide chain of α2(I) collagen, and has been shown to be a substrate for various MMPs (Knight *et al.*, 1992; Neumann *et al.*, 2004).

The chemical reaction

3-D Life hydrogels are formed by a Michael addition reaction (Mather *et al.*, 2006). This reaction is particularly useful for forming hydrogels in the presence of cells, because Michael donor and complementary acceptor groups are available, which allow the reaction to proceed under physiological conditions with very little interference from cellular or other cell culture components. As an addition reaction, no leaving group is produced, which might otherwise interfere with cells.

The *3-D Life* Hydrogel system employs maleimide groups as Michael acceptors and thiol groups as Michael donors (Figure 9.4c). Thiol groups are favoured for hydrogel formation, because of their biocompatibility and because the natural extracellular environment contains hardly any reactive thiol groups. The high specificity of the crosslinking reaction permits the presence of many additives during the gel-forming process. Thus, growth factors and cytokines can be added before the crosslinking reaction and thus be included in the gel. Since the thioether bonds formed by the Michael reaction and the polymer backbones themselves are stable at the conditions of cell culture (Majima *et al.*, 1995), hydrogels formed by these components are stable during long cultivation periods.

Formation of hydrogels using maleimide-functionalised polymers and thiol-functionalised crosslinkers

Crosslinking of the polymers works best if the molar amount of maleimide groups equals the amount of thiol groups. Since the molar concentrations of the reactive groups in the *3-D Life* reagents are known, it is easy to set up hydrogels of different strengths by combining the appropriate volumes of the polymer components.

If the maleimide groups are also used for covalently attaching peptides or other entities to the hydrogel matrix, the amount of the maleimide polymer should be adjusted to provide sufficient maleimide groups for both peptide modification and crosslinking.

Degradation of hydrogels and recovery of cells

Dextran hydrogels are degradable by the addition of the enzyme dextranase, which is included in *3-D Life* products containing Mal-Dextran. Due to its specificity for dextran, this enzyme does not affect cells, multicellular entities, cell surface proteins or ECM. As a macromelecule, dextranase will hardly enter the gel structure itself, but degrade the hydrogel from the outside by erosion. For a rapid dissolution of the hydrogel, it is helpful to maximise the ratio of gel surface to volume by slicing the gel in small pieces.

Since the *3-D Life* hydrogel structure does not contain amines, it is possible to fix the cells with amino-reactive fixation reagents like formaldehyde without affecting the gel chemistry. The gel can still be dissolved by the dextranase and cells can be recovered. However, the unreacted fixative should be removed before adding the dextranase to retain full enzyme activity. It is also possible to recover life cells for further cultivation or for extraction of cellular contents for downstream processes like polymerase chain reaction (PCR), Western blotting or other biochemical analyses. If the recovered cells are to be embedded in another dextran hydrogel, residual dextranase must be thoroughly removed by washing the cells 2–3 times before embedding. Otherwise, the hydrogel will be degraded rapidly by co-embedded traces of dextranase.

Covalent attachment of peptides to the hydrogel

Peptides carrying a single thiol, e.g. by inclusion of a cysteine residue during peptide synthesis, can be attached to the maleimide-functionalised polymers with a yield of over 90% (data not shown). Once the peptide is attached to the polymer and the polymer is crosslinked into the hydrogel network, the peptide is immobilised in this matrix.

Peptide concentrations of up to 5 mMol/L can be used in *3-D Life* hydrogels. Since most peptides work at much lower concentrations, the high capacity of the maleimide-functionalised polymers to bind peptides allows the attaching of multiple peptides in the same hydrogel. Multiple bioactive functionalities for mimicking the variety of ligands found in native matrix can thus be included in the gel to design suitable matrices for cell culture or to test the effect of different cell-binding motifs on cells.

Extracellular matrix proteins

Soluble ECM proteins (e.g. fibronectin, laminin, collagen) or protein fragments can be mixed into the hydrogel. For formation of a protein-loaded hydrogel, the protein is simply added to the maleimide polymer solution prior to crosslinking. Because of the high molecular weight of such proteins, their diffusion rate is quite low in the gel. Thus, they remain in the gel for an extended time where they can be bound by cells and used as matrix components.

Gel strength, viscoelastic properties and kinetics of gel formation of *3-D Life* hydrogels

The gel strength of *3-D Life* hydrogels can be adjusted over a shear storage modulus range of 20–5000 Pa by altering the concentration of the functionalised polymers and thereby altering the concentration of the resulting crosslinks (Figure 9.5). Since the polymers are offered as highly concentrated stock solutions, adjusting the gel strength is only a matter of different dilutions of the polymer stocks.

3-D Life hydrogels are highly elastic, as can be seen by rotational rheometry (Figure 9.6). After mixing the maleimide polymer and crosslinker, the shear storage modulus (G′), a measure of elasticity, increases, while the shear loss modulus (G″), which characterises viscosity, remains at the baseline.

On a macroscopic level, the consistency of the gels ranges from soft, almost water-like, to strong, similar to rubber balls. This is in the range of various native tissues such as the liver, which has a shear modulus of 900–3000 Pa (Palmeri *et al.*, 2008), breast tissue with a dynamic modulus of 140–400 Pa (Levental *et al.*, 2009) or tumour tissue with a dynamic modulus around 4000 Pa (Paszek *et al.*, 2005). Thus, the hydrogels can be used to approximate the viscoelastic properties of many native tissues.

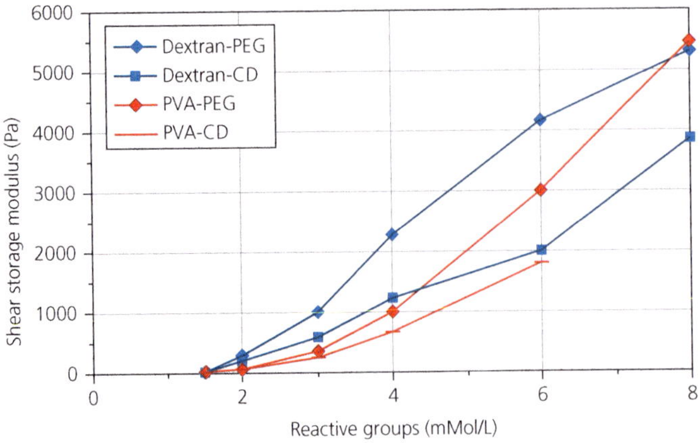

Figure 9.5 Gel strength of *3-D Life* hydrogels. Mal-PVA and Mal-Dextran were combined with PEG-Link and CD-Link at the indicated concentrations of reactive groups. After 20 minutes, the shear storage modulus of the resulting hydrogels was determined by rotational rheometry.

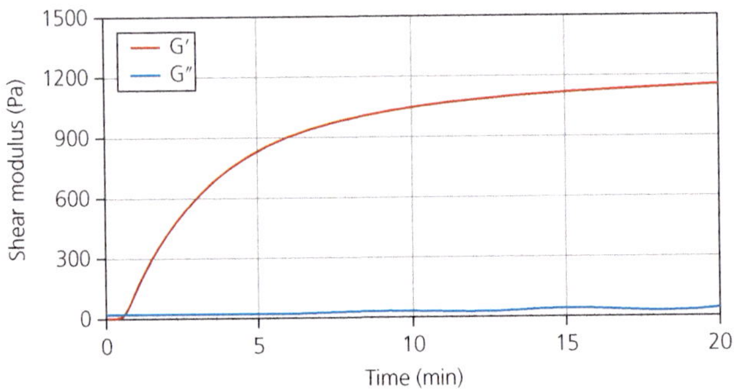

Figure 9.6 Development of shear storage modulus and shear loss modulus. Shear storage modulus (G′) and shear loss modulus (G″) were recorded during gel formation by Mal-Dextran (3 mMol/L maleimide groups) and PEG-Link (3 mMol/L SH groups). Immediately after mixing of the reagents, the solution was placed in a plate rheometer and the shear moduli G′ and G″ were recorded by rotational rheometry.

Kinetics of gel formation

In two-component gel systems, the crosslinking reaction starts as soon as the two components are combined. It is important that there is sufficient time to make a homogeneous mixture of the two solutions, and to place it in a culture dish before the gel forms. On the other hand, if the gel forms too slowly, cells may sink down to the surface of the culture dish and grow as a 2D monolayer.

The rate at which a gel forms after combining the *3-D Life* reagents depends on the rate constant of the Michael addition reaction between maleimide and thiol groups and on the concentration of those groups. The rate constant of the reaction can be controlled by the pH, since thiolate anions (R-S) are more

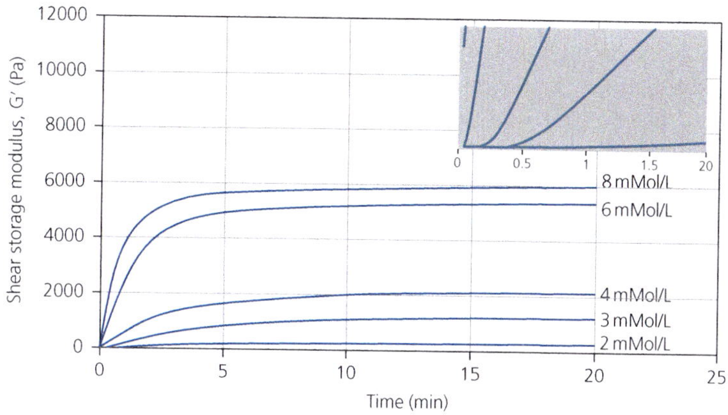

Figure 9.7 Development of gel strength. Mal-Dextran and PEG-Link were combined at the indicated concentrations of reactive groups. Immediately after mixing, the solution was placed into a plate rheometer, where the development of viscoelastic properties was recorded over time. Inset: G′ during the first two minutes.

reactive nucleophiles compared to thiols (R-SH) (Bednar, 1990). Because maleimide groups are very reactive, at neutral pH *3-D Life* hydrogels form almost immediately as soon as the two polymers are mixed. This is of great advantage, for example for automated set-up of gels (see below) or for bioprinting. In the latter application, rapid gelation facilitates the formation of 3D structures built by the continuous addition of nanodroplets of gel solution on a flat surface. However, when gels are prepared manually, rapid gelation leaves little time for homogeneous mixing of the two components. For this reason, *3-D Life* hydrogels are typically set up at pH 5.5, which reduces the reaction rate substantially, leaving enough time to generate homogeneous gels. The gels are usually solid after 2–5 minutes and can be covered by a neutralising cell culture medium. Most cell types tolerate this brief exposure to a slightly acidic pH. Changes in the concentration of the crosslinking groups also affect the rate of gel formation and need to be considered during gel set-up (Figure 9.7). The higher the concentration of the two reacting polymers, the faster the gelation proceeds.

The formation of *3-D Life* hydrogels usually occurs in three stages. Directly after combining the reactive groups, the elastic shear modulus remains at the baseline for a few seconds, indicating that the system is still fluid. In this first phase, the crosslinking reaction produces molecular aggregates of increasing size, but these aggregates are not yet crosslinked across the entire volume. In a second phase, a molecular network across the entire mixture is initiated and continuously reinforced by additional crosslinks. This phase is characterised by a linear increase of the elastic shear modulus, indicating increasing gel stiffness (see Figure 9.7). Finally, the reduced concentration of reactive groups and their reduced mobility due to constraints of the polymer network cause a third phase, where the development of gel strength slows down and eventually stops.

For some applications, it is important to have a longer time span between mixing the reactive polymers and the beginning of gelation. For example, placing the mixed solution into a specific cell culture device (e.g. microchannels) or the filling of syringes for injection may require more time. To this end, a new family of 3D reagents has recently been developed where polymer and crosslinker react more slowly and leave enough time for handling and manipulation of the mixed solution before it turns into a gel (data not shown). The slower gel formation also greatly facilitates the preparation of stiff gels.

Manual and automated set-up of hydrogels

The manual procedure for setting up biomimetic hydrogels containing embedded cells is very simple and does not require specific equipment or chemical expertise. Prior to crosslinking, all gel components are liquid. Essentially, all components except for the crosslinker are combined and mixed with a cell suspension (Figure 9.8). In a final step, this mixture is combined with the crosslinker to initialise gel formation. After a few seconds, a gel forms. After a few minutes, it is hard enough to be covered with culture medium and cultivation of the embedded cells can proceed.

All components have a low viscosity and can be handled at room temperature. Therefore, the entire gel-forming procedure is well suited for automation and applicable for 3D cell-based drug screening. Automated dispensation of liquid hydrogel reagents at room temperature is of great advantage compared to ECM-derived hydrogels like Matrigel, which has to be cooled to remain in a liquid state. Moreover, rapid gelation after the maleimide polymer is mixed with

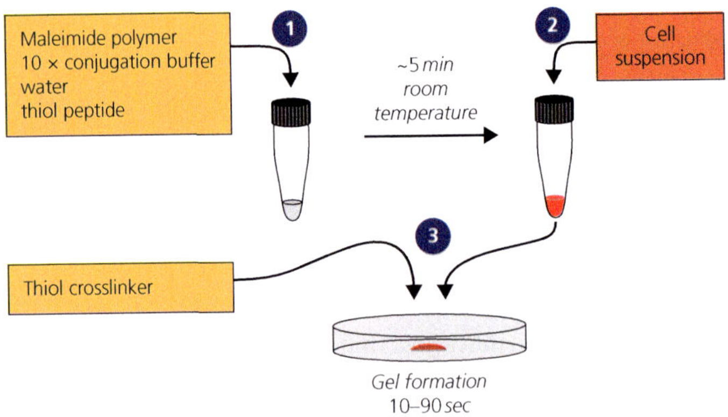

Figure 9.8 Schematic drawing of the steps required to embed cells in a *3-D Life* hydrogel. 1. Polymer, buffer and water are mixed. If desired, a thiol-functionalised peptide is added. The mixture is briefly incubated to covalently attach the peptide to the maleimide polymer. 2. A cell suspension is added to the peptide-polymer conjugate solution. 3. Cells and the peptide-conjugated polymer are mixed with crosslinker for gel formation and placed on a culture dish. After the gel is stable, culture medium is added and the culture dish is placed in an incubator for cell cultivation (not shown).

the crosslinker keeps cells from sinking to the bottom of the well. After concluding the cultivation period, gels can be dissolved by the robotic addition of dextranase to allow postculture analyses.

The automated set-up of gels for a tumour cell culture, drug treatment and biochemical analyses of these cultures is reported by Rimann *et al.* (2013). However, a caveat often cited for drug screening models using hydrogels compared to 2D cultures is the possibility that compounds adsorb to the hydrogel material, which may give a false drug-response relationship. We argue, though, that this is reduced in synthetic and chemically defined gels such as the *3-D Life* system compared to hydrogels made from natural basement matrix extracts. Since both structural polymers of the employed hydrogel are well known for their resistance to reagent binding (dextran is used as a stationary phase in size exclusion (e.g. Sephadex®) chromatography), it is likely that few compounds will interact with the hydrogel matrix. Nonetheless, this needs to be addressed in future experiments.

Hydrogel mesh size, molecular diffusion, cell labelling and growth factors

By calculation, the average mesh size of *3-D Life* hydrogels is approximately 5–10 nm. However, due to the random distribution of reactive groups along the polymers and imperfections during the gel-forming process, larger mesh sizes may also be present. Still, the mesh size is very small and prevents cells from migrating within the hydrogel. The mesh size is large enough, though, to allow for efficient diffusion of low molecular weight reagents. Chemical fixation of cells, labelling of live or fixed cells with small molecules (for example, fluorescently labelled phalloidin, nucleic acid stains, viability and cytotoxicity assay reagents, Ca^{2+} indicators, or proliferation assay reagents based on small molecule compounds) can be performed as in 2D cell culture. For immunostaining, however, gels should be made quite thin (approximately 200–400 μm thick) and the antibody treatment should be extended to overnight incubations to make sure that the antibodies have sufficient time for diffusing in and out of the gel. As an alternative, cells or multicellular entities which developed during the cultivation can be isolated from the gel by the addition of dextranase and subsequently immunostained in suspension.

Growth factors added to the medium will diffuse into the hydrogel more or less rapidly, depending on their molecular size. This is perfect for establishing gradients of chemoattractants or repellents across a hydrogel, and to observe directed cell migration in response to such gradients.

On the other hand, culture supernatants are often analysed for the presence of cell-secreted growth factors, cytokines and chemokines. Compared to 2D cultures, the slower diffusion of growth factors of higher molecular weight in hydrogels needs to be considered. Thus, the detection of growth factors in the culture medium during an experiment may be delayed compared to the real cell secretion time point. If the measurement is an endpoint analysis, the factors can be analysed after dissolution of the gel to circumvent this problem.

A tumour-stroma model: 3D co-culture of MCF-7 cells and primary fibroblasts

The development of cancer is accompanied by crosstalk between the cancer cells and their stromal microenvironment (Pietras & Ostman, 2010). *In vitro* 3D model systems that reflect cancerous tissue are in demand to study the inherent biology and to find effective drugs that target the interactions between tumour cells and stroma (Casbas-Hernandez *et al.*, 2011). The *3-D Life* hydrogel technology was used to build a tumour-stroma model in which a tumour cell line was co-cultured with fibroblasts, which are among the most important stromal cells forming the tumour stroma (Räsänen & Vaheri, 2010). Fibroblasts become reprogrammed into cancer-associated fibroblasts (CAFs) during tumour development and promote tumour progression by modulating composition and function of the extracellular matrix and the production of different growth factors and cytokines (Augsten, 2014).

The mammary tumour cell line MCF-7 was used for the observation of tumour progression in the presence of fibroblasts. MCF-7 cells have a luminal epithelial-like, non-aggressive phenotype (Kenny *et al.*, 2007). With the addition of growth factors, MCF-7 cells can be induced to go through an epithelial-mesenchymal transition (EMT) and acquire a more invasive phenotype (Vergara *et al.*, 2011; Walsh & Damjanovski, 2011). The co-culture model was developed to investigate how cancer cells change the fibroblast phenotype into the CAF phenotype and how CAFs will in turn promote tumour growth and tumour progression like the transition from a non-invasive into an invasive phenotype. Changes in proliferation, cell morphology, cell motility, matrix composition, production of growth factors and cytokines as well as changes in gene expression can be analysed in this system in order to shed more light on the tumour-fibroblast interactions.

Set-up of MCF-7 and fibroblast mono- and co-cultures

For the generation of 3D cell cultures, MCF-7 cancer cells and primary skin fibroblasts are embedded in hydrogels formed by crosslinking of Mal-Dextran with CD-Link at a crosslinking concentration of 3 mMol/L maleimide and SH groups. The stiffness of this gel is similar to breast tissue in a healthy state compared to stiffer transformed breast tissue (Levental *et al.*, 2009). Prior to crosslinking, Mal-Dextran is modified with RGD peptide at a final concentration of 0.5 mMol/L.

A thin PVA-PEG gel layer is placed at the bottom of the chamber (Figure 9.9). Because this gel contains the uncleavable linker PEG-Link and no cell adhesion

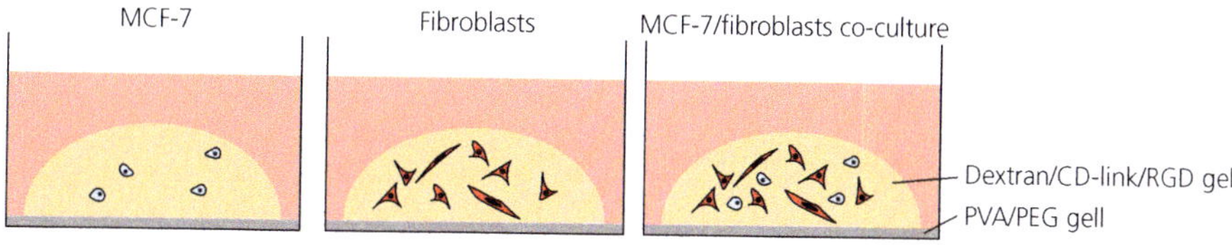

Figure 9.9 Schematic drawing of mono- and co-culture of MCF-7 cells and fibroblasts. Eight-well chamber slides with cover glass bottoms are coated with a thin layer of PVA-PEG gel (*grey*) to prevent adhesion of cells to the bottom of the well during cultivation. A RGD peptide-modified dextran gel containing cells and crosslinked with CD-Link is casted on top of this layer. After gelation occurs, gels are covered with growth medium (*red*). MCF-7 and primary human dermal fibroblasts are either grown alone or in co-culture. Cell nuclei are shown in black. Objects are not drawn to scale.

sites, it does not promote cells to enter the gel and thus acts as a barrier preventing cell attachment to the glass bottom and the formation of a 2D cell layer during long-term culture.

Reagents and materials

3-D Life hydrogel products

- *3-D Life* PVA-PEG hydrogel kit (Cellendes, Cat. No. 09-G-001): Mal-PVA, PEG-Link, 10x $CB_{pH\ 5.5}$, 10x $CB_{pH\ 7.2}$, thioglycerol, water
- *3-D Life* Dextran-CD hydrogel kit (Cellendes, Cat. No. G91-1): Mal-Dextran, CD-Link, 10x $CB_{pH\ 5.5}$, 10x $CB_{pH\ 7.2}$, thioglycerol, water, dextranase
- *3-D Life* RGD peptide, 1 µmol, lyophilised (Cellendes, Cat. No. 09-P-001)

Reagents and materials not included in the *3-D Life* products

- Nunc™ Lab-Tek™ chamber coverglass, eight chamber (Thermo Scientific, Cat. No. 155411)
- DMEM growth medium: Dulbecco's modified eagle medium high glucose (DMEM) w/o L-glutamine (PAA, Cat. No. E15-009) supplemented with 10% fetal bovine serum (PAA, Cat. No. A15-101), 4 mM L-glutamine (PAA, Cat. No. M11-004), 100 U/mL penicillin and 100 µg/mL streptomycin (PAA, Cat. No. P11-010)
- DMEM/Ham's F12 growth medium: DMEM/Ham's F12 (1:1) medium (PAA, Cat. No. E15-012) supplemented with 10% fetal bovine serum (PAA, Cat. No. A15-101), 2.5 mM L-glutamine (PAA, Cat. No. M11-004), 100 U/mL penicillin and 100 µg/mL streptomycin (PAA, Cat. No. P11-010)
- Co-culture medium: a 1:1 mixture of DMEM growth medium and DMEM/Ham's F12 growth medium
- PBS w/o Ca/Mg (PBS^-) (PAA, Cat. No. H15-002)
- Solution of trypsin (0.5 mg/mL) and ethylenediamine tetra-acetate (EDTA) 0.22 mg/mL in PBS^- (PAA, Cat. No. L11-004)
- One 10 cm culture plate of MCF-7 human breast cancer cells (Cell Line Services, Cat. No. 300273) grown in DMEM/Ham's F12 growth medium to 70% confluency

- Two 10 cm culture plates of primary human dermal fibroblasts (BioCat, Cat. No. aI001-f-DVB) grown in DMEM growth medium to 100% confluency
- Hemocytometer, 15 mL centrifuge tubes, 1.5 mL reaction tubes, pipette tips, micropipettor, serological pipettes, bucket of ice

Methods: 3D cell culture

The following protocol is for the preparation of a total of six 30 µL hydrogels, two gels for each monoculture and two gels for the co-culture.

Preparation of gel components

- Dissolve *3-D Life* Mal-Dextran, CD-Link and RGD peptide lyophilisates in water according to the instructions in the accompanying data sheets. Place Mal-Dextran on ice.
- Thaw all other hydrogel reagents (Mal-PVA, PEG-Link, 10x $CB_{pH\ 5.5}$) and water at room temperature. Place Mal-PVA on ice.
- Make sure that salts in the 10x $CB_{pH\ 5.5}$ are completely dissolved. Keep 10x $CB_{pH\ 5.5}$ at room temperature during use.

Note: Do not expose *3-D Life* thiol-containing reagents (RGD peptide, CD-Link, PEG-Link) to air longer than necessary to avoid oxidation of the thiol groups.

Coating of eight-chamber slides with a PVA-PEG hydrogel layer

- Generation of reagent mix: in a 1.5 mL reaction tube, mix 23.1 µL Mal-PVA, 256.9 µL water and 35 µL 10x $CB_{pH5.5}$.
- Place 5 µL PEG-Link in the first chamber.
- Add 45 µL of the reagent mix to the PEG-Link and pipette up and down three times to mix PEG-Link with reagent mix.
- Immediately spread the solution across the whole bottom of the chamber using the micropipette tip.
- Treat the other five chambers the same way.
- Cover the eight-chamber slide with its lid and let the chamber slide sit for at least 30 minutes in the sterile hood. The gel coating is stable for several hours.

Preparation of cell suspensions of MCF-7 cells and fibroblasts

- Prepare cell suspensions of MCF-7 cells at a density of 6×10^5 cells/mL and of fibroblasts at a density of 2×10^6 cells/mL in PBS^- using standard methods.
- Place cell suspensions on ice.

Preparation of cell-laden hydrogels

Except for CD-Link, which is placed separately in each chamber, the volumes of reagent mixes required for gel preparations of mono- and co-cultures are listed in Table 9.1.

Note: Reagent volumes are calculated for a total of three gels instead of only two, to avoid a potential shortage of reagent mix if slight inaccuracies in pipetting occur.

Table 9.1 Reagent mixes for MCF-7 and fibroblast mono- and co-cultures.

Reagent	Final concentration in the gel	Fibroblast mono-culture (μL)	MCF-7 monoculture (μL)	Co-culture (μL)
10x CB pH5.5	n.a.	6.0	6.0	6.0
Water	n.a.	27.6	27.6	27.6
Mal-Dextran (30 mMol/L maleimide groups)	3.5 mMol/L maleimide groups	10.5	10.5	10.5
RGD peptide (20 mMol/L SH groups)	0.5 mMol/L SH groups	2.4	2.4	2.4
MCF-7 cell suspension (6×10^5 cells/mL)	1×10^5 cells/mL	–	15.0	15.0
Fibroblast cell suspension (2×10^6 cells/mL)	3.3×10^5 cells/mL	15.0	–	15.0
Phosphate buffered saline		15.0	15.0	–

Prepare gels as follows.

1 Place three 1.5 mL reaction tubes in a rack and mark one reaction tube for the preparation of the fibroblast monoculture, one tube for the MCF-7 monoculture and one tube for the co-culture.

2 In these Eppendorf tubes, combine 10x $CB_{pH\ 5.5}$, water, Mal-Dextran and RGD peptide according to the volumes indicated in Table 9.1 for each type of culture. After the addition of RGD peptide, immediately mix well to ensure homogeneous modification of the Mal-Dextran with the peptide.

3 Incubate for 5–10 minutes at room temperature to allow time for the RGD peptide to attach to the maleimide groups of the dextran polymer.

4 In the meantime, mark the lid of the eight-chamber slide according to the respective cell culture to be placed in the wells underneath the lid.

5 Place 4.5 μL of crosslinker (CD-Link) in the middle of the first two wells designated for the fibroblast cell culture. Be careful not to damage the PVA-PEG gel layer at the bottom of the chamber.

6 Bring fibroblast cells back into homogeneous suspension by flicking the tube or by pipetting cells up and down. Add 15 μL of the fibroblast cell suspension and 15 μL PBS^- to the reaction tube containing the reagent mix designated for fibroblast gel culture and mix.

7 In a pipette tip, transfer 25.5 μL of this mixture to the 4.5 μL CD-Link in the first chamber and quickly mix both by pipetting up and down three times. Avoid the formation of air bubbles. Gel formation starts after a few seconds of mixing. Complete mixing step as fast as possible to avoid gel formation in the pipette tip.

8 Repeat step 7 with the second chamber for fibroblast monoculture.

9 Repeat steps 5–8 for the MCF-7 monoculture and the co-culture using the volumes indicated in Table 9.1.

10 Wait for five minutes after the last pipetting to let the gels form.
11 Optional: test gel formation by careful inspection of the gels with a pipette tip.
12 Once the gels have formed, add co-culture medium to each chamber until the gel is covered with medium (approximately 200 µL).
13 Add the lid to the eight-chamber slide and incubate gels at 37 °C and 5% CO_2.
14 Renew the co-culture medium after one hour.
15 Culture for 14 days and replace co-culture medium every 2–3 days.

After a culture period of 14 days, the 3D cultures are either treated with live-dead cell staining reagents (Live/Dead® Viability/Cytotoxicity Kit, Life Technologies, Cat. No. L-3224) or chemically fixed with 4% paraformaldehyde and subsequently labelled with phalloidin-TRITC (e.g. Sigma, Cat. No. 1951) for actin fibres and Syto® 24 Green (Life Technologies, Cat. No. S7559) for nuclei. Both labelling procedures are carried out as commonly done with 2D cell cultures. Chemically fixed cultures can be preserved up to several weeks when immersed in PBS^- and stored at 4 °C. After storing for more than one day, the staining may become diffuse due to dislocation of staining reagents from their target molecules. When stored for more than one day, seal the chamber slide with parafilm to avoid evaporation.

Results

MCF-7 and fibroblast monocultures versus co-culture in *3-D Life* hydrogels

Within 14 days of culture, MCF-7 cells divide and form cell clusters which reach an average size of up to approximately 200 µm in diameter. When compared with the phenotypic classification of tumour spheroids described by Kenny *et al.* (2007), the predominant phenotype is a round or mass phenotype (Figures 9.10a,b and 9.11a,a′), with the exception of a few acini- (see Figure 9.11a′), and grape-like phenotypes (Figure 9.10c,c′). The main phenotypes of MCF-7 spheroids in *3-D Life* hydrogels are very similar to the ones observed in Matrigel and corroborate the description of MCF-7 as a luminal, non-aggressive cancer cell type (Kenny *et al.*, 2007).

Live-dead cell staining of these cultures revealed that independent of the spheroid phenotype, most of the cells are alive, including in the inner core of the cell clusters (see Figure 9.10a′–c′). Dead cells are randomly dispersed and there is no indication of a specific area within the cluster that contains dead cells.

When co-cultured with fibroblasts, MCF-7 cluster morphology does not change much compared to the phenotype when cultured alone (see Figure 9.11c,c′). There is some increase in size of the spheroids (data not shown), which may indicate an increased proliferation in the presence of fibroblasts. We did not observe much tumour cell migration away from the cell mass that would indicate an invasive behaviour in the presence of fibroblasts.

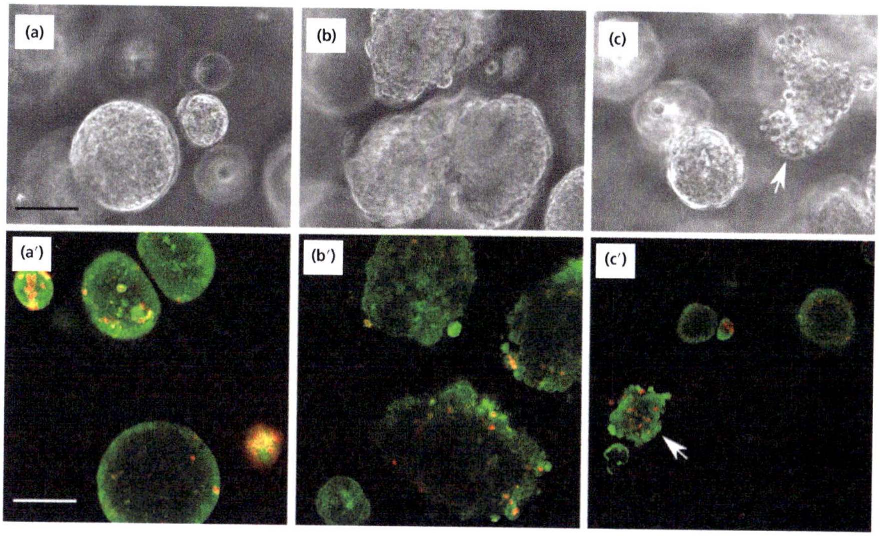

Figure 9.10 Growth phenotypes and survival of MCF-7 cells after cultivation for 14 days. MCF-7 cells are grown in dextran gel crosslinked with CD-Link and modified with RGD peptide. Pictures were taken by phase contrast microscopy (a–c) or by epifluorescence microscopy (a′–c′). MCF-7 cells grow in spheroids (a,a′), in a mass (b,b′) and in a grape-like (c,c′; *arrows*) phenotype. Independent of the growth phenotype, most of the cells are alive (green nuclei) and only a few dead cells (red nuclei) are randomly distributed throughout the cell clusters. Size bar, 100 μm.

Primary human skin fibroblasts grown in monoculture display a well-spread phenotype with randomly oriented protrusions and few cell-cell contacts (see Figure 9.11b,b′,e). However, fibroblasts change their morphological phenotype drastically when co-cultured with MCF-7 cells. They display a more slender appearance with fewer protrusions and aggregate in a parallel alignment to form dense cell clusters (see Figure 9.11c,c′,d). These clusters are distributed throughout the gel, independent from their distance to MCF-7 cells.

Discussion

Investigating tumour-stroma interactions in *3-D Life* hydrogels

The results show that *3-D Life* hydrogels provide the essential functions of a three-dimensional matrix for the cultivation of cancer cells and fibroblasts. MCF-7 cells form cell aggregates of the round or mass phenotype similar to the phenotype observed in basement membrane extracts (BME) (Kenny *et al.*, 2007) or collagen gels (Holliday *et al.*, 2009). Although BME is a 3D matrix often used for the culture of cancer cells, it does not seem to be a suitable 3D matrix for fibroblasts (Hakkinen *et al.*, 2011). Most of the cells do not spread and stay round. In contrast, fibroblasts in *3-D Life* hydrogels display a morphological appearance that is very similar to the one observed in collagen gels, often used for tumour cell fibroblast co-cultures.

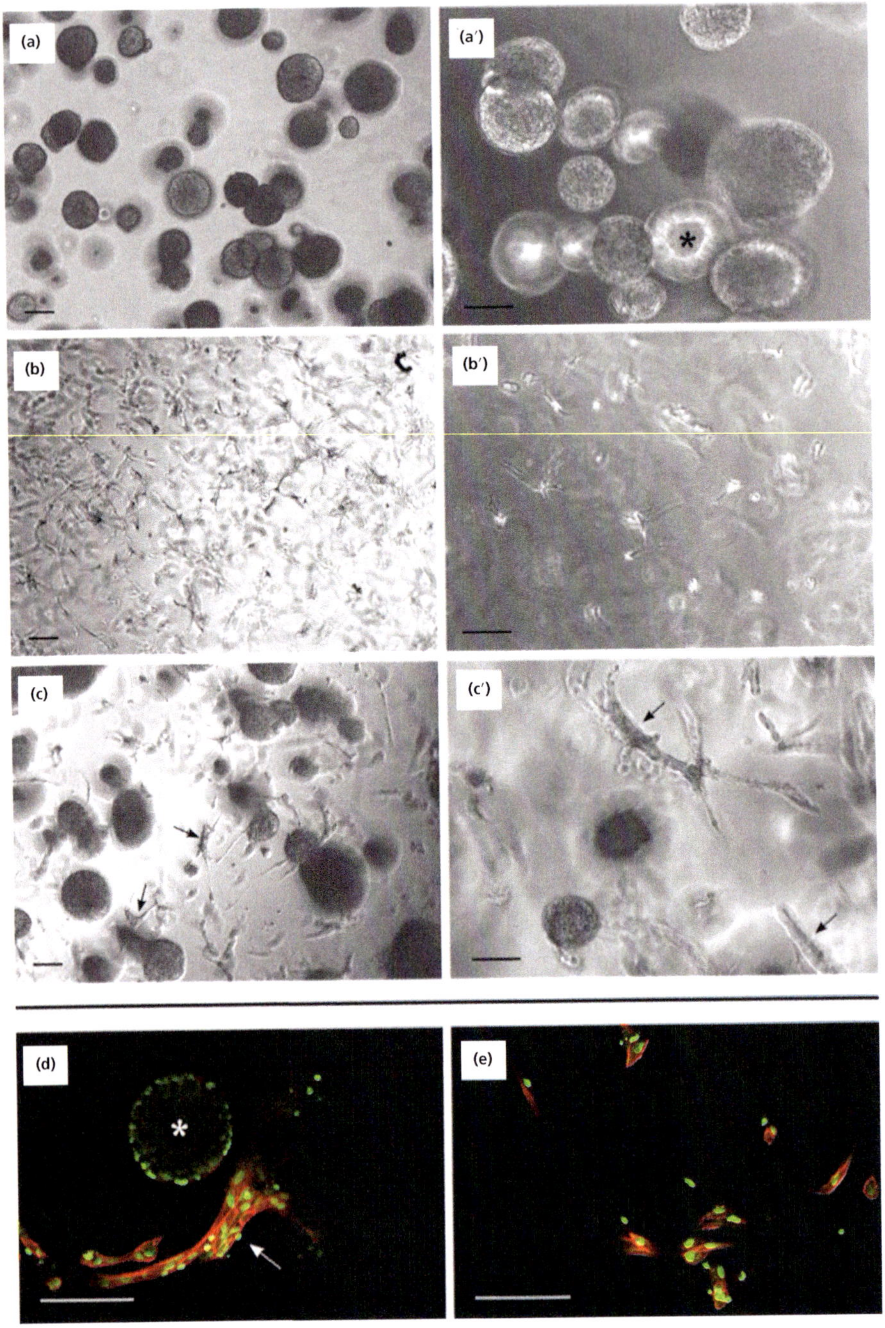

Figure 9.11 Mono- and co-culture of MCF-7 and fibroblasts. MCF-7 cells and primary human dermal fibroblasts are either grown alone (a,a′ and b,b′, respectively) or in co-culture (c,c′) in dextran gels crosslinked with CD-Link and modified with RGD peptide. After 14 days in monoculture, MCF-7 cells form spheroids of varying sizes (a,a′). Some spheroids show a lumen inside the spheroid similar to an acinus (asterisk in a′). Primary human dermal fibroblasts grown in monoculture grow mostly as single cells or in loose clusters in a well-spread phenotype (b,b′). When MCF-7 cells and fibroblasts are grown together, fibroblasts form long-shaped aggregates with tightly packed cells (arrows in c and c′). A co-culture (d) and a fibroblast monoculture (e) fixed and stained for filamentous actin (*red*) and nuclei (*green*). Again, in the presence of MCF-7 spheroids (asterisk in d), fibroblasts aggregate into long-shaped clusters (arrow in d) compared to the random distribution of cells in monoculture (e). (a–c′) Phase contrast microscopy at lower (a,b,c) and higher (a′,b′,c′) magnification. (d,e) Epifluorescence microscopy. Size bars, 100 µm. Photographs a′, b′ and c′ originally published in Charwat *et al.* 2015. Source: Charwat, V. *et al.* 2015. Reproduced with permission of Elsevier.

The observed changes in fibroblast morphology and aggregation when co-cultured with MCF-7 cells suggest that the tumour cells affect the fibroblasts within this hydrogel matrix. In tumour tissue, one result of such an interaction is an increased production of collagen by fibroblasts and the reorganisation of the extracellular matrix into a stiffer scaffold when compared to normal tissue (Otranto *et al.*, 2012). A quantitative analysis of ECM proteins, especially collagen, may be quite difficult to perform in collagen gels due to the large amount of this protein already contained in the 3D scaffold (Holton *et al.*, 2011). In contrast, the *3-D Life* dextran gel constitutes a protein-free scaffold where every matrix protein deposited in the gel is produced by the cells and thus can be identified and quantified as such with various non-invasive label-free detection methods. An increase of collagen around fibroblasts in co-culture when compared with monocultures was shown by Raman spectroscopy with a BioRam® instrument (Cell Tool GmbH) (Charwat *et al.*, 2015). Other non-invasive methods, such as multiphoton laser scanning microscopy (MPLSM), can also be used for label-free imaging of collagen fibres (Provenzano *et al.*, 2006) and might be an additional option for analysing cultures in dextran gels.

In vivo, increased collagen crosslinking stiffens the ECM and is believed to promote tumour progression due to mechanotransductional mechanisms (Levental *et al.*, 2009). One approach to test the influence of ECM stiffness on tumour progression would be to increase the stiffness of the gel itself by using increased concentrations of polymer. This allows an analysis of the mechanical effect on tumour progression independent from biochemical cues in the hydrogel. It has to be considered, though, that the architecture of a crosslinked polymer is different from a collagen network. The *3-D Life* hydrogel constitutes a tight polymer network that is not penetrable by cells unless they produce the appropriate MMPs to cleave the crosslinker. In collagen gels, cell-mediated formation of collagen fibres provides a stiff 3D context for cells while the whole matrix can still be wide-meshed and let cells migrate without encountering much of a physical barrier.

Modelling stages of tumour progression

Tumour cell spheroids generated in hanging drops are often used in cancer research. The technology has the advantage of a very reproducible production of the same composition and size of a cancer microtissue. Although the formed tumour spheroids are not all the same size in our model, the embedding in a matrix makes it possible to establish a situation reflecting the ductal carcinoma *in situ* (DCIS) stage (Polyak & Kalluri, 2010), where the tumour has not yet been infiltrated by stroma cells but interacts with the stroma by the exchange of molecular factors. In this model, transition to the invasive ductal carcinoma (IDC) stage, where tumour cells break loose from their cell mass and migrate through the stroma, can be assessed. An interesting step further into a more complex and controlled cancer tissue model would involve embedding spheroids that have been generated by the hanging drop technique in a hydrogel matrix.

Variations on the tumour-stroma co-culture model in *3-D Life* hydrogels

The *3-D Life* system offers the possibility to combine two types of hydrogels: the dextran-based gel that can be dissolved by dextranase and the PVA-based gel that withstands the dextranase treatment. Combining these two gels in cell culture, two adjacent 3D cell compartments can be created that contain two different cell types. For example, tumour cells can be seeded in a bottom PVA gel and covered with a second layer of dextran gel containing fibroblasts or vice versa. During culture, cells can exchange their paracrine signalling factors but stay physically separated. Cells of the dextran gel can be released by dextranase treatment for analysis without contamination from the second cell type, which is still contained in the intact PVA gel. Other variations to further modify the cultures are possible. For example, the two types of gels can be modified with different molecular functions, e.g. choice of incorporation of a MMP-cleavable linker or of different adhesion peptides or matrix proteins. Another set-up would be the incorporation of diffusible growth factors and cytokines in one compartment for a slow release and the analysis of effects these factors have on cells in the other compartment.

Other combinations of cell types can be applied in this set-up to explore tumour-stroma interactions. For example, CAFs isolated from tumour tissue could be combined with normal epithelial cells and provide a way to study the effect of the diseased fibroblasts on healthy cells. Moreover, other stromal cells can be added to the system to mimic a more complex tumour-stroma microenvironment (Margareta Mueller, personal communication).

Troubleshooting

Although the *3-D Life* hydrogel system is a simple set of liquid reagents, there are potential error sources. For example, it is necessary to store the reagents appropriately to prevent deterioration of the reactive groups. Also, the cellular response to a specific biomimetic microenvironment may be different from that anticipated. Possible causes and remedies for these kinds of problems, however, are beyond the scope of the simple troubleshooting guide listed in Table 9.2.

Conclusion

The *3-D Life* hydrogel system is based on the crosslinking of biologically inert polymers, offering the choice of forming completely synthetic or chemically defined hydrogels. The adaptation of this polymer network by use of MMP-cleavable crosslinkers, covalent attachment of adhesion peptides and variation of stiffness facilitates the mimicking of many functions of native ECM.

Table 9.2 Troubleshooting *3-D Life* hydrogels.

Observation	Possible causes	Possible remedies
No gel formation	Maleimide groups and thiol groups used for crosslinking are not near equimolar concentrations	Check the balance of maleimide groups and thiol groups of your components
	Deteriorated reactive groups	Use a fresh sample of reaction components; adhere to recommended storage conditions
Gel forms too fast	Concentration of reactive components is too high (>5 mMol/L)	Use lower equimolar concentration of maleimide and thiol groups for crosslinking
	pH too high	Use lower pH (CB pH 5.5) for performing the crosslinking reaction
Vitality of cells decreases rapidly	Cell type requires presence of an adhesion factor or of a different adhesion factor	Add appropriate cell adhesion peptides or extracellular adhesion proteins
	Cell may be sensitive to low pH during gel set-up	Use neutral pH during gel set-up
Dissolution of dextran-based hydrogel is not complete after 1 h at 37 °C	High gel strength; dextranase activity reduced due to long storage at 4 °C or higher	Mechanically disrupt gel prior to digestion. Prolong digestion up to 2 h; add more volume of dextranase-medium solution
Cells do not behave as intended	Biomimetic microenvironment not suitable	Vary microenvironment, e.g. reduce gel strength, alter adhesion peptide concentration, utilise CD-Link

The independent variation of these modifications provides the user with a very high flexibility in designing biomimetic matrices for many types of applications.

The system is supplied as a set of defined individual reagents. All reagents are liquid and can be easily handled using conventional pipetting techniques or robots.

The gel-forming chemical reaction does not require UV or radicals and does not produce any chemical by products which may harm cells. The gels can be moulded in any shape or containment. They are completely translucent and thus suitable for most types of light microscopy.

Gels based on the polymer dextran can be enzymatically degraded without adverse effects on cells or proteins. This allows the complete recovery of live or chemically fixed cells for downstream processing or for further culturing.

Acknowledgements

The authors wish to acknowledge the contributions of Nils Clausen, Claudia Hoffmann and Claudia Weber. This work was supported by grants from the German Federal Ministry of Education and Research and Federal Ministry for Economic Affairs and Energy.

References

Augsten, M. (2014) Cancer-associated fibroblasts as another polarized cell type of the tumour microenvironment. *Frontiers in Oncology*, **4**, 62.

Bednar, R. A. (1990) Reactivity and pH dependence of thiol conjugation to N-ethylmaleimide: detection of a conformational change in chalcone isomerase. *Biochemistry*, **29**, 3684–3690.

Casbas-Hernandez, P., Fleming, J. M. & Troester, M. A. (2011) Gene expression analysis of in vitro cocultures to study interactions between breast epithelium and stroma. *Journal of Biomedicine and Biotechnology*, **2011**, 520987.

Charwat, V., Schütze, K., Holnthoner, W. *et al.* (2015) Potential and limitations of microscopy and Raman spectroscopy for live-cell analysis of 3D cell cultures. *Journal of Biotechnology*, **205**, 70–81.

De Groot, C. J., van Luyn, M. J., van Dijk-Wolthuis, W. N. *et al.* (2001) In vitro biocompatibility of biodegradable dextran-based hydrogels tested with human fibroblasts. *Biomaterials*, **22**, 1197–1203.

Devolder, R. & Kong, H. J. (2012) Hydrogels for in vivo-like three-dimensional cellular studies. *Wiley Interdisciplinary Reviews: Systems Biology and Medicine*, **4**, 351–365.

Draye, J. P., Delaey, B., van de Voorde, A. *et al.* (1998) In vitro and in vivo biocompatibility of dextran dialdehyde cross-linked gelatin hydrogel films. *Biomaterials*, **19**, 1677–1687.

Ferreira, L. S., Gerecht, S., Fuller, J. *et al.* (2007) Bioactive hydrogel scaffolds for controllable vascular differentiation of human embryonic stem cells. *Biomaterials*, **28**, 2706–2717.

Hakkinen, K. M., Harunaga, J. S., Doyle, A. D. *et al.* (2011) Direct comparisons of the morphology, migration, cell adhesions, and actin cytoskeleton of fibroblasts in four different three-dimensional extracellular matrices. *Tissue Engineering Part A*, **17**, 713–724.

Hern, D. L. & Hubbell, J. A. (1998) Incorporation of adhesion peptides into non-adhesive hydrogels useful for tissue resurfacing. *Journal of Biomedical Materials Research*, **39**, 266–276.

Holliday, D. L., Brouilette, K. T., Markert,.*et al.* (2009) Novel multicellular organotypic models of normal and malignant breast: tools for dissecting the role of the microenvironment in breast cancer progression. *Breast Cancer Research*, **11**, R3.

Holton, S. E., Walsh, M. J., Kajdacsy-Balla, A. *et al.* (2011) Label-free characterization of cancer-activated fibroblasts using infrared spectroscopic imaging. *Biophysical Journal*, **101**, 1513–1521.

Hynds, R. E. & Giangreco, A. (2013) Concise review: The relevance of human stem cell-derived organoid models for epithelial translational medicine. *Stem Cells*, **31**, 417–422.

Kenny, P.A., Lee, G. Y., Myers, C. A. *et al.* (2007) The morphologies of breast cancer cell lines in three-dimensional assays correlate with their profiles of gene expression. *Molecular Oncology*, **1**, 84–96.

Knight, C. G., Willenbrock, F. & Murphy, G. (1992) A novel coumarin-labelled peptide for sensitive continuous assays of the matrix metalloproteinases. *FEBS Letters*, **296**, 263–266.

Levental, K. R., Yu, H., Kass, L. *et al.* (2009) Matrix crosslinking forces tumour progression by enhancing integrin signaling. *Cell*, **139**, 891–906.

Lévesque, S. G. & Shoichet, M. S. (2007) Synthesis of enzyme-degradable, peptide-cross-linked dextran hydrogels. *Bioconjugate Chemistry*, **18**, 874–885.

Liu, Y. & Chan-Park, M. B. (2010) A biomimetic hydrogel based on methacrylated dextran-graft-lysine and gelatin for 3D smooth muscle cell culture. *Biomaterials*, **31**, 1158–1170.

Lutolf, M. P. & Hubbell, J. A. (2005) Synthetic biomaterials as instructive extracellular microenvironments for morphogenesis in tissue engineering. *Nature Biotechnology*, **23**, 47–55.

Majima, E., Goto, S., Hori, H. *et al.* (1995) Stabilities of the fluorescent SH-reagent eosin-5-maleimide and its adducts with sulfhydryl compounds. *Biochimica et Biophysica Acta*, **1243**, 336–342.

Massia, S. P., Stark, J. & Letbetter, D. S. (2000) Surface-immobilized dextran limits cell adhesion and spreading. *Biomaterials*, **21**, 2253–2261.

Mather, B. D., Viswanathan, K., Miller, K. M. *et al.* (2006) Michael addition reactions in macromolecular design for emerging technologies. *Progress in Polymer Science*, **31**, 487–531.

Mroue, R. & Bissell, M. J. (2013) Three-dimensional cultures of mouse mammary epithelial cells. *Methods in Molecular Biology*, **945**, 221–250.

Neumann, U., Kubota, H., Frei, K. *et al.* (2004) Characterization of Mca-Lys-Pro-Leu-Gly-Leu-Dpa-Ala-Arg-NH2, a fluorogenic substrate with increased specificity constants for collagenases and tumour necrosis factor converting enzyme. *Analytical Biochemistry*, **328**, 166–173.

Nuttelman, C. R., Mortisen, D. J., Henry, S. M. *et al.* (2001) Attachment of fibronectin to poly(vinyl alcohol) hydrogels promotes NIH3T3 cell adhesion, proliferation, and migration. *Journal of Biomedical Materials Research*, **57**, 217–223.

Otranto, M., Sarrazy, V., Bonte, F. *et al.* (2012) The role of the myofibroblast in tumour stroma remodeling. *Cell Adhesion & Migration*, **6**, 203–219.

Palmeri, M. L., Wang, M. H., Dahl, J. J. *et al.* (2008) Quantifying hepatic shear modulus in vivo using acoustic radiation force. *Ultrasound in Medicine & Biology*, **34**, 546–558.

Paszek, M. J., Zahir, N., Johnson, K. R. *et al.* (2005) Tensional homeostasis and the malignant phenotype. *Cancer Cell*, **8**, 241–254.

Pietras, K. & Ostman, A. (2010) Hallmarks of cancer: interactions with the tumour stroma. *Experimental Cell Research*, **316**, 1324–1331.

Polyak, K. & Kalluri, R. (2010) The role of the microenvironment in mammary gland development and cancer. *Cold Spring Harbor Perspectives in Biology*, **2**, p.a003244.

Provenzano, P. P., Eliceiri, K. W., Campbell, J. M. *et al.* (2006) Collagen reorganization at the tumour-stromal interface facilitates local invasion. *BMC Medicine*, **4**, 38.

Räsänen, K. & Vaheri, A. (2010) Activation of fibroblasts in cancer stroma. *Experimental Cell Research*, **316**, 2713–2722.

Rimann, M., Angres, B., Patocchi-Tenzer, I. *et al.* (2013) Automation of 3D cell culture using chemically defined hydrogels. *Journal of Laboratory Automation*, **19**, 191–197.

Ruoslahti, E. & Pierschbacher, M. D. (1987) New perspectives in cell adhesion: RGD and integrins. *Science*, **238**, 491–497.

Schmedlen, R. H., Masters, K. S. & West, J. L. (2002) Photocrosslinkable polyvinyl alcohol hydrogels that can be modified with cell adhesion peptides for use in tissue engineering. *Biomaterials*, **23**, 4325–4332.

Seliktar, D., Zisch, A. H., Lutolf, M. P. *et al.* (2004) MMP-2 sensitive, VEGF-bearing bioactive hydrogels for promotion of vascular healing. *Journal of Biomedical Materials Research Part A*, **68**, 704–716.

Stack, M. S. & Gray, R. D. (1989) Comparison of vertebrate collagenase and gelatinase using a new fluorogenic substrate peptide. *Journal of Biological Chemistry*, **264**, 4277–4281.

Vergara, D., Valente, C. M., Tinelli, A. *et al.* (2011) Resveratrol inhibits the epidermal growth factor-induced epithelial mesenchymal transition in MCF-7 cells. *Cancer Letters*, **310**, 1–8.

Walsh, LA. & Damjanovski, S. (2011) IGF-1 increases invasive potential of MCF 7 breast cancer cells and induces activation of latent TGF-β1 resulting in epithelial to mesenchymal transition. *Cell Communication and Signaling*, **9**, 10.

Woerly, S., Laroche, G., Marchand, R. *et al.* (1995) Intracerebral implantation of hydrogel-coupled adhesion peptides: tissue reaction. *Journal of Neural Transplantation and Plasticity*, **5**, 245–255.

Zajaczkowski, M. B., Cukierman, E., Galbraith, C. G. *et al.* (2003) Cell-matrix adhesions on poly(vinyl alcohol) hydrogels. *Tissue Engineering*, **9**, 525–533.

PART III
Scaffolds

CHAPTER 10

Alvetex®, a highly porous polystyrene scaffold for routine three-dimensional cell culture

Antonio Romo-Morales[1], Eleanor Knight[1] and Stefan Przyborski[1,2]
[1] *Department of Biosciences, Durham University, Durham, UK*
[2] *ReproCELL Europe, NETPark, Sedgefield, UK*

Introduction

The majority of research in cell biology is currently carried out *in vitro* in two-dimensional (2D) culture environments where cells are grown on flat surfaces. Cells grown in these conditions adopt an unnatural geometry, remodel their cytoskeleton and change their growth characteristics to enable survival (Figure 10.1). This in turn affects their ability to function and differentiate, which has major implications for research discovery (Baker & Chen, 2012). Despite being cheap and convenient, monolayer cultures are a reductionist approach to cell culture. These cell-based assays modify normal cell behaviour, depicting an artificial biological phenomenon, which ultimately impairs their use as predictive assays in drug discovery. Three-dimensional (3D) cell culture seeks to eliminate this aberrant response by providing suitable surroundings for cells to maintain their normal 3D shape and freely form complex interactions with neighbouring cells. Replicating the natural environment *in vivo* is necessary to escape the limitations of monolayer cultures and thus enhance the biological relevance of cell culture research.

In recent years, multiple platforms for 3D cell culture have been developed, but the idea of a 3D cell culture environment is not new. It has been around for at least 40 years; for example, fibroblasts were grown in collagen gels where they attained a more *in vivo*-like phenotype (Elsdale & Bard, 1972). Materials such as collagen gels and biodegradable polymers in scaffolds degrade during long-term culture. Whilst this is a distinct advantage to aid their integration when transplanted into host tissues, biodegradability is detrimental in many *in vitro* studies (Griffith & Swartz, 2006). Scaffold degradation can lead to a loss of mechanical properties and the local release of by-products

Technology Platforms for 3D Cell Culture: A User's Guide, First Edition. Edited by Stefan Przyborski.

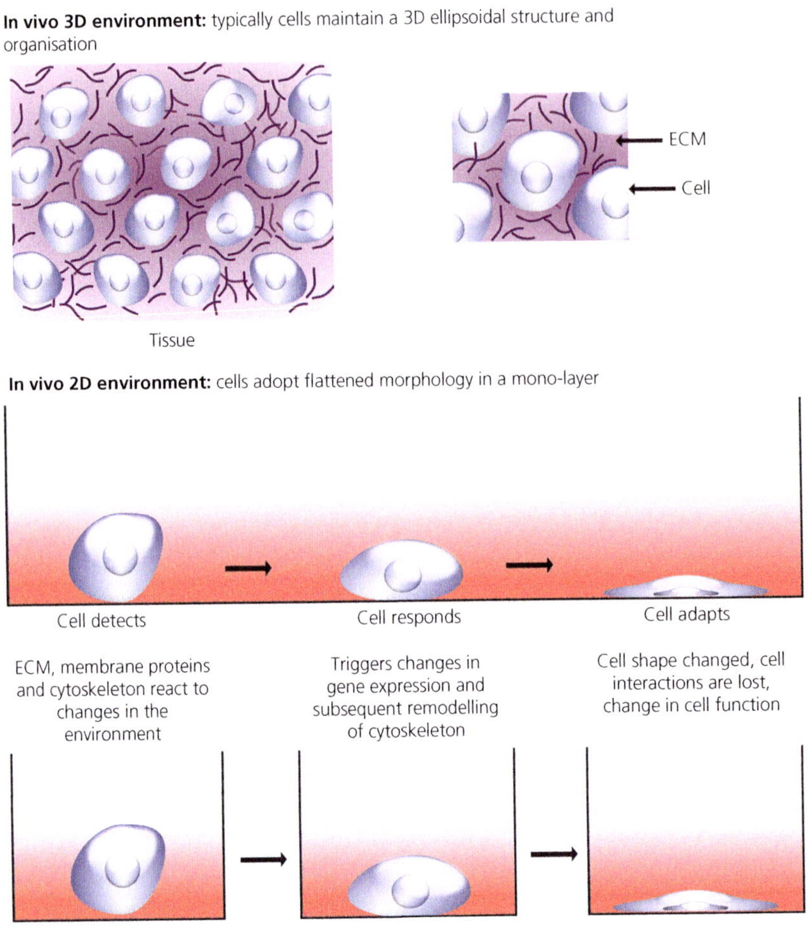

Figure 10.1 The effects of changing the growth environment on cells. Cells in 2D culture are forced to modify their morphology to survive. This alters the intracellular structure, which in turn triggers changes in gene expression and cell function.

that can subsequently influence cell behaviour. Moreover, such materials are not practical for routine 3D cell culture.

For the production of a reproducible 3D environment, variables such as shelf-life, storage and quality control need to be taken into consideration. Inert non-degradable synthetic polymer-based materials overcome these issues and can be engineered into highly porous materials suitable for cell infiltration, which could potentially support the production of *in vivo*-like structures. These types of 3D cultures hold the potential to replace some conventional cell culture assays; for example, it has been demonstrated that hepatocytes grown on polystyrene scaffolds display enhanced performance and sensitivity to toxic insult (Schutte *et al.*, 2011). This in turn could lead to a reduction in the need for animal models for testing toxicity. Additionally, there is evidence to show that

3D culture models can be used to study human tissue physiology in health and disease where tissue-like function and homeostasis are maintained (Schmeichel & Bissell, 2003). Three-dimensional scaffolds also provide a unique tool for developmental biologists to study early developmental processes without interference from early patterning information. This information is still present in alternative 3D systems such as the formation of embryoid bodies.

An advantage of scaffold-based technologies is their ability to provide support in three dimensions. This environment helps to create organised arrangements of cells in a controllable and reproducible manner, leading to natural 3D tissue-like structures *in vitro* (Knight & Przyborski, 2015). The development of artificial skin constructs is a classic example of the success of scaffold technologies. Scaffolds have been used routinely to recreate the layered structure of dermis and epidermis. Porous scaffolds have been shown to support epidermal-like structures from equine keratinocytes (Sharma *et al.*, 2016), and to produce co-cultures that can be used for the study of tumour invasion into the stromal layer (Fischbach *et al.*, 2007). The increasing number of applications in the literature reflects the rapid growth of scaffold-based technologies used to create native tissue architecture, enhancing the biological relevance of cell culture research.

Despite the availability of many promising examples of new scaffold-based technologies enabling tissue-like structures *in vitro* (Knight & Przyborski, 2015) or enhancing the efficacy of *in vitro* testing (Kimlin *et al.*, 2013), few can be manufactured as a commercial product for routine use. Most commercially viable 3D culture products are synthetic and free from animal-based components, supplied sterile and manufactured according to high-quality control procedures, hence with minimal batch-to-batch variation. Commercially available scaffolds need to maintain these standards and overcome the impracticalities and variation associated with user-prepared scaffolds. Polystyrene scaffolds are an attractive approach that can meet these requirements to support routine 3D cell culture. Polystyrene is an inert non-biodegradable material, already familiar in traditional cell culturing techniques. Additionally, it is inexpensive to mass-produce to a high-quality standard, making it a consistent and reproducible platform to model *in vivo* biological phenomena. With this in mind, Alvetex®, an inert non-degradable scaffold, has been developed from a porous polystyrene high internal phase emulsion (HIPE).

Description of technology

Alvetex provides a 3D physical space where cells can infiltrate and grow in three dimensions. It faithfully captures the microenvironment experienced by cells, making it a more physiologically accurate model of biological processes (Table 10.1). This polystyrene scaffold has a highly porous architecture consisting

Table 10.1 Summary: changing the physical dimension of the cell culture environment affects cell behaviour.

	Traditional 2D cell culture	Alvetex 3D cell culture	Normal *in vivo* environment
General dimensions and physical differences			
Maximum distance of cell from the source of nutrients	0	0–100 μm	0–200 μm
Resemblance to *in vivo* cellular environment	Low	High	n.a.
Appearance of cell morphology	Flattened	3D shape	3D shape
Potential for 3D cell-to-cell interactions	Very low	High	High
Ability to form complex three 3D cellular structures	Very low	High	High
Upon initial seeding of cells onto plasticware			
Degree of cellular stress placed upon cell structure	High	Low	n.a.
Changes to protein and gene expression	High	Low	n.a.
Cell surface area in contact with plastic	At least 50%	0–50%	n.a.
Post cell seeding phase			
Ongoing changes to cellular shape	High	Low	n.a.
Need for remodelling of cytoskeletal architecture	High	Low	n.a.
Deviation from normal *in vivo* morphology	High	Low	n.a.
Cell surface area in contact with plastic	At least 50%	0–50%	n.a.
Opportunity for enhanced *in vitro* cell functionality	Low	High	n.a.

of several voids interconnected by pores. This network of interconnects can be clearly visualised on the SEM micrograph (Figure 10.2c). Alvetex has an open surface structure (Figure 10.2b) which allows cells to infiltrate into the internal structures of the scaffold and which assists in the mass transfer of nutrients and waste products and the exchange of gases (Carnachan *et al.*, 2006). Cells growing and dividing within Alvetex Scaffold are free to migrate throughout the matrix, behaving as they would in their natural surroundings. Cultured cells are able to lay down extracellular matrix and self-organise into complex *in vivo*-like structures, extending throughout the substrate, forming tissue-like layers (Figure 10.3).

Alvetex undergoes quality control procedures to ensure a consistent structure is produced (Table 10.2). Void size is assessed as previously described

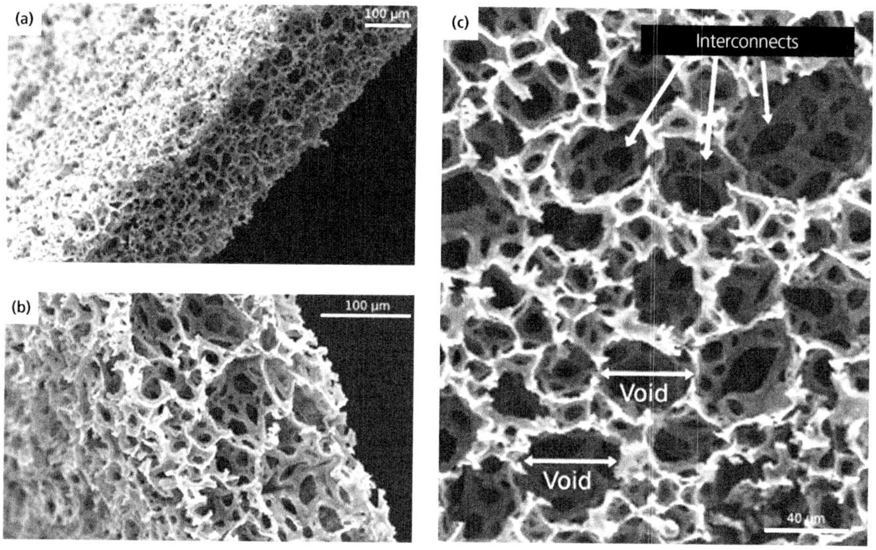

Figure 10.2 Alvetex viewed by scanning electron microscope to highlight its porous structure. (a) A 200 μm thick polystyrene scaffold. (b) Alvetex porosity of >90%. (c) Close-up of voids with dimensions of approximately 36–40 μm in diameter and interconnects of approximately 12–14 μm in diameter.

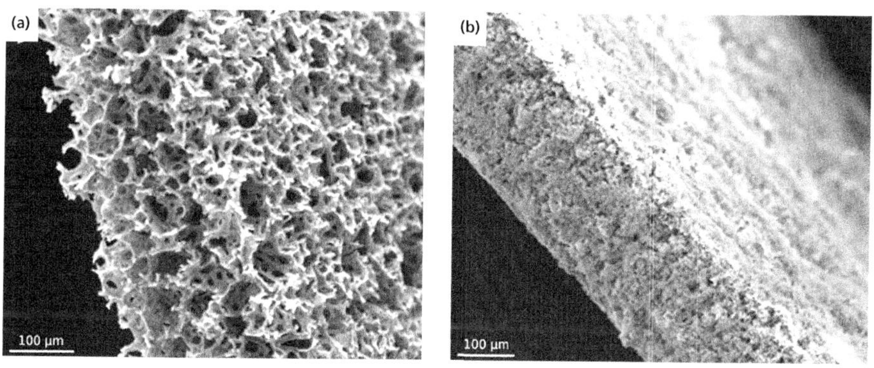

Figure 10.3 3D growth of cultured cells form 200 μm thick tissue-like layers. (a) Naked Alvetex Scaffold before cell seeding viewed under scanning electron microscope to show the highly porous scaffold. (b) Cells have grown throughout the scaffold to the point where the Alvetex Scaffold is no longer clearly visible.

(Bokhari *et al.*, 2007b). It is possible to tailor the morphology of these emulsion-templated porous polymers (Carnachan *et al.*, 2006) and to manufacture the membranes for a specific void and interconnect size. Void size has previously been shown to affect cell adhesion, cell viability (Zeltinger *et al.*, 2001), ingrowth, distribution (Wake *et al.*, 1994) and formation of extracellular matrix (Aronin *et al.*, 2009) during 3D culture. Alvetex is manufactured as two products with different void size. Both materials are highly porous but Alvetex Scaffold has a

Table 10.2 Alvetex – unique scaffold dimensions ideally suited to 3D cell culture.

Alvetex feature	Benefits of using Alvetex for 3D cell culture
Same polystyrene as conventional cell culture plasticware	• Easily switch between 2D and 3D protocols • Inert – no effect on cell growth or function – no new experimental variables • Stable – does not degrade, no change throughout long-term studies • Can be precoated with ECM proteins
Consistent scaffold structure – extremely low batch-to-batch variability (ISO Certificate)	• Reproducible, consistent results, low batch-to-batch variability • Genuine and homogeneous 3D cell growth
Entire scaffold is only 200 µm thick	• No cell is ever more than 100 µm away from nutrients and gases – mimics *in vivo* conditions • Cells can feed and excrete via passive diffusion – mimics *in vivo* conditions
>90% Porosity	• Cells can easily penetrate scaffold and lay down ECM to more closely mimic *in vivo* conditions • Cells and media move freely through the matrix • Nutrients and waste exchanged by passive diffusion • Partial cell retrieval is possible
Void dimensions 36–40 µm	• Typically up to 75–100 cells may occupy a single void

ECM, extracellular matrix.

void size of 36–40 µm, which allows a broad range of cells to infiltrate into the scaffold interior, whereas Alvetex Strata has a much smaller void size of 13–15 µm. This allows some very small cells to infiltrate into the Strata interior but the majority of cell types form a dense layer of 3D cells at the surface of the membrane.

Alvetex is supplied as a sterile product ready for immediate use. Consideration has been given to the presentation of the scaffold for alternative uses. The 200 µm thick porous membranes have been adapted to fit a variety of conventional cell culture plasticware formats. Available as inserts or plates, Alvetex gives researchers the ability to set up experiments with varying configurations for specific durations of the assay or in which the degree of cell penetration can be controlled (Figure 10.4). For example, placement of the membrane at the bottom of a 12-well plate held in place by a clip bathes the 3D culture in medium only from above and is ideal for shorter term cultures, whereas the membrane in a well insert enables medium contact from both above and below the culture. These subtle differences influence cell behaviour in the cultures and care should be taken to choose the correct format for specific applications (Knight & Przyborski, 2014). In addition, utilising well inserts enables different media fill options (Figure 10.4c–e) for different applications described in Table 10.3.

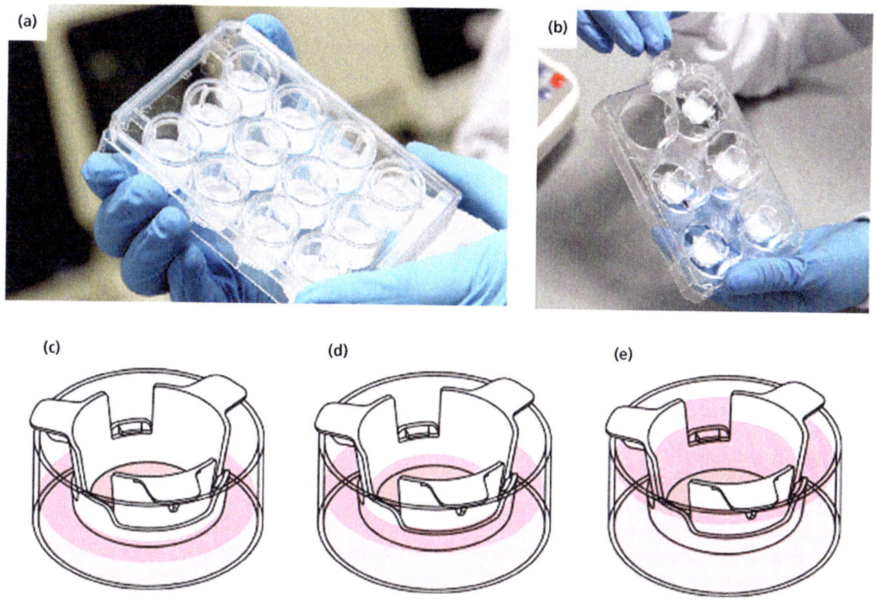

Figure 10.4 **Alvetex plate formats for different applications and medium fill options.** (a) Alvetex in a 12-well format (AVP002) comprising 22 mm discs of Alvetex (held by a removable clip). (b) Alvetex six-well inserts (AVP004). Schematics (c)–(e) illustrate three different filling options. (c) Medium from below only for cells grown in 3D at air-liquid interface. (d) Medium from above and below for routine 3D growth of cells with lower-average metabolic activity/proliferation rate or for experiments where cells are incubated with test substrate in top chamber only for permeability investigations. (e) Medium interconnected for routine 3D growth of cells with high metabolic activity/proliferation rate.

Table 10.3 Alvetex inserts enable three different medium fill options (see Figure 10.4c–e).

	Medium fill option A	Medium fill option B	Medium fill option C
Description	Medium in contact from below only	Medium in contact above and below – independent medium compartments	Medium in contact above and below – connected compartments
Enables	3D growth at the air-liquid interface	3D growth with two different media constituents	3D growth with two uniform media constituents
Example of application	Induction of epithelial stratification (e.g. skin epidermis)	Barrier penetration assay of test compounds Evaluation of compound permeability Growth with different media either side of the scaffold	Create optimal conditions for maximising cell growth and increased viability

Both of the Alvetex membranes are compatible with a range of biochemical assays, and are readily processed for cell visualisation using standard histological techniques, immunohistochemistry and electron microscopy (Figure 10.5) (Knight *et al.*, 2011).

The scaffolds have been used to culture a range of cell types, including hepatocytes (Bokhari *et al.*, 2007a,b; Schutte *et al.*, 2011), osteoblasts (Bokhari *et al.*, 2007c) and pluripotent stem cell-derived neurons (Hayman *et al.*, 2004, 2005) (Figure 10.6). Alvetex 3D cultures have demonstrated increased functionality compared with conventional 2D cultures (Burkard *et al.*, 2012) and cytotoxic

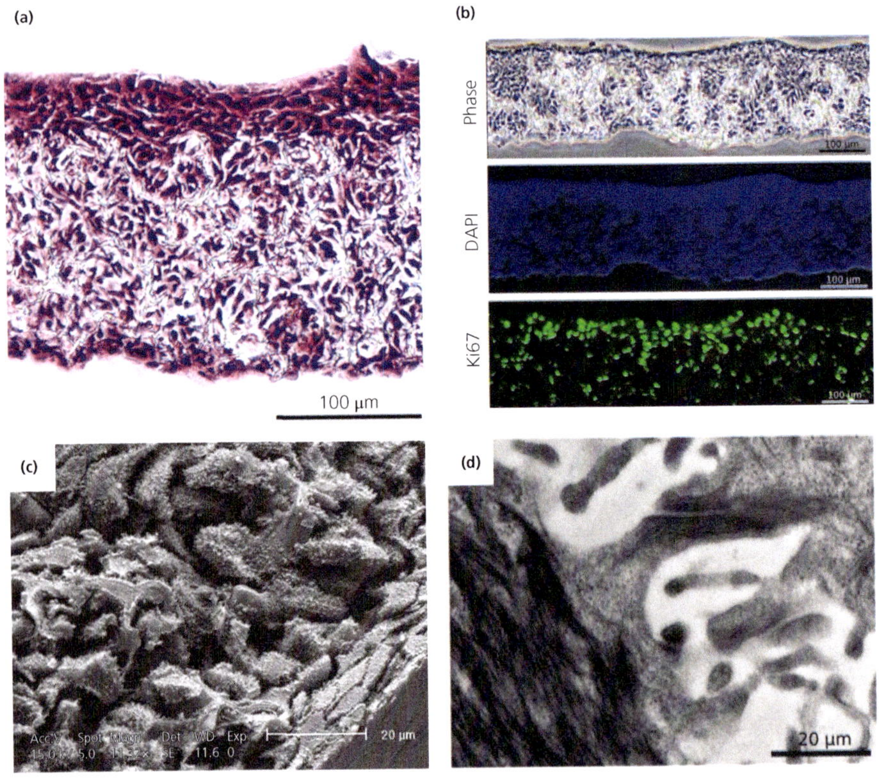

Figure 10.5 Alvetex is compatible with downstream applications involving tissue processing. (a) Cells fixed in 4% paraformaldehyde, embedded in paraffin wax, sectioned (7 micron) before staining with haematoxylin & eosin (H&E) and cover-slipped. Frozen and paraffin-embedded samples can be sectioned and stained to reveal the native cellular structures inside Alvetex. (b) Cell culture fixed in 4% paraformaldehyde, embedded in paraffin wax and sectioned (10 micron). Antigen retrieval followed by immunocytochemical analysis with the proliferation marker Ki67 (*green*) and the nuclear stain DAPI (*blue*) was performed following standard immunocytochemical methods. (c) Visualisation of the structure of skin cells that have penetrated throughout the scaffold, showing some that have stratified on the surface. (d) Standard transmission electron microscopy analysing the ultrastructure of cells grown in Alvetex.

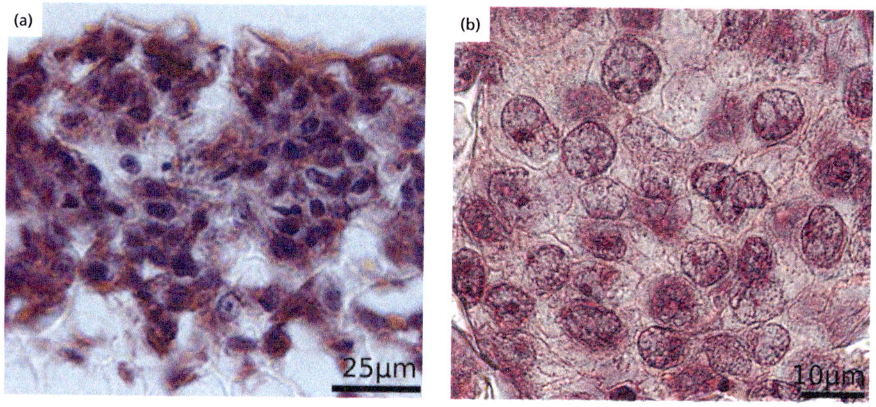

Figure 10.6 Cell structure. (a) Cells grown within Alvetex maintain their natural shape and 3D organisation. (b) 3D cell culture of liver hepatocytes grown within Alvetex.

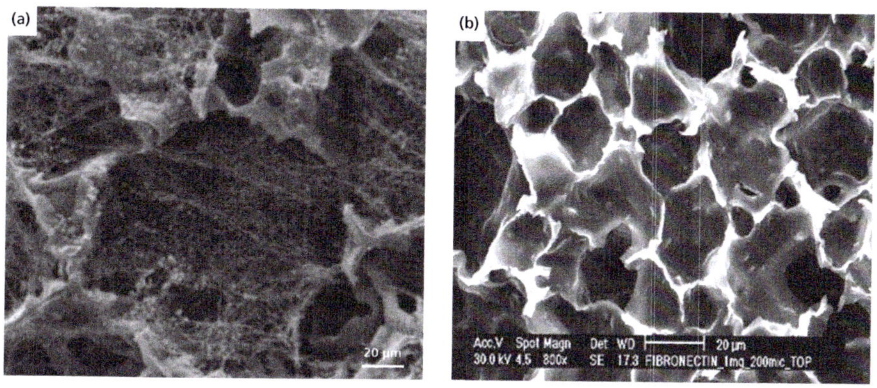

Figure 10.7 Alvetex can easily be coated with extracellular matrix proteins. (a) Scaffold preloaded with collagen IV. (b) Coating Alvetex with fibronectin. The ECM proteins form a web of fibres spanning voids into which cells can grow and migrate in 3D.

agents have displayed lower apparent cytotoxicity (Alayoubi *et al.*, 2013; Schutte *et al.*, 2011). Most recently, Alvetex Scaffold has been used to create 3D tissues suitable as positive controls for *in situ* hybridisation (MacDonald *et al.*, 2014). To address the inert nature of Alvetex Scaffold, this polystyrene substrate can easily be coated with extracellular matrix (ECM) proteins and other reagents, for example collagen I, collagen IV (Figure 10.7a), fibronectin (Figure 10.7b), laminin, poly-D-lysine, poly-L-lysine, poly-L-orthinine, Matrigel® and PuraMatrix®. Coating the scaffold with these components increases cellular interactions, enhancing cell attachment and migration within the scaffold, particularly when using a serum-free medium. In addition, these bioactive elements can support the crosstalk between cell populations and facilitate co-culture set-ups. Here we describe a few validated 3D cell culture protocols using Alvetex.

Methods for coating Alvetex Scaffold

Poly-D or L-lysine coating

1 Prepare Alvetex Scaffold for coating by first treating with 70% ethanol followed by two PBS washes as described in the relevant product information leaflet. Leave Alvetex Scaffold in the second PBS wash until ready to apply the poly-L-lysine solution.
2 Aspirate the second PBS wash and add 500 μL of poly-D or L-lysine per well. Replace plate lids and leave to stand for one hour at room temperature.
3 Tilt the 12-well plate and gently aspirate any excess fluid from the edge of the wells. If using well insert Alvetex Scaffold formats, remove excess fluid from Alvetex Scaffold by gently tapping the plate or petri dish on the worktop. Check that no residual fluid is hanging from the base of the well inserts. Aspirate to remove any residual coating agent from the bottom of the wells.
4 Prepare cells for seeding in the appropriate culture medium and seed directly on the wet poly-L-lysine coated Alvetex Scaffold membrane in the volumes relevant for the Alvetex Scaffold product format. Allow the cells to settle for 30–90 minutes in an incubator (5% CO_2, 37 °C) before flooding with medium.

Collagen I coating

1 Prepare Alvetex Scaffold for coating by first treating with 70% ethanol followed by two PBS washes as described in the relevant product information leaflet. Leave Alvetex Scaffold in the second PBS wash until ready to apply the collagen solution.
2 Dilute rat tail collagen I (BD Biosciences, Cat. No. 354236) to a concentration of 0.8 mg/mL using cell culture-grade water. Handle the reagents on ice, using prechilled pipette tips to perform the dilution and subsequent application onto Alvetex Scaffold.
3 Aspirate the second PBS wash from the Alvetex Scaffold discs and carefully pipette 500 μL of the diluted collagen solution onto each disc. Replace plate lids and leave to stand for one hour at room temperature.
4 Remove excess fluid from Alvetex Scaffold in well insert format by gently tapping the plate or petri dish on the worktop. Check that no residual fluid is hanging from the base of the well inserts. Aspirate to remove any residual coating agent from the bottom of the wells. If using Alvetex® Scaffold in 12-well plate format, tilt the plate and gently aspirate any excess fluid from the edge of the wells.
5 Prepare cells for seeding in the appropriate culture medium and seed directly on the wet collagen-coated Alvetex Scaffold membrane in the volumes relevant for Alvetex Scaffold product format. Allow the cells to settle for 30–90 minutes in an incubator (5% CO_2, 37 °C) before flooding with medium.

Fibronectin coating

1 Prepare Alvetex Scaffold for coating by first treating with 70% ethanol followed by two PBS washes as described in the relevant product information leaflet. Leave Alvetex Scaffold in the second PBS wash until ready to apply the fibronectin solution.
2 Reconstitute fibronectin (BD Biosciences, Cat. No. 356008) to a concentration of 0.5 mg/mL using PBS.
3 Aspirate the second PBS wash from the Alvetex Scaffold discs and carefully pipette 300 μL of the diluted fibronectin solution onto each disc. Replace plate lids and leave to stand for one hour at room temperature.
4 Remove excess fluid from Alvetex Scaffold in well insert format by gently tapping the plate or petri dish on the worktop. Check that no residual fluid is hanging from the base of the well inserts. Aspirate to remove any residual coating agent from the bottom of the wells. If using Alvetex Scaffold in 12-well plate format, tilt the plate and gently aspirate any excess fluid from the edge of the wells.
5 Prepare cells for seeding in the appropriate culture medium and seed directly on the wet fibronectin-coated Alvetex Scaffold membrane. Seed cells in volumes relevant for the specific Alvetex Scaffold format being used (see product information booklet for volume details). Allow the cells to settle for 30–90 minutes in an incubator (5% CO_2, 37 °C) before flooding with medium.

Matrigel coating

1 Prepare Alvetex Scaffold for coating by first treating with 70% ethanol followed by two PBS washes as described in the relevant product information leaflet. Leave Alvetex Scaffold in the second PBS wash until ready to apply the Matrigel solution.
2 Dilute Matrigel (BD Biosciences, Cat. No. 356234; prethawed overnight on ice) to a concentration of 0.8 mg/mL (1 in 10 dilution) using appropriate cell culture medium (e.g. MEM for the example below). Handle the reagents on ice, using prechilled pipette tips to perform the dilution and subsequent application onto Alvetex Scaffold.
3 Aspirate the second PBS wash from the Alvetex Scaffold discs and carefully pipette 350 μL of the diluted Matrigel solution onto each disc. Replace plate lids and leave to stand for 1–2 hours at room temperature.
4 Remove excess fluid from Alvetex Scaffold in well insert format by gently tapping the plate or petri dish on the worktop. Check that no residual fluid is hanging from the base of the well inserts. Aspirate to remove any residual coating agent from the bottom of the wells. If using Alvetex Scaffold in 12-well plate format, tilt the plate and gently aspirate any excess fluid from the edge of the wells.

5 Prepare cells for seeding in the appropriate culture medium and seed directly on the wet Matrigel-coated Alvetex Scaffold membrane in the volumes relevant to the Alvetex Scaffold product format. Allow the cells to settle for 30–90 minutes in an incubator (5% CO_2, 37 °C) before flooding with medium.

Example applications of Alvetex Scaffold technology

This section describes the use of Alvetex Scaffold for advanced routine cell culture methods. Co-culture of cells in 3D produces more relevant models for drug testing and understanding the *in vivo* microenvironment. Co-cultures using Alvetex have demonstrated both enhanced cell activity and differentiation. Using these principles, Alvetex technologies are also described for the development of a full-thickness human skin model and a model to study cancer cell invasion.

Construction of co-culture models using Alvetex Scaffold

A co-culture can be defined as the growth of at least two distinct cell types in a combined culture. *In vitro* co-culture models are particularly relevant in drug research as they provide a more physiologically relevant microenvironment. Such a microenvironment can be used during drug testing to monitor the effect on cell-cell interactions (Khetani & Bhatia, 2008). Co-cultures often improve the success of cells that are difficult to cultivate as monocultures and may lead to a more desired physiological behaviour (Amit *et al.*, 2000). For example, in previous studies rat primary hepatocytes have demonstrated enhanced functionality and maintained their viability for longer when co-cultured with 3T3 cells (Bhandari *et al.*, 2001). Moreover, the co-culture of carcinoma and intratumoural stromal cells simulated a breast cancer microenvironment, allowing cell-cell communication to be monitored for the initiation and progression of the cancer (Miki *et al.*, 2010). Finally, a study carried out using endothelial and smooth muscle cells co-culture mimicked the structure of a vessel wall, creating an improved model of the *in vivo* environment (Truskey, 2010).

Cell culture inserts containing semi-porous membranes have previously been used to assess the paracrine influence between two cell types (Miki *et al.*, 2012). Alvetex Scaffold in well inserts or in multiwell plates are an opportunity to investigate the interplay of two cell types in heterogeneous cultures within the same scaffold. In addition, 3D cell co-culture scenarios with an increased level of complexity can be set up to study different cell-cell interactions and the secretion of paracrine factors (Figure 10.8). One assembly option allows for two 3D cultures to be in direct contact to then establish layers of alternative cell types and thus investigate more complex biological phenomena such as cell invasion and migration. Here we demonstrate one application demonstrating the ability and advantages of using Alvetex Scaffold technology to support the co-cultures of dermal fibroblasts and primary human keratinocytes to form a terminally differentiated, cornified human skin equivalent.

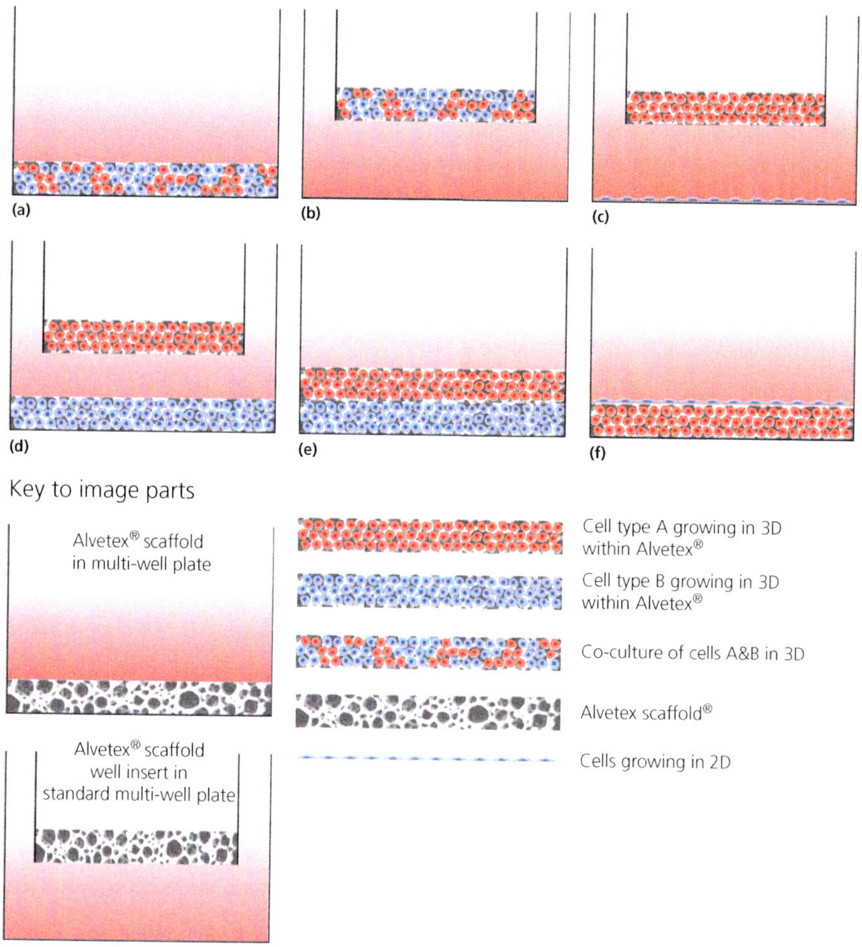

Figure 10.8 Using Alvetex to create co-culturing experiments with different assembly options. (a,b) 3D/3D co-culture in multiwell or well insert to emulate the structure of a tissue composed of more than one cell type. (c) 3D/2D co-culture in multiwell plate and well insert combined to study the secretion of factors and signalling molecules between two independent cell cultures. (d) 3D co-culture in multiwell plate and well insert combined to study the secretion of factors and signalling molecules between two independent cell cultures. (e) 3D/3D co-culture in multiwell plate to study the direct interaction of cells in contact with one another, to mimic tissue structures and to investigate invasion and migration of different cell types. (f) 2D/3D co-culture in multiwell plate to study the direct interaction of cells in contact with one another, to mimic tissue structures and to investigate invasion and migration of different cell types.

Development of full-thickness human skin model using Alvetex Scaffold technology

Human skin is a complex organ with a multilayered architecture. Broadly, it can be divided into the upper epidermis, composed of keratinocytes interacting with melanocytes, and the lower dermis with fibroblasts synthesising the extracellular matrix. Between these two layers, there is a basement membrane, rich in matrix molecules such as laminin isoforms and type IV, VII and XVII collagens (Fleischmajer *et al.*, 1998). Sitting on top of the basement membrane, keratinocytes form a proliferative basal layer (stratum basale) and become more differentiated as they move towards the surface of the skin.

Keratinocytes are a key cell type, forming these stratified layers through a highly organised adhesive network. These cells migrate upwards from the basal compartment to the cornified layers, constantly reorganising their cytoskeleton and adhesive junctions (Simpson *et al.*, 2011). Initially, keratinocytes in the stratum basale are anchored to the bottom of the basement membrane through hemidesmosomes and integrin-based adhesions. These cell adhesions extend across the basement membrane into the dermis, allowing keratinocytes to sense external cues transduced into intracellular signals. This in turn can affect keratinocytes' activity in the upper epidermal layer. For example, impairing cell attachment is known to trigger apoptosis in basal cells (Dowling *et al.*, 1997) and affect keratinocyte proliferation (Murgia *et al.*, 1998).

Daughter keratinocytes undergoing differentiation can migrate to the stratum spinosum, where they become larger and create more cellular adhesions and junctions, which ultimately increase the robustness of this layer. Along this journey, epidermal cells also undergo cytoskeletal modifications that tune their function to each layer of the epidermis. In this way, this tissue exhibits a paradoxical behaviour in which it appears stable and tightly bound, yet the cellular components are highly dynamic. In the upper layers of the epidermis, for example, keratinocytes flatten and create a cornified envelope, underlying the plasma membrane, whilst in the stratum corneum they secrete lysosomal enzymes, degrading organelles and becoming completely squamous. These differences in function arise from their position in the stratified epidermis as well as from the cascade of signals, transmitted along this tissue.

For these reasons, a full-thickness *in vitro* skin model must reproduce the layered architecture and its native cell populations to achieve a true representation of the normal human skin microenvironment (Figure 10.9). Recapitulating the cellular topography can provide a suitable environment for cells to self-organise into the *in vivo* anatomy and facilitate the modelling of the intricate physiology and network of signalling cues for routine cellular assays. Alvetex is a platform that has the potential to meet such demands. Its porous structure can enable 3D growth and the colonisation of primary human dermal fibroblasts (Hill *et al.*, 2015). By enabling the formation of cell-cell interactions, dermal fibroblasts can co-ordinate to lay out the ECM, extending throughout this polystyrene membrane. Primary human keratinocytes can then be introduced to the dermal component, where the crosstalk between these native cell populations will enhance tissue organisation and function. Co-culture of keratinocytes and dermal fibroblasts is a common starting point in the creation of skin equivalents. This possibility is available with Alvetex technology, enabling the formation of a stratified epithelium displaying the morphological and functional features of the epidermis *in vivo*.

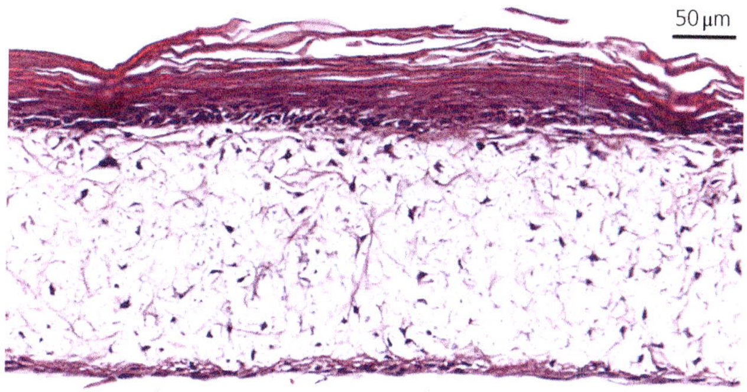

Figure 10.9 **Full-thickness human skin model using Alvetex Scaffold technology.**

Method for developing a full-thickness human skin model using Alvetex Scaffold technology (Hill *et al.*, 2015)

Cell culture

1 Primary human neonatal foreskin fibroblasts (CellnTec) were cultured in Media A for up to seven passages. Immortalised mouse embryonic 3T3 fibroblasts (ATCC-CCL-92) were cultured in Media D.
2 Primary human keratinocytes derived from surplus skin, obtained from patients between the ages of 20 and 55 years and undergoing routine surgery, were isolated by incubating the skin in dispase (Scientific Laboratory Supplies) for 12–18 hours at 4 °C to separate the epidermis with trypsin/EDTA (Scientific Laboratory Supplies) for five minutes at 37 °C. The primary human keratinocytes were subsequently cultured in Media E for up to two passages.
3 A co-culture with keratinocytes and mitomycin C (Sigma-Aldrich)-treated 3T3 feeder cells at 1:1 ratio in Media B was set up and passaged up to three time, changing the medium every day.
4 Following detachment with trypsin/EDTA, keratinocytes were then incubated with an equal volume of soybean trypsin inhibitor (Sigma-Aldrich) and centrifuged at 300× g for five minutes before resuspension in fresh culture medium and subsequent culture.

Human skin equivalent preparation

1 Twelve-well format Alvetex scaffolds (AVP005) (Reinnervate Ltd., Reprocell Group) were pretreated with 70% ethanol in a six-well plate according to protocols for preparation of Alvetex Scaffold for cell culture.
2 2.0×10^6 primary neonatal foreskin fibroblasts were seeded onto Alvetex in 100 μL Media A, which were subsequently incubated at 37 °C, in a humidified atmosphere of 5% CO_2 in air for 1.5 hours.
3 Nine millilitres of Media A +100 μg/mL ascorbic acid (Sigma-Aldrich) were subsequently added to the bottom of each well, gently covering the insert prior to incubation for a further 18 days. Medium changes are required every 3.5 days, to allow the formation of dermal equivalent.

4 Subsequently, the dermal equivalents were washed with 10 mL phosphate-buffered saline (PBS; Sigma-Aldrich) before adding 4 mL Media B to the outer side of the insert such that the bottom of each dermal equivalent was in contact with the medium.
5 Primary human keratinocytes were harvested by differential trypsinisation, discarding the 3T3 feeder cells, and 2.0×10^6 keratinocytes seeded onto dermal equivalents in 100 µL Media B and incubation continued for a further three hours. Five millilitres of Media B were then added to the outer side of each well to gently flood the inside of the insert prior to further incubation at 37 °C for three days, changing the medium every day.
6 On day 21, the insert was removed from the six-well plate and placed into a well insert holder in a deep petri dish (Reinnervate Ltd., Reprocell Group) on the middle rung of the stand. Thirty millilitres of Media C were then added to the disc, such that the bottom of the equivalent was in contact with the medium but the upper surface remained exposed to the air and incubation continued at 37 °C in 5% CO_2 for 14 days, changing the medium every 3.5 days, to allow the formation of a full-thickness skin equivalent.

After one month of culture at the air interface, an organised skin construct formed which showed several cornified layers and the stratification of keratinocytes in the upper layers of the epidermis (Figure 10.10d). Morphological features such as the development of the strata basale, spinosum and corneum indicate the ability to reproduce the integrity of *in vivo* structure and organisation of skin with Alvetex Scaffold technology. The uppermost layer, also known as the stratum corneum, is particularly important since it acts as the principal barrier of skin (Figure 10.10a). It is therefore essential to reproduce this component for allergen penetration studies as well as assessment of barrier function.

Regarding molecular markers, the dermis and the multilayered epidermis display different protein expression profiles. Whilst the dermal layer has ECM components, such as type II and III collagens, the epidermal layers show different expression of cytokeratins according to the stages of keratinocyte differentiation (Hill *et al.*, 2015). Histological analysis indicates the presence of the stratum corneum in this skin equivalent, confirming complete keratinocyte differentiation (Figure 10.10b). Utilising immunofluorescent staining can further assess the differences in protein expression and demonstrate the benefits of co-culture systems. For example, expression of type IV and VII collagens at the epidermal-dermal junction arises from the crosstalk between fibroblasts and keratinocytes (Hill *et al.*, 2015).

An alternative model of skin that is highly suited to test epidermal barrier function can be achieved through the 3D culture of HaCaT keratinocytes using Alvetex Scaffold. The HaCaT lineage is a spontaneously immortalised human keratinocyte cell line that despite showing a transformed phenotype with disordered tissue organisation (Boelsma *et al.*, 1999), still has capacity for differentiation (Maas-Szabowski *et al.*, 2003). This application of Alvetex technology allows for keratinocyte stratification by growing this cell line inside the

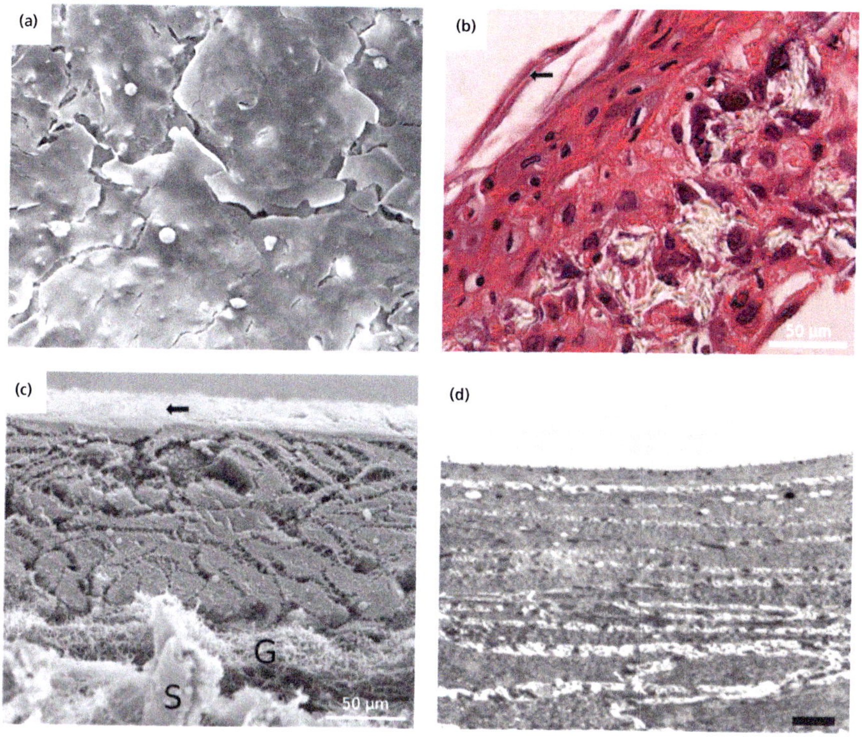

Figure 10.10 **3D culture of skin keratinocytes on Alvetex resulting in formation of 3D epidermis and maturation of the stratum corneum.** (a) Scanning electron microscope image illustrating the formation of the stratum corneum of a 21-day culture. (b) Alvetex Scaffold sectioned and stained with H&E and viewed by light microscopy after 35 days. (c) Scanning electron micrograph image showing full-thickness skin construct grown on a layer of collagen on top of Alvetex which includes an upper cornified layer. (d) Transmission electron microscopy demonstrating stratification of keratinocytes in upper layers of the culture. Source: Courtesy of Ross Carnachan.

scaffold at the air-liquid interface without underlying collagen or fibroblasts (Figure 10.10a). In this way, this method removes the inherent variability of using dermal substrates and facilitates reproducibility, given that the scaffold is completely inert. In addition, the length of the culture is shortened since no collagen gel formation or de-epidermised dermis isolation procedures are required. Culturing HaCaT keratinocytes in Alvetex Scaffold results in the formation of a basic stratum corneum regardless of whether a dermal equivalent is utilised (Figure 10.11). We propose that the described HaCaT model without collagen or fibroblasts is a useful tool to test epidermal barrier function.

Model of colon cancer cell invasion

Specific to cancer cells, a change from 2D to 3D culture leads to evidence of higher metastatic potential (Rhee, 2001) and differences in drug sensitivity (Padron & Peters, 2006; Weaver *et al.*, 2002). Co-cultures of cancer cells and

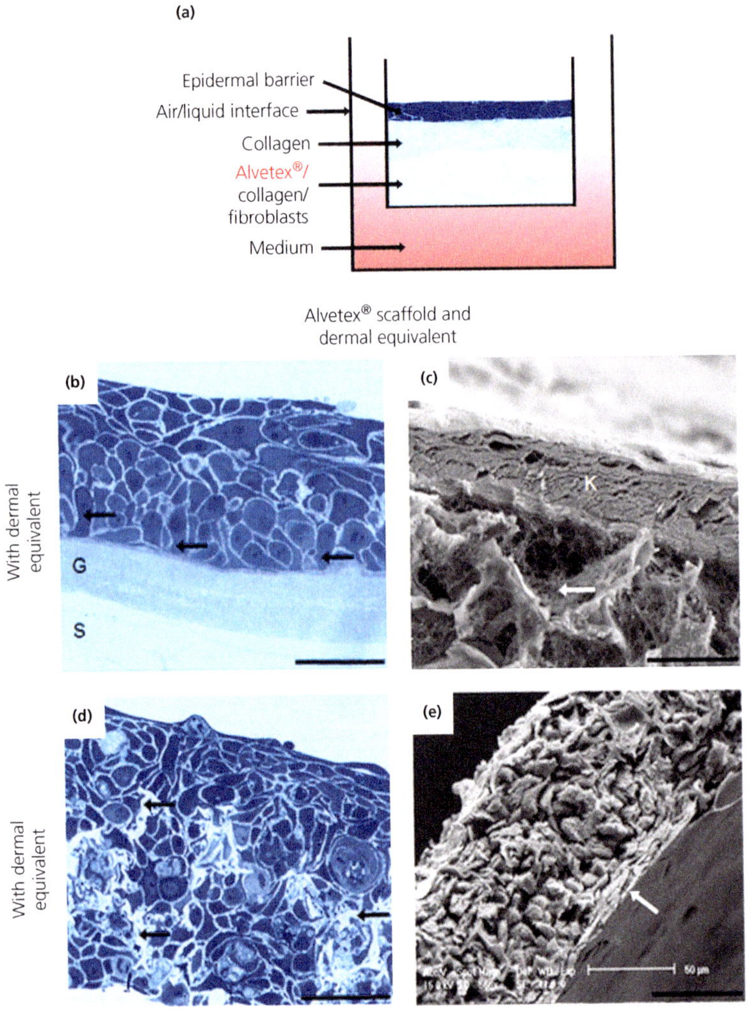

Figure 10.11 Growth of HaCaT keratinocytes on Alvetex Scaffold. (a) Schematic diagram illustrating how HaCaT keratinocytes cultures can be set up at the air-liquid interface in a well insert containing Alvetex Scaffold. There are several alternative permutations. For example, fibroblasts can be seeded into the scaffold prior to seeding keratinocytes on the surface. Alternatively, a collagen gel can be layered on the surface of the scaffold and keratinocytes seeded onto the surface of the gel. It is also feasible to use a combination of the gel and the fibroblasts. Collagen gels are classically used in RAFT cultures where keratinocytes are seeded onto the gel and brought to the air-liquid interface. However, such gels are notoriously difficult to work with and often shrink or tear. Working with a collagen gel layer on Alvetex Scaffold helps stabilise the gel by providing it with support during handling and reduces shrinkage. The growth of HaCaT keratinocytes was evaluated on Alvetex Scaffold in the presence (b,c) and absence (d,e) of a collagen gel coating. Fibroblasts were initially cultured in Alvetex prior to seeding of HaCaT cells (b,c). 3D cultures were imaged using either bright-field microscopy of resin-sectioned, toluidine blue-stained samples (b,d) or by scanning electron microscopy (SEM) (c,e). Image panels in detail: (b) toluidine blue-stained L R White

their surrounding stromal tissues are useful for the study of cancer cell invasion and hold potential for the testing of antimigratory drugs (Eccles *et al.*, 2005). The use of *in vitro* models to study cancer cell biology is preferred for both ethical and economical reasons. The tumour stroma is thought to influence the invasive potential of cancer cells (Kramer *et al.*, 2013). By culturing cancer cells together with fibroblasts in 3D, the crosstalk between the tumour stroma and cancer cells can be reproduced, providing a more accurate representation of the events occurring *in vivo*. Here we demonstrate the formation of stromal-tumoural co-cultures to study the invasion of SW480 colon carcinoma cells into the fibroblast-filled scaffold.

Establishing fibroblast cultures to mimic the stromal layer

1 Alvetex Scaffolds in six-well inserts are ideal for this purpose as they provide a suitable cell growth area to medium ratio for highly proliferative cell types.
2 To establish the fibroblast layer, seed NIH 3T3 cells onto the scaffolds. Trypsinise cells from existing 2D cultures and count using a haemocytometer. Produce a single cell suspension at a density of 5.0×10^6 cells per mL.
3 Immediately prior to cell addition, remove the final wash from the scaffold preparation process. Add 10 mL of complete growth medium to each well containing a scaffold.
4 Add 100 µL of cell suspension directly to the medium in the centre of the scaffold insert. Allow a few minutes for the cells to settle onto the scaffold.
5 Carefully transfer the plate to a cell culture incubator at 37 °C and 5% CO_2.
6 Culture the fibroblast-containing scaffolds for seven days with a complete medium change every 2–3 days.

Construction of co-cultures to assess cancer cell invasion

1 Transfer the established 3D fibroblast cultures to a well insert in a deep petri dish. Three inserts can be housed in one petri dish.

Figure 10.11 (Continued)
resin thin section (1 µm) of organotypic culture with 3D scaffold filled with collagen and fibroblasts cultured at the air-liquid interface for seven days (scale bar 100 µm). The scaffold (S) and layer of collagen gel (G) are labelled and black arrows indicate basal keratinocytes of columnar morphology. (c) SEM image of keratinocytes cultured for seven days at air-liquid interface supported by 3D Alvetex Scaffold prefilled with collagen and fibroblasts (scale bar 50 µm). The layer of keratinocytes (K) is labelled and the white arrow indicates presence of the collagen filaments in the 3D polymer scaffold. (d) Toluidine blue-stained L R White resin thin section (1 µm) of keratinocytes cultured within the 3D scaffold with no collagen or fibroblasts for seven days (scale bar 100 µm). The black arrow indicates the 3D scaffold. (e) SEM image showing keratinocytes cultured within the 3D scaffold with no collagen or fibroblasts at air-liquid interface for seven days (scale bar 50 µm). The upper surface of flattened stratified cells is indicated with a white arrow. Source: Courtesy of Ross Carnachan.

2 Add 50 mL of medium to the outside of the inserts to allow the medium to rise inside the insert to cover the substrate but not to go over the sides of the inserts (see notes).
3 Trypsinise the SW480 cells from their existing 2D culture substrates and count using a haemocytometer. Produce a single cell suspension at a density of 1×10^7 cells per mL.
4 Add 100 µL of cell suspension directly to the medium in the centre of the scaffold insert. Allow a few minutes for the cells to settle onto the scaffold.
5 Incubate the petri dishes overnight at 37 °C and 5% CO_2 to allow the cells to settle onto the stromal layer.
6 The following day, top up the medium to 70 mL, taking care to not dislodge cells from the scaffold.
7 Co-cultures are maintained for a further 10 days with a medium change every 2–3 days. After seven days, cultures are established and may be used to test antimigratory compounds. Additionally, a chemical gradient can be achieved across the 3D culture, depending on the level of medium surrounding the scaffolds (see notes).

Another method of 3D co-culture involves establishing a 3D layer of cells prior to addition of another cell type in a single cell suspension. Here we demonstrate its use as a cell invasion model. For this purpose, it is necessary to produce a uniform fibroblast-filled scaffold prior to the addition of the other cell type (Figure 10.12a). Once the fibroblast culture is established, a carcinoma line can be added. Figure 10.12 shows co-cultures of two cell lines – SW480 and SW620. These carcinoma cells are derived from early and late-stage tumours respectively and therefore demonstrate the migratory potential of cells either side of the metastatic progression of a tumour. Once these co-cultures are established, it is possible to measure the migration of the carcinoma cells into the stromal layer by image analysis of scaffold sections. Identification of the carcinoma cells and fibroblasts can be achieved via immunochemistry or by a fluorescent-tagged cell line.

Troubleshooting

Choosing the right format to suit your needs

In deciding which Alvetex format to use, the type of cell to be cultured, the length of the culture, the depth of cell penetration and the type of assay read-out should all be considered. Well inserts are designed to assist in the production of chemical gradients and connecting medium reservoirs after cell seeding. The well insert housing Alvetex consists of polystyrene walls with three equally spaced openings. This allows for different medium fill levels during and after cell seeding as detailed (Knight & Przyborski, 2014).

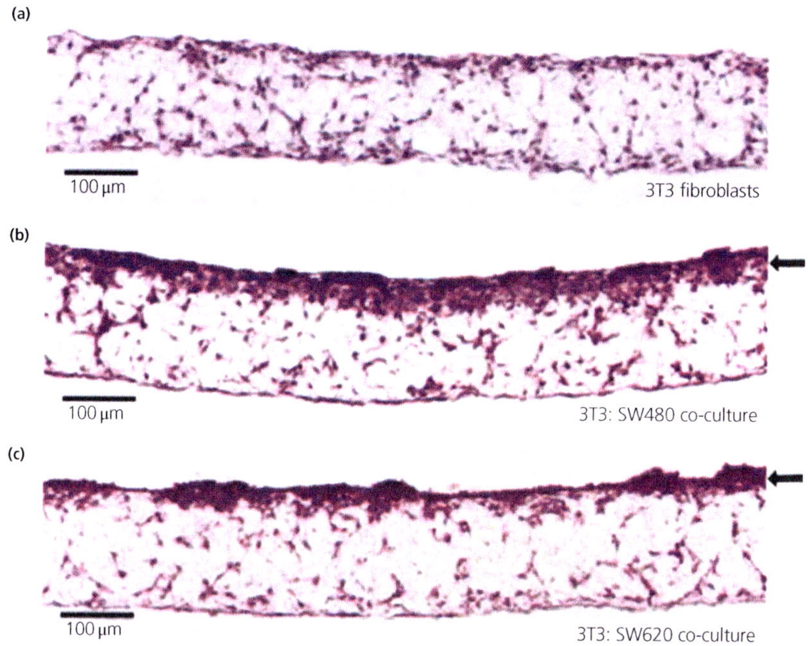

Figure 10.12 Development of 3D co-culture cell invasion models. Alvetex Scaffold provides an ideal platform to create layers of alternative cell types. In this example, we show how fibroblasts can be used to mimic the stromal layer to study cancer cell invasion through associated tissues. Scaffolds were paraffin embedded, sectioned and stained with H&E. (a) Scaffold packed with NIH 3T3 fibroblasts prior to the addition of cancer cells. Seeding carcinoma cells onto the surface of fibroblast-filled scaffolds holds potential as a 3D cell invasion assay. Here we demonstrate such co-cultures using the early and late-stage colon adenocarcinoma cells SW480 (b) and SW620 (c). Arrows indicate approximate location of co-cultured adenocarcinoma cells on the surface with the 3D fibroblast culture beneath. Scale bar 100 µm. Source: Rosie L. Adams.

Cell seeding optimisation

1 For most applications, initial cell seeding densities of $0.5–2.0 \times 10^6$ cells in 100–150 µL per membrane is suitable. Seeding in low volumes enables cells to attach to the disc and avoid cell loss.

2 When seeding, remove the wash medium thoroughly from the plate and carefully dispense cells onto the middle of the discs without touching the membrane.

3 Replace the lid and place the plate in a humidified incubator at 37 °C with 5% CO_2 for approximately three hours. This will facilitate cell attachment to the Alvetex membrane.

4 Gently flood the wells with medium. The amount of medium will depend on the format and the purpose of the experiment (see Figure 10.4c–e). For

experiments such as the creation of artificial skin constructs, in which cells must be grown at the air-liquid interface, it is important that the medium is only in contact with the Alvetex membrane from below.

Medium changes

To maintain medium quality, frequent changes are necessary, but they may disturb the culture. For these reasons, there are Alvetex components that enable users to grow cells in larger volumes of medium.

Conclusion

Alvetex provides a polystyrene environment for robust 3D culture. Three-dimensional cultures on Alvetex produce cells which more closely mimic cell physiology *in vivo* by avoiding cell flattening which can lead to remodelling of cell cytoskeleton and nuclear shape. Changes in cytoskeleton assembly have been shown to alter gene expression (Vergani *et al.*, 2004) and nuclear shape changes also lead to differences in gene expression and protein synthesis (Thomas *et al.*, 2002).

Alvetex Scaffold provides a versatile platform that enables cells to infiltrate the scaffold and maintain their natural 3D shape and organisation, thereby allowing the investigation of more *in vivo*-like cell behaviour and function than with conventional 2D model systems. The ability to grow more than one cell type simultaneously enables the study of complex intercellular interactions by more accurately simulating tissue microenvironments. These co-culture experiments can be valuable for understanding and observing cell behaviour and differentiation, for studying cell signalling and migration dynamics, and for recapitulating tissues in both health and diseased states, with the potential for developing novel therapeutics. Alvetex Scaffold is simple to use and enables the routine generation of genuine 3D cell co-cultures.

Acknowledgements

We would like to acknowledge Rosie Adams for her contribution in producing the model of colon cancer cell invasion used in this chapter, Ross Carnachan (Reinnervate Ltd) for developing the HaCaT cultures on Alvetex and Neil Robinson (Durham University, UK) for his work in the development of the full-thickness skin equivalent using Alvetex Scaffold technology described in this chapter.

We thank Simon Padbury for his help in procuring and designing the figures utilised in this chapter. Lastly, we would like to extend our thanks to Reinnervate Ltd, part of the ReproCELL Group and in part support from the Biotechnology and Biological Research Council (BBSRC) UK (awards BB/K019260/1; BB/I015825/1).

References

Alayoubi, A., Alqahtani, S., Kaddoumi, A. & Nazzal, S. (2013) Effect of PEG surface conformation on anticancer activity and blood circulation of nanoemulsions loaded with tocotrienol-rich fraction of palm oil. *Aaps Journal*, **15**(4), 1168–1179.

Amit, M., Carpenter, M. K., Inokuma, M. S. *et al.* (2000) Clonally derived human embryonic stem cell lines maintain pluripotency and proliferative potential for prolonged periods of culture. *Developmental Biology*, **227**(2), 271–278.

Aronin, C. E. P., Sadik, K. W., Lay, A. L. *et al.* (2009) Comparative effects of scaffold pore size, pore volume, and total void volume on cranial bone healing patterns using microsphere-based scaffolds. *Journal of Biomedical Materials Research Part A*, **89A**(3), 632–641.

Baker, B. M. & Chen, C. S. (2012) Deconstructing the third dimension – how 3D culture microenvironments alter cellular cues. *Journal of Cell Science*, **125**(13), 3015–3024.

Bhandari, R. N. B., Riccalton, L. A., Lewis, A. L. *et al.* (2001) Liver tissue engineering: a role for co-culture systems in modifying hepatocyte function and viability. *Tissue Engineering*, **7**(3), 345–357.

Blanpain, C. & Fuchs, E. (2009) Epidermal homeostasis: a balancing act of stem cells in the skin. *Nature Reviews Molecular Cell Biology*, **10**, 207–217.

Boelsma, E., Verhoeven, M. C. H. & Ponec, M. (1999) Reconstruction of a human skin equivalent using a spontaneously transformed keratinocyte cell line (HaCaT). *Journal of Investigative Dermatology*, **112**, 489–498.

Bokhari, M., Carnachan, R. J., Cameron, N. R. & Przyborski, S. A. (2007a) Culture of HepG2 liver cells on three dimensional polystyrene scaffolds enhances cell structure and function during toxicological challenge. *Journal of Anatomy*, **211**, 567–576.

Bokhari, M., Carnachan, R. J., Cameron, N. R. & Przyborski, S. A. (2007b) Novel cell culture device enabling three-dimensional cell growth and improved cell function. *Biochemical and Biophysical Research Communications*, **354**(4), 1095–1100.

Bokhari, M., Carnachan, R. J., Przyborski, S. A. & Cameron, N. R. (2007c) Emulsion-templated porous polymers as scaffolds for three dimensional cell culture: effect of synthesis parameters on scaffold formation and homogeneity. *Journal of Materials Chemistry*, **17**, 4088–4094.

Burkard, A., Dahn, C., Heinz, S. *et al.* (2012) Generation of proliferating human hepatocytes using upcyte (R) technology: characterisation and applications in induction and cytotoxicity assays. *Xenobiotica*, **42**(10), 939–956.

Carnachan, R. J., Bokhari, M., Przyborski, S. A. & Cameron, N. R. (2006) Tailoring the morphology of emulsion-templated porous polymers. *Soft Matter*, **2**(7), 608–616.

Cukierman, E., Pankov, R., Stevens, D. R. & Yamada, K. M. (2001) Taking cell-matrix adhesions to the third dimension. *Science*, **294**(5547), 1708–1712.

Dowling, J., Yu, Q. C. & Fuchs, E. (1997) Beta4 integrin is required for hemidesmosome formation, cell adhesion and cell survival. *Journal of Cell Biology*, **134**, 559–572.

Eccles, S. A., Box, C. & Court, W. (2005) Cell migration/invasion assays and their application in cancer drug discovery. *Biotechnology Annual Review*, **11**, 391–421.

Elsdale, T. & Bard, J. (1972) Collagen substrata for studies on cell behavior. *Journal of Cell Biology*, **54**(3), 626–637.

Fischbach, C., Chen, R., Matsumoto, T. *et al.* (2007) Engineering tumors with 3D scaffolds. *Nature Methods*, **4**(10), 855–860.

Fleischmajer, R., Utani, A., MacDonald, E. D. *et al.* (1998) Initiation of skin basement membrane formation at the epidermo-dermal interface involves assembly of laminins through binding to cell membrane receptors. *Journal of Cell Science*, **111**, 1929–1940.

Griffith, L. G. & Swartz, M. A. (2006) Capturing complex 3D tissue physiology in vitro. *Nature Reviews Molecular Cell Biology*, **7**(3), 211–224.

Hayman, M. W., Smith, K. H., Cameron, N. R. & Przyborski, S. A. (2004) Enhanced neurite outgrowth by human neurons grown on solid three-dimensional scaffolds. *Biochemical and Biophysical Research Communications*, **314**(2), 483–488.

Hayman, M. W., Smith, K. H., Cameron, N. R. & Przyborski, S. A. (2005) Growth of human stem cell-derived neurons on solid three-dimensional polymers. *Journal of Biochemical and Biophysical Methods*, **62**(3), 231–240.

Hill, D. S., Robinson, N. D. P., Caley, M. P. *et al.* (2015) A novel fully humanized 3D skin equivalent to model early melanoma invasion. *Molecular Cancer Therapeutics*, **14**, 2665–2673.

Khetani, S. R. & Bhatia, S. N. (2008) Microscale culture of human liver cells for drug development. *Nature Biotechnology*, **26**(1), 120–126.

Kimlin, L. C., Casagrande, G. & Virador, V. M. (2013) In vitro three-dimensional (3D) models in cancer research: an update. *Molecular Carcinogenesis*, **52**(3), 167–182.

Knight, E. & Przyborski, S. (2014) Method for simple and routine three-dimensional cell culture, in *Cellular In Vitro Testing* (eds J. Haycock, A. Ahluwalia & J. M. Wilkinson), Pan Stanford Publishing, Abingdon, UK, pp. 149–163.

Knight, E. & Przyborski, S. (2015) Advances in 3D cell culture technologies enabling tissue-like structures to be created in vitro. *Journal of Anatomy*, **227**, 746–756.

Knight, E., Murray, B., Carnachan, R. & Przyborski, S. (2011) Alvetex (R): polystyrene scaffold technology for routine three dimensional cell culture, in. *3D Cell Culture: Methods and Protocols* (ed. J. Haycock), Humana Press Inc., Totowa, New Jersey.

Kramer, N., Walzl, A., Unger, C. *et al.* (2013) In vitro cell migration and invasion assays. *Mutation Research/Reviews in Mutation Research*, **752**(1), 10–24.

MacDonald, C., Finlay, D. B., Jabed, A. *et al.* (2014) Development of positive control tissue for in situ hybridisation using Alvetex scaffolds. *Journal of Neuroscience Methods*, **238**, 70–77.

Maas-Szabowski, N., Starker, A. & Fusenig, N. E. (2003) Epidermal tissue regeneration and stromal interaction in HaCaT cells is initiated by TGF-α. *Journal of Cell Science*, **116**, 2937–2948.

Miki, Y., Suzuki, T., Abe, K. *et al.* (2010) Intratumoral localization of aromatase and interaction between stromal and parenchymal cells in the non-small cell lung carcinoma microenvironment. *Cancer Research*, **70**(16), 6659–6669.

Miki, Y., Ono, K., Hata, S. *et al.* (2012) The advantages of co-culture over mono cell culture in simulating *in vivo* environment. *Journal of Steroid Biochemistry and Molecular Biology*, **131**(3-5), 68–75.

Moody, J. (2013) Feeder-independent culture systems for human pluripotent stem cells. *Methods in Molecular Biology*, **946**, 507–521.

Murgia, C., Blaikie, P., Kim, N. *et al.* (1998) Cell cycle and adhesion defects in mice carrying a targeted deletion of the integrin beta4 cytoplasmic domain. *European Molecular Biology Organization Journal*, **17**, 3940–3951.

Neofytou, E. A., Chang, E., Patloia, B. *et al.* (2011) Adipose tissue-derived stem cells display a proangiogenic phenotype on 3D scaffolds. *Journal of Biomedical Materials Research Part A*, **98A**(3), 383–393.

Padron, J. M. & Peters, G. J. (2006) Cytotoxicity of sphingoid marine compound analogs in mono- and multilayered solid tumor cell cultures. *Investigational New Drugs*, **24**(3), 195–202.

Rhee, H. W. (2001) Permanent phenotypic and genotypic changes of prostate cancer cells cultured in a three-dimensional rotating-wall vessel. *In Vitro Cellular & Developmental Biology-Animal*, **37**(9), 127–140.

Schmeichel, K. L. & Bissell, M. J. (2003) Modeling tissue-specific signaling and organ function in three dimensions. *Journal of Cell Science*, **116**(12), 2377–2388.

Schutte, M., Fox, B., Baradez, M. O. *et al.* (2011) Rat primary hepatocytes show enhanced performance and sensitivity to acetaminophen during three-dimensional culture on a polystyrene scaffold designed for routine use. *Assay and Drug Development Technologies*, **9**(5), 475–486.

Sharma, R., Barakzai, S. Z., Taylor, S. E. & Donadeu, F. X. (2016) Epidermal-like architecture obtained from equine keratinocytes in three-dimensional cultures. *Journal of Tissue Engineering and Regenerative Medicine*, **10**, 627–636.

Simpson, C. L., Patel, D. M. & Green, K. J. (2011) Deconstructing the skin: cytoarchitectural determinants of epidermal morphogenesis. *Nature Reviews Molecular Cell Biology*, **12**, 565–580.

Thomas, C. H., Collier, J. H., Sfeir, C. S. & Healy, K. E. (2002) Engineering gene expression and protein synthesis by modulation of nuclear shape. *Proceedings of the National Academy of Sciences USA*, **99**(4), 1972–1977.

Truskey, G. A. (2010) Endothelial cell vascular smooth muscle cell co-culture assay for high throughput screening assays for discovery of anti-angiogenesis agents and other therapeutic molecules. *International Journal of High Throughput Screening*, **2010**(1), 171–181.

Vergani, L., Grattarola, M. & Nicolini, C. (2004) Modifications of chromatin structure and gene expression following induced alterations of cellular shape. *International Journal of Biochemistry & Cell Biology*, **36**(8), 1447–1461.

Wake, M. C., Patrick, C. W. & Mikos, A. G. (1994) Pore morphology effects on the fibrovascular tissue-growth in porous polymer substrates. *Cell Transplantation*, **3**(4), 339–343.

Weaver, V. M., Lelievre, S., Lakins, J. N. *et al.* (2002) Beta 4 integrin-dependent formation of polarized three-dimensional architecture confers resistance to apoptosis in normal and malignant mammary epithelium. *Cancer Cell*, **2**(3), 205–216.

Zeltinger, J., Sherwood, J. K., Graham, D. A. *et al.* (2001) Effect of pore size and void fraction on cellular adhesion, proliferation, and matrix deposition. *Tissue Engineering*, **7**(5), 557–572.

CHAPTER 11

CelluSponge™ and Go Matrix as innovative three-dimensional cell culture platforms

Bramasta Nugraha[1,2]

[1] *Department of Biosystem Science and Engineering, Swiss Federal Institute of Technology (ETH Zürich), Basel, Switzerland*

[2] *Roche Pharmaceutical Research and Early Development, Basel, Switzerland*

Introduction

Advances in 3D cell culture technology have sparked interest in introducing complexity into cultured cells *in vitro* for various biomedical and pharmaceutical applications, ranging from bioartificial-assisted devices to drug and antiviral screening platforms (Gautier *et al.*, 2011; Molina-Jimenez *et al.*, 2012; Ploss *et al.*, 2010; Xia *et al.*, 2009). This new concept of culturing cells in 3D was initiated by the notion that cells behave differently outside their native niche, especially when cultured on conventional flat tissue culture flasks, known as 2D culture (Pampaloni *et al.*, 2007). Since 2D culture is unable to reproduce the complexity of *in vivo* tissue in *in vitro* models, there is an inherent discrepancy between how real tissue behaves in its physiological microenvironment and in cultured cells (Pampaloni *et al.*, 2007). The possibility of culturing cells in a 3D microenvironment, mimicking the events occurring in real organs, has the potential to enhance the investigation of various biological responses that cannot be directly performed in organs due to ethical concerns and source scarcity. Culturing cells in a 3D culture and having a stronger correlation with the studied organ could then minimise the use of expensive and labour-intensive animal studies, especially in disease modelling and drug screening.

When developing 3D cell culture technologies, it is important to mimic how cells interact with neighbouring cells and their own native matrices *in vivo*. Reproducing these cell-cell and cell-matrix interactions *in vitro* can make the study of cell behaviour more physiologically accurate. In this way, 3D cell culture offers the possibility to incorporate what previous researchers had ignored by simply plating cells on a flat tissue culture flask.

Technology Platforms for 3D Cell Culture: A User's Guide, First Edition. Edited by Stefan Przyborski.

Some important factors in the scaffold design and fabrication that have to be strictly controlled include pore size and porosity to maintain proper diffusion of nutrients and gas exchange for cellular growth, and scaffold surface chemistry to present proper cues to the cells grown in the region. When the pore size of the scaffolds is larger than 500 μm, cells can no longer migrate and reorganise, as they do not recognise the surface any more. Pore size with high interconnectivity should ideally be in the range of 50–150 μm for hepatocyte culture (Wu *et al.*, 2007). Within these dimensions, hepatocytes' cell-cell contact would be ensured in order to maintain their differentiated functions when compared to 2D systems. When cultured on 2D collagen monolayer, hepatocytes tend to dedifferentiate rapidly and lose their liver-specific functions (Wen *et al.*, 2009). The surface chemistry of the scaffold is known to influence cell adhesion, proliferation and functionality. When hepatocytes were cultured on an arginine-glycine-aspartate (RGD) modified polyethylene glycol gel and type I collagen-modified polysulfonate sponges, long-term enhancement of liver differentiated functions was observed, but not on the unmodified respective substrates (Kinasiewicz *et al.*, 2007; Underhill *et al.*, 2007). Thus, the incorporation of cell adhesion-specific ligands on the scaffold surface is paramount to promote cell adhesion.

The raw materials to make a scaffold can be natural materials such as extracellular components (collagen, fibronectin, fibrin, laminin, Matrigel®) or decellularised liver matrices. However, scaffolds made with these natural materials are not cost-effective, impose batch-to-batch variation and may induce immunogenic reactions. On the other hand, synthetic materials are advantageous in their chemistry, making them easier to control and tune for multiple purposes. The majority of scaffold-based tissue engineering technologies utilise synthetic polymers such as polyglycolic acid (PGA), polylactic acid (PLA), or polyhydroxyl alkanoate (PHA). These polymer scaffolds are designed to guide cell organisation and growth, allowing *in vivo*-like diffusion of nutrient to the cells.

The fundamental development of our in-house cellulosic sponge scaffold (CelluSponge™) was inspired by the absence of a platform which combines the advantages of having macroporous diffusible networks for good mass transfer, soft mechanical stiffness needed to culture soft tissues, i.e. hepatocytes, and the ease of functionalisation with cellular ligands/modifiers. Hydroxypropyl cellulose is an FDA-approved cellulose derivative, which has been used as a material for commercial artificial tears (Lacrisert) (Palakuru *et al.*, 2008; Prause, 1986). The mixture of hydroxypropyl cellulose and water shows an interesting temperature-sensitive behaviour and forms a biphasic stable state beyond its liquid crystal solution temperature (LCST) (Spontak & Hirsch, 2002). This temperature sensitivity is due to the response of hydroxypropyl cellulose colloidal nanoparticles to increasing temperature. In 2000, Hu *et al.* introduced the primary concept of making porous cellulose hydrogel from cellulose nanoparticle networks. The advantages of this process are that it has two levels of structural hierarchy: primary network of polymer chain and secondary network of cellulose nanoparticles. In comparison with other porous gels, other advantages of this nanoparticle network hydrogel are high uniformity and easily tunable mesh sizes.

CelluSponge allows for the formation of three-dimensional microtissues, commonly known as multicellular aggregates or spheroids. 3D spheroid culture has been extensively used *in vitro* for various cell types, such as cancer, primary cells and stem cells (Abu-Absi *et al.*, 2002; Baraniak & McDevitt, 2012; Fischbach *et al.*, 2007; Ong *et al.*, 2007; Pampaloni *et al.*, 2007). The extensive cell-cell interactions in spheroids provide a closer reflection of the *in vivo* environment, which can enhance the function to successfully mimic the original tissue. These spheroids can be maintained without vascularisation when the size of spheroid diameter is within diffusible dimensions (less than 200 μm) to allow penetration of sufficient nutrients and oxygen (Curcio *et al.*, 2007).

Macroporous 3D CelluSponge development as 3D cell culture platform

Macroporous scaffolds (pore size larger than 50 μm) are beneficial to support the cells, as these pores facilitate migration and growth of the cells inside the pores, and allow sufficient nutrient access and removal of metabolites (Yue *et al.*, 2012). As depicted in Figure 11.1, initially the macroporous CelluSponge was synthesised through side chain chemical partial substitution of hydroxypropyl cellulose with allyl isocyanate, a chemical with an unsaturated double bond, which acts as a crosslinking group. The macroporous scaffolds produced have high interconnected porosity (50% or higher) and total porosity of 80% or higher. The macropores have pore size distribution peaking at above 50 μm. i.e. 90–100 μm, 85% water content and Young's modulus of about 10–20 kPa in a hydrated state.

The clear polymer solution (Figure 11.2a) was heated up to 313 K to induce formation of a meta-stable colloidal system whose structures are pictured in Figure 11.2b. There are micro/nano-spheres present in the colloidal system as shown in the dynamic light scattering (DLS) and scanning electron microscope (SEM) images at this temperature (Figure 11.2c). Gamma (γ) irradiation was selected to crosslink the colloidal system due to its deep penetration ability and its clean chemistry for uniform activation and fixation of the colloidal structure without using chemical initiators.

Abundant hydroxyl groups present in hydroxypropyl cellulose facilitate side chain modification to design both the non-cleavable and cleavable CelluSponge. Side chain modification with allyl isocyanate, as described previously, creates a physically and chemically stable cellulosic sponge, which can be useful for 3D cell culture that does not require harvesting of cultured cells. On the other hand, side chain modification with dithiodipropionic acid and 2-amino ethyl methacrylate creates a chemically cleavable CelluSponge (CelluSponge CB) that can

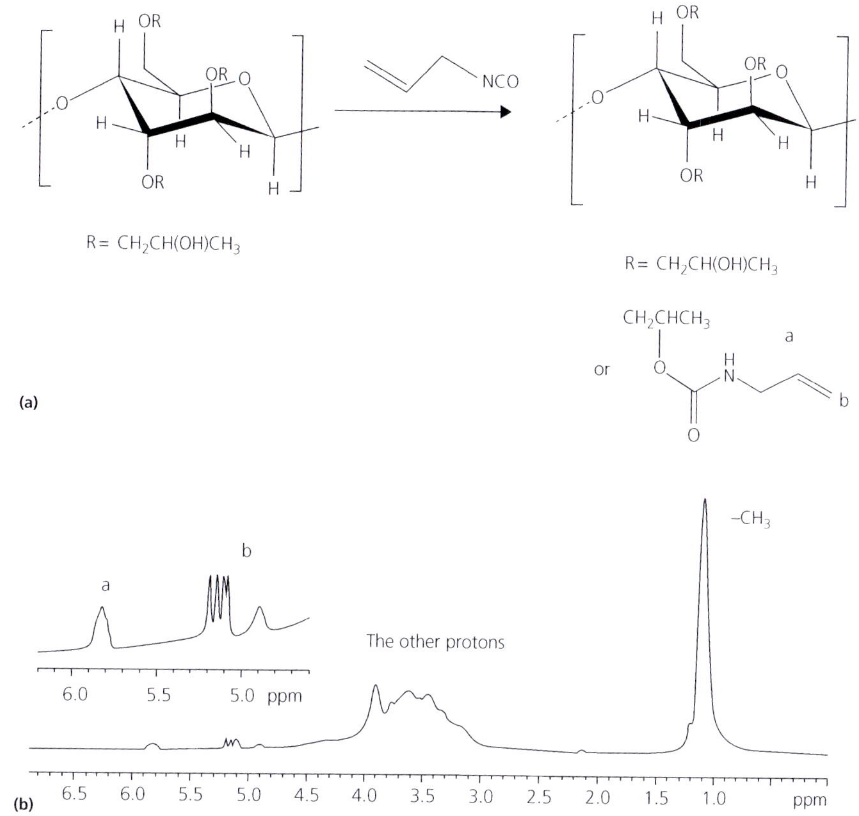

Figure 11.1 CelluSponge synthesis with side chain modification. (a) Hydroxypropyl cellulose side chain modification with allyl isocyanate and its NMR spectra (b). Source: Yue, Z., Wen, F., Gao, S. *et al.* 2010. Reproduced with permission of Elsevier.

be used for 3D cell culture, which requires only temporary scaffolding construct (Figure 11.3). The sponge can be easily removed upon demand with a mild reaction that occurs in the cell's physiological conditions (Nugraha & Yu, 2014).

Both types of sponges have been fabricated with γ-ray irradiation that helps to create uniform crosslinking and ensure reduce batch-to batch variability in the mass production of CelluSponge. The dimension and shape of the sponge can also be tailored depending on its use, making it suitable as a high-throughput platform. In addition, both types of sponge can be further ligand modified for the growth of cells with specific needs.

Go Matrix development as a 3D cell culture platform

The major advances of modern cell biology have mostly been established through 2D culture because of its convenience in operation and facility for observation. However, more and more studies demonstrate that cells cultured on 2D substrates do not provide a true physiological reflection of what occurs *in vivo*. Thus, it is important to develop 3D cell culture to capture these physiological

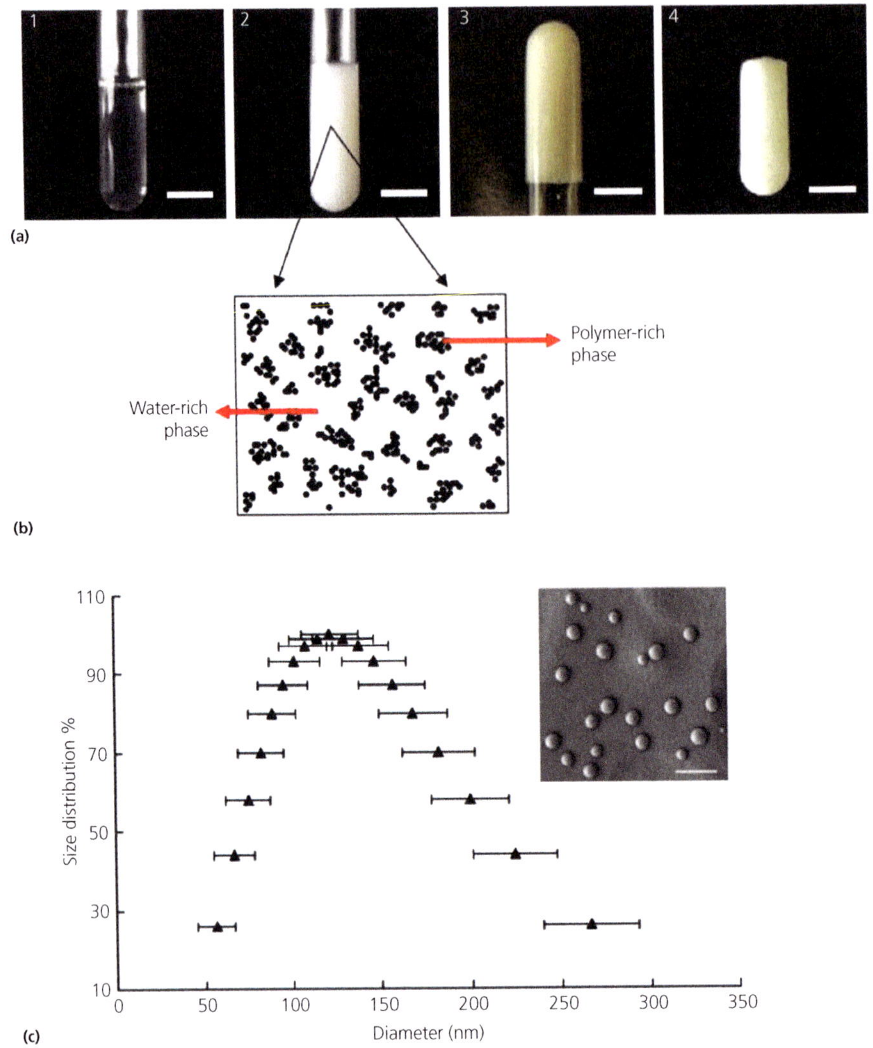

Figure 11.2 CelluSponge preparation. (a) Preparation of CelluSponge (scale bar 1.0 cm) from a homogeneous solution of 10 wt% uncrosslinked polymer solution in water at room temperature (1), to phase separated colloidal fluid at ~ 313 K (2), crosslinked heterogeneous gel at room temperature (3) and freeze-dried scaffold (4). (b) Schematic illustration of the phase separated colloidal network. (c) Size distribution of the biphasic colloidal fluid measured by dynamic light scattering; insert image is the SEM micrograph of colloidal nano/microparticles (scale bar 1.0 μm). Results are shown as the average values ± standard deviation (n = 4). Source: Yue, Z., Wen, F., Gao, S. *et al.* 2010. Reproduced with permission of Elsevier.

conditions *in vitro* for applications such as tissue engineering, disease modelling and drug screening.

Adherent cells need to attach to their surroundings for growth. Scaffolds are a 3D culture support, consisting of porous materials of either natural origin or man-made polymers. Different approaches to fabricating culture matrices have

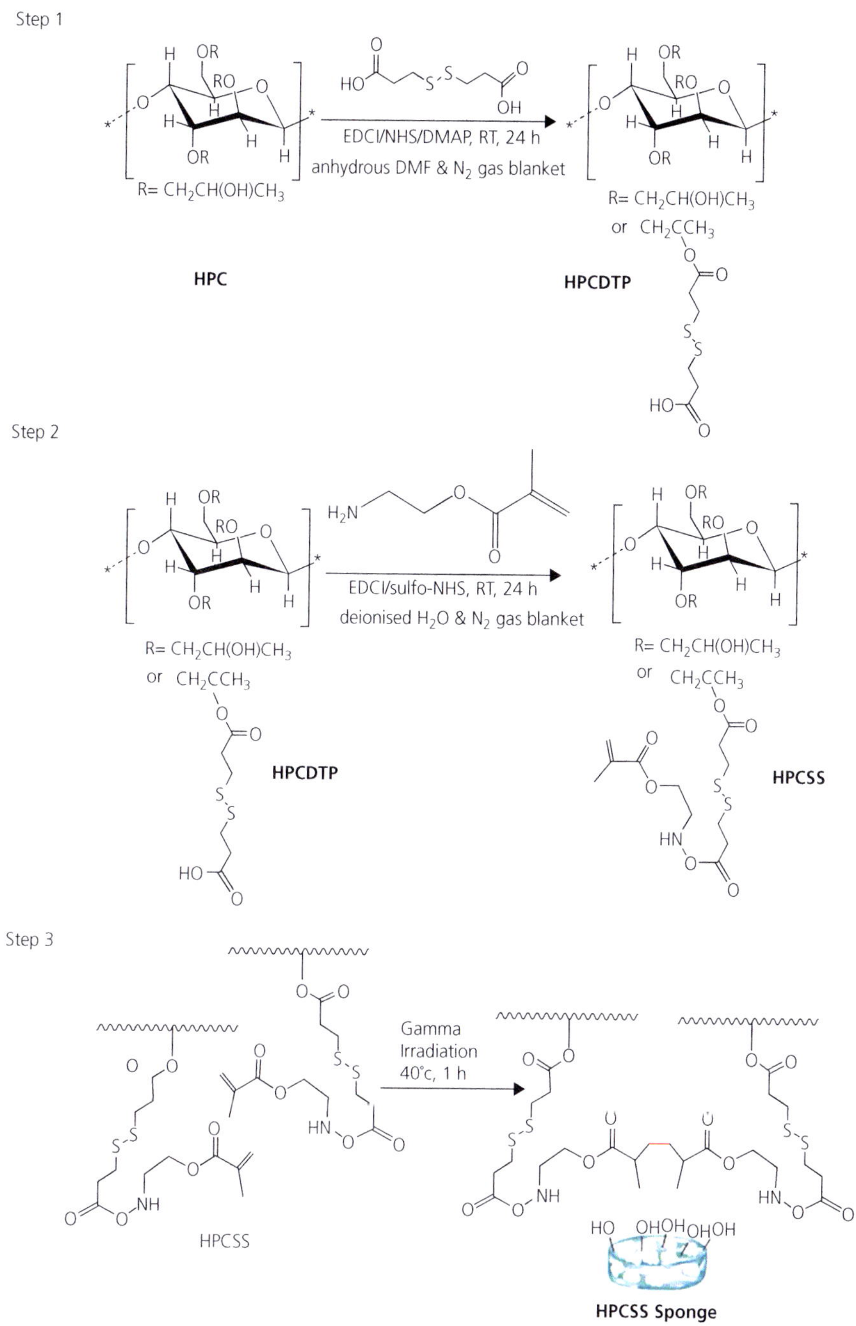

Figure 11.3 **Schematic diagram of cleavable cellulosic sponge synthesis and fabrication (CelluSponge-CB/HPCSS).** Source: Adapted from Nugraha, B. & Yu, H. 2014.

their pros and cons. For example, though Matrigel is a widely used extracellular matrix (ECM), it is also known for its uncharacterised composition, low permeability and weak mechanical support. While tuning gel concentration, the stiffness and permeation change at the same time. The purpose is to incorporate shape, structure, biochemical cues and mechanical parameters to culture models *in vitro* to form a microenvironment with controlled porosity, stiffness and adhesive ligands.

Bio-Byblos Biomedical adapted techniques developed by Lin's lab to build a platform for 3D cell culture consisting of gelatin scaffolds with uniform and ordered pores commercialised under the name Go Matrix. The fabrication scheme involves utilising a microfluidic device to generate monodispersed foam crystal as a template. The liquid foam then congeals before the foam coarsens. After the liquid foam turns solid, it is immersed in buffer under vacuum to be degased. The solid foam becomes open and all the pores are interconnected. At this point, many chemical steps can be carried out such as quenching and conjugation. Finally, the open solid foam is readily used as a cell culture scaffold.

The gelatin used in Go Matrix originated from porcine skin. The stiffness of solidified gelatin gel ranges from 120 kPa to 420 kPa. The porosity of Go Matrix is kept around 74%. The biodegradability is adjustable for the duration of experiment and fluorescence can be modified for imaging purposes. Currently, we have product lines of 60, 90, 130 μm pore Go Matrix. By adopting an appropriate pore size, we can optimise the seeding efficiency and growth curves for different cell types loaded in Go Matrix can be established. For example, with NIH-3T3 as the cell model, 60–90 μm pore Go Matrix will readily retain fibroblasts' 3D morphology in their spindle-like shape, whilst 150–200 μm pore Go Matrix (the exact dimension could be customised; pore size up to 220 μm is now available) is more suitable for mouse cardiomyocytes, which possess larger cell volume than other cell types. Larger interconnected pores ensure better seeding density. With its high interconnectivity and orderliness provided by its isotropic structure, Go Matrix provides a uniform microenvironment for cell-matrix and cell-cell interactions and thus facilitates observation.

The mechanical stiffness of the microenvironment affects cellular behaviour. Our company has another product line of polyacrylamide (PA) scaffolds under development. The stiffness of the PA gel can be tuned from 1 to 70 kPa. The surface can be coated with different ECM proteins such as collagen and fibronectin. This product line enables researchers to choose the appropriate mechanical properties for their cells to facilitate purposes such as differentiation.

CelluSponge applications

3D aggregate culture and stem cell differentiation

Fibroblast cell line NIH3T3 was cultured in the sponge to test cell compatibility. The increase in alamarBlue® percentage reduction (reduced %) from day 1 to day 5 indicates that the cells are not only viable but also proliferative over this period (Figure 11.4a). The majority of the cells are viable even in the core of the aggregate (Figure 11.4b). The SEM image shows 3D cell morphology, different from what was normally observed in conventional 2D culture (Figure 11.4c–f). Other cell types such as C3A, MCF-7 and HFF also show similar phenotypes (data not shown).

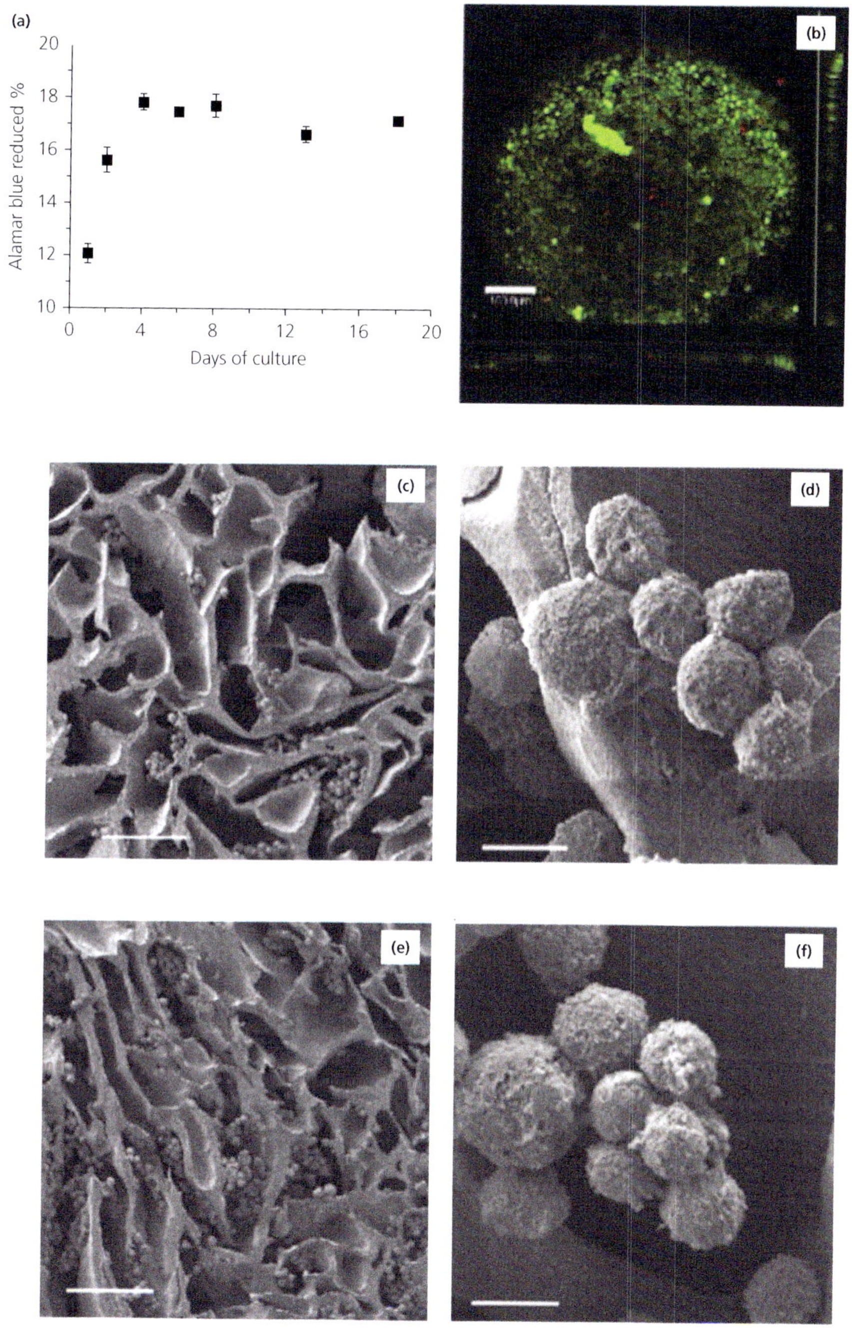

Figure 11.4 **Culture of NIH3T3 in CelluSponge**. (a) Cell proliferation is monitored by alamar Blue assay; data are shown as the average values ± standard deviation (n = 3). (b) Live (green)/dead (red) staining of NIH3T3 after four weeks culture (scale bar 100 μm). (c,d) SEM images of NIH3T3 cultured for one day at low (scale bar 100 μm) and high magnification (scale bar 10 μm), respectively. (e,f) SEM images for day 5 at low (scale bar 100 μm) and high magnification (scale bar 10 μm), respectively. Source: Yue 2010. Reproduced with permission of Elsevier.

To demonstrate the versatility of the sponge surface modification with extracellular cues, the sponge was activated by 1,1′-carbonyldiimidazole (CDI) and conjugated with collagen type I (Figure 11.5). X-ray photoelectron spectroscopy (XPS) detected the increased level of atomic nitrogen. Live/dead staining of HFF cultured in the collagen-modified sponge showed spreading cell morphology due to enhanced cell-matrix interaction.

CelluSponge surface modified with collagen (CelluSponge-Collagen) was used to systematically study the growth and neural differentiation of human mesenchymal stem cells (hMSCs). As shown in Figure 11.6, hMSCs began to differentiate into neural cells after being cultured in neural differentiation

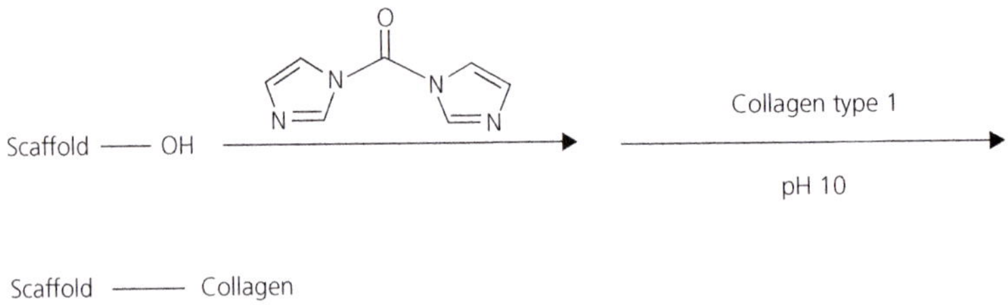

(a)

XPS analysis of atomic ratio (%)

Samples	C%	O%	N%
Unmodified scaffold	66.23	33.10	0.67
Scaffold treated with CDI	76.83	20.30	2.87
Scaffold treated with CDI and collagen 1	66.53	32.20	1.27

(b)

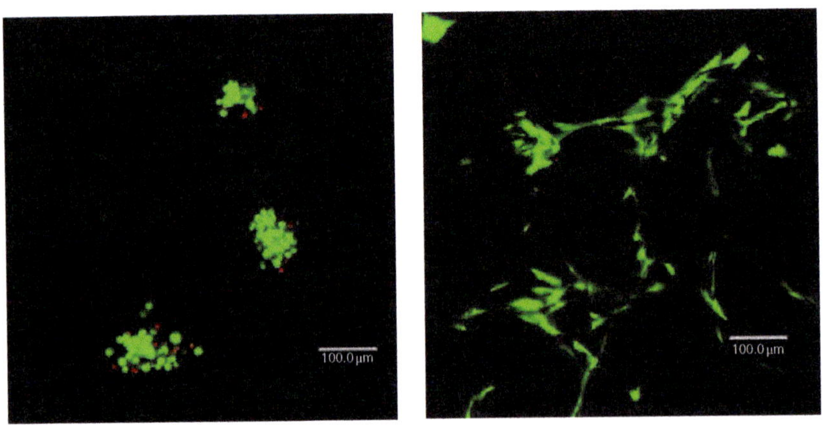

(c)

Figure 11.5 Surface modification of CelluSponge with type I collagen (CelluSponge-Collagen). (a) Surface modification synthesis scheme. (b) XPS analysis of surface-modified scaffold surfaces. (c) Fluorescent live/dead staining of HFF in unmodified (*left*) and collagen conjugated (*right*) scaffolds. Scale bar 100 μm. Source: Yue 2010. Reproduced with permission of Elsevier.

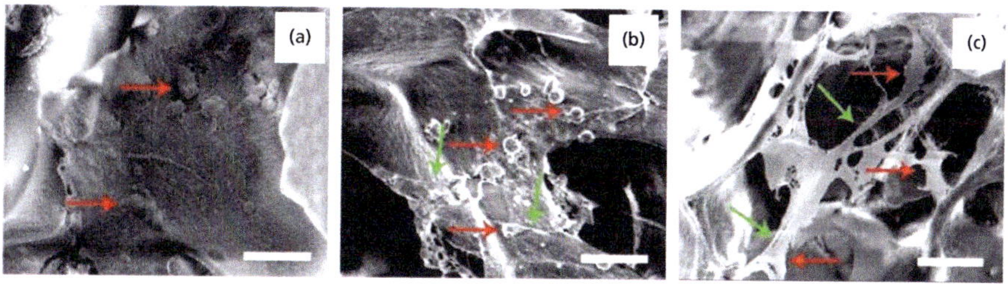

Figure 11.6 **SEM images of neural differentiation of hMSCs in CelluSponge-Collagen at various time intervals.** (a) Two days, (b) seven days and (c) 14 days. The red and green arrows indicate cell body and neurites, respectively. Scale bar 30 μm. Source: Adapted from Gua *et al.* 2010.

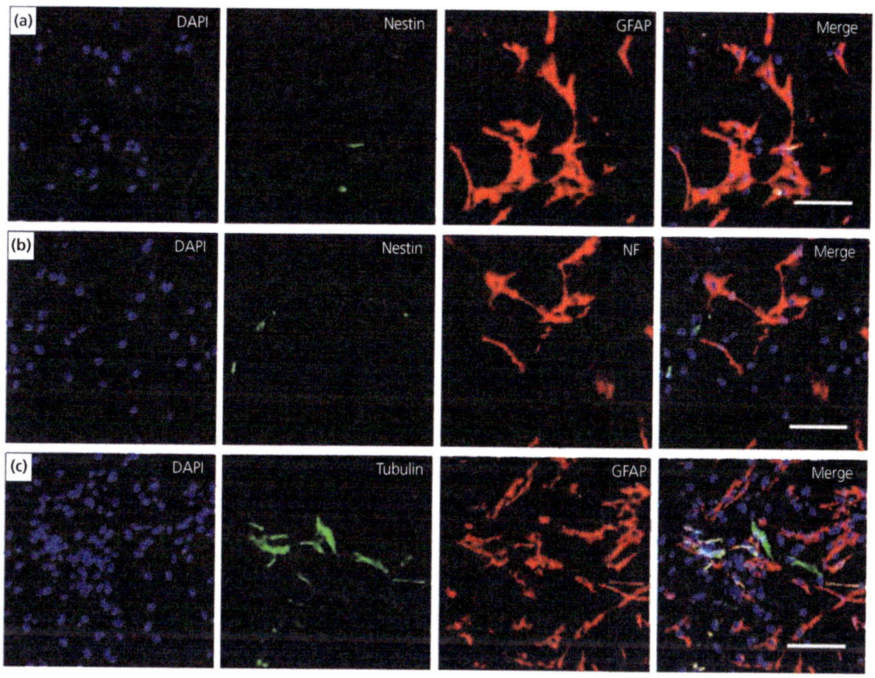

Figure 11.7 **Confocal images of the complete differentiation of hMSCs into neurons or glial cells in CelluSponge-Collagen after 14 days' culture.** (a) Some induced hMSCs only show GFAP-positive staining. (b) Some induced hMSCs only show NF-positive staining. (c) Some induced hMSCs in sponge showing tubulin of GFAP-positive signal. Scale bar 100 μm. Source: Adapted from Gua *et al.* 2010.

medium for two days; however, the neurites of induced cells were still short. Seven days after induction, the induced cells on the 3D sponge showed neuron-like or glial-like morphologies and neurites were longer. The image from day 14 after induction shows the longest neurites.

After 14 days of cultivation in neural differentiation medium, all the induced cells were nestin negative and some were stained with β-tubulin III, NF or GFAP antibodies (Figure 11.7). This indicates successful hMSC differentiation into

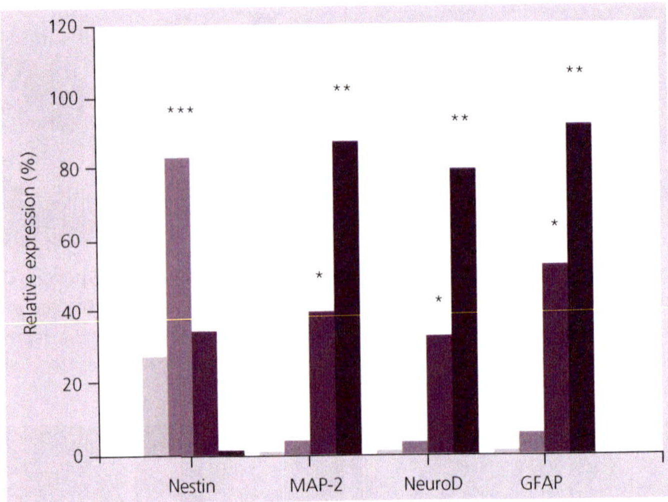

Figure 11.8 mRNA transcript levels of neural cells differentiated from hMSCs. Comparison of nestin, microtubule-associated protein 2, NeuroD and glial fibrillary acidic protein mRNA transcript levels of neural cells differentiated from hMSCs cultured in CelluSponge-Collagen at different time points by RT-PCR (light to dark colour gradients range from day 0 (undifferentiated hMSCs), 2, 7 and 14). Results represent the average ± s.e.m. (n = 3). Significant differences determined by one-way analysis of variance. *P < 0.05 compared with two days, **P < 0.05 compared with seven days, ***P < 0.01 compared with undifferentiated hMSCs. GFAP, glial fibrillary acidic protein; MAP, microtubule-associated protein. Source: Adapted from Gua *et al.* 2010.

mature neuron or glial cells in CelluSponge-Collagen, also confirmed by real-time polymerase chain reaction (RT-PCR) in Figure 11.8.

3D primary hepatocyte culture

In vitro platforms for drug safety testing

In the pharmaceutical industry, new drugs are being developed through rigorous processes that involve compound target validation, governing absorption, distribution, metabolism, excretion and toxicity (ADME Tox). ADME Tox is critical to determine whether a compound target is safe enough to enter the clinical stage. This remains the major challenge for the pharmaceutical industry, as evidenced by early drug withdrawal or severe use limitation of marketed drugs due to unexpected adverse effects (lack of efficacy, toxicity and unfavourable pharmacological properties) (Gebhardt *et al.*, 2003). Advances in system biology, genomics, epigenetics, bioinformatics and proteomics have led to a swift increase in lead compound production. This has increased the importance of ADME Tox prediction before the preclinical stage, contrary to what was previously done. Screening lead compounds before the preclinical stage will tremendously reduce the cost of these trials and predict the toxicity of the compounds more accurately

By knowing the target compound's metabolic profile, predicted by hepatocyte culture models, it can be predicted whether orally administered drugs will be metabolised by cytochrome P450 enzymes in the liver before moving to the systemic circulation. Additionally, if the liver-metabolised target compound resulted in toxic by-products, it would be possible to find an alternative and administer the drugs via another route with comparable pharmacological activity. Finally, it can also be investigated which cytochrome P450 is important in metabolising certain compounds (Ananthanarayanan *et al.*, 2011). Elucidating the metabolic mechanism can extend analysis of the target toxicity from various dose concentrations or interspecific toxicity responses, and from acute to chronic exposure (Ananthanarayanan *et al.*, 2011). The importance of these hepatocyte-based *in vitro* platforms is to bridge the gap between laboratory animal models and human tissue, as well as to closely predict compound toxicity.

Rat hepatocytes cultured on three different platforms (collagen monolayer, galactosylated polyethylene terephthalate (PET) membrane and galactose-modified CelluSponge (CelluSponge-Galactose) revealed platform-dependent cell behaviours (Figure 11.9). Hepatocytes cultured on collagen monolayers were relatively flat and spread on day 3 onwards, which correlated with loss of differentiated functions (Du *et al.*, 2006). Hepatocyte spheroid formation on galactosylated PET membranes (2D PET Gal membrane) took three days, and often collided with adjacent spheroids, thus forming larger spheroids. In CelluSponge-Galactose cultures, hepatocytes immediately organised into 3D spheroids within one day of culture, and remained stable in this configuration until at least day 6. Hepatocyte spheroids formed in CelluSponge-Galactose were smaller than those formed on 2D PET Gal membrane, with spheroid diameter $60.7 \pm 15.9\,\mu m$ and $108.1 \pm 19.2\,\mu m$, for each platform, respectively. In addition, spheroids formed in CelluSponge-Galactose were constrained by the sponge pores and thus did not easily detach as did those plated on 2D PET Gal membranes.

Hepatocyte spheroids cultured between days 1 and 7 showed gradual increases in surface smoothness and disappearance of the cell-cell boundaries (Figure 11.10a). On day 1, cell morphology was spherical. The nature of the hydrophilic and soft hydrogel sponges would prevent the cells from spreading, which normally occurs on hard substrates (Wells, 2008). Tethered spheroids on the sponge surface were observed; the arrows in the figure show the contact points where the spheroids adhered to the sponge surface. Hepatocyte spheroid viability, which was assessed by co-staining live and dead cells using CellTracker™ Green (CTG) and propidium iodide (PI), respectively, showed no PI signal which revealed good viability maintenance from day 1 to day 7 in culture (Figure 11.10b). The CTG signals illustrated indistinguishable borders between single cells in the spheroids, which reflected the tightness of the cell-cell contacts.

Immunofluorescence staining of F-actin, E-cadherin and MRP2/CD147 in the hepatocyte spheroids 48 hours post seeding, in comparison to collagen

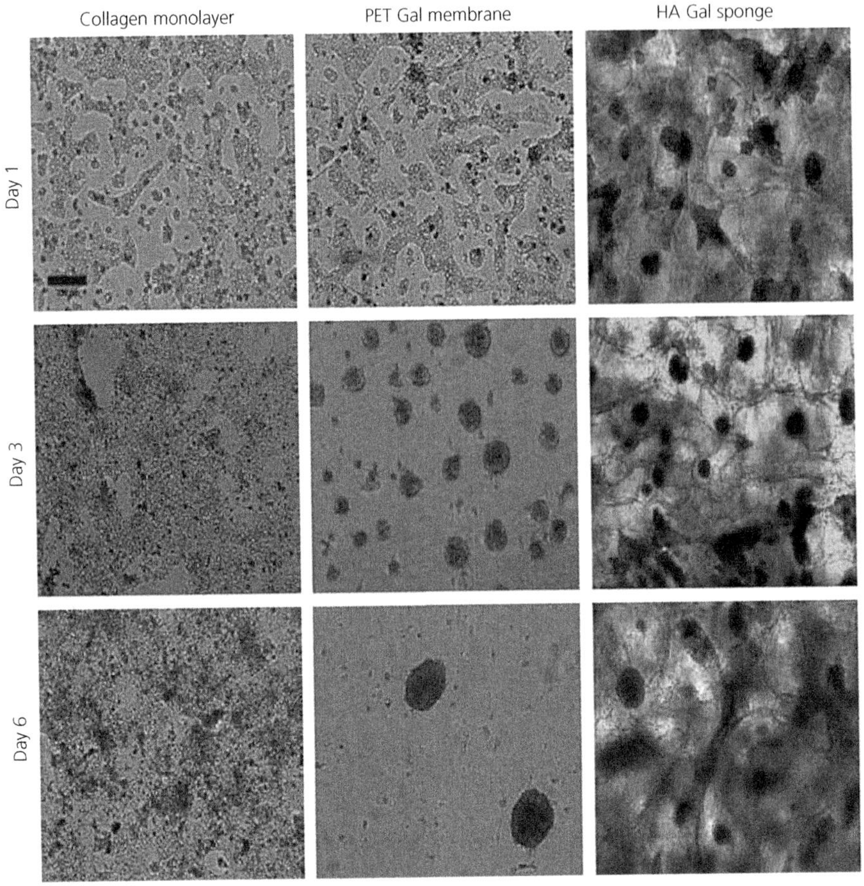

Figure 11.9 Phase contrast images of rat hepatocyte cultured in three different platforms. Scale bar 100 μm. Diameter of hepatocyte spheroids formed on PET Gal membrane on day 3: 39.8 ± 8.1 μm, day 6: 108.1 ± 19.2 μm, and on CelluSponge-Galactose/HA Gal sponge day 1: 46.1 ± 9.3 μm, day 3: 55.7 ± 21.1 μm, day 6: 60.7 ± 15.9 μm. Source: Nugraha *et al.* 2011. Reproduced with permission of Elsevier.

sandwich control, showed localisation of these markers (Figure 11.11a). As would be expected in non-spreading cells, F-actin staining revealed that the actin cytoskeleton had a predominant cortical localisation in both sponge and collagen sandwich cultures and an absence of stress fibres. E-cadherin staining, a marker of cell-cell junctions, demonstrated that cells in the hepatocyte spheroids have tight associations with neighbouring cells. E-cadherin expression also supports the maintenance of cell viability during long-term culture (Luebke-Wheeler *et al.*, 2009). MRP2/CD147 staining marked the apical and basolateral domains of the hepatocytes, respectively. In the MRP2/CD147 staining image (see Figure 11.11a, rightmost panel), the signals showed a comparable and non-co-localised signal as observed in the collagen sandwich culture.

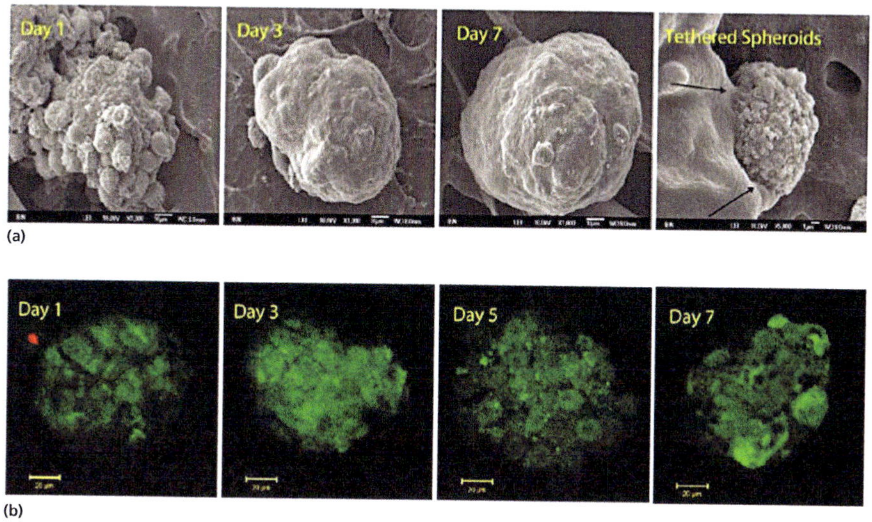

Figure 11.10 SEM of spheroids in CelluSponge. (a) SEM images of hepatocyte spheroids formed in CelluSponge-Galactose/HA Gal sponge. (b) Hepatocyte spheroids cultured in CelluSponge-Galactose (scale bar 20 μm). Source: Nugraha *et al.* 2011. Reproduced with permission of Elsevier.

Ultrastructural views observed by transmission electron microscopy illustrated the subcellular microstructures located inside the spheroids. The images of hepatocyte spheroids cultured for 48 hours demonstrated a space between neighbouring cells, reminiscent of the bile canaliculi with the presence of microvilli (Figure 11.11b) (Harada *et al.*, 2003; Talamini *et al.*, 1997). Tight junctions between the two cells were also clearly observed.

Several hepatocyte functions are thought to be dependent on the polarised phenotype of the cells, including biliary excretion, albumin secretion and urea synthesis. Since early formation of the spheroid, polarity of hepatocytes in the cellulosic sponges was observed, but it is important to determine if these functions were also enhanced. Biliary excretion was examined by the addition of fluorescein diacetate (FDA) dye at various time intervals, including 16, 24 and 48 hours post seeding (Figure 11.11c). Polarised hepatocytes formed bile canaliculi structures between neighbouring cells that contain the MRP2 transporter. Viable cells in the spheroids will cleave FDA into a fluorescein dye by intracellular esterases; the dye is then excreted by MRP2 into the bile canaliculi. FDA staining in the hepatocyte spheroids formed in CelluSponge showed an accumulation in the bile canaliculi between two cells, starting from 16 hours post seeding, significantly faster than has been reported in collagen sandwich, which normally occurs between 48 and 72 hours post seeding (Ng *et al.*, 2006). Morphology of the FDA signal resembled the mouth-like shape of bile canaliculi described elsewhere (Sidler *et al.*, 2004; Xia *et al.*, 2009), confirming its proper excretion.

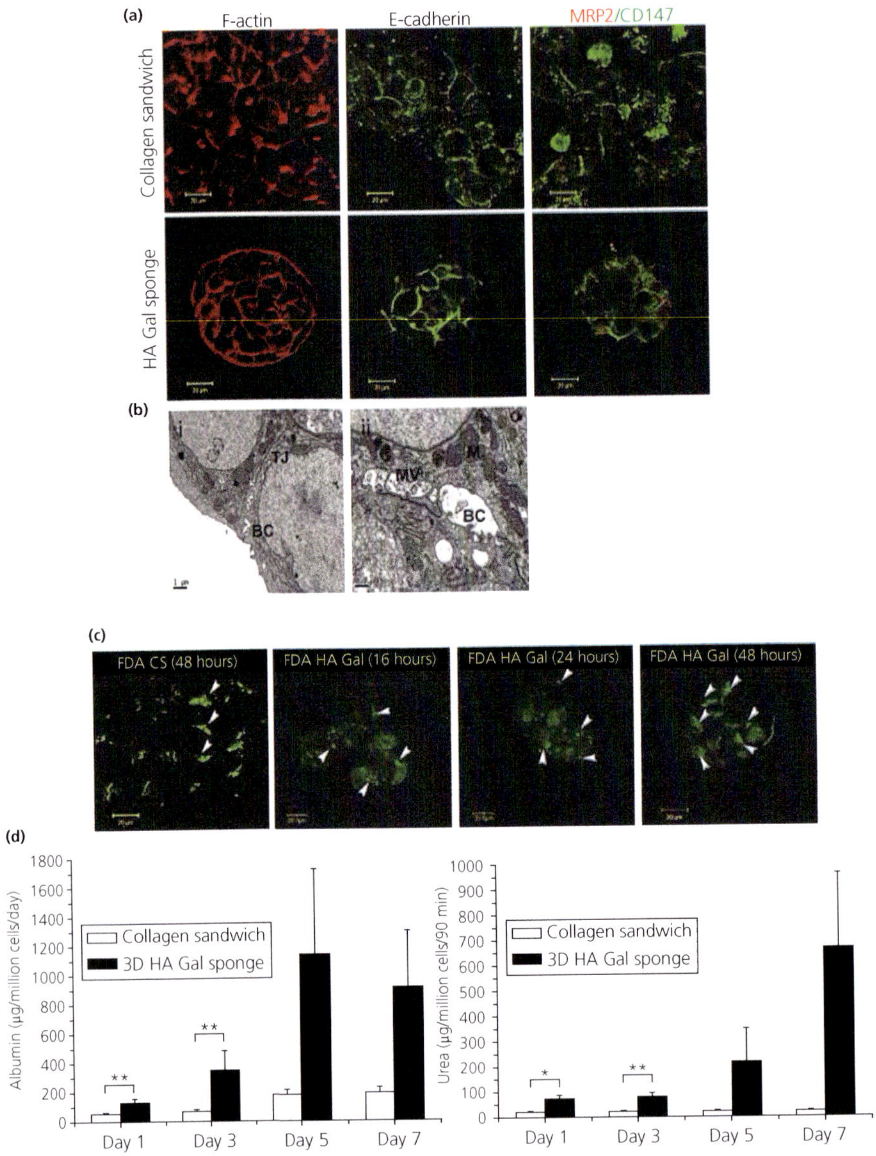

Figure 11.11 Cellular characterisations of hepatocyte spheroids. (a) Immunofluorescence staining of mature hepatocyte markers (scale bar 20 μm). (b) Transmission electron microscopy images at 48 hours post seeding (scale bars for (i) and (ii) are 1 mm and 0.5 μm, respectively). BC, bile canaliculi; M, mitochondria; Mv, microvili: TJ, tight junction. (c) Fluorescein diacetate excretion of hepatocytes in collagen sandwich (CS) and CelluSponge-Galactose/HA Gal at different time intervals (scale bar 20 μm). (d) Albumin secretion and urea synthesis function of hepatocyte in the sponges and collagen sandwich. *$P<0.1$. **$P<0.05$. Source: Nugraha *et al.* 2011. Reproduced with permission of Elsevier.

Maintenance of albumin secretion and urea synthesis, markers of mature differentiated hepatocytes, during extended culture is a prerequisite for drug safety testing applications (Nussler *et al.*, 2001). Figure 11.11d demonstrates that these functions were generally better maintained in hepatocytes cultured in CelluSponge-Galactose compared to the collagen sandwich for at least seven days.

Primary human hepatocyte culture models for hepatitis C antiviral screening

Hepatitis C virus (HCV) is a major cause of acute and chronic hepatitis in the world. Most cases of infection with HCV result in chronic hepatitis, with 60–70% of all cases developing active liver disease and the remaining ones leading to cirrhosis, end-stage liver disease and liver cancer. In estimation, 3% of the world's population is chronically infected with HCV (Centers for Disease Control, 2010; Salloum & Tai, 2012). Within 20 years of infection, chronic HCV patients who develop cirrhosis have an increased risk of developing hepatocellular carcinoma and liver failure. This complication of HCV infection has significantly increased mortality and thus the need for liver transplantation (Wise *et al.*, 2008). It has been demonstrated that there are six major HCV genotypes, and approximately three-quarters of patients with chronic HCV have genotype 1 infection (Simmonds *et al.*, 2005). This genotype is more resistant to current available treatments, peginterferon α-2a and ribavirin (Salloum & Tai, 2012). There are still no effective preventive vaccines, which is why there is a strong need to develop new therapies against HCV genotype 1 infection.

The host specificity of HCV, exclusive to chimpanzees and the human liver, has stalled the study of HCV over the past decades (Wilson & Stamaki, 2012). Moreover, it has been hampered by the lack of adequate human hepatocyte culture and small animal models (Zhu *et al.*, 2008). Since the use of chimpanzees as HCV infection models is not cost-effective, or in line with efforts to reduce animal use in scientific investigation, research is shifting towards the culture of human liver cells (Ananthanarayanan *et al.*, 2011). Huh7, as a cell source for an HCV infection model, has been investigated but found to lack effective RIG-I-mediated interferon production, rendering it unsuitable for innate immunity to HCV infection studies (Sumpter *et al.*, 2005). Previously reported cell lines outside the Huh7 family that are able to support HCV infection are derived from tumour tissues or immortalised, making them incompatible with any research intended to determine potential oncogenic effects of HCV infection (Wu *et al.*, 2012). Therefore, primary human hepatocytes are still regarded as the gold standard in HCV infection elucidation.

Primary human hepatocytes need to be cultured in a proper system to be polarised over an extended culture period, because viral entry into the cells is controlled by cell polarisation and cellular localisation of different markers such as cluster of differentiation 81 (CD81), claudin1 and SCARB1 (Harris *et al.*, 2008;

Mee *et al.*, 2008). This presents a challenge to create a robust 3D cell culture platform to maintain primary human hepatocyte polarity. There is a need to recapitulate hepatocyte polarity in the form of cell-dense constructs such as spheroids, which act like artificial human liver tissue. By achieving polarised human hepatocyte models *in vitro*, HCV infection can be studied as well as viral replication over a period of time and thus screen new candidate compounds for antiviral drugs. The *in vitro* human hepatocyte model is considered to be more economic compared to other methods of repopulating mouse liver ectopically with human hepatocytes, which can also show the replication of HCV (Chen *et al.*, 2011).

The ability of the galactosylated cellulosic sponge in inducing human hepatocytes and Huh7.5 cells to form spheroids was tested and it was found that the cells formed spheroids within the first 24 hours post seeding (Figure 11.12). Both human hepatocytes and Huh7.5 cells formed tight spheroids within three days and maintained their spheroid configuration over prolonged culture of up to 40 days and 21 days, respectively.

The size distribution of primary human hepatocyte spheroids formed in culture was 50–80 µm from day 3 until day 14 in culture (Figure 11.13a). On the other hand, due to the proliferative nature of Huh7.5 cells, most Huh7.5 spheroids were between 80 and 120 µm on day 3 of culture and there was an increase in the number of spheroids between 150 and 200 µm in diameter over 14 days in culture (Figure 11.13b). This is equivalent to 1300 spheroids in a 10 mm

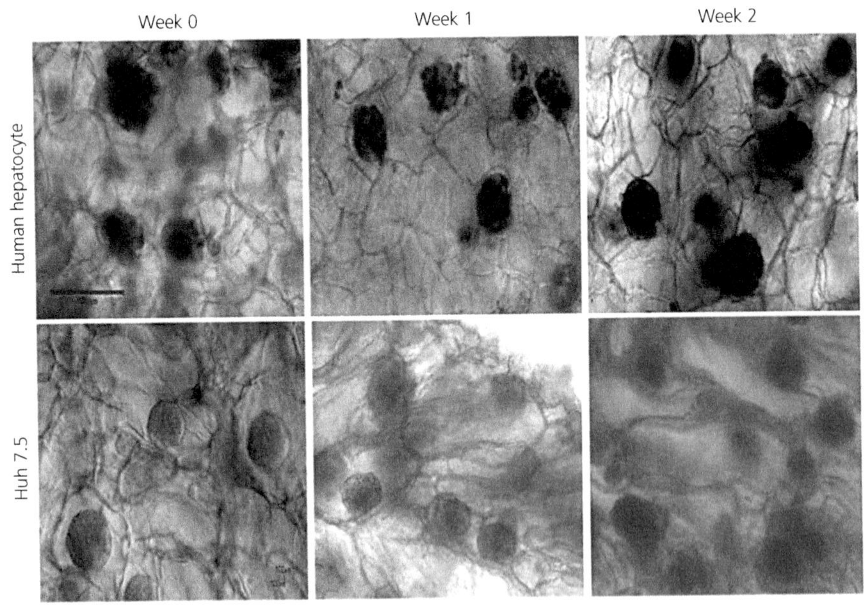

Figure 11.12 Phase contrast images of human hepatocyte spheroids and Huh7.5 spheroids. Upper panel shows human hepatocytes and lower panel Huh7.5 spheroids cultured in CelluSponge-Galactose at different weeks of culture (scale bar 100 µm).

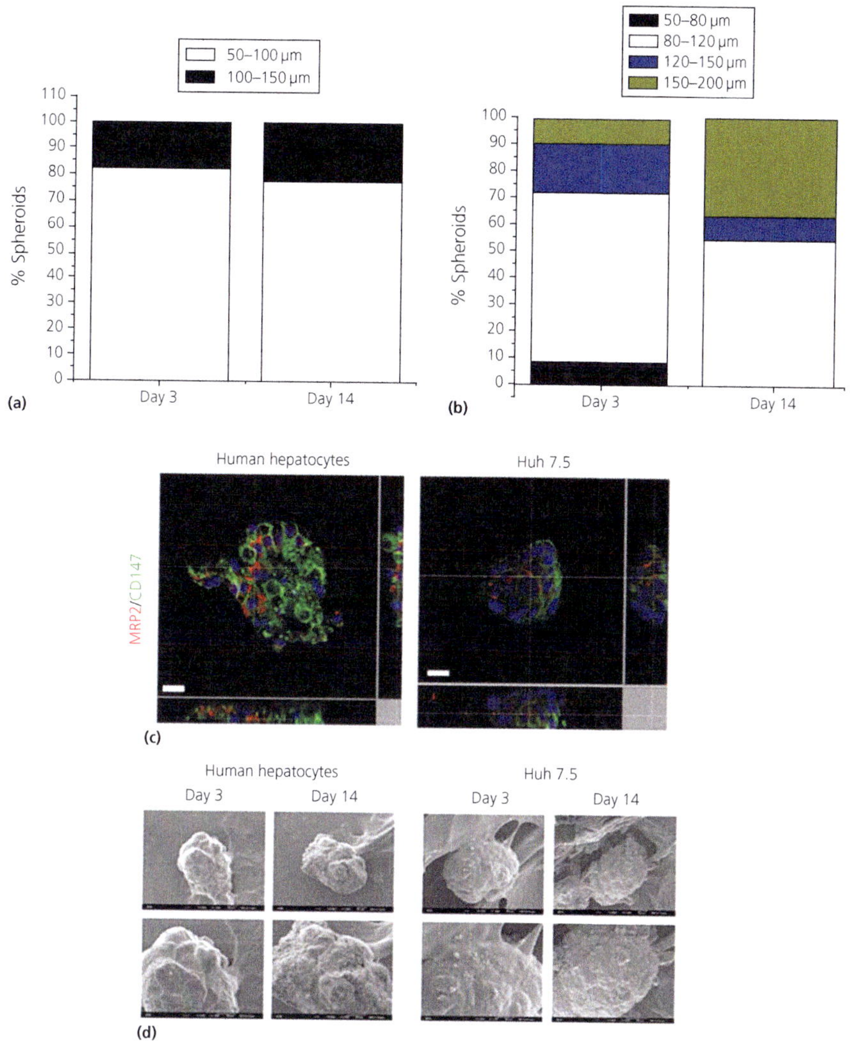

Figure 11.13 **Spheroid formation of human hepatocytes and Huh7.5 cells.** Human hepatocytes (a) formed spheroids of predominantly 50 – 100 µm size and maintained over time while Huh7.5 (b) initially formed 80 – 120 µm spheroids that grew larger over time. (c) Immunofluorescence staining of MRP2 (*red*) and CD147 (*green*) co-localisation in spheroids indicates spatial segregation of these apical and basolateral markers as a sign of cell polarisation. Percentage of co-localisation in primary human hepatocyte and Huh7.5 is 0% and 13% respectively. Scale bar 20 µm. (d) SEM images at magnifications of 1000 Å ~ and 2000 Å ~ indicate that both primary human hepatocytes and Huh7.5 formed tight aggregates and maintained spheroid integrity over 14-day culture while they remain tethered to the sponge materials. There are micropores on the surface of primary human hepatocytes as seen in typical hepatocyte spheroids. Scale bar 10 µm. Source: Nugraha, B. & Yu, H. 2014.

diameter sponge and 500 spheroids in a 6 mm sponge. Each spheroid had approximately 100 cells, and the 1.3×10^5 cells per 6 mm sponge allowed for sensitive detection of downstream endpoints and assays.

Spheroids of both Huh7.5 cells and primary human hepatocytes expressed markers of apical and basolateral domains such as multidrug-resistant associated protein 2 (MRP2) and CD147 (Figure 11.13c). In Huh7.5 spheroids, MRP2 protein was found in the apical domain, forming punctuated structures between cells that did not extend to form tube-like structures resembling the bile canaliculi in the liver. Staining of CD147 indicated the basolateral domain. There was 13% co-localisation of MRP2 and CD147 staining as determined by image analysis, which indicated the formation of spatially segregated apical and basolateral domains as non-polarised cells. Primary human hepatocytes in spheroids formed defined apical domains indicated by MRP2 staining which marks the spatial localisation of the apical transporter that is necessary for maintenance of hepatocyte function and viability in *in vitro* cultures. The primary human hepatocytes also exhibited spatially segregated basolateral and apical domains as indicated by the staining of CD147 and MRP2. There was no overlap or co-localisation of MRP2 and CD147 markers in spheroid culture, supporting the enhanced cell polarisation in these spheroids.

Scanning electron microscopic images of both cell types' spheroids on day 3 and day 14 portrayed tight spheroid formation and maintenance of the spheroids' compact morphology over prolonged culture (Figure 11.13d). Interestingly, SEM images of human hepatocyte spheroids showed 1–2 μm size holes on the spheroid surface, previously reported as bile canaliculi-like structures extending from the surface to the interior of the spheroid (Figure 11.13d) (Peshwa *et al.*, 1996).

Analysing the liver-specific genes over 14 days of culture assessed phenotypic differentiations of the two kinds of spheroids. Primary human hepatocytes cultured as spheroids exhibited minimal dedifferentiation compared to freshly thawed cryopreserved human hepatocytes for various lots (Figure 11.14a). Mature human hepatocyte genes such as cytochrome P450, family 1, subfamily A, polypeptide 1 (CYP1A1), cytochrome P450 3A4 (CYP3A4), HFN4α (hepatocyte nuclear factor 4 α) and albumin were upregulated in the spheroid culture by 1024-, 4-, 5700- and 90-fold, respectively. α1 Antitrypsin (A1AT) gene was downregulated 45-fold. For Huh7.5 spheroids, a significant increase in the transcript levels of CYP3A4 and albumin genes (256- and 16384-fold, respectively after seven days of culture) was observed (Figure 11.14b). These important mature hepatocyte genes showed significant increase over seven days of culture and maintained similar levels of expression until 14 days in culture. The other genes A1AT, CYP1A1 and HNF4α were maintained from day 1 to day 14 of culture.

Various HCV entry markers, namely CD81, claudin1, occludin and SCARB1, were also characterised in the spheroids. CD81 was visualised at the basolateral domain of the cells. Claudin1, SCARB1 and occludin were visualised at the tight junction regions (Figure 11.15). These markers' expression was observed in the spheroids over extended culture of up to 14 days.

Spheroid culture of human hepatocyte and Huh7.5 cells showed higher uptake of HCV pseudoparticles (HCVpp) compared to the monolayer culture

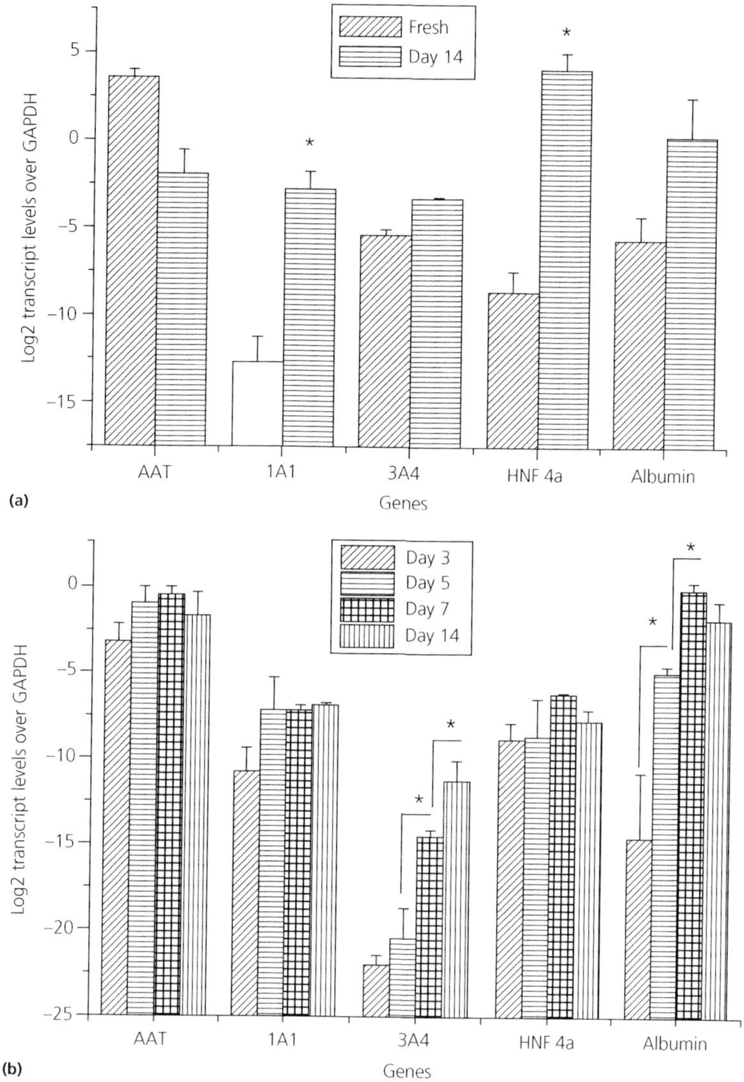

Figure 11.14 Gene expression profiles of human hepatocytes and Huh7.5 cells in spheroids over 14-day culture. (a) Human hepatocytes and (b) Huh7.5 cells. Hepatocyte-specific genes were maintained at similar or higher levels. $^{*}P<0.05$. Data expressed as mean ± s.e.m. (n = 3). GAPDH, glyceraldehyde 3-phosphate dehydrogenase. Source: Nugraha, B. & Yu, H. 2014.

(Figure 11.16a,b). To demonstrate the utility of our spheroids model to screen antiviral compound, co-incubation of HCVpp with CD81 antibody was performed. CD81 antibody was previously reported to inhibit HCV entry by reducing the availability of CD81 receptors to mediate HCV entry (Lindenbach *et al.*, 2005). The presence of CD81 antibody inhibited HCVpp entry in a dose-dependent manner (Figure 11.16c,d). The CD81 antibody concentration when

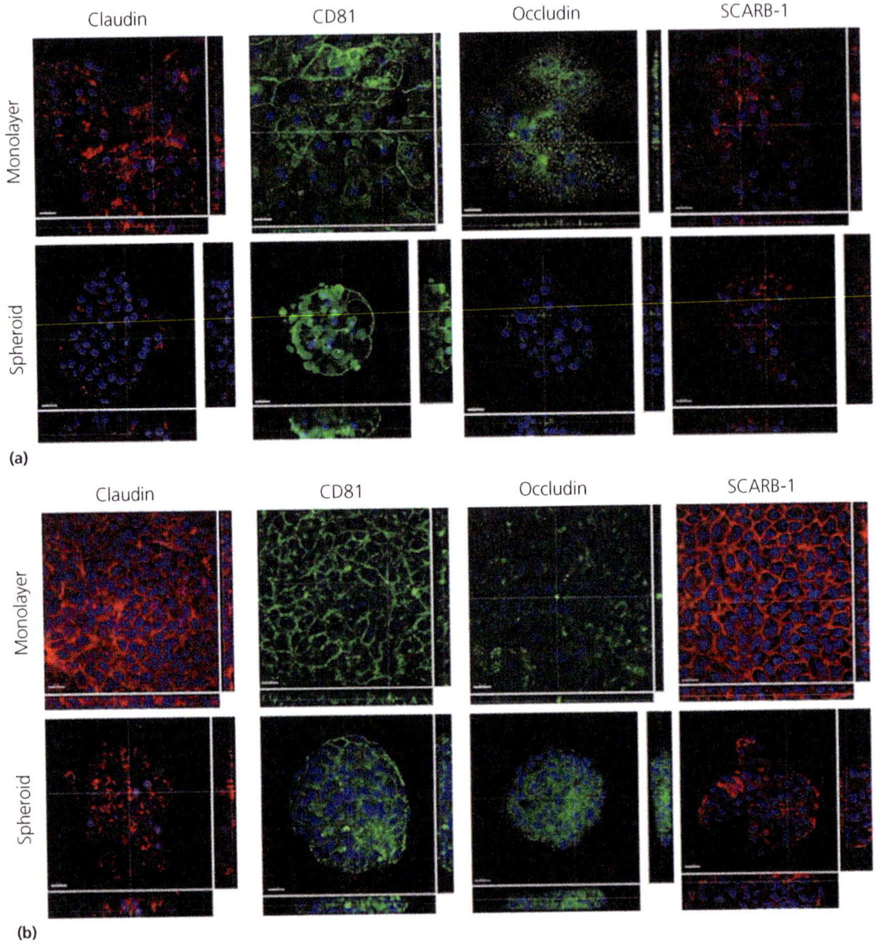

Figure 11.15 Immunofluorescence stainings of viral entry markers. Viral entry markers: claudin-1 (*red*), CD 81 (*green*), occludin (*green*) and SCARB-1 (*red*) in (a) primary human hepatocytes and (b) Huh7.5 cells. Orthogonal sections (xy, yz and xz) were visualised. All markers were expressed in both spheroids and the control 2D monolayer cultures, while the spatial localisation patterns in spheroids are distinctly different from the control but more reminiscent of the pattern seen *in vivo*. Scale bar 20 μm. Source: Nugraha, B. & Yu, H. 2014.

50% of HCVpp entry was inhibited, labelled as IC_{50}, was 500 ng/mL for human hepatocytes and 100 ng/mL for Huh7.5 cells. These IC values were close to the previously reported values, i.e. < 1 μg/mL (Ploss *et al.*, 2010).

To continue the investigation on whether spheroid cultures can support the replication of live full-length virus in culture, spheroids were infected with the JFH-1 (genotype 2a) strain of the virus and replication over eight days of culture was monitored. The cryopreserved primary human hepatocytes, however, did not support HCV live viral replication as detected using RT-PCR, consistent with previous reports (Ploss *et al.*, 2010; Zeisel *et al.*, 2007). On the other hand, Huh7.5

cell spheroids supported HCV replication with a threefold increase in viral titres from day 2 to day 4 in culture, after which there was a twofold increase from day 4 to day 6 and a 1.5-fold increase from day 6 to day 8 (Figure 11.17a). The spheroid cultures also supported fourfold and eightfold higher copy number of the JFH-1 virus than the conventional monolayer cultures on day 4 and day 8, respectively (Figure 11.17b). Although spheroids of primary human hepatocytes

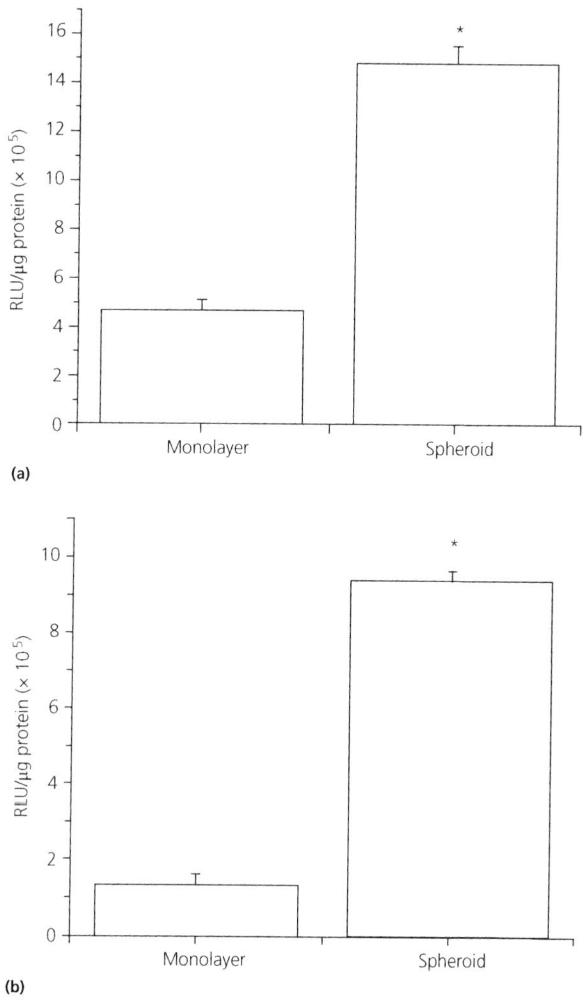

Figure 11.16 The observation of HCV entry and its inhibition. Comparison of HCVpp entry between 2D monolayer control and spheroid cultures of (a) primary human hepatocytes and (b) Huh7.5 cells indicates respectively four- and ninefold higher levels of HCVpp entry into spheroids than the control. (c,d) Dose-dependent inhibition of HCVpp entry by JS-81 antibody against CD81 receptor in (c) human hepatocytes and (d) Huh7.5 cells. $*P<0.05$. Data expressed in relative luminescence units (RLU) and normalised RLU as mean ± s.e.m. (n = 3). Source: Nugraha, B. & Yu, H. 2014.

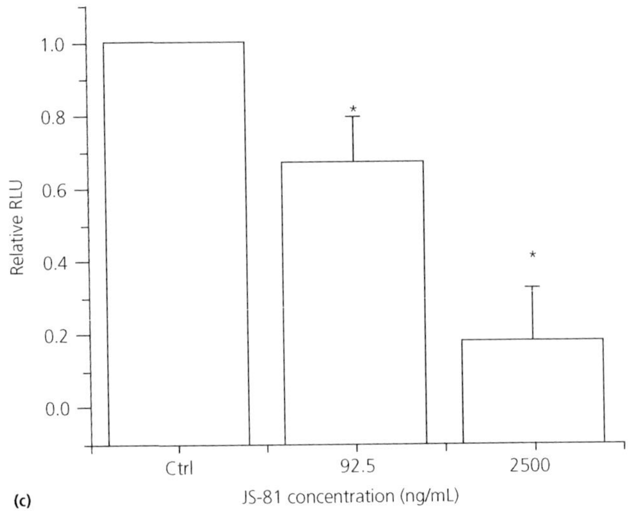

(c)

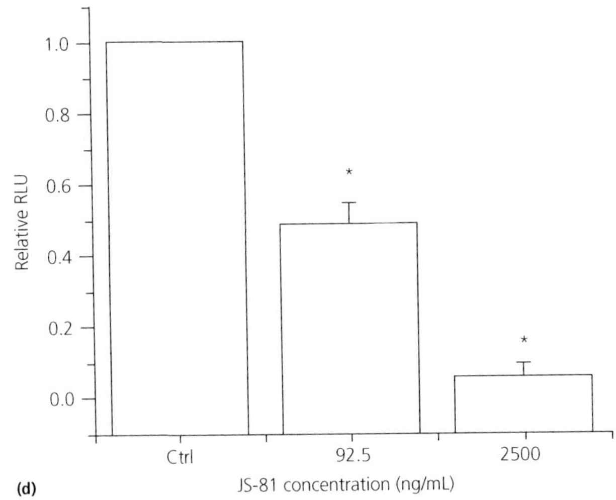

(d)

Figure 11.16 (Continued)

support higher levels of HCV entry compared to monolayer cultures, they do not support its persistent infection or replication. This could be due to the effects of cryopreservation on primary human hepatocytes or the activation of interferon pathways upon viral entry and consequent inhibition of robust HCV replication, which might lead to undetectable viral titres.

3D spheroids culture with retrieval-ability

In tissue engineering, it is ideal to have a biodegradable scaffold because the 3D scaffold only serves as a temporary support for cell growth (Chien *et al.*, 2012). Additionally, the biodegradability of the scaffold can facilitate further

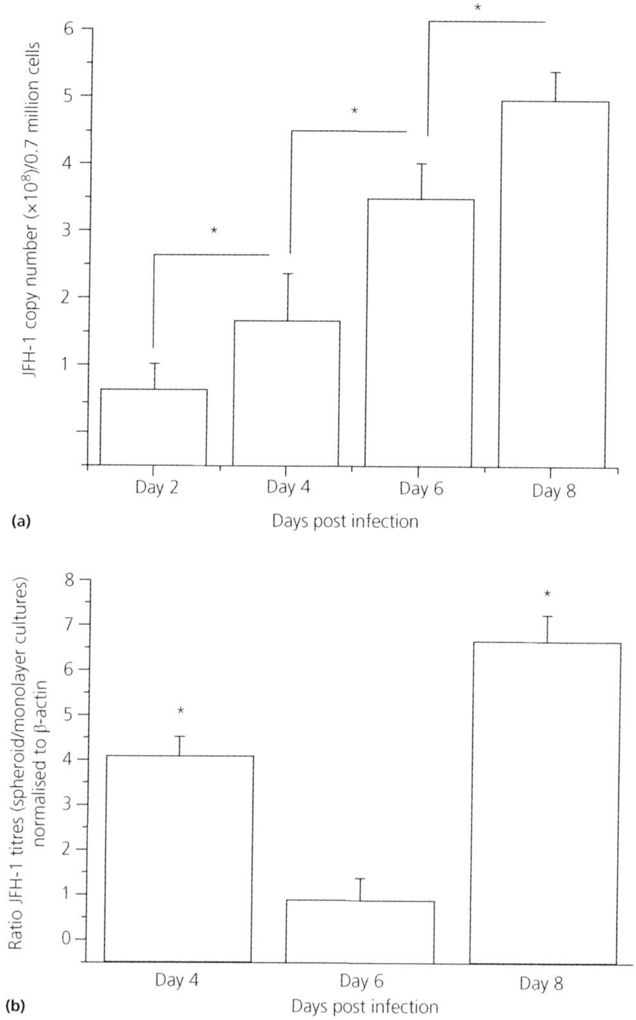

Figure 11.17 HCV clone JFH-1 replication in Huh7.5 cells spheroids. (a) Increase in JFH-1 copy number in spheroids at eight days indicates viral replication. (b) Ratio of JFH-1 titres in spheroids over 2D monolayer control cultures as measured by ΔΔCt in RT-PCR normalised to β-actin shows preferential increase in viral replication in spheroids compared to the control over time. *$P<0.05$. Data expressed as mean ± s.e.m. (n = 3). Source: Nugraha, B. & Yu, H. 2014.

downstream cell assays. However, scaffold degradation by-products should not be cytotoxic or elicit immunological responses. Preferably, these products should be water-soluble for easy removal. CelluSponge can entrap cells, induce spheroid formation and maintain these constructs within the macroporous networks over an extended culture period. However, it is a non-cleavable scaffold, which imposes challenges for retrieving intact spheroids from the sponge for further downstream analysis and use. Therefore, the biodegradation or cleavage capability of this novel cellulosic sponge needs to be investigated further.

The process of disulfide bond cleavability has been known for years in protein synthesis. When cysteine is oxidised it forms cystine, which is actually two cysteine residues joined by a disulfide bond (cysteine-S-S-cysteine) between the -SH group (Park & Raines, 2001). In recent years, there has been increasing interest in preparing scaffold matrices containing disulfide crosslinked bonds, since cleaving these bonds can be controlled under physiological conditions by simply altering the concentration of reductant used (Bromberg *et al.*, 2005; Hisano *et al.*, 1998; Liu *et al.*, 2005; Plunkett *et al.*, 2005). Through chemical cleaving, disulfide bonds (-S-S-) can be readily cleaved into thiol groups (-SH HS-). These thiol groups do not decrease the local pH as commonly observed in polyester hydrolysis mechanisms (Yoshida *et al.*, 2009). Various chemical reductants are known to cleave disulfide bonds, e.g. dithiothreitol (DTT), glutathione (GSH), L-cysteine (Cys) and tris (2-carboxyethyl) phosphine (TCEP) (Burns *et al.*, 1991; Hisano *et al.*, 1998; Kakizawa *et al.*, 2001; Lees & Whitesides, 1993; Shu *et al.*, 2003). The insertion of this disulfide bond into the hydrogel construct has been known to induce cleavage under physiological conditions by the simple incubation with reductants. Soluble decomposed products were easily removed by washing with excess cell culture medium or buffer (Matsusaki *et al.*, 2007). Various disulfide-containing hydrogels were already reported as a temporary hydrogel template, ECM mimicry matrix and 3D cell encapsulation platform (Chien *et al.*, 2012; Hisano *et al.*, 1998; Matsusaki *et al.*, 2007; Shu *et al.*, 2003); however, the cleavable macroporous disulfide-containing cellulosic sponge as a hydrogel template for spheroids culture and retrieval has not been explored.

Careful polymer synthesis, design and choice of a less cytotoxic reductant were performed. Dissolving buffer-CBX was chosen as the reductant to cleave disulfide bonds in the cellulosic macroporous sponge. Compared to a known and stronger reductant, DTT, dissolving buffer-CBX was found to be more effective in cleaving disulfide bonds at wider pH range, i.e. pH 1.5–8.5. To develop a disulfide-containing cellulosic sponge, the ultimate aim is to synthesise a sponge that can be cleaved in less than one hour in order to reduce the exposure time of the cells to the dissolving buffer-CBX.

Comparing our disulfide-containing cellulosic sponge (CelluSponge CB) to the other disulide-containing hydrogel systems, our technology shows the importance of having macroporous networks with mechanically and chemically tunable properties for 3D cell culture, within a diffusible construct rather than cells encapsulated in a hydrogel system. The presence of macroporous networks is hypothesised to accelerate the cleavage. Sponge macroporosity helps in the formation of cell-dense constructs, i.e. spheroids. Our disulfide-containing sponge also shows a significantly faster cleavage rate compared to other disulfide-containing hydrogels (which cannot form spheroids due to blockage of cell-cell interactions), i.e. spheroids can be easily retrieved from CelluSponge CB in as little as 30 minutes (Chien *et al.*, 2012; Matsusaki *et al.*, 2007).

Similarly, as was shown previously in the non-cleavable galactosylated CelluSponge (CelluSponge-Galactose/HA Gal sponge), in the cleavable galactosylated CelluSponge (CelluSponge-GalactoseCB/HPCSS Gal sponge), primary rat hepatocytes also immediately reorganised into 3D spheroids within one day of culture and remained stable in this configuration until at least day 7 (Figure 11.18). Hepatocyte spheroids formed in the CelluSponge-GalactoseCB were 88.77 μm in average diameter. The spheroids formed in

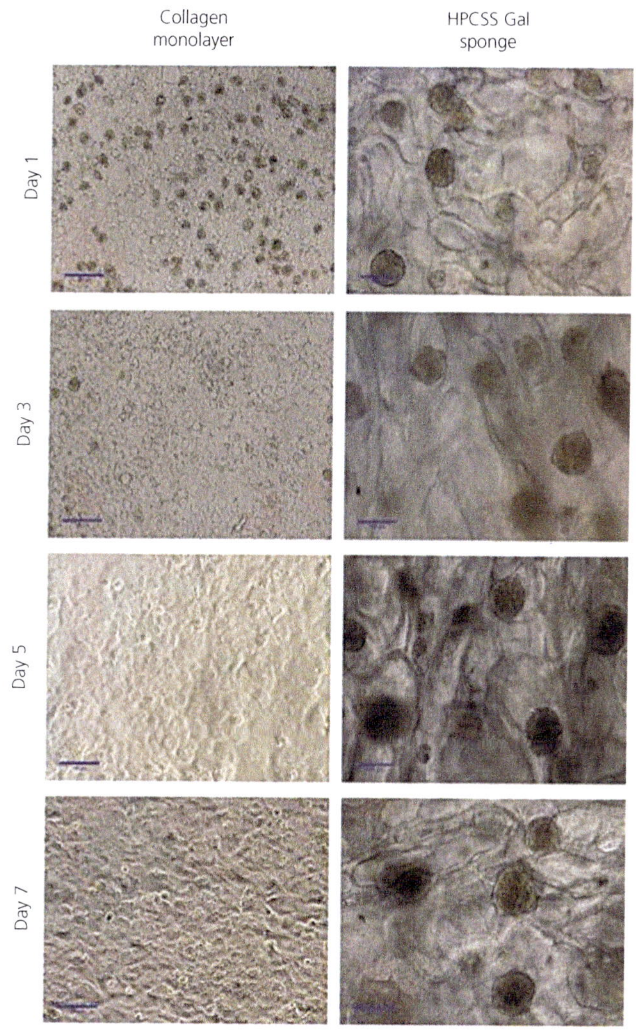

Figure 11.18 Rat hepatocytes cultured in cleavable HPCSS Gal sponge (CelluSponge-Galactose CB). Compact spheroids are formed within one day post seeding (scale bar 100 μm). Average spheroid size on day 1: 75.58 ± 16.61 μm, day 3: 84.68 ± 17.64 μm, day 5: 96.95 ± 13.52 μm and day 7: 97.88 ± 11.84 μm. Source: Nugraha, B. & Yu, H. 2014.

CelluSponge-GalactoseCB were constrained in the sponge macropores and thus did not detach as easily as those on 2D PET Gal membranes.

The intended applications of this cleavable cellulosic sponge are for 3D spheroid culture and particularly for easy retrieval of spheroids for further manipulations. This proposed mechanism is shown in Figure 11.19. During cleavage, disulfide bonds exchanged into thiol bonds loosen up the crosslinked

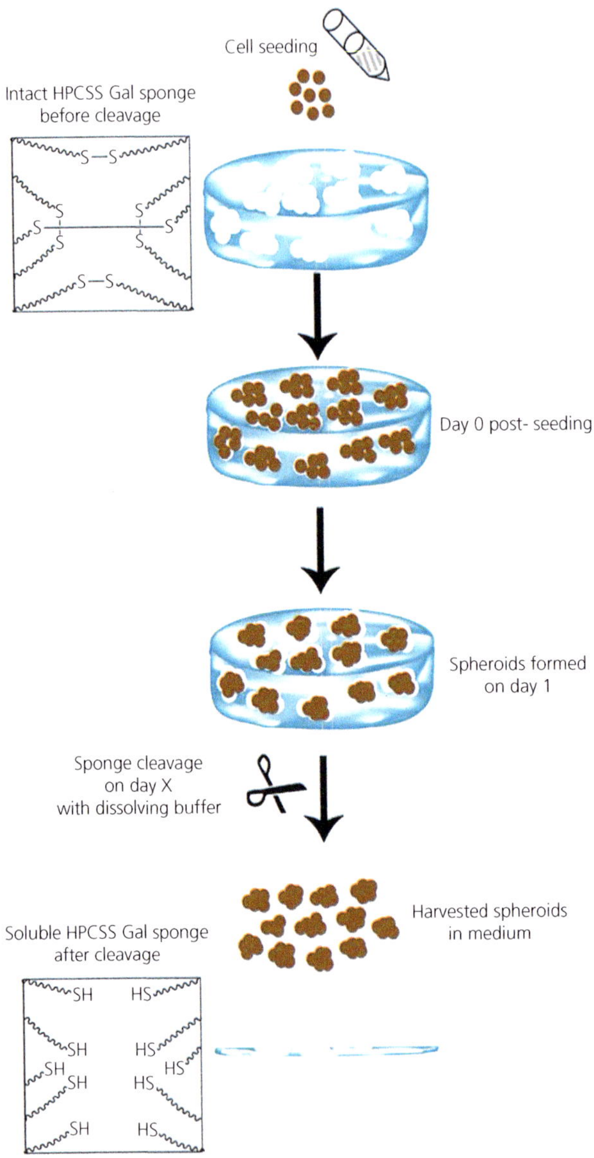

Figure 11.19 Schematic drawing of cleavable cellulosic sponge (CelluSponge-Galactose CB) cleavage mechanism. Source: Nugraha, B. & Yu, H. 2014.

bond, making the sponge soluble. The cleavage should only occur upon adding the disulfide bond reductant.

Immunofluorescence staining of F-actin, E-cadherin and MRP2 in the hepatocyte spheroids 72 hours post seeding retrieved from cleavable (HPCSS Gal/CelluSponge-Galactose CB) and non-cleavable (HA Gal/CelluSponge-Galactose) sponges showed comparable localisation of these markers (Figure 11.20).

Upon confirming the maintenance expression of MRP2 (apical domain marker as well as hepatocyte efflux transporter) in the retrieved spheroids, the excretory function of these spheroids was studied by incubating them with FDA. Viable cells in the spheroids will cleave FDA into a fluorescein dye by intracellular esterases; the dye is then excreted by MRP2 into the bile canaliculi. FDA staining of the retrieved hepatocyte spheroids showed dye accumulation in the bile canaliculi region between the two cells (Figure 11.21). This excretion exhibited similarity to what had been observed in the intact hepatocyte spheroids in the non-cleavable CelluSponge (Nugraha *et al.*, 2011).

The ability to replate the retrieved spheroids obtained after cleaving the sponge with dissolving buffer was demonstrated on collagen-coated dishes. The spheroids settled at the bottom of the dish within one hour. Four hours post replating, the spreading pattern of spheroids was clearly observed at the periphery (Figure 11.22) and 16 hours later, the spheroids exhibited almost

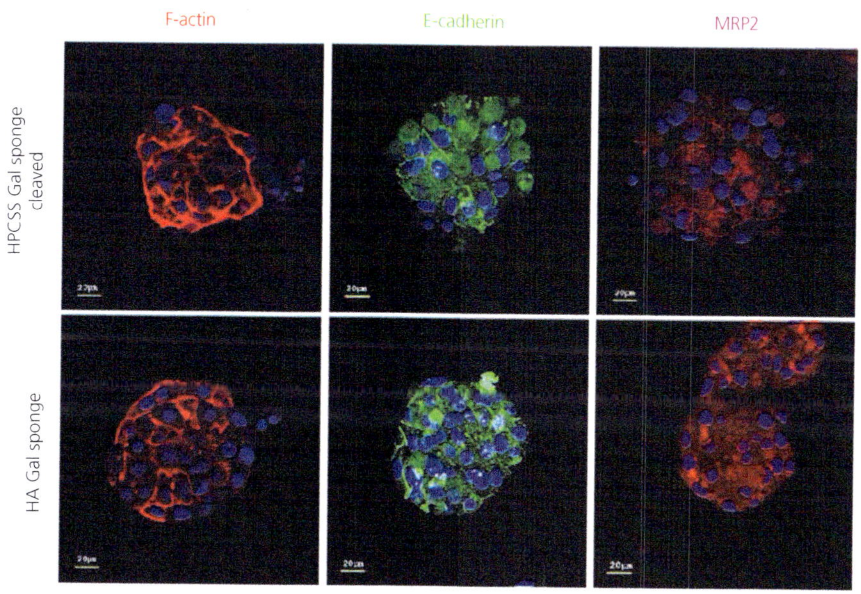

Figure 11.20 **Immunofluorescence staining of polarity markers and cell-to-cell junction of hepatocyte spheroids.** Spheroids were retrieved from cleavable HPCSS Gal sponge (CelluSponge-Galactose CB) and cultured in non-cleavable HA Gal sponge (CelluSponge-Galactose). The incubation of spheroids with dissolving buffer-CBX to cleave the sponge does not show a harmful effect on these markers (scale bar 20 μm). Source: Nugraha, B. & Yu, H. 2014.

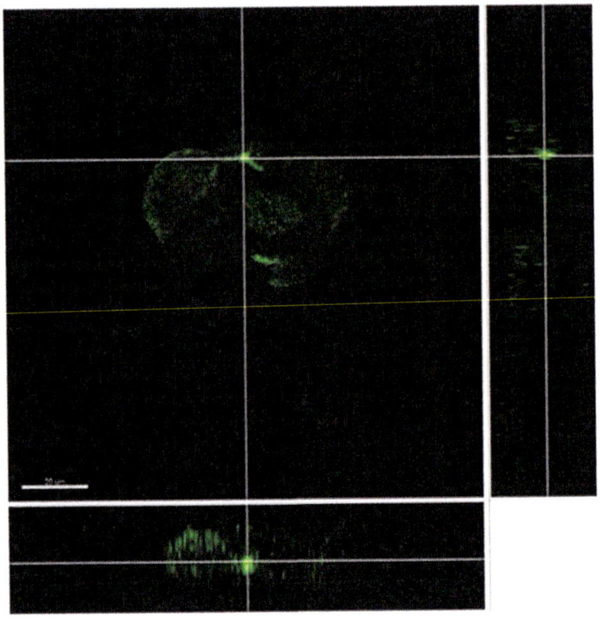

Figure 11.21 **FDA staining of the retrieved rat hepatocyte spheroids from cleavable HPCSS Gal sponge (CelluSponge-Galactose CB).** Staining indicates putative accumulation of dye at the bile canaliculi region (scale bar 20 μm). Source: Nugraha, B. & Yu, H. 2014.

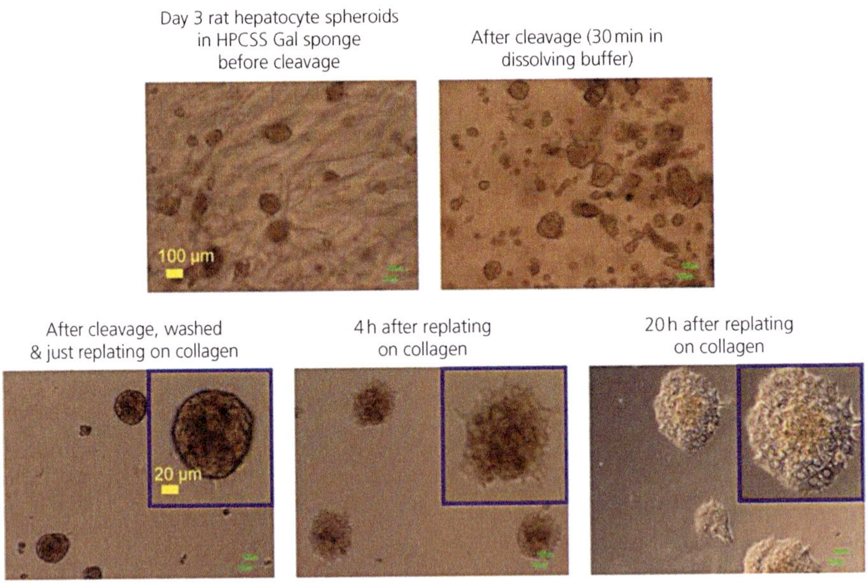

Figure 11.22 **Rat hepatocyte spheroids.** After retrieval, these spheroids were replated on a collagen dish, indicating the ability of spheroids to respread on a collagen-coated plate (scale bar 100 μm). Insert is the zoomed-in view of the spheroid, scale bar 20 μm). Source: Nugraha, B. & Yu, H. 2014.

complete spreading. Cells that were spreading out of the spheroids showed the clear bile canaliculi-like structure at the cell-cell boundary, which was normally seen on hepatocyte collagen monolayer culture.

Not limited to hepatocyte-based applications, our cleavable cellulosic sponge (CelluSponge CB series) also shows utility for 3D cancer cell and stem cell cultures (Figure 11.23). Both E1 (metastatic derivative human colorectal cancer cells) and mESCs seeded into our cleavable CelluSponge CB reorganised into compact spheroids within 24 hours post seeding. The spheroids remained compact up to at least five days of culture, were easily retrieved through sponge cleavage and maintained good cell viability during culture and post retrieval.

Go Matrix applications

Several representative cells from different origins, (3T3, C2C12, MDCK, rabbit chondrocytes, mouse cardiomyocytes) have been shown to be viable and retain their 3D morphology in Go Matrix. The life span of cells in this scaffold extends to more than a month. Cells also show more physiological responses in Go Matrix. For example, chondrocyte expression more closely resembles that of healthy cartilage. Furthermore, other factors can be incorporated to this scaffold, including surface modification for cell mobility test, materials of high compatibility to actualise implantation or controlled degradability for mini-tissue harvest, and drug screening.

Conclusion

A novel class of hydrogel-based macroporous sponges for soft tissue culture has been generated from hydroxypropyl cellulose with a range of side chain chemical functionalisation. CelluSponge provides a relatively easy-to-synthesise, low-cost platform which can be fabricated in large scale as well as having the option to be non-cleavable (stable) and cleavable (degradable). Cell seeding is quite simple, with possible manipulation for robotic liquid handler-assisted seeding for high-throughput cell culture.

Acknowledgements

The author would like to acknowledge the guidance of Professor Hanry Yu and Dr Zhilian Yue in the development of cellulosic sponge technology. This work was supported by the Institute of Bioengineering and Nanotechnology A*STAR Singapore for a PhD thesis performed at A*STAR and the National University of Singapore. Currently, Bio-Byblos Biomedical Co. Ltd has licensed exclusively two patent applications of non-cleavable and cleavable cellulosic sponges.

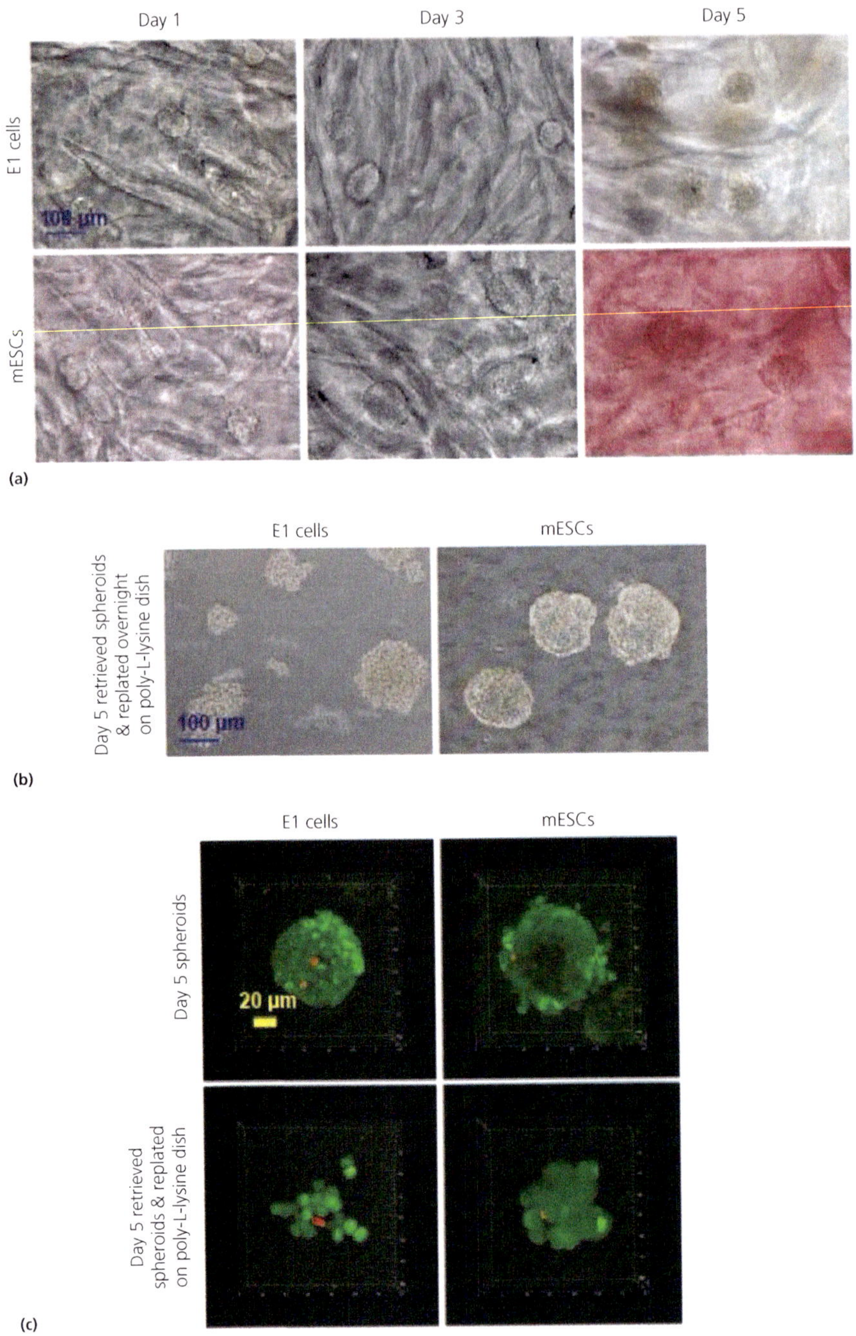

Figure 11.23 E1 and mESCs cultured in cleavable CelluSponge-CB (HPCSS sponge). (a) Phase contrast images. (b) Spheroid retrieval. (c) Live/dead staining. Source: Nugraha, B. & Yu, H. 2014.

References

Abu-Absi, S.F. Friend, J.R., Hansen, L.K. *et al.* (2002) Structural polarity and functional bile canaliculi in rat hepatocyte spheroids. *Experimental Cell Research*, **274**, 56–67.

Ananthanarayanan, A., Narmada, B.C., MO, X. *et al.* (2011) Purpose-driven biomaterials research in liver-tissue engineering. *Trends in Biotechnology*, **29** 110–118.

Baraniak, P. R. & McDevitt, T.C. (2012) Scaffold-free culture of mesenchymal stem cell spheroids in suspension preserves multilineage potential. *Cell and Tissue Research*, **347**, 701–711.

Bromberg, L., Temchenko, M., Alakhov, V. *et al.* (2005) Kinetics of swelling of polyether-modified poly(acrylic acid) microgels with permanent and degradable cross-links. *Langmuir*, **21**,1590–1598.

Burns, J. A., Butler, J. C., Moran, J. *et al.* (1991) Selective reduction of disulfides by tris(2-carboxyethyl)phosphine. *Journal of Organic Chemistry*, **56**, 2648–2650.

Centers for Disease Control and Prevention (2010*) Sexually Transmitted Diseases Treatment Guidelines, Hepatitis C, 2010*. Available at: www.cdc.gov/std/treatment/2010/hepC.htm>.

Chen, A. A., Thomas, D. K., Ong, L. L. *et al.* (2011) Humanized mice with ectopic artificial liver tissues. *Proceedings of the National Academy of Sciences USA*, **108**, 11842–11847.

Chien, H. W., Tsai, W. B. & Jiang, S. (2012) Direct cell encapsulation in biodegradable and functionalizable carboxybetaine hydrogels. *Biomaterials*, **33**, 5706–5712.

Curcio, E., Salerno, S., Barbieri, G. *et al.* (2007) Mass transfer and metabolic reactions in hepatocyte spheroids cultured in rotating wall gas-permeable membrane system *Biomaterials*, **28**, 5487–5497.

Du, Y., Chia, S. M., Han, R. *et al.* (2006) 3D hepatocyte monolayer on hybrid RGD/galactose substratum. *Biomaterials*, **27**, 5669–5680.

Fischbach, C., Chen, R., Matsumoto, T. *et al.* (2007) Engineering tumors with 3D scaffolds. *Nature Methods*, **4**, 855–860.

Gautier, A., Carpentier, B., Dufresne, M. *et al.* (2011) Impact of alginate type and bead diameter on mass transfers and the metabolic activities of encapsulated C•A cells in bioartificial liver applications. *European Cells & Materials*, **21**, 94–106.

Gebhardt, R., Hengstler, J. G., Müller, D. *et al.* (2003) New hepatocyte in vitro systems for drug metabolism: metabolic capacity and recommendations for application in basic research and drug development, standard operation procedures. *Drug Metabolism Reviews*, **35**, 145–213.

Gua, H., Yue, Z., Leong, W. S. *et al.* (2010) Control of in vitro neural differentiation of mesenchymal stem cells in 3D macroporous, cellulosic hydrogels. *Regenerative Medicine*, **5**, 245–253.

Harada, K., Mitaka, T., Miyamoto, S. *et al.* (2003) Rapid formation of hepatic organoid in collagen sponge by rat small hepatocytes and hepatic nonparenchymal cells. *Journal of Hepatology*, **39**, 716–723.

Harris, H. J., Farquhar, M. J., Mee, C. J. *et al.* (2008) CD81 and claudin 1 coreceptor association: role in hepatitis C virus entry. *Journal of Virology*, **82**, 5007–5020.

Hisano, N., Morikawa, N., Iwata, N. *et al.* (1998) Entrapment of islets into reversible disulfide hydrogels. *Journal of Biomedical Materials Research*, **40**, 115–123.

Hu, Z., Lu, X., Gao, J. *et al.* (2000) Polymer gel nanoparticle networks. *Advanced Materials*, **12**, 1173–1176.

Kakizawa, Y., Harada, A. & Kataoka, K. (2001) Glutathione-sensitive stabilization of block copolymer micelles composed of antisense DNA and thiolated poly(ethylene glycol)-block-poly(l-lysine): a potential carrier for systemic delivery of antisense DNA. *Biomacromolecules*, **2**, 491–497.

Kinasiewicz, A., Dudzinski, K., Chwojnowski, A. *et al.* (2007) Three-dimensional culture of hepatocytes on spongy polyethersulfone membrane developed for cell transplantation. *Transplantation Proceedings*, **39**, 2914–2916.

Lees, W. J. & Whitesides, G. M. (1993) Equilibrium constants for thiol-disulfide interchange reactions: a coherent, corrected set. *Journal of Organic Chemistry*, **58**, 642–647.

Lindenbach, B. D., Evans, M. J., Syder, A. J. *et al.* (2005) Complete replication of hepatitis C virus in cell culture. *Science*, **309**, 623–626.

Liu, Y., Zheng Shu, X. & Prestwich, G. D. (2005) Biocompatibility and stability of disulfide-crosslinked hyaluronan films. *Biomaterials*, **26**, 4737–4746.

Luebke-Wheeler, J. L., Nedredal, G., Yee, L. *et al.* (2009) E-cadherin protects primary hepatocyte spheroids from cell death by a caspase-independent mechanism. *Cell Transplantation*, **18**, 1281–1287.

Matsusaki, M., Yoshida, H. & Akashi, M. (2007) The construction of 3D-engineered tissues composed of cells and extracellular matrices by hydrogel template approach. *Biomaterials*, **28**, 2729–2737.

Mee, C. J., Grove, J., Harris, H. J. *et al.* (2008) Effect of cell polarization on hepatitis C virus entry. *Journal of Virology*, **82**, 461–470.

Molina-Jimenez, F., Benedicto, I., Dao Thi, V. L. *et al* (2012) Matrigel-embedded 3D culture of Huh-7 cells as a hepatocyte-like polarized system to study hepatitis C virus cycle. *Virology*, **425**, 31–39.

Ng, S., Han, R., Chang, S. *et al.* (2006) Improved hepatocyte excretory function by immediate presentation of polarity cues. *Tissue Engineering*, **12**, 2181–2191.

Nugraha, B. & Yu, H. (2014) Cleavable cellulosic sponge development for 3 dimensional cell culture and spheroids retrieval. US Patent 2A1.

Nugraha, B., Hong, X., Mo, X. *et al.* (2011). Galactosylated cellulosic sponge for multi-well drug safety testing. *Biomaterials*, **32**, 6982–6994.

Nussler, A. K., Wang, A., Neuhaus, P. *et al.* (2001) The suitability of hepatocyte culture models to study various aspects of drug metabolism. *ALTEX*, **18**, 91–101.

Ong, S. M., He, L., Thuy Linh, N.T. *et al.* (2007) Transient inter-cellular polymeric linker. *Biomaterials*, **28**, 3656–3667.

Palakuru, J. R., Wang, J. & Aquavella, J. V. (2008) Effect of blinking on tear volume after instillation of midviscosity artificial tears. *American Journal of Ophthalmology*, **146**, 920–924.

Pampaloni, F., Reynaud, E. G. & Stelzer, E.H. (2007) The third dimension bridges the gap between cell culture and live tissue. *Nature Reviews Molecular Cell Biology*, **8**, 839–845.

Park, C. & Raines, R. T. (2001) Adjacent cysteine residues as a redox switch. *Protein Engineering*, **14**, 939–942.

Peshwa, M. V., Wu, F. J., Sharp, H. L. *et al.* (1996) Mechanistics of formation and ultrastructural evaluation of hepatocyte spheroids. *In Vitro Cellular & Developmental Biology-Animal*, **32**, 197–203.

Ploss, A., Khetani, S. R., Jones, C. T. *et al.* (2010) Persistent hepatitis C virus infection in micro-scale primary human hepatocyte cultures. *Proceedings of the National Academy of Sciences USA*, **107**, 3141–3145.

Plunkett, K. N., Berkowski, K. L. & Moore, J. S. (2005) Chymotrypsin responsive hydrogel: application of a disulfide exchange protocol for the preparation of methacrylamide containing peptides. *Biomacromolecules*, **6**, 632–637.

Prause, J. U. (1986) Treatment of keratoconjunctivitis sicca with Lacrisert. *Scandinavian Journal of Rheumatology*, **61(suppl)**, 261–263.

Salloum, S. & Tai, A. W. (2012) Treating hepatitis C infection by targeting the host. *Translational Research*, **159**, 421–429.

Shu, X. Z., Liu, Y., Palumbo, F. *et al.* (2003) Disulfide-crosslinked hyaluronan-gelatin hydrogel films: a covalent mimic of the extracellular matrix for *in vitro* cell growth. *Biomaterials*, **24**, 3825–3834.

Sidler Pfandler, M. A., Hochli, M. & Inderbitzin, D. (2004) Small hepatocytes in culture develop polarized transporter expression and differentiation. *Journal of Cell Science*, **117**, 4077–4087.

Simmonds, P., Bukh, J., Combet, C. *et al.* (2005) Consensus proposals for a unified system of nomenclature of hepatitis C virus genotypes. *Hepatology*, **42**, 962–973.

Spontak, R. J. & Hirsch, S. G. (2002) Temperature-dependent property development in hydrogels derived from hydroxypropylcellulose. *Polymer*, **43**, 123–129.

Sumpter, R. Jr, Loo, Y. M., Foy, E. *et al.* (2005) Regulating intracellular antiviral defense and permissiveness to hepatitis C virus RNA replication through a cellular RNA helicase, RIG-I. *Journal of Virology*, **79**, 2689–2699.

Talamini, M. A., Kappus, B. & Hubbard, A. (1997) Repolarization of hepatocytes in culture. *Hepatology*, **25**, 167–172.

Underhill, G. H., Chen, A. A., Albrecht, D. R. *et al.* (2007) Assessment of hepatocellular function within PEG hydrogels. *Biomaterials*, **28**, 256–270.

Wells, R. G. (2008) The role of matrix stiffness in regulating cell behavior. *Hepatology*, **47**, 1394–1400.

Wen, Y. A., Liu, D., Xiao, Y. Y. *et al.* (2009) Enhanced glucose synthesis in three-dimensional hepatocyte collagen matrix. *Toxicology in Vitro*, **23**, 744–747.

Wilson, G. K. & Stamaki, Z. (2012) *In vitro* systems for the study of hepatitis C virus infection. *International Journal of Hepatology*, **2012**:292591.

Wise, M., Bialek, S., Finelli, L. *et al.* (2008) Changing trends in hepatitis C-related mortality in the United States, 1995–2004. *Hepatology*, **47**, 1128–1135.

Wu, C., Pan, J., Bao, Z. *et al.* (2007) Fabrication and characterization of chitosan microcarrier for hepatocyte culture. *Journal of Materials Science: Materials in Medicine*, **18**, 2211–2214.

Wu, X., Robotham, J. M., Lee, E. *et al.* (2012) Productive hepatitis C virus infection of stem cell-derived hepatocytes reveals a critical transition to viral permissiveness during differentiation. *PLoS Pathogens*, **8**, e1002617.

Xia, L., Ng, S., Han, R. *et al.* (2009) Laminar-flow immediate-overlay hepatocyte sandwich perfusion system for drug hepatotoxicity testing. *Biomaterials*, **30**, 5927–5936.

Yoshida, H., Klinkhammer, K., Matsusaki, M. *et al.* (2009) Disulfide-crosslinked electrospun poly(γ-glutamic acid) nonwovens as reduction-responsive scaffolds. *Macromolecular Bioscience*, **9**, 568–574.

Yue, Z., Wen, F. & Yu, H. (2012) Forming porous scaffold from cellulose derivatives. US Patent 8283028 B2.

Yue, Z.,Wen, F., Gao, S. *et al.* (2010) Preparation of three-dimensional interconnected macroporous cellulosic hydrogels for soft tissue engineering. *Biomaterials*, **31**, 8141–8152.

Zeisel, M. B., Koutsoudakis, G., Schnober, E. K. *et al.* (2007) Scavenger receptor class B type I is a key host factor for hepatitis C virus infection required for an entry step closely linked to CD81. *Hepatology*, **46** 1722–1731.

Zhu, H., Elyar, J., Foss. R. *et al.* (2008) Primary human hepatocyte culture for HCV study. *Methods in Molecular Biology*, **510**, 373–382.

CHAPTER 12
Mimetix® electrospun scaffold: An easy-to-use tool for 3D cell culture in drug discovery and regenerative medicine

Robert J. McKean and Elena Heister
The Electrospinning Company Ltd, Rutherford Appleton Laboratory, Harwell Oxford, Didcot, UK

Introduction

Cell-based assays are routinely used in life science research for applications including testing the safety, toxicity and efficacy of a drug, as well as assessing the safety and toxicity of chemicals in the cosmetic and chemical industries.

The need for *in vitro* cell-based models that generate accurate information is becoming an increasing challenge. The cost of assays and limitations of being able to use real tissues and animal models, as well as the inconsistent predictive value of existing *in vitro* assays, are all drivers for developing more physiologically relevant cell culture systems. *In vivo*, cells have a three-dimensional (3D) geometry and are surrounded by other cell types in a 3D extracellular matrix (ECM). This ECM allows cell-cell communication via physical cues and a network of signalling molecules. Cells cultured in a flat two-dimensional (2D) monolayer lack these communication methods and therefore their growth, differentiation, morphology and metabolisms function differently from how they would *in vivo* (Dvir & Fleischer, 2013; Dvir *et al.*, 2011). A good example of the practical impact of this is how tumour cells in particular show a different cellular response to drugs and growth factors in 2D versus 3D cultures, indicating that 2D cells may not accurately mimic the natural tumour microenvironment (Ekert *et al.*, 2014), which can be hypoxic.

In vitro, cells are typically grown as 2D monolayers. These are easy to set up and maintain, and provide good cell viability. However, 2D cultures are often monocultures (consisting of only one cell type) without ECM or interstitial fluid, and they can represent a rather poor approximation of tissue organisation and function in living organisms (*in vivo*). This causes major limitations for applications such as *in vitro* efficacy/toxicity testing for drug discovery or *in vivo/ex vivo* growth of tissues for regenerative medicine.

Technology Platforms for 3D Cell Culture: A User's Guide, First Edition. Edited by Stefan Przyborski.

In contrast, cells cultured in 3D are more representative of real cells in the body than those cultured in 2D models, especially with respect to the cell-cell and cell-matrix interactions that influence molecular targets, uptake and cellular metabolism of drugs. Cells in 2D have most of their surface area exposed to fluid and the flat culture surface, whereas cells cultured in 3D have a more natural shape and structure (ellipsoids of 10–30 μm instead of flat cells of 4–10 μm height), and nearly all of their surface area is exposed to other cells or matrix. Physiological activities such as morphogenesis, angiogenesis and tissue invasion occur exclusively in 3D cell culture and some cell types, for example hepatocytes, respond differently when cultured in 3D than when cultured in 2D monolayers (Aly *et al.*, 2009).

When cells grow in 3D clusters, compared to 2D monolayers, the level of oxygenation differs depending on whether a cell is situated toward the inside or the outside of the 3D structure. The difference in level of oxygenation appears to produce areas of hypoxic microenvironments that influence the growth of some cells for example, cancer stem cells (Lin & Yun, 2010). These hypoxic microenvironments cannot be reproduced in a 2D monolayer and it is thought that this is one reason, along with changes in electrophysiology (Mulhall *et al.*, 2013), why some antitumour drugs show different responses in 2D and 3D cell culture.

One of the main and ongoing challenges has been to develop 3D cell culture protocols that are easy to use in both laboratory testing and high-throughput screening assays, compatible with automation. Recent 3D cell culture methodologies that have been developed to meet these issues have included the use of scaffolds made of various materials for cell adherence in culture. Natural materials such as alginate, collagen and fibrin have been used as 3D cell culture scaffolds for over a decade (Ahmed *et al.*, 2008; Becker *et al.*, 2006; Perets *et al.*, 2003). These are used because they are composed of proteins and polysaccharides found in the ECM and as a result contain binding sites for cell adhesion to support cell culture. Additionally, scaffolds made from natural materials can be modified using a variety of methods to incorporate controlled release systems for delivering therapeutic factors. They can also be used to successfully culture embryonic stem cells (Kolehmainen & Willerth, 2012), where the scaffold culture can be analysed to determine the effects of various growth factors on the differentiation of the cells seeded inside.

Natural scaffolds

Alginate is derived from seaweed and recently, alginate microcapsules have been used to produce 3D hepatocyte cultures for use in preclinical studies as a cell therapy to treat liver damage (Capone *et al.*, 2013; Vaghjihani *et al.*, 2014). The advantage of using alginate as a scaffold is that the cells can be encapsulated without losing viability or function and the alginate protects the cells from being destroyed by the immune system, thus preventing immune cell-mediated rejection. An alginate scaffold has been developed commercially, known as AlgiMatrix® (Life Technologies), and this is being utilised for culturing *in vitro* tumour models for anticancer drug screening (Godugu *et al.*, 2013).

Fibrin scaffolds, produced by polymerising the protein fibrinogen obtained from plasma, have been widely investigated for a variety of 3D cell culture applications. These include the generation of mimetic bone for use in preclinical studies on bone regeneration (McDuffee *et al.*, 2014), as a method of generating an artificial ovary (Luyckx *et al.*, 2013) and as a cardiac tissue mimic (Liau *et al.*, 2011).

Collagen is the most abundant protein in the body and is an important component of the ECM. Scaffolds manufactured with collagen provide physical support for cells and have appropriate incorporation sites for mineral substances and cytokines. They also exhibit good biocompatibility and biodegradability profiles (Geiger *et al.*, 2003). Therefore using scaffolds generated from collagen for 3D cell culture is a popular choice and several commercial systems are marketed. These include Optimaix-3D (Matricel) which uses collagen harvested from rat tails as the scaffold material. Collagen scaffolds have been used for a number of 3D cell culture applications including bone regeneration (Lee *et al.*, 2013; Miyazaki *et al.*, 2009) and for the production of knee menisci tissue mimics (Croutze *et al.*, 2013).

While scaffolds made of natural materials offer many benefits, they also have a number of disadvantages. Natural scaffolds may exhibit immunogenicity and contain pathogenic impurities, which is not desirable if the 3D cell culture is to be used in human clinical or veterinary applications. There is also less control over the mechanical properties of natural scaffolds with batch-to-batch inconsistency making a reproducible product for use in a Good Manufacturing Practice (GMP) environment problematic. Additionally, batch variation can lead to assay-to-assay changes if the scaffolds are being used in high-throughput screening, affecting the quality of the data. Another issue with using natural materials as a scaffold is that the pore sizes are difficult to control and this can lead to an uneven distribution of cells throughout the scaffold, causing problems with transferring drugs or chemicals throughout the 3D matrix which again could generate inconsistent results.

When using 3D cell culture constructs for clinical or drug screening applications, the reproducibility and consistency of the scaffolds are paramount. For clinical applications, a scaffold which retains the same physical and chemical components from batch to batch, that can be sterilised, does not cause immunogenic reactions and is biodegradable has the ideal features for successful use as a 3D tissue or bone mimic.

In drug or chemical screening applications, the optimal scaffolds to use for 3D cell culture have to allow cell lines (sometimes these have to be layers of different cells) of the researchers' choice to adhere throughout the scaffold. The scaffold has to be simple to process, flexible enough to be used on the bench or in an automated screening system, and be compatible with a broad range of analytical techniques, including microscopic imaging, biochemical and histological testing.

Synthetic polymer scaffolds

Substituting natural materials for synthetic materials such as polystyrene overcomes some of these manufacturing and immunogenicity issues and fulfils many of the criteria stated for 3D cell culture for use in clinical and drug and chemical screening applications. Polystyrene is attractive to use as a scaffold material because it is chemically inert, stable and similar to conventional 2D tissue culture plasticware such as microplates. Other benefits of using polystyrene scaffolds are that they are inexpensive to produce in large quantities and because many are non-biodegradable, they have a long shelf-life. Commercially available polystyrene scaffolds include Alvetex® (Reinnervate), consisting of a porous sponge-like membrane containing interconnecting pores, and 3D Insert™PS (Biotek), a fibrous mesh-type scaffold.

Polystyrene scaffolds have been used for a number of 3D cell culture applications, including culturing lymphoma cells (Caicedo-Carvajal *et al.*, 2011). One disadvantage of using a polystyrene scaffold for 3D cell culture is that many are around 200 microns in thickness and form a relatively inflexible matrix. For 3D cell culture to be used routinely in drug discovery assays, the scaffolds have to be thin enough to permit diffusion of drug candidates and also, in the absence of a functioning vascular system, to allow nutrients and oxygen to maintain the cells as true phenotypic mimics. Scaffolds that are too dense make visualisation of cells using conventional and inexpensive confocal microscopy techniques unreliable, which in some cases means that the assays cannot be analysed using automated high-content screening systems that rely on this type of imaging. Additionally, polystyrene is also known to exhibit autofluorescence (Young *et al.*, 2013), especially in the 300–400 nm wavelength range, which overlaps with the excitation and emission spectra of many commonly used fluorophores. Therefore, polystyrene scaffolds may not be suitable for some fluorescence or luminescence cell-based assays because of background light emissions. Not being able to automate assay analysis and having to use alternative and sometimes more costly assays can make using 3D cell culture in polystyrene scaffolds prohibitively expensive for high-throughput assay applications.

For 3D tissue culture to be used in a regenerative medicine context, polystyrene is not considered a desirable option because it is non-biodegradable, and if particles of polystyrene break away from the scaffold and begin circulating, they could cause tissue deposition and granuloma formation (Vandooren *et al.*, 2013). Polystyrene is also relatively inflexible so cannot easily be used in applications where 3D cell cultures are required to curve around a surface, for example as a corneal tissue mimic.

Using synthetic polymers as scaffolds in the correct format can overcome diffusion issues and offers advantages with 3D cell culture for use in clinical and drug and chemical screening applications. Synthetic polymer scaffolds were developed in the 1990s (Mikos *et al.*, 1993) and their use has been widespread. They are versatile and have been used in a range of formats for the 3D cell culture of osteogenic stem cells (Kim *et al.*, 2008), hepatocytes (Chou *et al.*, 2013) and pluripotent stem cells (Gao *et al.*, 2010).

Many scaffolds made of synthetic polymers are produced in a matrix format by patterning polymers into configurations such as porous films and fibrous mats. With these formats, it is difficult to control pore sizes and batch-to-batch variation can occur. Pore size and porosity are of paramount importance in the design of scaffolds for 3D cell culture. Pore characteristics play an important role in directing tissue formation and function (Zhang *et al.*, 2014). Scaffold porosity and adequate pore size are required for homogeneous cell distribution and interconnection throughout tissue mimics. Additionally, increased porosity can have a beneficial effect on the diffusion of nutrients and oxygen, especially in the absence of a functional vascular system (Lee *et al.*, 2008).

An alternative method of producing a 3D cell culture scaffold from synthetic polymers, which addresses these issues, is to produce a matrix generated from electrospun fibres. When compared to other fibre-forming processes such as self-assembly and phase separation techniques, electrospinning provides a simple and cost-effective means to produce fibrous scaffolds with an interconnected pore structure and fibre diameters in the sub-micron range.

The next section of this chapter explains how a scaffold made of an electrospun synthetic polymer can be used for 3D cell culture, describing the electrospinning technique and how it can be utilised to generate a commercially available scaffold range known as Mimetix. It will also discuss a protocol for using Mimetix with specific examples which demonstrate why Mimetix is ideally suited to 3D cell culture for both drug screening and tissue engineering applications.

Description of technology

Electrospinning is a well-established technique used for over two decades in which fibres are electrospun from a wide variety of natural and synthetic polymers. To date, a range of polymers such as polylactic acid (Yang *et al.*, 2005) and polystyrene (Sun *et al.*, 2005) have been fabricated into electrospun matrices for use as 3D cell culture scaffolds.

The electrospinning process involves injecting a polymer solution at a constant feed rate through a nozzle or needle that is charged to a high voltage, typically 10–30 kV. The applied voltage induces a charge on the surface of the liquid droplet and when this is sufficiently high, the hemispherical surface of the fluid elongates and a Taylor cone is established. On increasing the applied voltage further, a charged liquid jet is ejected from the Taylor cone and attracted to the earthed collector, which is positioned at a fixed distance from the needle. During this process, the solvent evaporates from the polymer solution, leaving dry polymer fibres on the collector.

Using electrospinning, a linear relationship between three principal process variables (applied voltage, working distance and polymer concentration) allows fibres with micro- and nanometre diameters (Reneker & Chun, 1996) to be consistently produced. The architecture of electrospun scaffolds is similar to the natural ECM and this helps electrospun fibres to mimic the natural tissue environment. Electrospinning can be applied to a range of natural or synthetic polymers to generate fibres with certain characteristics, such as stiffness or elasticity, to be hydrophilic or hydrophobic and biodegradable or non-biodegradable. The process can even be designed to incorporate functional molecules, such as drugs (Canton *et al.*, 2010), and the degradation time of some biodegradable polymers can be tuned to fit the application.

Electrospun fibres offer several benefits which make them highly suited to 3D cell culture for both regenerative medicine and drug or chemical screening applications. These include having a high surface-to-volume ratio, providing more cell attachment space and a high porosity – over 80% porosity has been demonstrated (Soliman *et al.*, 2011). The porosity also helps to mimic the ECM by facilitating efficient nutrient exchange and allows extraction of expression markers. With electrospun fibres, the porosity of the scaffolds can be defined by changing the fibre diameter and packing density; the fibres can be formed into a wide variety of sizes and shapes and can be configured from different material compositions to suit the required physical properties and function of the scaffold.

Electrospun fibres in 3D cell culture

Scaffolds constructed of electrospun fibres used in regenerative medicine research typically contain electrospun synthetic, biodegradable polymers. Synthetic polymers are preferred because they can be sterilised and engineered to biodegrade in specific time periods, making these scaffolds more suitable for use in clinical applications. Electrospun synthetic polymers that have been successfully used for 3D cell culture and tissue engineering applications include poly-D,L-lactide-co-glycolide (PGLA) as a bone (Ni *et al.*, 2011), skin (Blackwood *et al.*, 2008) and corneal tissue mimic (Deshpande *et al.*, 2010); poly-ε-caprolactone (PCL) for cardiac grafts (Guex *et al.*, 2014); and poly-L-lactide (PLLA) for nerve cells (Murray-Dunning *et al.*, 2011). One advantage of using electrospun fibre scaffolds is that fibres can be aligned (Wang *et al.*, 2009) and this is particularly useful to direct the orientation of, for example, nerve cells.

Electrospun fibres for scaffolds used in drug and chemical screening applications have been produced from a variety of materials. To date, these include collagen scaffolds for 3D cancer models (Hartman *et al.*, 2009) and PCL to support 3D hepatocyte cultures (Hashemi *et al.*, 2009). Electrospun fibre scaffolds are used in these applications because their porosity allows drugs and chemicals to diffuse across the scaffold more easily and because the scaffolds are thin, and the polymers they are made of generally do not show autofluorescence, they allow microscopic analysis of the cells using a range of fluorophores.

Ultra-Web™ (Corning) was the first commercial electrospun scaffold available for 3D cell culture but was discontinued in 2009 due to a supply problem with the electrospun polyamide fibres. This left a gap in the market in which demand for electrospun scaffolds remains high. To ensure this need is met, the Electrospinning Company is using state-of-the-art electrospinning equipment and proprietary manufacturing and polymer chemistry know-how to produce a commercially available range of highly consistent, reproducible 3D electrospun scaffolds.

The company can develop electrospun synthetic polymers with different degradation properties (Deshpande *et al.*, 2013), and is able to tune the fibre diameter to create scaffolds with pore sizes to suit different cell types. The company can also incorporate bioactive molecules into the fibres (Millas *et al.*, 2014) and, through collaboration with the company Orla Protein Technologies, any of the electrospun fibre scaffolds can be coated with a choice of 29 different peptide motifs to promote cell attachment, growth and differentiation.

The Electrospinning Company produces a commercially available range of scaffolds made from electrospun fibres, known as the Mimetix scaffold. These are suitable for use in both drug screening and regenerative medicine applications and their structural features and protocols for use are described in the next section.

Mimetix electrospun polymer scaffolds

Mimetix scaffolds are fabricated from randomly orientated electrospun fibres (Figure 12.1) of consistent fibre diameter made from the medical-grade polymer PLLA. This polymer was chosen as scaffold material because as well as being biocompatible and biodegradable, it is US Food and Drug Administration (FDA) approved for use in humans, which means Mimetix could potentially be used for clinical applications.

The resulting scaffold is highly porous and mimics the ECM, providing an ideal environment to support the growth of cells in 3D. The technology guarantees batch-to-batch reproducibility and the precise control of important fibre parameters, such as diameter and surface texture. The Mimetix scaffold comes in ready-to-use standard-size plates, which are compatible with laboratory liquid handling robotics and automated image analysis equipment, such as plate readers. The scaffold is available in fibre diameters of 2 μm and 4 μm, corresponding to pore sizes of 10–18 μm and 15–31 μm and at a thickness of 50 or 100 μm. The thickness of the Mimetix scaffold offers the benefit that cell morphology and

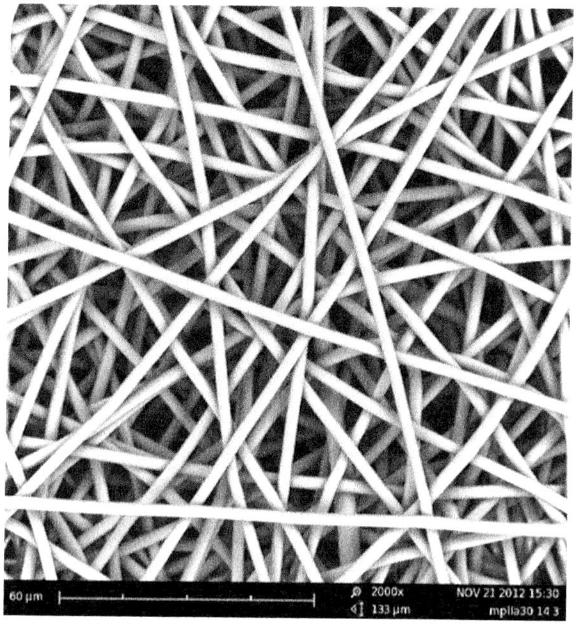

Figure 12.1 **Scanning electron micrograph of the fibrous Mimetix scaffold.**

behaviour can be observed inside the scaffold using (confocal) fluorescence microscopy imaging, utilised in many high-content screening imagers. Additionally, because the Mimetix scaffold is constructed of PLLA which does not exhibit autofluorescence, it can be used for fluorescence- and luminescence-based assays without causing background light emission. This means that for drug or chemical screening applications, the Mimetix scaffold can be used in microplates that can be analysed in standard high-throughput automation without making changes to existing screening protocols.

For added utility in automated screening systems, the Mimetix scaffold has been designed for a number of different plate formats. The 96-well plate contains the Mimetix scaffold laser-welded by a proprietary process into the base (Figure 12.2). These scaffolds cannot be removed from the wells, although the cells can be separated from the scaffold by trypsinisation and gentle agitation. The Mimetix scaffold is also available as 12-well plate discs that are held in place by removable rings, which allows them to be removed for imaging.

A hanging insert Mimetix scaffold format that allows for cells to be cultured with an air-liquid interface is available. Alternatively, it is suited to experiments with longer culturing periods or for cell migration experiments, as the cells can be exposed to medium from above and below. These scaffold discs can also be removed for imaging, or for seeding with a second cell type on the reverse side. The Mimetix scaffold can also be provided as a sheet (220 × 350 mm) for customised cutting, or as precut discs with diameters between 2 mm and 50 mm.

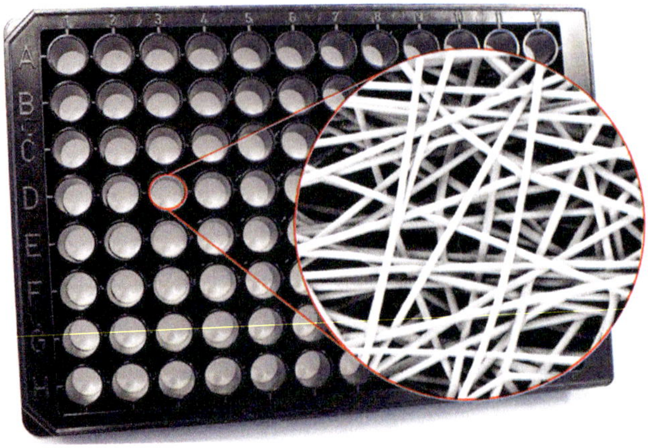

Figure 12.2 **Image of the Mimetix 96-well microplate format of the Mimetix scaffold.**

Figure 12.3 **Scanning electron micrograph of a Mimetix aligned fibre scaffold.**

Additionally, there is a Mimetix scaffold with aligned fibres (Figure 12.3), to support the differentiation of cells that require a particular orientation, such as peripheral nerves and tendons. These scaffolds are available with a fibre diameter of 2 µm, mounted on 12-well Cell Crown® inserts; a 96-well plate is currently under development.

Mimetix scaffolds can be easily sterilised via γ-irradiation and have been used successfully to culture a range of cell types for use in 3D cell-based assays.

These include human primary hepatocytes, retinal pigment epithelium, breast, ovarian and prostate cancer cells, keratinocytes, neural cells, dermal and lung fibroblasts, mesenchymal and embryonic stem cell lines.

For regenerative medicine applications, the Electrospinning Company and the primary cell products firm Medicyte are partners in the European Union Seventh Framework Programme project Re-Liver which aims to build a bioartificial liver organoid. To date, within the Re-Liver project, upcyte® hepatocytes (Medicyte) have been shown via confocal fluorescence microscopy to migrate into a 50 μm-thick Mimetix scaffold and proliferate into three layers of cells, demonstrating 3D morphology. These 3D cells also exhibit higher levels of basal and induced cytochrome P450 enzymes than 2D cell culture of upcyte hepatocytes (donor 151). This research indicates that the Mimetix scaffold provides a tissue-like environment for the generation of a basic, functioning 3D hepatic mimic, which the company continues to develop.

The choice of which Mimetix scaffold format to use for 3D cell culture depends on a variety of factors, such as cell type, duration of culture, desired penetration depth into the scaffold and application of the 3D cell culture, and an example 3D cell culture protocol is detailed below using one of the available Mimetix formats.

Application of Mimetix for 3D breast cancer cells

The following method describes how to seed and maintain HMT3909S8 human epithelial breast cancer cells as 3D cultures in the 96-well format Mimetix electrospun scaffold. This 3D cell culture can then be used in drug screening assays to determine efficacy or toxicity of lead compounds.

Materials

Cell line and media

The cell line used was an HMT3909S8 human breast cancer cell line (AvantiCell Science Ltd). The cells were thawed in a water bath (37 °C) and transferred to a sterile conical-bottom tube containing breast cancer cell culture medium CDM3 (10 mL) (AvantiCell Science Ltd). The cell suspension was centrifuged (240 g, 20 °C, five minutes) and the supernatant was removed. The cell pellet was resuspended in CDM3 medium (1 mL) and the cells plated out at $12–15 \times 10^3$ cells/cm^2. The cells were incubated (37 °C, 5% CO_2, 24 hours).

3D scaffold

Two Mimetix 3D scaffolds fabricated from 2 μm and 4 μm electrospun fibres both in 96-well format (Electrospinning Company) were prewetted with a 30% ethanol solution (200 μL) in each well to soak the scaffolds and allow cell culture medium to access the pores.

Method

Cell proliferation

The ethanol in each well of the 96-well Mimetix scaffolds was removed and each well was washed twice with CDM3 medium (1 mL). The HMT3909S8 cell suspension (200 μL, 1.5–6 × 10^4 cells/mL) was added to each well of both scaffolds. Human breast cancer cells cultured in the 4 μm fibre diameter scaffold were stained with a nuclear dye (TO-PRO-3) and analysed for cell penetration with confocal microscopy.

Cell seeding density

To determine optimum cell seeding density, three wells in each of the Mimetix 3D scaffolds fabricated from 2 μm and 4 μm fibres were seeded at 1.5 × 10^4, 3 × 10^4 and 6 × 10^4 cells/mL. The cells were incubated (37 °C, 5% CO_2 for 10 days) and culture medium was replaced by demi-dilution every 2–3 days. From three wells, cells at densities of 1.5 × 10^4, 3 × 10^4 and 6 × 10^4 cells/mL in both 2 μm and 4 μm Mimetix scaffolds were lysed after 72 hours in culture and then tested for basal cytotoxicity and total cell number using a lactate dehydrogenase (LDH) assay according to the manufacturer's instructions. The LDH assay was analysed with a plate reader.

Culture duration

To determine optimum cell culture duration, three wells in each of the Mimetix 3D scaffolds, fabricated from 2 μm and 4 μm fibres, were seeded at 1.5 × 10^4 cells/mL. The cells were incubated (37 °C, 5% CO_2 for 10 days) and medium replaced by demi-dilution every 2–3 days. From three wells, cells cultured in both 2 μm and 4 μm Mimetix scaffolds were sampled and lysed at days 2, 4 and 10 after inoculation and cells were cultured for up to 10 days. The samples were then tested for basal cytotoxicity and total cell number using an LDH assay according to the manufacturer's instructions. The LDH assay was analysed with a plate reader.

Apoptosis

To determine if the HMT3909S8 cell line is resistant to apoptosis in 3D culture, a HMT3909S8 cell suspension (200 μL, 1.5–6 × 10^4 cells/mL) was added to wells in 96-well plates containing no scaffold, or a 2 μm or 4 μm Mimetix scaffold. The cells were incubated (37 °C, 5% CO_2 for 10 days) and medium replaced by demi-dilution every 2–3 days. In cells cultured without a scaffold or either a 2 μm or 4 μm Mimetix scaffold, apoptosis was induced by adding staurosporine (1 μM), 24 hours prior to read-out at day 10 (3D cultures) or day 3 (2D cultures). The cells were then tested for apoptosis using the Caspase-Glo® 3/7 according to manufacturer's instructions. The assay was analysed with a plate reader.

Results

Cell penetration

The 3D reconstruction image (Figure 12.4) shows that the human breast cancer cells migrated into the Mimetix scaffold (4 μm fibre diameter) and were able to grow in different planes, forming a true 3D cell matrix.

Cell seeding density

Analysis of the results of the LDH assay (Figure 12.5) indicate that signals obtained for total cell numbers for 3×10^4 and 6×10^4 seeded cells were not much higher than for 1.5×10^4 seeded cells, but basal cytotoxicity increased at the higher seeding densities. Total cell numbers were slightly higher for the Mimetix 2 μm fibre scaffold at the two lower densities and similar to the Mimetix 4 μm scaffold at the highest cell density. Based on these results, a seeding density of 1.5×10^4 cells is considered to be optimal for this cell line. Since the cell density and cytotoxicity results were comparable for both fibre diameter types of Mimetix scaffold, this indicates that either scaffold is equally suitable for culturing this cell type.

Culture duration

Total cell number increased over the 10-day cell culture period (Figure 12.6), showing that Mimetix scaffolds with electrospun fibre diameters of 2 μm and 4 μm support 3D cell culture proliferation. As Figure 12.6 shows, the measurements were very consistent (coefficients of variation (CVs) of <10%) up to day 10. On days 2 and 4, basal cytotoxicity was low and cells sensitive to induced cytotoxicity, but basal cytotoxicity/CVs were significantly increased by day 10. The results suggest that culture duration of 4–6 days is optimal and is consistent with published data of breast cancer cell lines cultured on polymer scaffolds for 3–5 days (Saha, 2012).

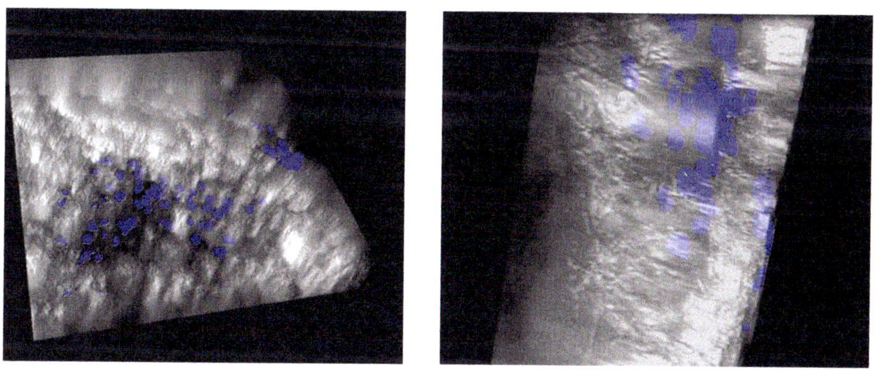

Figure 12.4 3D confocal microscopy reconstruction image showing human breast cancer cells. Cells are seen proliferating into Mimetix scaffold (4 μm fibre diameter) with cell nuclei counterstained in blue (TO-PRO-3).

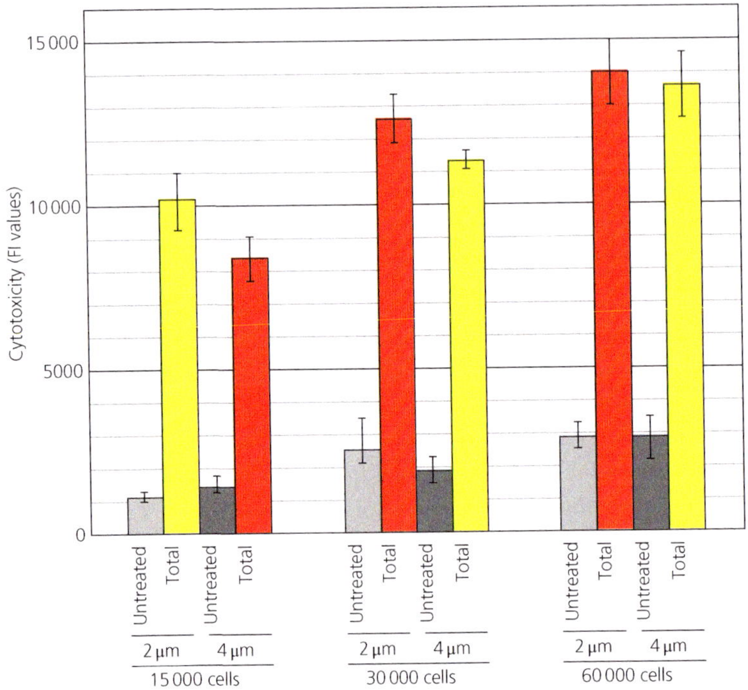

Figure 12.5 Cytotoxicity and cell density of HMT3909S8 cells. Values were taken after 72 hours culture in Mimetix scaffold 2 µm and 4 µm fibre diameter.

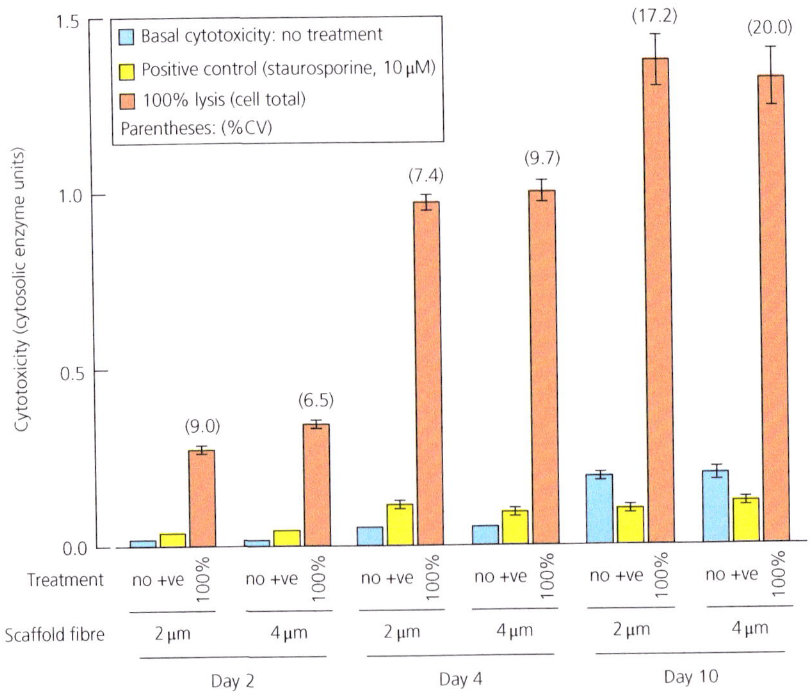

Figure 12.6 Cell viability and proliferation of HMT3909S8 cells in Mimetix scaffolds over 10 days. CV, coefficient of variation.

Apoptosis

The cancer cell line HMT3909S8 cultured in 3D Mimetix scaffolds of both fibre diameters became more resistant to drug-induced apoptosis compared with those cells grown in 2D culture (Figure 12.7). Breast cancer cells *in vivo* can become resistant to chemotherapy and in turn to apoptosis. These results demonstrate the suitability of the Mimetix scaffold as a tool for modelling the tumour behaviour human breast cancer cells, which could be used in screening the effectiveness of chemotherapeutic drug candidates.

Summary

The Mimetix electrospun scaffolds of both 2 μm and 4 μm fibre diameter support and allow the HMT3909S8 breast cancer cell line to proliferate and penetrate the scaffold over a 10-day period. The optimum cell seeding density of 1.5×10^4 cells/mL allowed the cells to proliferate well without significant cytotoxicity for up to six days in Mimetix scaffolds. In addition, cells became more resistant to drug-induced apoptosis when cultured in the scaffold. This is consistent with published results showing that breast cancer cells cultured in 3D are more resistant to apoptosis (Weaver *et al.*, 2002). These results show that Mimetix scaffolds can successfully support 3D cell culture and allow simple microscopy or plate reader analysis of a breast cancer cell line *in situ*, which indicates that

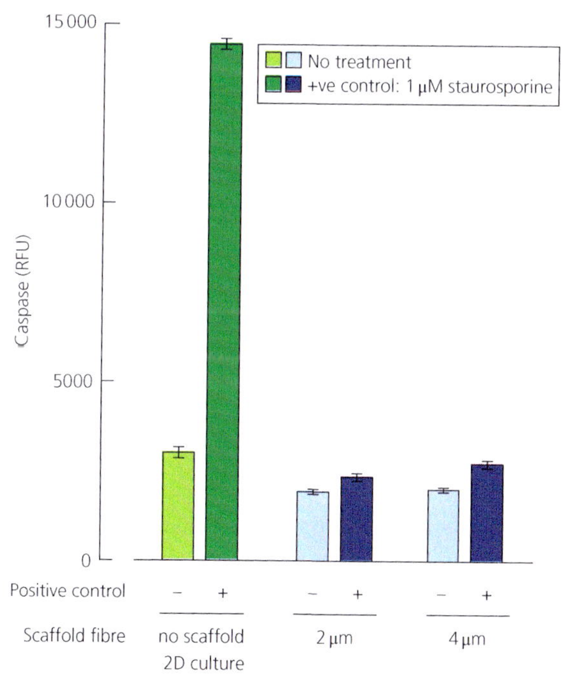

Figure 12.7 Staurosporine-induced apoptosis of HMT3909S8 cells cultured in Mimetix scaffolds and 2D cell culture microplates. RFU, relative fluorescence unit.

this scaffold type could be easily implemented and provide more realistic tumour models for use in either laboratory-scale or automated high-throughput drug discovery assays.

Troubleshooting

In specific circumstances, the Electrospinning Company has noted issues in routine use of the Mimetix scaffold and this is a guide to solving some of those commonly observed.

Scaffold preparation

Insufficient wetting of scaffold

A solution of 20% ethanol (20% v/v) in water should normally wet the scaffold within less than a minute. If incomplete wetting is experienced, a slightly higher concentration of ethanol (25–30% vv) in water should be used. If that does not wet the scaffold, then there is a fault with the scaffold and the technical support at the Electrospinning Company should be contacted via email or phone.

Air bubbles trapped under the scaffold

This happens especially for the 12-well Mimetix format with removable scaffold discs. The cell medium should be pipetted up and down to remove the bubble. If this does not help, a suction pump is required for air displacement and this should remove the bubbles.

Loose retaining rings

This occurs especially with the removable scaffold discs for the Mimetix 12-well and six-well formats with removable scaffold discs. These can become loose during shipping and handling. To put them back in place, press the retaining ring down on two sides using tweezers or pipette tips. Please note: the retaining rings for the six-well plates with hanging inserts are very slightly conical. If they have fallen out completely, make sure that they are replaced in the correct orientation.

Cell culture and assays

No cell proliferation

It is well established in the literature that cells generally proliferate more slowly in a 3D culture than they do in 2D culture. The Electrospinning Company has generated data which indicates that cells cultured in 3D, including in Mimetix scaffolds, take longer to reach a specific density than they do in 2D culture. However, because of this slower proliferation in 3D (and the greater space available), cells can be maintained in culture for much longer. If cells appear to be proliferating incorrectly, then they might require a coating to improve attachment and the technical support at the Electrospinning Company can be contacted via email or phone for advice on which coating is appropriate for use with Mimetix.

Difficulty determining cell confluence

Confluence and cell viability can be estimated with optical or fluorescence microscopy using appropriate dyes for cell staining (for example, neutral red). Please note that some hydrophobic dyes may adsorb onto the Mimetix scaffold and users should contact technical support at the Electrospinning Company for advice on which dyes are appropriate for use with the Mimetix scaffold. Quantitative, colorimetric assays are also compatible with the scaffold technology (supernatants have to be transferred to a new, clear 96-well plate prior to read-out to avoid background interference of the scaffold) and can provide a more accurate measurement of cell numbers within a scaffold. It is recommended to record a growth curve before using a new cell line in the Mimetix scaffold to determine its growth behaviour in 3D and how long cells take to become half-confluent or confluent.

Analysis

High background signal from plate

The Mimetix scaffolds gives rise to a background signal in absorbance-based assays and the Electrospinning Company recommends the use of fluorescence- or luminescence-based assays to avoid this problem. Alternatively, the well contents can be transferred to a new 96-well plate for an absorbance-based read-out.

Difficulty recovering cells from scaffold

Cells can be partly removed from the Mimetix scaffold by trypsinisation and gentle agitation. As a general rule, slightly longer incubation times are required for 3D-cultured cells than for those grown in 2D, but care needs to be taken that cell performance and viability are not compromised by performing a longer incubation period. However, some cell lines are particularly adherent and hard to remove. In most cases, the Electrospinning Company recommends finding analytical techniques which do not rely on retrieving the cells from the scaffolds. Researchers should consider assays that measure analytes in the supernatant or assays that lyse the cells as a first step to determining DNA or protein content.

Conclusion

Using 3D cell culture scaffolds fabricated from electrospun fibres offers scientists a range of benefits. These scaffolds are inexpensive to produce and the flexible electrospun fibres, engineered with highly reproducible parameters, can be formed into scaffolds ranging in size and shape to fit standard 2D cell culture plates. Since the scaffolds produced are thin, they allow homogeneous diffusion and cell seeding. As demonstrated in this chapter, the Mimetix scaffolds made of electrospun polymer fibres allow good cell proliferation of breast cancer cells to form a tumour-like model. Additionally, the structure of electrospun scaffolds such as Mimetix allows researchers to image cells *in situ* using standard inexpensive analysis methods including microscopy and plate reading.

The 3D nature of electrospun scaffolds provides an ideal structure in which cells can grow, differentiate and function in close proximity to neighbouring cells, forming networks of biomimetic tissue-like layers. Using electrospun scaffolds of medical-grade, flexible polymers which are biodegradable and biocompatible, and can be sterilised prior to use, means electrospun scaffolds like Mimetix could potentially be safely fabricated as a scaffold to support tissue or bone growth. Additionally, because electrospun scaffolds can be aligned, they could be used to drive directional growth of neural or tendon tissue.

These capabilities make electrospun scaffolds such as Mimetix an excellent tool for use in life science and clinical applications, which could help reduce the costs and numbers of animal tests required for drug and chemical screening, as well as offering safer and more effective tissue mimics for regenerative medicine.

Acknowledgements

The authors would like to thank Colin Wilde, Tracey Duff and Hazel McLuskey at AvantiCell Science Ltd for generation of the biological data used in Figures 12.5, 12.6 and 12.7.

References

Ahmed, T. A., Dare, E. V. & Hincke, M. (2008) Fibrin: a versatile scaffold for tissue engineering applications. *Tissue Engineering Part B: Reviews*, **14**, 199–215.

Aly, H. H., Shimotohno, K. & Hijikata, M. (2009) 3D cultured immortalized human hepatocytes useful to develop drugs for blood-borne HCV. *Biochemical and Biophysical Research Communications*, **379**, 330–334.

Becker, C., Olde-Damink, L., Laeufer, T. *et al.* (2006) 'UroMaix' scaffolds: novel collagen matrices for application in tissue engineering of the urinary tract. *International Journal of Artificial Organs*, **29**, 764–771.

Blackwood, K. A., McKean, R., Canton, I. *et al.* (2008) Development of biodegradable electrospun scaffolds for dermal replacement. *Biomaterials*, **29**, 3091–3104.

Caicedo-Carvajal, C. E., Liu, Q., Remache, Y. *et al* (2011) Cancer tissue engineering: a novel 3D polystyrene scaffold for *in-vitro* isolation and amplification of lymphoma cancer cells from heterogeneous cell mixtures. *Journal of Tissue Engineering*, **2011**, 362326.

Canton, I., McKean, R., Charnley, M. *et al.* (2010) Development of an Ibuprofen-releasing biodegradable PLA/PGA electrospun scaffold for tissue regeneration. *Biotechnology and Bioengineering*, **105**, 396–408.

Capone, S. H., Dufresne, M., Rechel, M. *et al.* (2013) Impact of alginate composition: from bead mechanical properties to encapsulated HepG2/C3A cell activities for *in vivo* implantation. *PLoS One*, **8**, 62032.

Chou, M. J., Hsieh, C. H., Yeh, P. L. *et al.* (2013) Application of open porous poly (D, L-lactide-co-glycolide) microspheres and the strategy of hydrophobic seeding in hepatic tissue cultivation. *Journal of Biomedical Materials Research Part A*, **101**, 2862–2869.

Croutze, R., Jomha, N., Uludag, H. & Adesida, A. (2013) Matrix forming characteristics of inner and outer human meniscus cells on 3D collagen scaffolds under normal and low oxygen tensions. *BMC Musculoskeletal Disorders*, **14**, 353–377.

Deshpande P., McKean, R., Blackwood, K.A. *et al.* (2010) Using poly (lactide-co-glycolide) electrospun scaffolds to deliver cultured epithelial cells to the cornea. *Regenerative Medicine*, **5**, 395–401.

Deshpande, P., Ramachandran, C., Sefat, F. *et al.* (2013) Simplifying corneal surface regeneration using a biodegradable synthetic membrane and limbal tissue explants. *Biomaterials*, **34**, 5088–5106.

Dvir, T. & Fleischer, S. (2013) Tissue engineering on the nanoscale: lessons from the heart. *Current Opinion in Biotechnology*, **24**, 664–671.

Dvir, T., Timko, B. P., Kohane, D. S. & Langer, R. (2011) Nanotechnological strategies for engineering complex tissues. *Nature Nanotechnology*, **6**, 13–22.

Ekert, J. E., Johnson, K., Strake, B. *et al.* (2014) Three-dimensional lung tumor microenvironment modulates therapeutic compound responsiveness in vitro – implication for drug development. *PLoS One*, **9**, 92248.

Gao, S. Y., Lees, J. G., Wong, J. C. *et al.* (2010) Modeling the adhesion of human embryonic stem cells to poly(lactic-co-glycolic acid) surfaces in a 3D environment. *Journal of Biomedical Materials Research Part A*, **92**, 683–692.

Geiger, M., Li, R. H. & Friess, W. (2003) Collagen sponges for bone regeneration with rhBMP-2. *Advanced Drug Delivery Review*, **55**, 1613–1629.

Godugu, C., Patel A. R., Desai, U. *et al.* (2013) AlgiMatrix™ based 3D cell culture system as an in vitro tumor model for anticancer studies. *PLoS One*, **8**, 53708.

Guex, A. G., Frobert, A., Valentin, J. *et al.* (2014) Plasma-functionalized electrospun matrix for biograft development and cardiac function stabilization. *Acta Biomaterialia*, **10**, 2996–3006.

Hartman, O., Zhang, C., Adams, E. L. *et al.* (2009) Microfabricated electrospun collagen membranes for 3-D cancer models and drug screening applications. *Biomacromolecules*, **10**, 2019–2032.

Hashemi, S. M., Soleimani, M., Zargarian, S. S. *et al.* (2009) *In vitro* differentiation of human cord blood-derived unrestricted somatic stem cells into hepatocyte-like cells on poly(epsilon-caprolactone) nanofiber scaffolds. *Cells Tissues Organs*, **190**, 135–149.

Kim, S., Kim, S. S., Lee, S. H. *et al.* (2008) *In-vivo* bone formation from human embryonic stem cell-derived osteogenic cells in poly (d, l-lactic-co-glycolic acid)/hydroxyapatite composite scaffolds. *Biomaterials*, **8**, 1043–1053.

Kolehmainen, K. & Willerth, S. M. (2012) Preparation of 3D fibrin scaffolds for stem cell culture applications. *Journal of Visualized Experiments*, **61**, 3641.

Lee, M., Wu, B. M. & Dunn, J. C. (2008) Effect of scaffold architecture and pore size on smooth muscle cell growth. *Journal of Biomedical Materials Research Part A*, **15**, 1010–1016.

Lee, S. S., Huang, B. J., Kaltz, S. R. *et al.* (2013) Bone regeneration with low dose BMP-2 amplified by biomimetic supramolecular nanofibers within collagen scaffolds. *Biomaterials*, **34**, 452–459.

Liau, B., Christoforou, N., Leong, K. W. & Bursac, N. (2011) Pluripotent stem cell-derived cardiac tissue patch with advanced structure and function. *Biomaterials*, **32**, 9180–9187.

Lin, Q. & Yun, Z. (2010) Impact of the hypoxic tumor microenvironment on the regulation of cancer stem cell characteristics. *Cancer Biology & Therapy*, **9**, 949–956.

Luyckx, V., Dolmans, M. M., Vanacker, J. *et al.* (2013) First step in developing a 3D biodegradable fibrin scaffold for an artificial ovary. *Journal of Ovarian Research*, **6**, 83.

McDuffee, L. A. Esparza Gonzalez, B. P., Nino-Fong, R. & Aburto, E. (2014) Evaluation of an *in vivo* heterotopic model of osteogenic differentiation of equine bone marrow and muscle mesenchymal stem cells in fibrin glue scaffold. *Cell Tissue Research*, **355**, 327–335.

Mikos, A. G., Sarakinos, G., Leite, S. M., Vacanti, J. P. & Langer, R. (1993) Laminated three-dimensional biodegradable foams for use in tissue engineering. *Biomaterials*, **14**, 323–330.

Millas, A. L. G., McKean, R., Stevens, M. *et al.* (2014) Fabrication of electrospun scaffolds incorporating an Amazonian therapeutic oil from the *Copaifera* sp. for wound care applications. *Journal of Biomaterials and Tissue Engineering*, **4**, 217–220.

Miyazaki, M., Morishita, Y., He, W. *et al.* (2009) A porcine collagen-derived matrix as a carrier for recombinant human bone morphogenetic protein-2 enhances spinal fusion in rats. *Spine Journal*, **9**, 22–30.

Mulhall, H. J., Hughes, M. P., Kazmi, B., Lewis, M. P. & Labeed, F. H. (2013) Epithelial cancer cells exhibit different electrical properties when cultured in 2D and 3D environments. *Biochimica et Biophysica Acta*, **1830**, 5136–5141.

Murray-Dunning, C., McArthur, S. L., Sun, T. *et al.* (2011) Three-dimensional alignment of Schwann cells using hydrolysable microfiber scaffolds: strategies for peripheral nerve repair. *Methods in Molecular Biology*, **695**, 155–166.

Ni, P., Fu, S., Fan, M. *et al.* (2011) Preparation of poly (ethylene glycol)/polylactide hybrid fibrous scaffolds for bone tissue engineering. *International Journal of Nanomedicine*, **6**, 3065–3075.

Perets, A., Baruch, Y., Weisbuch, F. *et al.* (2003) Enhancing the vascularization of three-dimensional porous alginate scaffolds by incorporating controlled release basic fibroblast growth factor microspheres. *Journal of Biomedical Materials Research*, **65A**, 489–497.

Reneker, D. H. & Chun, I. (1996) Nanometre diameter fibres of polymer, produced by electrospinning. *Nanotechnology*, **7**, 216–223.

Saha, S., Duan, X., Wu, L. *et al.* (2012) Electrospun fibrous scaffolds promote breast cancer cell alignment and epithelial-mesenchymal transition. *Langmuir*, **28**, 2028–2034.

Soliman, S., Sant, S., Nichol, J. W. *et al.* (2011) Controlling the porosity of fibrous scaffolds by modulating the fiber diameter and packing density. *Journal of Biomedical Materials Research Part A*, **96**, 566–574.

Sun, T., Mai, S., Norton, D. *et al.* (2005) Self-organization of skin cells in three-dimensional electrospun polystyrene scaffolds. *Tissue Engineering*, **11**, 1023–1033.

Vaghjiani, V., Vaithilingam, V., Saraswati, I. *et al.* (2014) Hepatocyte-like cells derived from human amniotic epithelial cells can be encapsulated without loss of viability or function *in vitro*. *Stem Cells and Development*, **23**, 866–876.

Vandooren J., Berghmans, N., Dillen, C. *et al.* (2013) Intradermal air pouch leukocytosis as an *in vivo* test for nanoparticles. *International Journal of Nanomedicine*, **8**, 4745–4756.

Wang, H. B., Mullins, M. E., Cregg, J. M. *et al.* (2009) Creation of highly aligned electrospun poly-L-lactic acid fibers for nerve regeneration applications. *Journal of Neural Engineering*, **6**, 016001.

Weaver, V. M., Lelièvre, S., Lakins, J. N. *et al.* (2002) beta4 integrin-dependent formation of polarized three-dimensional architecture confers resistance to apoptosis in normal and malignant mammary epithelium. *Cancer Cell*, **2**, 205–216.

Yang, F., Maurugan, R., Wang, S. & Ramakrishna, S. (2005) Electrospinning of nano/micro scale poly (L-lactic acid) aligned fibers and their potential in neural tissue engineering. *Biomaterials*, **26**, 2603–2610.

Young, E. W. K., Berthier, E. & Beebe, D. J. (2013) Assessment of enhanced autofluorescence and impact on cell microscopy for microfabricated thermoplastic devices. *Analytical Chemistry*, **85**, 44–49.

Zhang, Q., Lu, H., Kawazoe, N. & Chen, G. (2014) Pore size effect of collagen scaffolds on cartilage regeneration. *Acta Biomaterialia*, **10**, 2005–2013.

PART IV
3D bioreactor technologies

CHAPTER 13
Quasi Vivo® bioreactor technology

J. Malcolm Wilkinson
Kirkstall Ltd, Rotherham, UK

Introduction

It is now widely accepted that current 2D *in vitro* cell culture systems are poorly representative of human or animal physiology (Kirkpatrick *et al.*, 2007). This has been generally attributed to the fact that the complexity of the physiological environment is not replicated in petri dishes or microplates (Zhang, 2004). Cells are exquisitely sensitive to their microenvironments, which are rich in molecular signals from the extracellular matrix, other cells and from mechanical stimuli due to flow, concentration gradients and movement. These mechanical and biochemical signals are largely absent in static culture in well plates. For this reason, there is growing interest in seeding cells at higher density on scaffolds in an attempt to recreate the 3D environment. However, at the higher cell density in 3D culture, the supply of nutrients and oxygen becomes more critical, particularly for culture experiments lasting for several days. Flow of media can be introduced to overcome this problem but then the cell culture chamber design is far more complex in order to predict and control flow-induced stress. With flow systems, practical issues such as the ability to avoid leakage and blockages become important. However, once flow is introduced, the capability to couple together chambers allows more sophisticated co-culture models to be considered. This allows the crosstalk between different tissues to be investigated as a way of modulating and enhancing cell function.

A variety of new cell culture chamber designs are now being offered to refine the simplistic *in vitro* representations which are currently still used in cell culture laboratories. In parallel with the interest in flow and co-culture, there is a fascination with the concept of 'organ-on-a-chip' using microfluidics. Michael Shuler of Cornell University was one of the first to pioneer this technology for DPMK studies as early as 2004. The company he founded, Hurel, is still striving to bring a product to the market. It has proved exceptionally difficult to translate experimental methods from the millimetre to the micrometre scale because of practical problems such as blockages, air bubbles and difficulties with loading cells into

Technology Platforms for 3D Cell Culture: A User's Guide, First Edition. Edited by Stefan Przyborski.

microscopic chambers. Many other groups worldwide are still working to overcome these fundamental challenges.

For biologists and technicians to embrace new, physiologically more relevant culture methods, the transition from wells and dishes to other tools has to be as smooth as possible. For this reason, Kirkstall Ltd has developed an easy-to-use modular cell culture system known as Quasi Vivo®, which enables microwell protocols to be transferred directly to the connected chamber modules without the need to redesign cell culture experiments. The new system offers controlled mechanical stimuli from flow and biochemical stimuli from cells placed in connected modules, thus offering a physiologically realistic microenvironment to cells *in vitro*. Its design principles are based on the allometric scaling of cell numbers and the mean residence times of molecules in metabolic tissues, as well as consideration of oxygen tension and shear stress, which together can be combined to establish organ and system models. These principles have been most thoroughly developed by the multidisciplinary team at the Centro Piaggio at the University of Pisa (Sbrana & Ahluwalia, 2012). The Pisa group has been responsible for the original concepts and techniques which now form the foundation for the Quasi Vivo system described in this chapter.

Following the licensing of technology from the University of Pisa, Kirkstall has been developing a range of products which can be simply described as Lego blocks for cell biologists. A wide range of cell types have been cultured in these chambers by many of Europe's leading cell culture laboratories. These include hepatocytes (Vinci *et al.*, 2011), Caco-2 gut cells (Ucciferri *et al.*, 2013), adipocytes and endothelial cells (Vinci *et al.*, 2012). Work in progress is extending the range of applications and cell types to skin, kidney, respiratory epithelium and the blood–brain barrier. The practical feedback from over 36 laboratories has now been incorporated in improvements and new chamber designs.

Description of technology

Chamber design

There are currently three different chamber designs in production by Kirkstall. A brief description and the important design features are given below.

QV500 (Figure 13.1)

The QV500 cell culture chamber is a highly flexible research tool, ideal for academic use. Produced from medical-grade silicone, the chamber provides a leak-proof seal, allowing flow of cell culture media across cells. The modular design allows the flow system to be set up in a wide range of configurations to best replicate the *in vivo* environment. The internal diameter of the chamber is equal to the well diameter of a 24-well plate, making it easy to transfer protocols

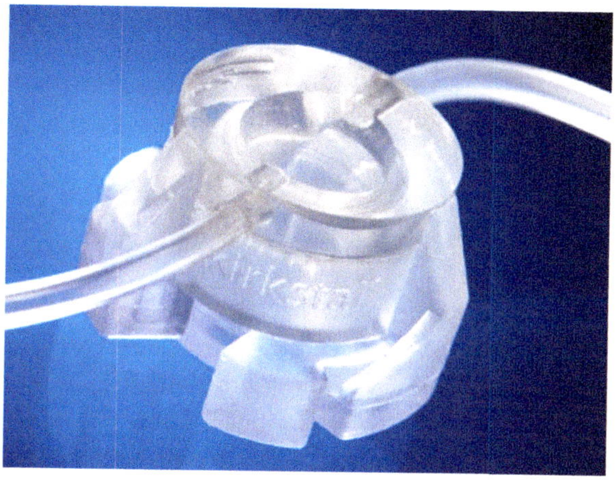

Figure 13.1 **QV500 chamber.**

developed earlier using static culture. The chamber is compatible with glass and plastic coverslips as well as a range of 3D scaffolds.

The chamber has a sloping roof which routes bubbles towards the outlet tube during filling. The dimensions of the chamber were optimised to minimise flow stress on the surface of a cell culture on a coverslip or scaffold at the base of the chamber (Mazzei *et al.*, 2010). The two parts of the chambers snap together and are locked by a twisting action. The compliance of the silicone material creates an excellent seal against leaks without the need for a gasket.

The overall scale of the chamber is based on solid engineering principles that give it clear advantages over a 'microfluidics' approach where the chamber volume is perhaps 10 or even 100 times smaller. These advantages are detailed below.

- There is space in the chamber to culture enough cells to create a tissue-like 3D culture from adherent cells. The chamber is also large enough to culture tissue slices. The chamber can contain enough cells to complete certain endpoint assays such as the Western blot or polymerase chain reaction (PCR) expression analysis. Microfluidics chambers cannot provide enough cells.
- Edge effects dominate at the small scale; cells are in contact with glass, plastic or nutrient media instead of being embedded in an extracellular matrix as they are in *in vivo* or in the Quasi Vivo chambers.
- It is difficult, if not impossible, in most microfluidics devices to grow differentiated 3D tissue-like models of skin, lung, etc. This is because the depth of the chambers is not great enough to allow more than 2–3 layers of cells. Skin and lung models need layers of cells 10 or more deep.
- It is difficult to seed cells at the beginning of experiments or remove cells from microfluidics devices at the end of the experiment for analysis.
- Air bubbles are extremely difficult to remove from the microcapillary tubes in microfluidic chips because of the dominance of surface tension forces. This creates problems for researchers using microfluidics. (Air bubbles can prevent the smooth flow of media or even stop flow completely.)

Whatever the scale of the cell culture chambers, complex modelling is required to establish whether the oxygen supply can be maintained at an appropriate level to ensure cell viability throughout the chamber and the thickness of a scaffold. Scaffold and bioreactor design are both critical in defining the oxygen profile in a cell construct (Mattei *et al.*, 2014).

QV600 (Figure 13.2)

This chamber was developed as a tool for culturing cells that need to be positioned at an air-liquid interface (ALI) to reach maturity (for example, skin, respiratory epithelia and cornea). The QV600 chamber uses an innovative design (protected by patents) to solve a very important problem: how to maintain a constant level of liquid in a chamber when the flow rate needs to be varied over a wide range. Most other commercial designs are very sensitive to the relative height of the chamber and the pumping system. As a result, they are liable to overflow or leak. The QV600 has a leak-proof sealed chamber.

The QV600 air-liquid interface chamber is a highly adaptable tool which allows a wide range of cell culture experiments to be conducted with media flow. The chamber maintains a constant surface level for the medium, allowing the cultivation of cells which require exposure to both air and medium to function properly. The QV600 can also be fitted with a horizontal barrier, splitting the chamber into two, each with its own separate medium flow. This is used to study the function of membranes and create barrier models such as gut, blood–brain barrier or skin. It is designed with the same internal dimensions as a standard 24-well cell culture plate well and is compatible with hanging inserts designed for this size, so that cells in culture can be held at the air-liquid interface.

Benefits include the following.

- Automatic maintenance of a constant medium surface level.
- Separate control of basal and apical medium flow.

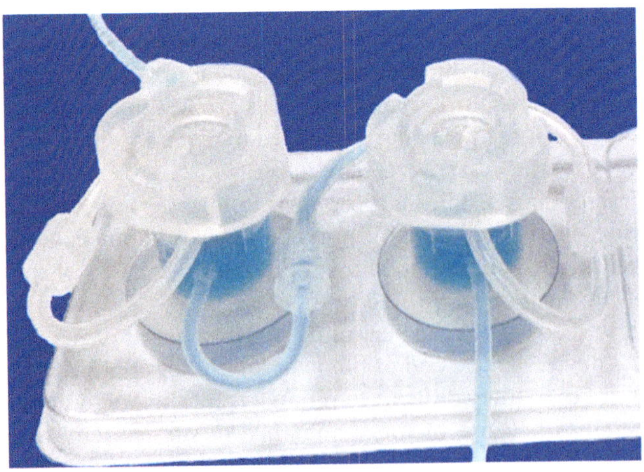

Figure 13.2 **Two interconnected QV600 chambers.**

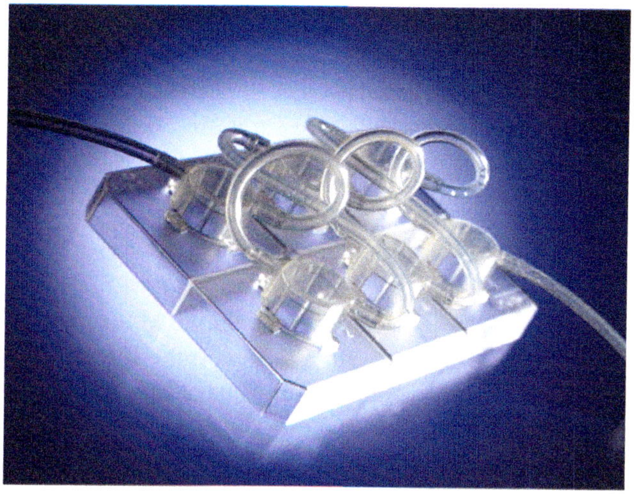

Figure 13.3 **QV900 chambers connected in a series configuration.**

- Recirculation allows cells to condition the medium whilst flow provides constantly refreshed nutrient supply to cells.
- Compatible with a wide range of inserts.
- Gases within the chamber can be controlled.
- Chambers can be connected to create systemic models.

QV 900 (Figure 13.3)

The QV900 six-chamber optical tray provides an ideal platform for higher-throughput testing in a compact format. Each tray contains six 24-well sized chambers within the same footprint as a standard well plate, all of which have optically clear floors and ceilings to allow imaging of cultures *in situ*. Chambers are connected using the same tubing as for our other Quasi Vivo products, allowing the system to be set up in a variety of configurations to meet user requirements within a compact, easy-to-handle format.

The flow of medium through the chambers provides a constantly refreshed nutrient supply to cells. Flow can be single pass from a reservoir bottle to waste collection or recirculating. Recirculation of media allows cells to condition the medium and helps to avoid shock to the cells when medium is replenished. The base of the chamber is optically clear and allows observation and imaging of cells *in situ* if the cells are cultured directly on the base of the chamber. Alternatively, cells can be cultured on coverslips or scaffolds. The chambers can also be used to culture tissue slices. The QV900 chambers can be connected in a variety of configurations to create systemic models.

Pumps for creating flow of medium

There are a variety of pumps that can be used to create flow of media: syringe pumps, peristaltic pumps and microfluidic pumps based on piezoelectric actuators.

Syringe pumps

Syringe pumps use a simple mechanical drive to push fluid out of a syringe. It is possible to create bidirectional flow by reversing the direction of mechanical movement. It is difficult to know what pressures are exerted by syringe pumps and this can be a hazard if cells are sensitive to damage from pressure. The syringes are typically single-use disposable components and are not easy to clean or sterilise.

Peristaltic pumps

A peristaltic pump is a positive displacement pump used for pumping a variety of fluids. The fluid is contained within a flexible tube fitted inside a circular pump casing (although linear peristaltic pumps have been made). A rotor with a number of 'rollers' attached to the external circumference of the rotor compresses the flexible tube. As the rotor turns, the part of the tube under compression is pinched closed, forcing the fluid to be pumped to move through the tube. Typically, there will be two or more rollers, or wipers, occluding the tube, trapping between them a body of fluid. The body of fluid is then transported, at ambient pressure, toward the pump outlet.

Because the tube in a peristaltic pump forms part of a closed flow loop, there is very little chance of contamination. Tubes can easily be cleaned and sterilised. The pump design prevents backflow and syphoning without valves.

There are some minor disadvantages of peristaltic pumps:

- The mechanical forces on the manifold tubing mean that it will degrade over time and eventually have to be replaced.
- The flow is not constant but slightly pulsatile as each of the rollers engages. However, this may be an advantage in mimicking physiological effects as the human body blood flow is varying.

Choosing the most appropriate pump will involve several criteria:

- Flow rates required.
- Space available for the pump.
- Does the pump need to withstand certain temperature or humidity conditions?
- Cost of pump/tubing cost.

Impact of pressure on cells

Creating media flow requires pressure gradients and these will influence the behaviour of cells. The important reference point is the flow and pressure conditions experienced by cells *in vivo*. The graph in Figure 13.4 shows the range of pressures and flows experienced by cells *in vivo*.

Micropumps

Several of the lab-on-a-chip or organ-on-a-chip technologies see the integration of a microfluidics pump as a potential advantage. However, caution must be exercised because microfluidic pumps can have serious performance limitations.

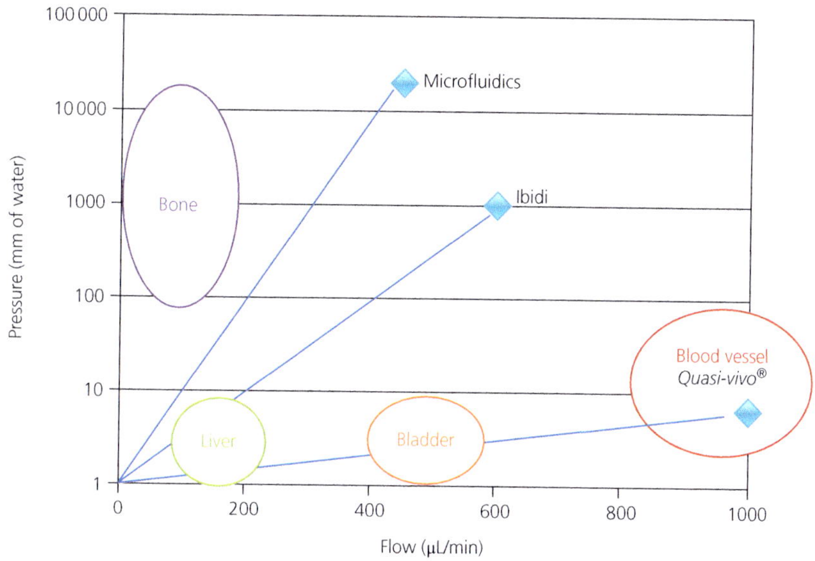

Figure 13.4 Graph of pressure versus flow.

Before selection of this option, several criteria for selection should be considered:

- Flow rate range
- Flow regime (pulsatile versus steady)
- Control and reproducibility
- Cost
- Reliability
- Backpressure.

This latter feature can be a major limitation for microfluidics pumps. Although microfluidics pumps can achieve moderate flow rates, they are very poor at sustaining this flow under conditions where there is significant backpressure. The plot in Figure 13.5 shows how the flow of one of the leading microfluidics pumps varies with backpressure. Micropumps are considered unsuitable for use with Quasi Vivo systems.

Setting up flow systems (Figure 13.6)

The QV500 and QV900 Quasi Vivo chambers can be used without flow of media and under these conditions the results will be very similar to using conventional 24-well plates. However, the main advantages of Quasi Vivo systems come into play when gentle flow of medium is introduced. This requires a flow circuit to be established. Examples of how this is done are given below after a short introduction that describes the typical way a cell culture experiment is designed. (For more information on tissue culture and sterile techniques, see the book by Ian Freshney (Freshney, 2005).)

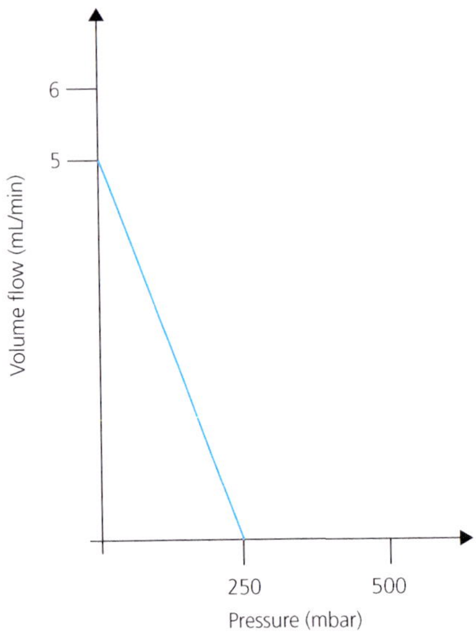

Figure 13.5 Effect of backpressure on micropump output.

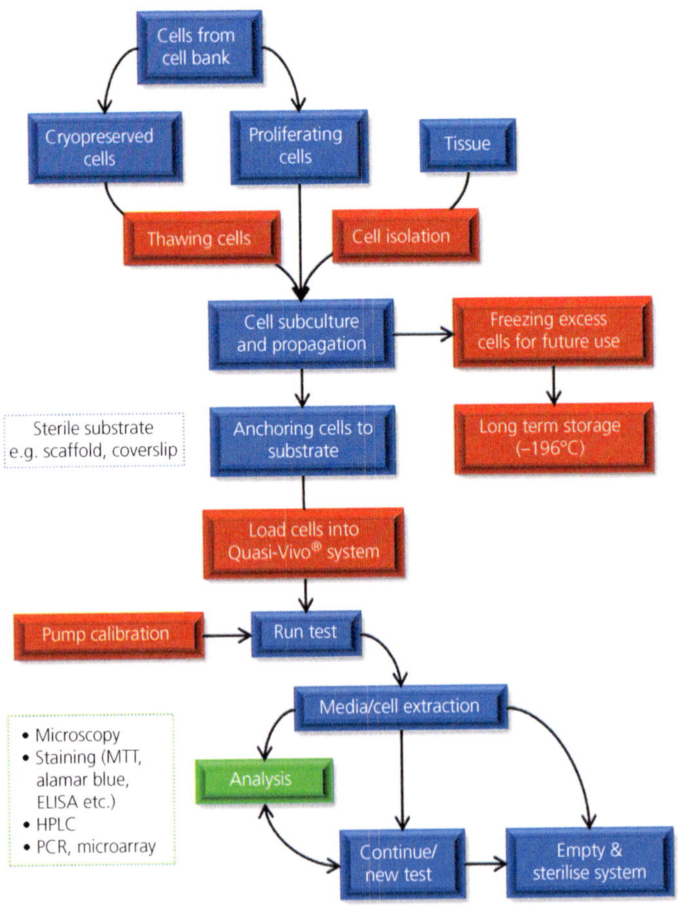

Figure 13.6 Quasi Vivo flowchart. This explains the practical implementation of the Quasi Vivo system, and how it fits into current tissue culture protocol.

Please note that the current QV500 and QV900 designs enable the culture of any primary cells or cell lines that grow as an adherent layer. They are not currently designed for culture of cells in suspension. The system has been tested using human and rat primary hepatocytes, HepG2 (human liver carcinoma), C3A (human hepatoblastoma), NIH 3T3 (mouse embryonic fibroblast), adipocytes, NCI H292, human dermal microvascular endothelial cells (HDMECs), human umbilical vein endothelial cells (HUVEC), human skin and oral fibroblasts, skin keratinocytes, osteoblasts and chondrocytes and many more cell types. These cells may be cultured on coverslips or scaffolds or in a gel matrix.

Selection of chambers

The QV500 silicone chamber (Figure 13.7) is the basic building block of the Quasi Vivo system. Both QV500 and QV600 chambers are made of polydimethyl-siloxane (PDMS), a biocompatible silicone polymer (Figure 13.8). For applications where non-specific binding of proteins or chemicals is important, the QV900 chamber should be selected as this is manufactured from a different plastic material. The external dimensions of the QV500 chambers are approximately 23 mm in height and extending to 37 mm in diameter. The internal volume of each chamber is approximately 2 mL and designed to accommodate 12 mm diameter coverslips.

Ethylene oxide gas-plasma sterilisation is used to sterilise the chambers at the end of the manufacturing process. This is the most efficient method which

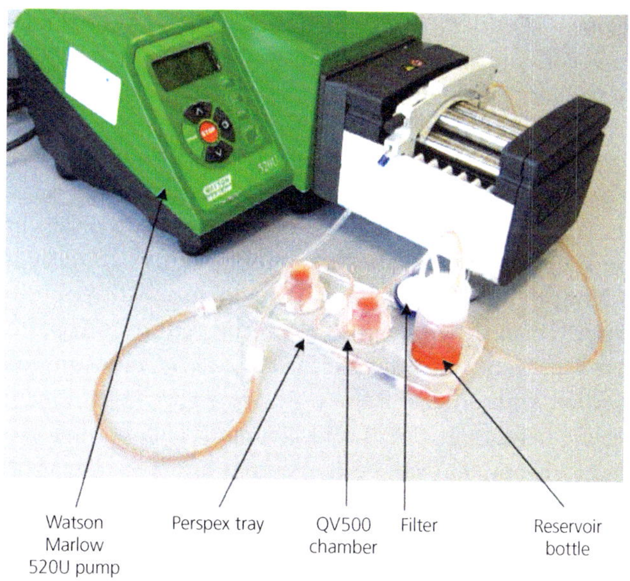

Figure 13.7 **QV500 chambers and reservoir bottle connected to a Watson Marlow 520U peristaltic pump.**

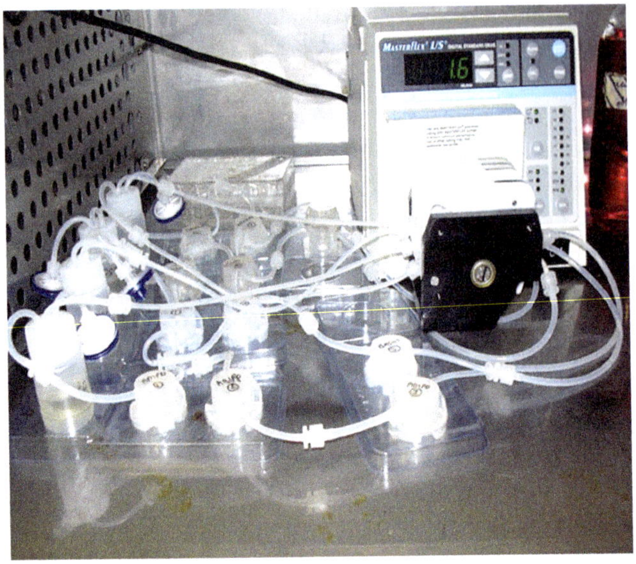

Figure 13.8 Experimental set-up with four flow loops running in parallel inside a cell culture incubator.

maintains the unique properties of silicone: flexibility and breathability. The chambers are supplied double-wrapped in sealed packs, each of which contains a colour change indicator which turns from purple to green when the sterilisation process is complete.

QV500 and QV600 chambers can be reused if washed with 70% ethanol and thoroughly rinsed with PBS after each experiment. The chambers are autoclave safe; however, autoclaving more than three times is not recommended as the silicone will lose flexibility and become discoloured. QV900 chambers are designed for single use and cannot be sterilised.

General comments and preparing the system for first use

All procedures which involve cell culture (seeding cells on coverslips, opening sterile chambers, opening and closing reservoir bottles, disconnecting and connecting the tubing) should be carried out in accordance with aseptic techniques. It is necessary to ensure that all cell culture procedures are performed to a standard that will prevent contamination from bacteria, fungi, mycoplasma and crosscontamination with other cell lines. All work should be done in a class II tissue culture hood, wearing gloves and laboratory coats. After assembly, the completed Quasi Vivo system can be transferred into a CO_2 incubator and left for the duration of the experiment.

The reservoir bottle and chambers are supplied sterile and with the Luer connectors attached. The extra Luer connectors and tubing are not sterile. Initially, the supplied Luer connectors should be attached to the tubing in the required configuration; for consistency, tube inlets should have female connectors and

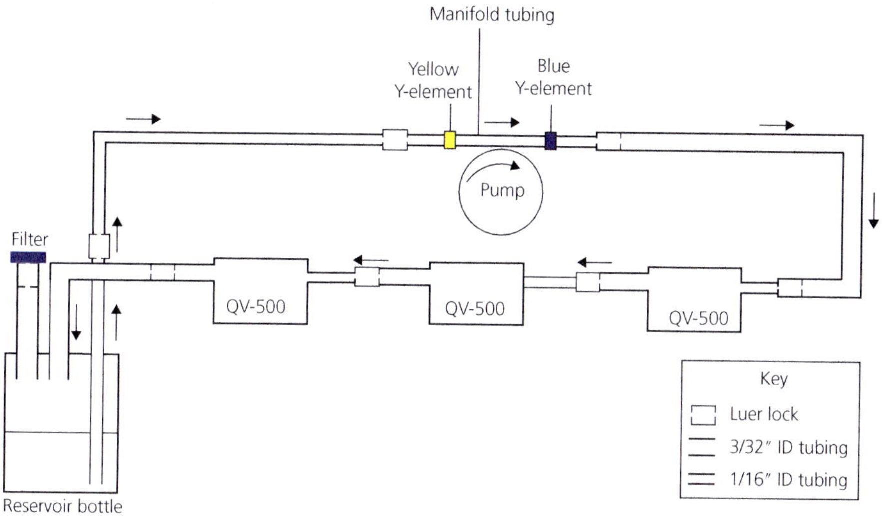

Figure 13.9 **Schematic diagram of typical flow system configuration using QV500.**

tube outlets should have male connectors. The thinner (1/16″ internal diameter (ID)) tubing requires the connectors of this size, while the thicker (3/32″ ID) tubing requires the connectors of this size. These assembled components should then be sterilised (we recommend autoclaving).

The whole system should now be connected under aseptic conditions in the configuration shown in Figure 13.9. The reservoir bottle has three tubes connected to it: the longest tube (1/16″ ID) is the outlet and should be submerged in medium, the middle length tube (3/32″ ID) is the reservoir inlet and should be just above the level of the medium to avoid foaming. Finally, the shortest tube (3/32″ ID) is for gas exchange and should not be touching the medium. A filter should be connected to this gas exchange tube.

Following the system round in a loop, the outlet tube of the reservoir is connected to 1/16″ ID extension tube (22 cm in length). This is connected to the pump manifold tubing, which is in turn connected to the 3/32″ ID extension tube (22 cm in length). This is connected to the chambers that are then connected to the reservoir inlet. Chamber inlets must be the 1/16″ ID tube and the outlets must be the 3/32″ ID tube (see Figure 13.7). Figure 13.9 shows three chambers connected in series but the number of chambers employed may be adjusted for the particular experiment. Once the system is connected in this way, it is recommended not to disconnect the tubes or chambers as this will increase the risk of infection.

The connected system should be washed through by filling the reservoir bottle with sterile PBS and connecting the closed system to the peristaltic pump at a flow rate of 1 mL/minute for at least 20 minutes. The PBS should be removed from the reservoir bottle under aseptic conditions using a sterile pipette and discarded. The PBS in the chambers and tubes should also be removed; this can be achieved by lifting and tilting the chambers to allow the liquid to return to the reservoir through the 1/16″ ID tube (normally this is the outlet from the reservoir), from where it can be removed and disposed of. Finally, the system should be primed with the culture medium to be used in the experiment, in the same manner as the PBS wash, and removed. The system is now ready to be used in an experiment.

Preparation of cells

Cells should be seeded 24 hours before the experiment on pretreated coverslips. The coverslips can be pretreated by coating with collagen I (from rat tail) or any other coating protein for a minimum of one hour in the incubator or overnight at 4 °C. The concentration of collagen depends on the type of cells and should be devised experimentally. As an example, for keratinocytes and fibroblasts, the optimum concentration of collagen I is 0.2 mg/mL and for human hepatocytes, 0.1 mg/mL. After incubation, the coverslips should be washed three times with PBS and either used immediately or stored in a fridge for up to one week.

Cells should be seeded on coverslips, typically in a 24-well plate. The number of cells depends on their type. Table 13.1 shows examples of seeding densities for different cell types.

Some cells, such as hepatocytes, need to be coated by a thin, permeable upper layer, which protects cells from damage. The most commonly used coatings are collagen I and alginate. To coat the cells with collagen, it is necessary to prepare rat tail collagen I solution (BD Bioscience, cat. no. 354236) at a concentration of 1 mg/mL for one hour before the experiment. For detailed instruction, refer to supplier's product specification sheet and 'Alternate Gelation Procedure for Rat Tail Collagen I'. 200 μL of this solution is applied to each glass slide to cover the seeded cells, and then incubated at 37 °C for 30 minutes to one hour. The coverslips are now ready to be used in an experiment. When removing the coverslips

Table 13.1 Guidelines for growing different cell types.

Type of cells	Number of cells	Confluence
Dermal keratinocytes	5×10^4	50%
	1×10^5	100%
Dermal fibroblasts	2.5×10^4	50%
	5×10^4	100%
Osteoblasts	1×10^5	100%
Chondrocytes	1×10^5	100%
Hepatocytes	2.4×10^5	100%
HepaRG	4.2×10^5	100%
HDMEC	1×10^4	100%

HDMEC, human dermal microvascular endothelial cell.

from the plate, ensure that the collagen layer on the top of the cells is left undisturbed, and transfer the coverslips to static plates or cell culture chambers.

Note: You should take care to prevent any contact of your cells with alcohol, which might be used to sterilise coverslips or forceps. If coverslips are sterilised with alcohol, a sufficient length of time should be allowed for these to dry completely. Forceps should be additionally rinsed with PBS before use.

Setting up an experiment

Having cleaned and primed the system, it is ready for use. Open each of the chambers under aseptic conditions in the class II tissue culture hood. Add approximately 1 mL of the required medium to each chamber. Gently lift the seeded coverslips from the wells of the static culture plate and put each one on the bottom of a chamber, with the cells uppermost (see Figure 13.7). Lock the chamber as previously described.

Fill the reservoir bottle with the medium of your choice – approximately 4 mL per chamber plus some extra medium. For example, for two chambers we suggest adding 10–20 mL of medium. The volume of medium used depends on the length of the culture period and can be optimised for the specific cell type. Carefully screw on the top of the reservoir bottle. If highly proliferative cells are used, it is advisable to use medium with no serum added, to prevent excessive cell proliferation. The system is now closed and the experiment may begin; the system should be placed inside a CO_2 incubator and connected to a peristaltic pump.

Flow rates

The flow rate for each cell type and system configuration must be carefully chosen. We have found that 25–1000 μL/min is the 'safe range' for the cells used in our experiments as listed in Table 13.1. However, every new cell line should be tested in order to find the optimum flow rate.

Therefore, it is strongly advised to calibrate and measure the flow rate of the liquid circulating in the system. The flow rate will vary depending on the number of chambers, the diameter of the tubing (Figure 13.10), the length of the tubing and the position of the chambers and reservoir in relation to the pump (for instance, placing on a higher shelf in the incubator) and, most importantly, on the type of pump used. The flexibility of the system allows a wide range of options, e.g. attachment of different combinations of chambers and tubing. Any modification should be followed by calibration of the system. Figure 13.10 shows the relationship between the volume flow rate and pump speed depending on the diameter of the tubing used.

Calibrating the system

To calibrate the system, run the required configuration with sterile PBS at various pump speed settings and collect the liquid output over one minute from the final chamber. At each speed setting, measure the volume of liquid

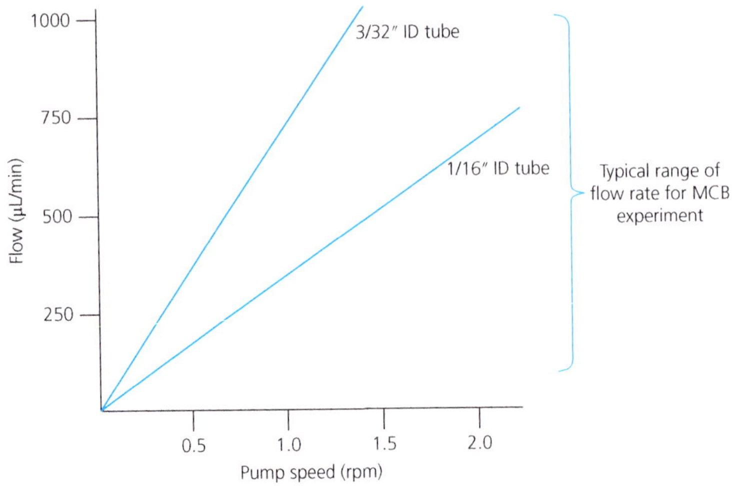

Figure 13.10 The relationship between the volume flow rate and pump speed for the two diameters of tubing used.

circulated by the increase in mass of the collection container after one minute. It is important that the air is expelled from the system before starting to measure the flow rate. Repeat this procedure three times for each setting to obtain a mean value.

Disassembling the system

Remove the system from the incubator and switch off the pump. Remove the medium from the reservoir, chambers and connecting tubes. Open the chambers and remove the coverslips, placing them in a fresh 24-well plate for analysis. Lock the chambers so the system is ready for cleaning.

Cleaning the system

Ensure that all the chambers and tubing are securely connected together, as for an experiment, and fill the reservoir bottle with sterile PBS which can be supplemented with antibiotics. Circulate the PBS through the whole system at a high speed (1–2 mL/min) for a few hours. It is important that the PBS should circulate a minimum of twice throughout the whole system. The time required for this will depend on the number of chambers and hence the volume of the system.

Sterilisation and cleaning procedures

QV500 and QV600 chambers can be used for more than one experiment without autoclaving, if the rules of aseptic techniques to prevent contamination are strictly followed. In this case, it is advisable to circulate 70% ethanol through the whole system and rinse very thoroughly with sterile PBS (ideally overnight) to remove any traces of ethanol which might affect cell viability. Tubing, connectors, chambers and reservoir bottles can be stored after opening for longer periods of time. In this case, tubing and connectors should be stored submerged in 70% ethanol; chambers and reservoir bottles should be rinsed with 70% ethanol followed by sterile PBS (long exposure to ethanol can damage the surface of silicone chambers), emptied and stored until needed.

Chambers, tubing and reservoir bottles are gas-plasma sterilised which is the optimum and least harmful sterilisation method for silicone chambers and tubing. Silicone parts can also be sterilised by using standard autoclaving procedure (121 °C, 15 psi, 15 minutes) and also γ-irradiation. However, to maintain reliability of the system, we advise no more than three sterilisation cycles for any component. Also, we advise users *not to wash whole system with sodium azide* which can accumulate on the surfaces of chambers and leak out during the experiment. The translucent polypropylene connectors supplied with the kit are autoclave safe, but can also be sterilised by γ-irradiation.

Co-culture: Using the Quasi Vivo system to build systemic models

Co-culture of cells or tissues is an essential part of building more representative *in vitro* models of *in vivo* systems. No cell in the body exists in isolation – cells are constantly communicating with each other, and this stimulation informs their function. Simple or complex co-cultures can be easily set up using Kirkstall's Quasi Vivo (QV) cell culture system, which includes various chamber types designed to best reflect the environment a cell experiences *in vivo*.

Co-culture is especially important for drug, medical and cosmetic testing, where the response of any given tissue may be determined by interactions with other tissues, or is dependent on metabolism of a compound in another part of the body. Perfusion culture using the QV system enables cells to communicate through a common recirculated medium, in addition to other aspects (nutrient supply, laminar flow stimulation) which work to recapitulate the *in vivo* environment *in vitro*. Using QV enables co-culture to be combined with 3D and flow conditions, creating a realistic alternative to *in vivo* in animals.

Co-culture challenges

- Static culture without flow causes cells to behave differently, which will influence interactions between different cell types if they are cultured on a single plate.
- Replicating cell-cell communication between cells or tissues from different organs.
- Finding a compatible medium that suits all cell types in the model.
- Restriction of complex modelling by standard static culture plasticware.

Benefits of co-culture in a QV system

- Co-culture of cells from different organs can be carried out with a single optimised medium.
- Can be used to create fully functioning models of complex *in vivo* systems involving multiple tissues, for example free fatty acid homeostasis (Vinci *et al.*, 2012) for the study of diabetes, or modelling of nanoparticle ingestion and adsorption, and associated toxicity in target organs.
- Modularity of the system allows for allometric scaling – differences in organ size can be replicated, for example modelling obesity by introducing excess adipose tissue into a system.
- Adjustable laminar flow rate and chamber pressure to suit the specific requirements of different tissue types.
- The *in vivo*-like stimulation of cells and tissues encourages them to behave with metabolic competency, resulting in improved function, viability and cell-cell signalling.
- Easy medium sampling for biochemical analysis.
- Variety of chambers available to suit different tissues.
- Compatible with 3D or 2D culture, suspended culture, cell culture inserts.

QV600 chamber for culture at air-liquid interface

The QV600 chamber provides an air-liquid interface for culture of cells that require contact with both air and medium to function properly, such as those in lung or skin tissues (Figure 13.11). The easy-to-use device is connected to a circulating medium supply through tubing and a peristaltic pump. Exposure to gentle medium flow stimulates gene expression and replenishes nutrients,

Figure 13.11 QV600 chamber schematic side view. This shows the placement of a Transwell insert (*orange*).

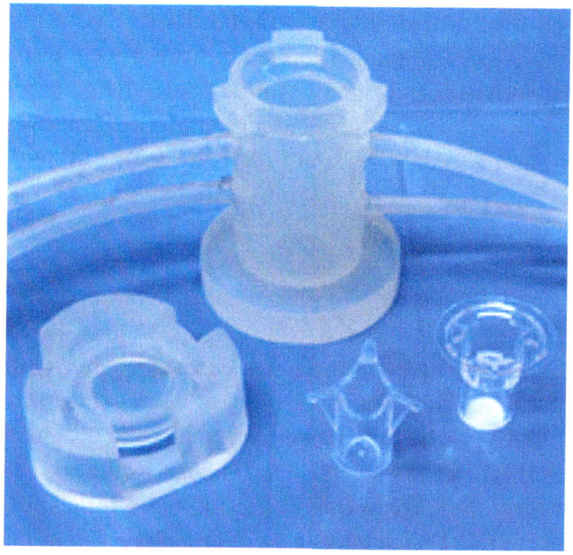

Figure 13.12 **QV600 chamber with lid removed, alongside 24-well Transwell and ThinCert™.**

maintaining peak cell viability and generating cell functionality close to that seen *in vivo*. The QV600 is compatible with cell culture inserts, scaffolds or tissue explants, and can be connected to other QV chambers for formation of systemic co-culture models with other cell types (Figure 13.12).

Benefits of QV600

- Easy to use.
- Same internal width as a standard 24-well plate well, making it easy to change from static to flow culture.
- Large circulating volume of medium reduces the frequency of medium changes; reservoir bottles hold up to 30 mL of media, enough to maintain a culture for 7–12 days without need for feeding.
- Removes the risk of disturbing or shocking cultures on feeding.
- Replicates flow, nutrient replenishment and other key aspects of the *in vivo* environment.
- Air-liquid interface is maintained at a constant level.
- Control of the gaseous phase to introduce a toxic component or an aerosol to the air in the chamber if required.

Applications

- Skin inflammation models
- Wound healing assays
- Inhalation exposure to chemical entities or infectious agents
- Eye drop safety testing

QV600 system configuration with an ALI chamber

The QV600 ALI chamber is designed to operate with a weir system, maintaining the liquid level at the level of the liquid outlet. The chamber has been designed to hold inserts at this level, maintaining cell cultures at the ALI, and thus the system needs to be set up in such a way as to allow this weir to be maintained.

Basic set-up with one pump

With a single pump in the system, it is recommended to locate it after the chamber, so that the liquid is pulled out of the chamber (Figure 13.13a). When the chamber is empty, the pump will remove air from the chamber, reducing the internal air pressure and pulling in liquid from the reservoir bottle. Once the fluid reaches the level of the liquid outlet, the pump will pull liquid from the chamber. Once the liquid has reached this level, the level will not change, as the pump cannot reduce the liquid level below the level of the outlet, and the internal pressure will not allow more liquid to enter the chamber than is pumped out.

It is recommended not to set the system up with the pump driving liquid into the chamber (Figure 13.13b). In this formation, the internal pressure has to be greater than atmospheric pressure in order to force the liquid back to the

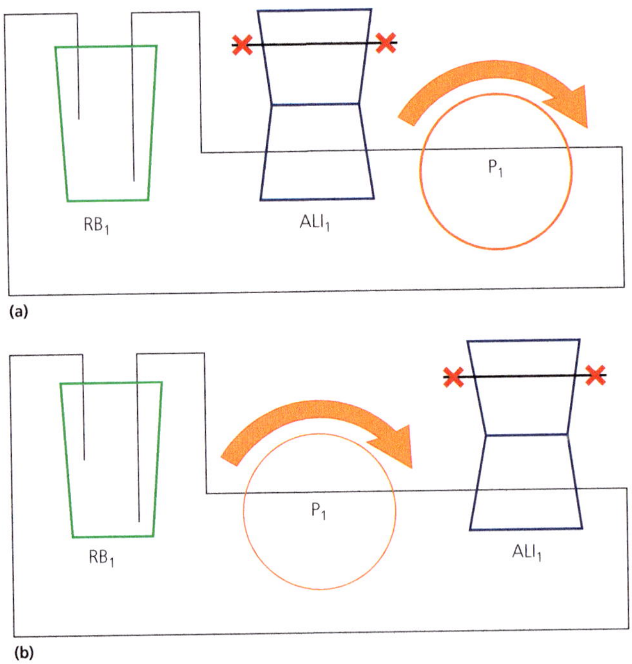

Figure 13.13 Configuration diagram for ALI system with a single pump. Note that the upper inlets of the chamber are sealed. (a) Recommended set-up, with pump after chamber. (b) Set-up with chamber after pump. This arrangement is not recommended. ALI$_1$, QV600 ALI chamber; P$_1$, pump; RB$_1$, reservoir bottle.

reservoir, and the liquid level can rise above the weir level until the pressure is great enough to increase the outlet flow to match the inlet flow.

Results obtained with Quasi Vivo

Viability

Work at Reinnervate using a proprietary scaffold material has demonstrated the improved viability and proliferation achieved under flow conditions compared to simple static conditions (www.reinnervate.com/) (Figures 13.14 and 13.15).

Function

Work at INSERM Laboratories in Montpellier has demonstrated that gene expression in primary hepatocytes can be maintained at levels close to those in freshly isolated cells (Vinci *et al.*, 2011) (Figure 13.16).

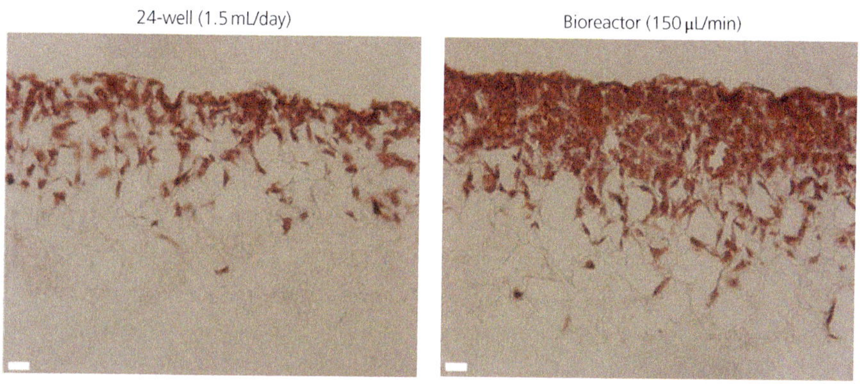

Figure 13.14 **Enhanced 3D cell growth** (HepG2, six-day culture on Alvetex scaffold).

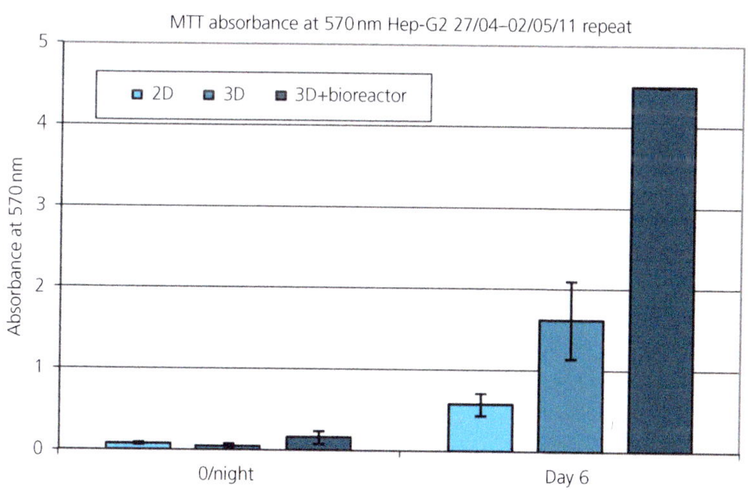

Figure 13.15 **Increased numbers of viable HepG2 cells.**

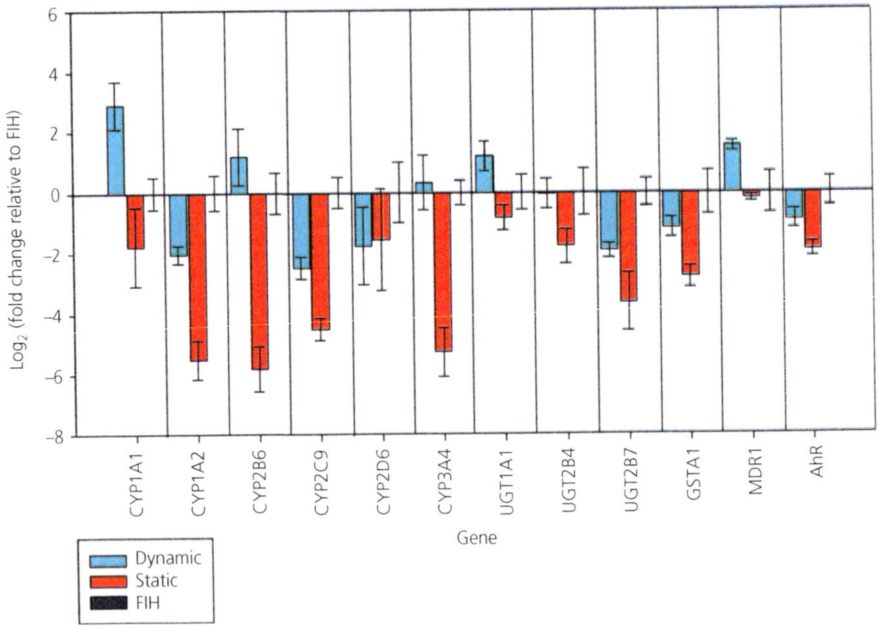

Figure 13.16 Changes in gene expression in human primary hepatocytes. These cells were cultured under flow or static, compared to freshly isolated hepatocytes.

Co-culture

Research at Pisa University (Vinci *et al.*, 2012) has demonstrated a model where nutrient balance in the human body is maintained through systemic signalling between different cells and tissues. Breaking down this circuitry to its most basic elements and reconstructing the metabolic network *in vitro* provides a systematic method to gain a better understanding of how crosstalk between the organs contributes to the whole-body metabolic profile and the specific role of each different cell type. To this end, a three-way connected culture of hepatocytes, adipose tissue and endothelial cells, representing a simplified model of energetic substrate metabolism in the visceral region, was developed. It managed to maintain glucose and fatty acid homeostasis *in vitro*.

Barrier properties

Several groups have demonstrated improvements in barrier function as a result of media flow.

- Caco-2 cells in a co-culture model of the gut and circulatory system as part of an EU Framework 7 project (Ucciferri *et al.*, 2013).
- Respiratory tract epithelium using commercially available culture in Transwells from Epithelix in QV600 ALI chamber swas characterised by TEER measurements, cell viability and cilia beating frequency (unpublished results).
- A lung tissue model at Nottingham University confirms tight junction formation (paper submitted for publication).

Extracellular matrix (ECM) production in chondrocytes

The Quasi Vivo bioreactor gave the greatest ECM deposition and alkaline phosphatase expression during PhD research at the University of Sheffield Dental School (paper in preparation).

Troubleshooting

Issue	Possible causes	Fixes
Quasi Vivo chambers leak	The chambers have not been closed properly	Ensure that chambers are fully locked until you feel a click
Cells show reduced viability compared with static controls	Flow rates have not been optimised for your cell type	Conduct optimisation experiments using different flow rates
QV900 chamber bottoms are cloudy	QV900 is made of acrylic which goes cloudy after contact with IMS	Take care not to use too much IMS when spraying for sterilisation
Peristaltic pump is not drawing medium from the reservoir bottle	Tubing over pump rollers is worn or damaged	Replace with new tubing, following instructions of manufacturer/ supplier

IMS, industrial methylated spirits.

Conclusion

Quasi Vivo systems have very quickly established themselves as a useful tool for academic research that is easy to use and provides a route to building more physiologically relevant models of 3D cell culture. By sponsoring research at leading cell culture laboratories, and encouraging publication and sharing of protocols, Kirkstall is speeding up the adoption of the new methods and hoping to facilitate the reduction and replacement of animal models as a method for testing the safety and efficacy of new drugs.

References

Freshney, R. I. (2005) *Culture of Animal Cells: A Manual of Basic Technique and Specialized Applications*. Wiley Blackwell, Chichester.

Kirkpatrick, J., Fuchs, S., Hermanns, I. *et al.* (2007) Cell culture models of higher complexity in tissue engineering and regenerative medicine. *Biomaterials*, **28**, 5193–5198.

Mazzei, D., Guzzardi, M. A., Giusti, S. *et al.* (2010) A low shear stress modular bioreactor for connected cell culture under high flow rates. *Biotechnology and Bioengineering*, **106**, 127–137.

Mattei, G., Giusti, S., Ahluwalia, A. (2014) Design criteria for generating physiologically relevant in vitro models in bioreactors. *Processes*, **2**, 548–569.

Sbrana, T. & Ahluwalia, A. (2012) Engineering Quasi-Vivo in vitro organ models. *Advances in Experimental Medicine and Biology*, **745**, 138–153.

Ucciferri, N., Collnot, E. M., Gaiser, B. K. *et al.* (2013) In vitro toxicological screening of nanoparticles on primary human endothelial cells and the role of flow in modulating cell response. *Nanotoxicology*, **8**, 697–708.

Vinci B. & Duret, C. *et al.* (2011) Modular bioreactor for primary human hepatocyte culture: medium flow stimulates expression and activity of detoxification genes. *Biotechnology Journal*, **6**, 554–564.

Vinci, B., Murphy, E., Iori, E. *et al.* (2012) An in vitro model of glucose and lipid metabolism in a multicompartmental bioreactor. *Biotechnology Journal*, **7**, 117–126.

Zhang, S. (2004) Beyond the petri dish. *Nature Biotechnology*, **22**, 151–152.

CHAPTER 14

Three-dimensional cell-based assays in hollow fibre bioreactors

John J. S. Cadwell[1] and William G. Whitford[2]
[1] *FiberCell Systems Inc., Frederick, Maryland, USA*
[2] *GE Healthcare, Life Sciences, Cell Culture, Logan, Utah, USA*

Introduction

In many ways, the hollow fibre bioreactor may be considered the original three-dimensional (3D) cell culture system (Figure 14.1). Hollow fibre-based cell culture was first developed by Richard Knazek at the NIH in 1972. He was searching for a way to culture adrenal tumour cells under *in vivo*-like conditions to study hormone secretion in response to drug stimuli. His reports that response curves can be generated by assaying the amount of hormone secreted in response to biochemical stimuli were the original dynamic *in vitro* cell-based assay (Knazek *et al.*, 1972). Hollow fibre bioreactors offer a method by which cells can be cultured at tissue-like densities over long periods of time. Hollow fibres act as 'artificial capillaries' and act much as capillaries do in the human body. Hollow fibre bioreactors hit their peak of popularity in the late 1970s to 1980s when they were employed in the bio-manufacturing of monoclonal antibodies (Tharakan & Chau, 1986).

The controlled molecular weight cut-off (MWCO) of the fibres allowed inhibitory cytokines such as TGF-β to diffuse away from the hybridoma cells, while entrapping the secreted antibody to high concentration within the small volume of the extracapillary space. Poor flux in the particular fibres available at the time, coupled with a lack of available technology for the scale-up of hollow fibre bioreactors, limited their application and the method fell into disfavour, except for small-scale research applications.

In the past 10 years, advances such as new high-flux fibre materials, better system engineering and an improved understanding of hollow fibre cell culture methods have given new life to the use of hollow fibre bioreactors. It had been thought that the benefits of hollow fibre bioreactors were restricted to simply higher secreted protein concentrations, the ability to reduce serum levels or more easily adapt cultures to commercially available serum free media. Much

Technology Platforms for 3D Cell Culture: A User's Guide, First Edition. Edited by Stefan Przyborski.

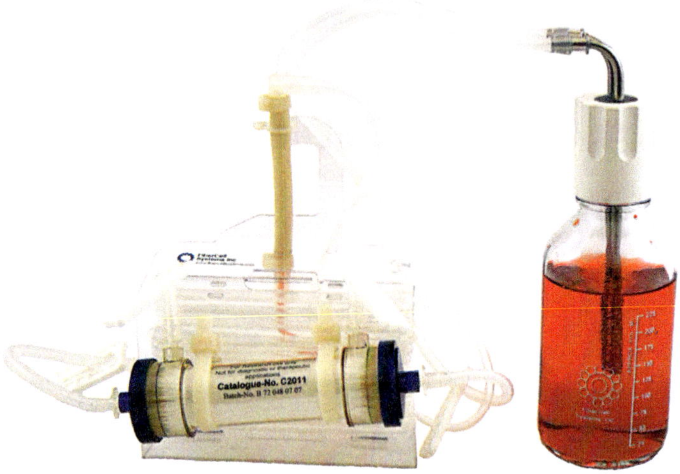

Figure 14.1 **Hollow fibre cartridge** A hollow fibre cartridge represents a single-use, bio-safe closed system for culturing cells, bacteria and viruses. The standard sized cartridge has an extracapillary space volume of 20 mL and 3000 cm^2 of area.

new information has been gathered regarding hollow fibre bioreactors and it is now understood that they represent a fundamentally different and more *in vivo*-like way to culture cells. There is no question that the conditions under which cells are cultured have a profound effect on their behaviour, and cell culture conditions are fundamentally different in a hollow fibre bioreactor (Stanness *et al.*, 1997).

The hollow fibre bioreactor

The biomimetic hollow fibre (HF) bioreactor is a high-density continuous perfusion culture system. It presents many unique distinctions from the commonly employed non-porous plastic surfaces, for example flasks, microcarrier beads and discs or roller bottles. A HF bioreactor includes a cartridge containing thousands of semi-permeable hollow fibres in a parallel array within a tubular housing fitted with inlet and outlet ports (Figure 14.2). These fibre bundles are potted at each end so that any liquid entering the ends of the cartridge will necessarily flow through the interior of the fibres. Cells are generally seeded within the cartridge but outside the hollow fibres in what is referred to as the extracapillary space (ECS).

In this configuration, culture media is pumped inside the hollow fibres, allowing nutrients and waste products to diffuse both ways across the fibre walls. Having passed through the cartridge, the culture medium is oxygenated and recirculated to the cartridge. There are, however, a few other possible configurations and alterations of this basic implementation and flow circuit. Endothelial cells, the cells

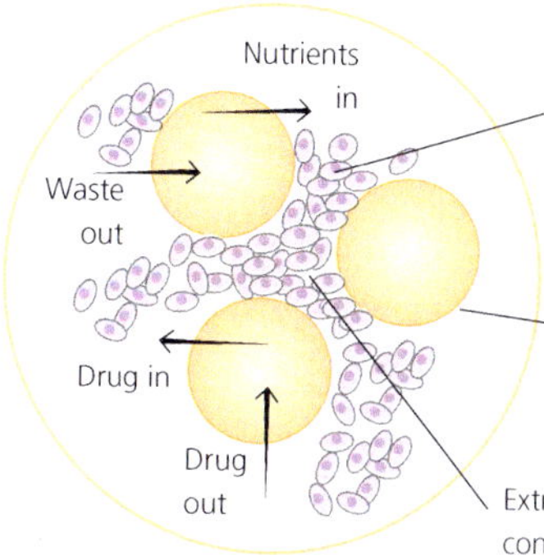

Figure 14.2 Cross-section of the hollow fibre cartridge. Medium is recirculated through the insides of the fibres and cells are typically grown on the outside of the fibres. The cells will attach to the porous support of the fibres and therefore do not require splitting or periodic subculturing.

Figure 14.3 Lymphocytes cultured in a hollow fibre bioreactor. Lymphocytes are growing at very high densities, $1–2 \times 10^8$ cells per mL.

that line all blood vessels, can be seeded on the insides of the hollow fibres with the culture medium circulated over them. This flow provides hydrodynamic force (or shear stress) similar to what the cells would experience *in vivo*, and the cells respond to this shear.

Hollow fibre bioreactors offer a unique environment for a more biomimetic cell cultivation and disparate cell co-cultivation. They present a 3D environment similar to the conditions found in the body, and support the continuous control of parameters such as oxygenation levels, medium composition and drug concentration. HF bioreactors are an effective means for the generation of a number of products, from secreted proteins or viruses to cells or conditioned medium. There are three fundamental characteristics that differentiate hollow fibre cell culture from any other type.

- Cells are bound to a porous, perfused matrix much as they are *in vivo*.
- The MWCO of the support matrix can be controlled.
- A very high surface area to medium volume ratio (>150 cm^2 per mL) (Figure 14.3).

Cells bound to a porous support

The fact that the cells are bound to a porous support provides a number of distinct features. One is the support of a continuous system where there is no requirement to split the cells. Cultures in this system can maintain viability and production-relevant metabolism in a postconfluent manner for extended periods of time – months or longer. For example, one hybridoma cell line was reported to maintain productivity in a HF bioreactor-based culture for over one year. It has been reported that a contributing factor here is that more *in vivo*-like growth conditions afforded by HF bioreactors result in significantly reduced apoptosis (Hirschel *et al.*, 2011). Another advantage is that, due to the extremely low shear generated with the cartridge, the majority of cells that become necrotic will not release significant cytoplasmic proteins or DNA into the culture medium. This provides features such as an improved culture environment, more accurate culture parameter monitoring and a raw harvest that is cleaner and easier to handle for downstream assays and purification.

Controlled molecular weight cut-off

Through the selection of fibre porosity, desired products can be retained to significantly higher concentrations and the location/effects of cytokines can also be controlled. This is well illustrated in the case of hybridoma culture where the inhibitory cytokine TGF-β can be selectively removed from the culture while secreted antibody is concurrently retained (Dennler *et al.*, 2002). Secreted recombinant proteins can be selectively retained and concentrated and cytokines and other factors that facilitate cell-cell interactions can be concentrated as well. Small molecule drugs can easily exchange across the fibre and rapidly reach equilibrium.

High surface-to-volume ratio

The small diameter of the fibres (of the order of 200 microns) generates an extremely high surface area-to-cartridge volume ratio in the range of 100–200 cm^2/mL. In conjunction with the high gross filtration rate of polysulfone fibres, the exchange of primary and secondary metabolites is high enough to support cell densities of $1–2 \times 10^8/mL$, approaching *in vivo* tissue-like densities. A 20 mL cartridge will easily support the cell mass provided by a standard 2 L suspension culture or 20–40 roller bottles. High cell densities produce more protein per area of reactor footprint and facilitate adaptation to both lower serum concentrations and simplified serum-free culture, such as employing the FiberCell® Systems' CDM HD perfusion-optimised serum replacement. The use of such a defined and protein-free medium results in much cleaner product harvests and more simplified purification as well as a more defined culture environment compared to the presence of fetal bovine serum (FBS) (Figure 14.4).

Operation of a HF bioreactor, in its most simplified form, begins by seeding a prepared cartridge with either suspension or harvested adherent cells.

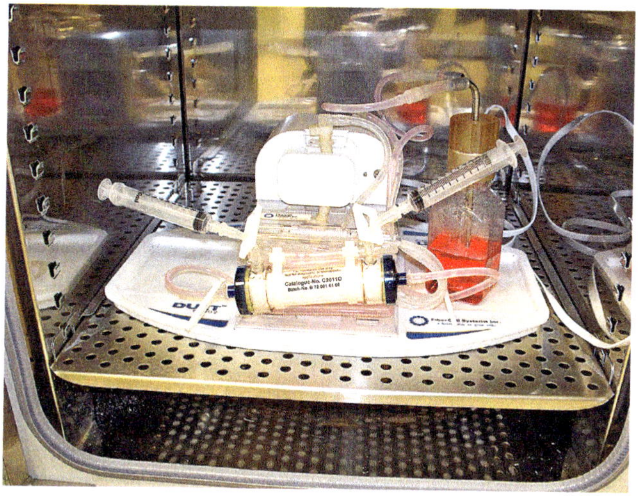

Figure 14.4 **FiberCell Duet pump in a CO_2 incubator.** Temperature and gas control is provided by the incubator and a thin cord is used to go through the incubator door without affecting gas composition.

The reactor cartridge is connected to an external reservoir and the medium recirculated from the reservoir through the cartridge. In one popular implementation, the recirculation is accomplished using a positive pressure displacement pump designed specifically for this purpose. Here, a piston compresses a short piece of pump tubing and one-way check valves on either side produce flow, in the same manner as the human heart. This frictionless pumping mechanism can generate 100 mL/minute of flow rate without wear on peristaltic pump tubing. Mass transfer of gases can be accomplished in a variety of ways, one being diffusive exchange through a loop of gas-permeable silicone tubing prior to the medium entering the bioreactor itself. The medium is constantly recirculated, providing a supply of oxygen and nutrients as well as removal of CO_2 and secondary metabolites. Medium recirculation rate and culture feeding can be linked to any number of culture parameters, and a number of control and automation options have been explored. A basic approach is to manually monitor the glucose and replace the medium as the concentration approaches 50% of the original.

Several cartridges sizes and fibre types are available, including those composed of polysulfone (PS) and polyvinylidene fluoride (PVDF). The MWCO in the PS fibres includes 5 KD or 20 KD and the pore size of PVDF is 0.1 micron. The PVDF fibre is of particular interest as various protein matrices, antibodies or growth factors can be readily bound to its surface via hydrophobic interaction. Cartridges come presterilised, assembled and ready to use and are intended for a single use.

During this renaissance of perfusion in general, and of the HF bioreactor in particular, several characteristics of hollow fibre cell culture have recently been identified.

- Reduction in apoptosis (Weeraphan *et al.*, 2012).
- Consistency of culture over long periods of time (G. Pavlakis, personal communication, 2013).
- More *in vivo*-like growth conditions resulting in improved cell function (Bennet *et al.*, 2007).
- Facilitation of the use of serum-free, protein-free and chemically defined media formulations (Whitford & Cadwell, 2011).

Hollow fibre bioreactors provide many particular (and some unique) culture characteristics due to a number of physical and ambient chemical conditions provided by the system, including:

- perfused medium flow and porous support permits long-term culture
- high cell density culture increases cell-cell contact
- selectable MWCO of fibres concentrates interactive cytokines
- selectable MWCO of fibres segregates cells, metabolites and products
- directional flow establishes gentle interstitial gradients within the cell mass
- hydrodynamic (shear) force on endothelial cells required for proper physiology
- long-term high-density culture on porous support facilitates development of cell-cell interactions over time.

Reduced apoptosis

It has long been recognised that reduced proteolysis and contamination by intracellular proteins and DNA are characteristics of the HF bioreactor. During extremely long culture periods of several months or longer, no degradation of secreted products has been reported, even when using serum-free and protein-free media. P.C. Liao's group has identified a potential mechanism for this (Wu *et al.*, 2009). Their research focused on the culture of primary cancer cells to identify potential secreted biomarkers. Cultures were initiated in FBS and then switched to a protein-free, chemically defined medium in order to reduce the amount of extraneous protein that might obscure the secretome of the cells. They found a significant reduction in apoptosis of up to 90%. The cells would die, but not lyse and go through the typical process of releasing proteolytic enzymes, intracellular proteins and DNA (Chang *et al.*, 2009). This, and the absence of significant shear forces inside the HF cartridge, results in dramatically reduced postsecretion alterations and cleaner harvests (Srisomsap *et al.*, 2010).

Culture consistency

In a HF bioreactor, the cells are bound to a porous support, not a non-porous plastic surface. Cell division rate and generation number are reduced, cultures do not require splitting, passage number is irrelevant, and cells grow in multiple layers in a 'postconfluent' fashion. Cultures can be maintained for many months and up to a year or longer (personal communication). Culture conditions remain

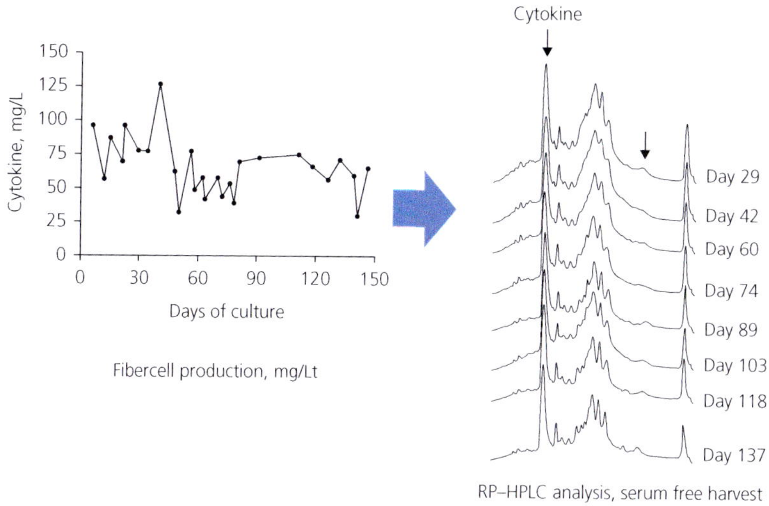

Figure 14.5 HEK 293 cells' cytokine product. HEK 293 cells expressing a heavily glycosylated protein in CDM HD protein-free medium. The HPLC demonstrates no changes in post-translational modifications over 137 days of continuous culture.

highly consistent during this time, as does cell physiology. In an example of this, provided by Dr George Pavlakis (NCI), 293T cells were transformed to produce a very complex protein. The cytokine product described was more than 40% carbohydrate and somewhat labile due to the fact that the two subunits are held together only with hydrophobic interactions. Attempts at producing this protein in standard flask or spinner cultures were unsuccessful. It was demonstrated that consistent production of dimerised product with complete and consistent post-translational modifications was maintained for over 140 days (Figure 14.5).

In vivo-like growth conditions

In vivo, most animal cells grow in 3D at very high density under tightly defined and highly controlled conditions. There is very little variation in oxygen tension, pH, glucose levels, etc. These parameters can have a wide range in flasks and other culture devices. It has been recently observed that 'The 2D cell culture systems used so far have several drawbacks: the morphology, proliferation, metabolism and expression profiles of cells grown in 2D systems are very different to cells in living tissues' (Tecan Group, 2012). It has been shown that in a HF bioreactor, consistency of culture conditions has a direct effect on cell physiology and product generation. One researcher found that transformed CHO cells expressing a hexamerised IgG responded favourably to HF bioreactor culture. The researcher first cultured the CHO cells in standard T150 flasks. Protein folding and post-translational modifications from this condition were incomplete, resulting in nearly 40% of the protein being expressed as an incompletely assembled monomeric subunit. When these same cells were removed from the

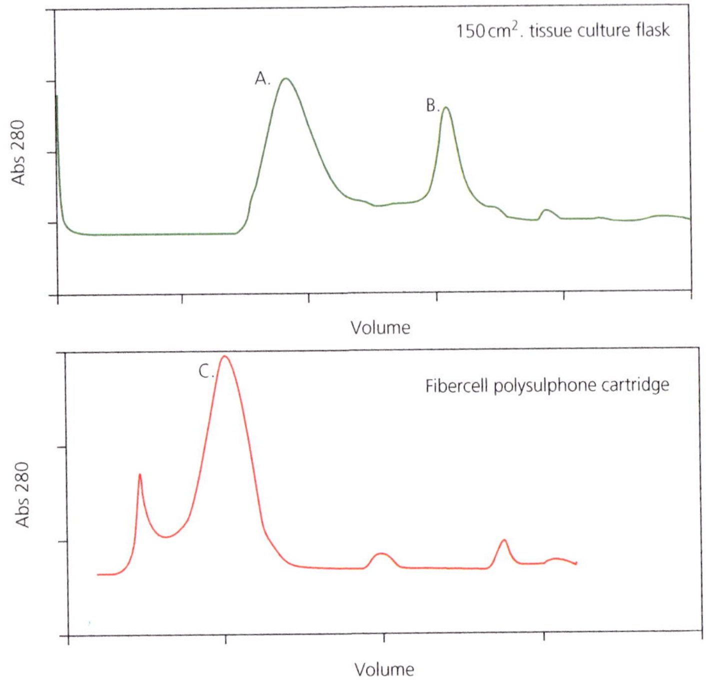

Figure 14.6 A CHO cell line expressing a hexamerised IgG protein construct. When these cells are cultured in flasks, 40% of the protein is expressed as a monomeric subunit. The exact same cells, when removed from the flask and seeded into the hollow fibre bioreactor, now express 95% of the protein as a properly folded hexamer.

flask and seeded into a HF bioreactor, within one week the expressed protein product was discovered to be 95% properly folded hexamer (J. Arthos, personal communication, 2013). The more biomemetic cell culture conditions had resulted in improved cell physiology and correct protein expression (Figure 14.6).

Single-use flow path systems

The culture-contact surfaces of many HF bioreactor instruments are composed of entirely single-use (SU) materials. These provide a number of benefits in the establishment of manufacturing and assay systems. They present reduced cross-contamination risks for many reasons, including that the materials have never seen living agents before and that they are sterilised by validated γ-irradiation systems. They provide lower initial investment costs as the SU housing apparatus is less expensive than full reuseable instrumentation. Lower facility and operating costs are supported by a reduced requirement for services such as autoclaves and process water as well as reduced operator requirements. Process efficiency and flexibility are provided by the ability to easily reroute the fluid flow path, and to order variant culture-contact materials or custom-designed assemblies from the manufacturer.

Protein-free, chemically defined culture medium

Besides providing many positive factors, animal serum presents a number of detrimental features including foaming, risk of contamination and variability in both production and cell-based assay performance. There are a number of non-culture-related limitations as well, including high and variable cost and the fact that high protein concentration can interfere with culture analysis. Cells *in vivo* are exposed to serum only during injury, and many cells then respond by particular activations. HF bioreactor culture characteristics, such as the very high cell density, allow for a reduction in serum concentration and facilitate adaptation to commercially available serum-free media. This has been taken one step further with the introduction of CDM HD, a commercially available serum replacement. CDM HD is a protein-free, chemically defined (CD), animal component-free serum replacement that is simplified and optimised for high-density cell culture.

Applications for cell-based assays

Many drug effects are dependent not only on concentration but also time. Some researchers believe cell-based assays and *in vitro* testing methods are a useful, time- and cost-effective tool for drug discovery. However, it is generally accepted that many of the available assays are not effective for examining the effects of both time and concentration, and do not closely mimic physiological kinetics. More specifically, such tests poorly report pharmacodynamic actions (what a drug does to the body) and pharmacokinetic actions (what the body does to a drug). Static cell-based assays in plates, flasks or other formats do not readily permit changes in drug concentration, as would be seen in humans from administration, uptake, distribution, distal metabolism and elimination effects. Animal models generally do not provide the same drug kinetics as would be found in humans, many infections cannot be supported with an animal model and often the bacterial load is not high enough to reveal the emergence of resistance. HF bioreactor cartridges have continuous medium circulation supporting dynamic control of drug concentration over time and resulting in the mimicking of naturally occurring gradients in tissue drug concentration. A high surface area-to-volume ratio permits extremely rapid exchange of metabolites and pharmacoactive molecules between the central reservoir and cells growing in the relatively small ECS of the cartridge. Furthermore, the volume of this central reservoir can be easily adjusted to permit rapid and reproducible changes in drug concentration (Figure 14.7).

Simulation of the kinetics of multiple drugs can also be accomplished so drug-drug interactions and combination therapies, as well as transport and efflux, can readily be modelled. The system is compact enough that multiple cartridges can be conveniently manipulated in a relatively small space, providing

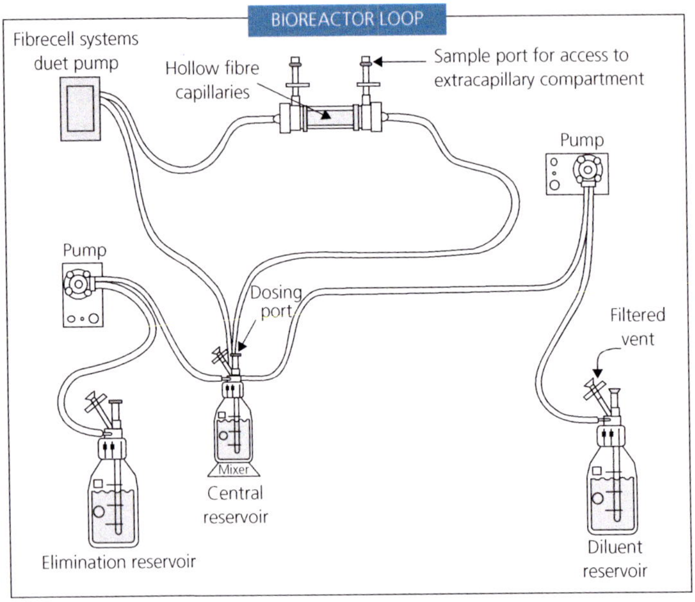

Figure 14.7 **The set-up for controlling drug concentration to mimic human PK/PD.** Medium or broth is constantly recirculating through the cartridge from the central reservoir. Drug can be introduced into the central reservoir at a defined rate representing the absorption curve. Diluent is then added and removed from the central reservoir at a controlled rate to model the elimination curve. Drug concentration in the extracapillary space of the cartridge equals the drug concentration in the central reservoir and the volume of the central reservoir remains constant.

multiplexed or parallel and higher throughput activities. Such systems can be configured for cell-based assays employing either a single-cell type or multicell in co-cultivation.

Examples of hollow fibre-based cell assays

Hollow fibre-based assays are inherently more complex and costly to design and set up than conventional cell-based assays. They require a higher degree of cell culture technical ability, more laboratory space and are more suited for higher value, more complex questions rather than smaller and truly high-throughput screening type assays. It can take up to a week before the culture is established and data can be generated. They are most effective when either using suspension cell types (permits the sampling of a small population of cells) or when measuring something the cells are secreting over time, such as a specific protein marker or virus particles.

However, these assays can generate data that is not available in any other manner, and can bridge an important gap between animal studies and phase I clinical trials. Large numbers of cells can be assayed over a pharmacologically relevant period of time. Drug concentrations can be controlled in a dynamic fashion and both adsorption and elimination curves can be modelled. Multiple tests can be performed on the same cell population. 3D cultures of multiple cell types can model complex processes such as virus infections in tissues, haematopoiesis, cancer cell propagation, cancer cell metastasis and the blood–brain barrier. Furthermore, nuclear magnetic resonance (NMR) and other imaging methods have demonstrated potential as non-invasive ways to monitor the culture (Mancuso *et al.*, 1990).

Assays using only one cell type

The simplest type of 3D cell-based assays performed in HF bioreactors consist of only one cell type, seeded in the ECS. A popular implementation of this approach is in the study of antimicrobial compounds on bacteria. While outside the purview of this chapter, it is mentioned because it is the most common application of this format (Inloes *et al.*, 2004). The primary cell-based assays based upon this method are virus culture using a mammalian host cell and cancer cell-based assays.

There are many advantages for assaying drug effects on viral activity in a HF bioreactor. The rapid delivery of nutrients, removal of waste products and equilibration of drugs across the fibres facilitates the culture of cells at high densities. It is the only cell culture method that can support cells at physiological densities and provide for *in vivo*-like viral infection and pharmacokinetics at these densities. While both adherent and suspension cells can be cultured in a HF bioreactor, it is notable that adherent cells bound to a porous support do not require periodic splitting and can be maintained for extended periods of time. Suspension cells can also be supported for extended periods of time due to the same constant feeding and removal of metabolites.

The ability to add medium with or without drugs is particularly important for dose fractionation pharmacodynamic studies where compounds are added to the system over a short period of time and then removed by dilution with drug-free medium without disturbing the cells or their environment. Both absorption and elimination can be simulated and the small volume of the ECS contributes to the rapidity and economy with which this equilibrium takes place. Lastly, and perhaps most importantly, because of their large size both viruses and virus-infected cells are retained in the small volume of the ECS. These infectious agents cannot cross the fibres into the medium. The system is completely closed to the external environment and provides an added biosafety component protecting laboratory personnel from exposure.

Antiviral pharmacodynamics

Pharmacodynamics is the science that links drug exposure to response. A key element here involves identification of the best dosing and scheduling. This is derived from the principle that the drug concentration-time curve is a factor in a drug's effectiveness (Eagle *et al.*, 2004). For example, the duration that a drug concentration remains above inhibitory levels may be closely linked to the antiviral effect. In this case, continuous or defined episodic intervals of drug exposure may lead to desired effects. On the other hand, peak concentrations of drug may be more important for antiviral activity. Here, brief doses at maximal concentrations may be most effective. It might also be demonstrated that the mode of dosing does not significantly affect antiviral activity. There are four essential pillars to the pharmacodynamic index (dose and schedule) of a particular entity for a particular virus:

1 The EC_{50}/EC_{95} values of the compound for the particular virus.
2 Binding (and therefore removal) of the compound to unreactive tissue and serum factors.
3 Identification of the compound's pharmacokinetics (and any variability of this within the population).
4 Identification of optimal exposure and administration schedule of the compound for the virus in question.

The high-density culture within HF reactors supports the efficient (biomimetic) cell-to-cell spread of virus in either matrixed or suspension cells. In either case, released virus and virus-infected cells accumulate in the ECS over time. Computer-controlled pumps administer drug and diluent into the central reservoir to model any schedule of drug exposure.

Antiviral effect can be determined by sampling the contents of the ECS through the sampling ports. This supports many divergent assays, beginning with virus-infected cell identification and counting by, for example, FACS analysis following treatment with a fluorochrome-labelled viral antigen-specific monoclonal antibody. The effect of drug on the yield of free infective virus particles can be determined by means such as plaque assay of harvested cell-free medium or by ELISA of viral antigen. The dosing regimen providing inhibition of viral replication and/or cell-to-cell spread of virus can be determined by sequential analysis of cell and ambient media viral titre and drug concentration from identified points later validated by, for example, LC/MS^n. Values for both drug exposure antiviral effects allow construction of an exposure–response relationship curve. Accuracy in this relationship is facilitated by the fact that eight or more evaluations can be performed simultaneously (McSharry *et al.*, 2011).

HIV

Nucleoside analogues

Nucleoside analogues are versions of the building blocks of DNA that almost act like normal nucleosides in DNA synthesis. But when these faulty building blocks are used, the new DNA cannot be built correctly. This in turn prevents the cell from producing new infective virus. Nucleotide analogue-based drugs also exist and are technically different from nucleoside analogues but act in much the same way. HF-based pharmacodynamic models have been employed in the examination of such HIV-1 antivirals for quite some time.

The nucleoside analogue 2′,3′-didehydro-3′deoxythymidine (d4T) was examined using HF infection models (HFIM) (Drusano *et al.*, 2002). Here, three HF units containing HIV-infected and uninfected CEM cells were set up. One unit was continuously infused with medium without drug and served as a no-drug control. A second unit received drug as a continuous infusion. A third unit received a bolus of d4T every 12 hours followed by a no-drug washout to model a one-hour pharmacokinetic half-life. In the absence of d4T there was a fourfold increase in p24 antigen over the nine days of the experiment. In the presence of d4T, delivered either as a continuous infusion or a twice-daily bolus, the amount of p24 antigen did not increase. In separate experiments, HF units infected with the same amount of virus and treated in the same way, but with d4T at half these doses, failed to completely inhibit virus replication. The results indicated the dosing required as either a bolus or continuous infusion to inhibit virus replication. The HFIM system predicted that the minimum effective dose of d4T to treat patients infected with HIV was approximately 0.5 mg/kg/day administered twice a day. This prediction was confirmed in a clinical study (Anderson *et al.*, 1992).

Protease inhibitors

The precursor Gag polyprotein is processed by a viral protease to many HIV structural proteins. A77003 impairs HIV-1 protease-mediated Gag processing and leads to an irreversible loss of the infectivity of the virus. A HF system was successfully used to determine the minimum concentration of the protease inhibitor A77003 that inhibits the replication of HIV in CEM cells (Preston *et al.*, 2003).

Vaccinia virus

Because routine smallpox vaccination among the public stopped after the disease was eradicated, the majority of the world's citizens are susceptible to infection with variola major virus. After the events of September and October 2001, however, the US government established goals for developing newer vaccines. Smallpox vaccine is made from vaccinia, a virus related to smallpox. Most smallpox vaccine contains live, replication-competent, vaccinia virus, not dead or attenuated virus like many other vaccines. For that reason, the vaccination site must be cared for to prevent the virus from spreading to other portions of the patient's body or other people. Also, the vaccine can cause undesirable side effects and complications.

The main challenge in developing a new, safer vaccine is that its effectiveness cannot be tested on humans, and other animals do not naturally contract smallpox. Monkeys or ferrets infected with other poxviruses are often used, but tests on animals that are artificially modelled with related disease often give false or misleading results. For improved pharmacodynamic studies on vaccinia virus, some researchers have employed HF-based models. One group developed such a model based upon used HeLa-S3 cells because they have been used to grow

Figure 14.8 **Example of a set-up to run multiple cartridges and multiple assays simultaneously.** The closed nature of the system permits different experiments to be run in the same incubator.

large amount of vaccinia virus and can be maintained as adherent or suspension cultures. Here, the effect of cidofovir on vaccinia virus replication in the HeLa-S3 cells was monitored by FACS analysis of virus-infected cells and by the production of infectious virus using a plaque assay (McSharry *et al.*, 2009a) (Figure 14.8).

Influenza virus

There are many distinct types of antiviral compounds effective against influenza virus. Licensed examples include the adamantanes and neuraminidase inhibitors. Amantadine has been used extensively for years, but recently almost all circulating influenza viruses have become resistant to it. Oseltamivir, a neuraminidase inhibitor, is now the most commonly used antiviral for uncomplicated influenza. The current recommendation for treatment with oseltamivir is to take two 75 mg tablets twice a day. Researchers wanted to determine if once-daily treatment was possible. They employed HF-based models to perform dose range and dose fractionation experiments (McSharry *et al.*, 2009b). As is the case for poxviruses, there are no cell lines that continuously produce influenza virus infection. The standard cell line used for influenza viruses is the Madin–Darby canine kidney (MDCK), which grows well on glass and plastic surfaces and is permissive to various strains of influenza virus, but not with high efficiency.

To perform pharmacodynamic studies for oseltamivir and influenza viruses, this group used a derivative of MDCK that expresses higher levels of a particular cell surface glycoprotein than the original line. It was determined that the R292 strain of influenza virus grows well in these cells in the HF units. To perform dose-ranging studies in the HFIM system, six HF units containing influenza A virus-infected AX-4 cells were mixed with uninfected AX-4 cells and continuously infused with various concentrations of oseltamivir. The effect of the drug on virus replication was determined daily by plaque and haemagglutination assay. This provided an EC_{50} value in the HFIM system for oseltamivir with this isolate of the R292 strain of influenza virus. To determine the pharmacodynamically linked index, a dose fractionation study was designed to mimic a variety of half-lives of the drug in application. Here, one HF unit received an appropriate exposure delivered by continuous infusion; a second unit received the same exposure once a day followed by a no-drug washout; a third unit received the same exposure delivered twice a day followed by a no-drug washout; and a fourth unit received the same exposure delivered three times a day followed by a no-drug washout.

The data showed that in the absence of drug, the virus grew well in the HFIM system. The continuous dose and all three fractionated doses gave the same amount of inhibition of virus replication. This group concluded that the pharmacodynamically linked index for oseltamivir for the R292 strain of influenza A virus is the $AUC/EC_{50}/_{95}$ ratio. This means that the model indicates that at the appropriate dose, oseltamivir could be given once a day.

The demonstration that adherent cells can be used to grow virus in the HFIM system opens this system up to the pharmacodynamic analysis of antiviral compounds for a wide variety of viruses. As long as the virus-infected cells produce virus that is released into the ECS, then the effect of antiviral compounds can be measured, such as in H1N1 influenza (Brown *et al.*, 2011).

Anticancer agents

Another example of a 3D cell-based assay using only one cell type in the ECS of the HF cartridge is for the analysis and characterisation of anticancer agents. Anticancer agents also exhibit both time- and concentration-dependent efficacy and the goal of cell-based assays can be twofold. The first is to ascertain the most effective dosage profile for cytotoxicity of cancer cells while leaving normal cells in an acceptable state. The second is to provide an indication of the compound's robustness and application breadth by testing different types of cancer cell lines under identical conditions.

Kirstein reported on the use of the HF model for anticancer drug evaluations. In this case, gemcitabine was examined in the anchorage-dependent MDA-MB-231 breast cancer cell line (Kirstein *et al.*, 2006). Longer-term exposures are possible in continuous culture HF systems since the cells do not require splitting when a confluence is reached. More accurate results are obtained for a few reasons. One is that because the multilayer, 3D organisation of these continually perfused cells more closely reflects the *in vivo* structure of the tumour, they have an increased relevance for assays of anticancer agents. Another is that the cell division rates in static cultures are commonly artificially high, and this can render them more sensitive to some chemotherapeutic agents then their natural counterpart (Kirstein *et al.*, 2008).

Assays using more than one cell type

Hollow fibre and 3D co-cultivation for cell-based assays

Historically, the science of cell biology was intended to reduce complex systems to their most fundamental components in order to better understand them. However, in this age of systems biology, we are appreciating the limitations to such reductionist research, and better understanding the implications of organisms and tissues being made up of more than one type of cell. From the consequences of complex physical interactions to the signalling between multiple cell types, we are seeing the behaviour of cells in the context of an organisation or ensemble of heterotypes. HF bioreactor-based culture is one of the few *in vitro* techniques providing large numbers of cells in close enough proximity and sufficiently high density to observe a number of tissue-like behaviours (Figure 14.9). These behaviours include:

- signalling and direct interactions between different cell types
- co-ordinate (or synergistic) activity upon the ambient media
- co-ordinate (or synergistic) activity upon a compound or drug
- mixed cell type-specific effect (+/−) upon viral infections.

Hollow fibre cell culture permits the recapitulation of natural structures containing more than one cell type in a defined, controlled and more biomimetic environment. Endothelial cells are the only cells that can be easily cultured on the insides of the fibres under flow. There are two types of cell co-cultivation that can be performed in a HF bioreactor. They are one or more cell types:

- on both the inside and the outside of the fibres (Davis, 2007)
- on only one side of the fibres (typically on the ECS).

An example of the first type of cellular co-cultivation is the use of HF bioreactors to culture endothelial cells on the insides of the fibres while culturing a different cell type on the outside to provide cell signalling. Endothelial cells comprise the inner lining of all blood vessels, and the endothelium can actually be considered a discrete organ, rather than simply a cell type. Endothelial cells were first

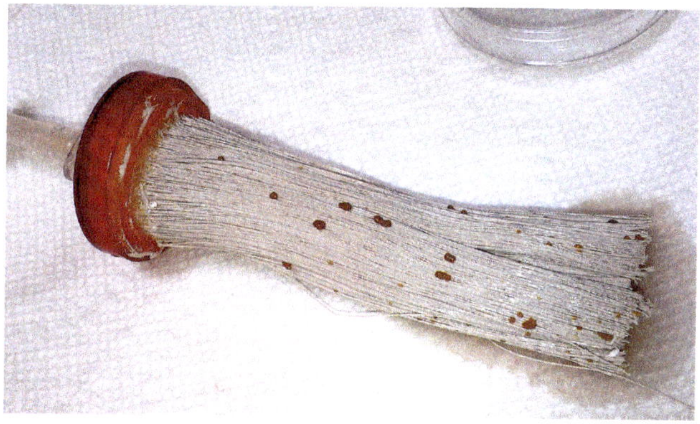

Figure 14.9 **Co-cultivation of mixed cell population derived from collagenase digestion of a human placenta.** This results in the formation of three-dimensional structures capable of generating suspension cells with stem-like phenotypes. Source: Courtesy of FiberCell Systems.

Figure 14.10 **Cross-section of a PVDF fibre with an endothelial cell monolayer.** This monolayer forms on the inside of the fibre under the influence of shear stress.

cultured on the insides of hollow fibres by Dr Barbara Ballerman (Ballerman & Ott, 1995). When cultured on the insides of a hollow fibre (the intracapillary space) and exposed to hydrodynamic shear stress from medium flow, endothelial cells behave differently (more *in vivo*-like) than they do in static, flask culture. Under shear, endothelial cells lay down flat as a monolayer, stop dividing, form tight junctions and express different genes than when cultured under static conditions in flasks (Figure 14.10). Differences in shear stress throughout the body may explain both the different functions and characteristics of endothelial cells in those areas (Redmond *et al.*, 1995).

The concept of 'organ recapitulation' in hollow fibres was first applied by Jorg Gerlach using liver tissue (Gridelli *et al.*, 2012). The system used was a complex HF bioreactor with two different fibre types and three separate bundles of fibres. Liver tissue was simply digested with collagenase and this mixture of cells seeded into the HF bioreactor. Primary function was maintained for four weeks. It appeared that endothelial cells provided a necessary component of the cell mixture and were involved in the formation of 3D structures (Hoffman *et al.*, 2012).

Bone marrow model

Perhaps the most rigorous application of HF systems for single-compartment cell co-cultivation is in the area of stromal cell/suspension cell interactions. Contrary to popular belief, it is possible under specific conditions to harvest large numbers of healthy, intact cells from a HF bioreactor. In fact, the most powerful applications for HF systems involve longer-term cultures with periodic harvesting over time. There are a number of recent examples of this effectiveness presenting some very interesting results. In demonstration of the second type of co-culture, Dr Mayasari Lim at University Hospital, Hong Kong, published an article on the co-cultivation of a human bone marrow stromal cell line with a leukaemic T cell line (Usuludin *et al.*, 2012). The standard approach to this is generally accomplished in flask culture, but here she reported the establishment of a stromal line cultured to fairly high density within one week of culture initiation in the HF cartridge. Once the glucose uptake rate of standard media in the cartridge reached 1 g per day, the leukaemic T cell line was seeded into the cartridge. When co-cultured in flasks, the T cells would typically undergo a 10× expansion. When cultured in the HF cartridge, the T cells underwent a 4000-fold expansion, much more in line with what occurs *in vivo* (Figure 14.11).

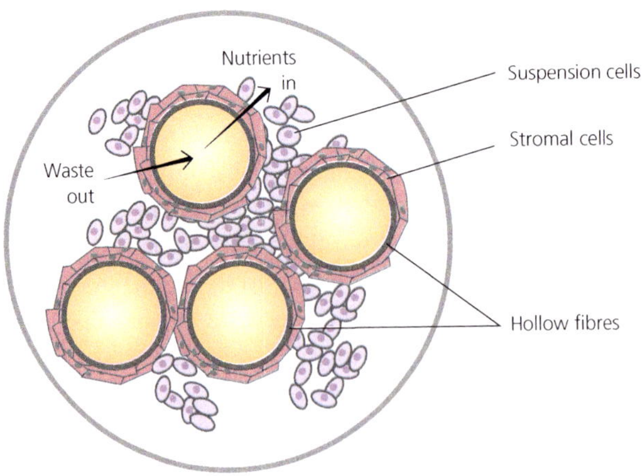

Figure 14.11 Schematic drawing of adherent stromal cells interacting with suspension cells. Since the cells are bound to a porous support, they do not require splitting and the culture can be maintained for several weeks to months.

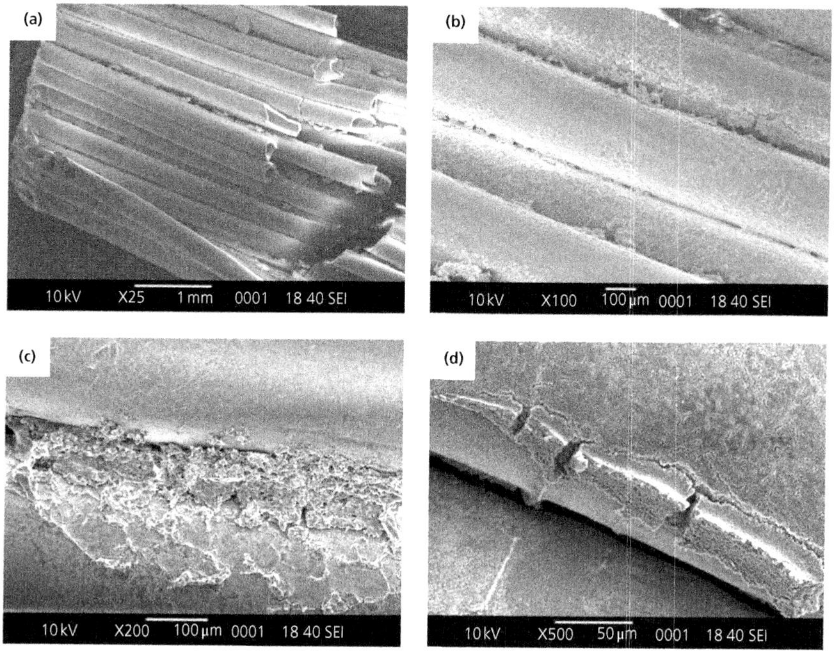

Figure 14.12 **Scanning electron micrographs of bone marrow stroma cells bound to the surface of the 5 KD MWCO polysulfone fibres.** These cells can interact with haematopoietic stem cells or leukaemic lymphocytes or other cell types. Images (a–d) are of increasing magnification showing the general structure of the polysulfone fibres with 3D cell growth in between.

The bone marrow microenvironment plays an integral role in the regulation of haematopoiesis. Residing stromal cells and the various extracellular matrixes in the bone marrow provide biological signals that control haematopoietic stem cell (HSC) function. In this study, a biomimetic co-culture platform using the HF bioreactor was developed for the *ex vivo* expansion of HSCs. The efficacy of such a platform was evaluated in comparison to standard cultures performed on tissue culture polystyrene (TCP). A human stromal cell line (HS-5) was employed as a co-cultured stromal support of lineage cell-depleted human cord blood cells. This was accomplished in serum-free medium supplemented with a cytokine cocktail (Xue *et al.*, 2014) (Figure 14.12).

Results showed that the performance of the HF bioreactor in supporting total cell and CD34+ progenitor cell expansion was comparable to that of cultures on TCP, while cells harvested from the HF bioreactor had a higher clonogenic ability. The performance of *ex vivo*-expanded cells from the HF bioreactor upon haematopoietic reconstitution in humanised mice was also comparable to that of the TCP control. Scanning electron microscopy revealed that stroma cell growth inside the HF bioreactor created a three-dimensional cell matrix. These findings demonstrate the feasibility of utilising an HF bioreactor for creating a complex matrix architecture, which may provide good *in vitro* mimicry of the bone marrow supporting large-scale expansion of HSCs.

Figure 14.13 Scanning electron micrograph of bovine aortic endothelial cells. These cells were cultured on the insides of the fibres and vascular smooth muscle on the outside of the fibre. Note the 3D structure formed by the vascular smooth muscle cells. Source: Courtesy of Dr. Paul Cahill and Dr. Eileen Redmond.

Asymmetric co-culture using endothelial cells

Three-dimensional cell-based assays utilising co-culture of multiple cell types in HF bioreactors can recapitulate more complex structures than those with a single cell type. One type of cellular co-cultivation, as described above, is the use of HF bioreactors to culture endothelial cells on the insides of the fibres and a different cell type on the outside. A number of such examples have been reported, including some with significantly more complex asymmetric cell co-cultivation. For example, Paul Cahill cultured bovine aortic endothelial cells on the insides of the fibres and vascular smooth muscle cells on the outside of the fibres. As altering the flow rate changed the shear stress, G-protein formation and endothelin receptor expression were directly modulated, even though there was no physical contact between the two cell types (Redmond *et al.*, 1995) (Figure 14.13).

Blood–brain barrier model

Endothelial cells at the blood–brain barrier (BBB) demonstrate tight intercellular junctions, the absence of fenestrations and reduced pinocytotic vesicles. The *in vitro* study of the BBB has progressed rapidly of late as improved monitoring technologies and new or enhanced cell culture models have become available. Developments in *in vitro* models promise the design of new drugs for treatment of CNS disorders. These models play an important role in investigating the ability of such compounds to cross the BBB. There are many *in vitro* approaches to modelling BBB physical and biochemical behaviour, but most fail to represent its three-dimensional nature and do not support the associated exposure of endothelial cells to such complex influences as exist *in vivo* (Janke *et al.*, 2013).

To answer this challenge, Janigro developed a new, dynamic, *in vitro* BBB model (NDIV-BBB) designed to allow for extensive pharmacological, morphological and physiological studies (Stanness *et al.*, 1999). He first reported that brain microvascular endothelial cells developed robust growth and differentiation when cultured alone. In the presence of glial cells, they developed measurable properties allowing monitoring of the system and experimental perturbations and physical manipulations. For example, it was noted that flow rates played an essential role in endothelial cell differentiation and concomitant decreased cell division. His new dynamic model of the BBB allows for longitudinal studies of the effects of flow and co-culture in a controlled and fully recyclable environment. It also permits visual inspection of the abluminal compartment and manipulation of individual capillaries. This model and testing services are available from FloCel (www.flocel.com).

In perhaps the ultimate embodiment of state-of-the-art HF-based cell assays, Dr Chris Pepper and co-workers have developed a model for primary investigations and assaying for cancer metastasis (Walsby *et al.*, 2014). It utilises endothelial cells on the inside of the fibre subjected to shear stress as well as their interaction with circulating leukaemic lymphocytes (Figures 14.14 and 14.15). There is

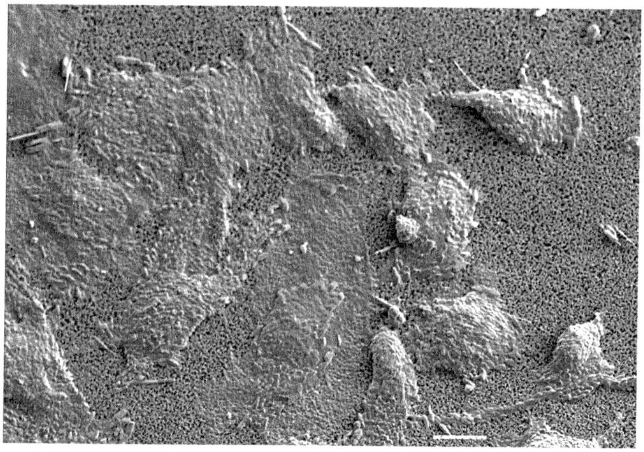

Figure 14.14 **Endothelial cells attached to the inside of a PVDF hollow fibre with a short period of shear stress.** Cells have begun to flatten in response to the shear.

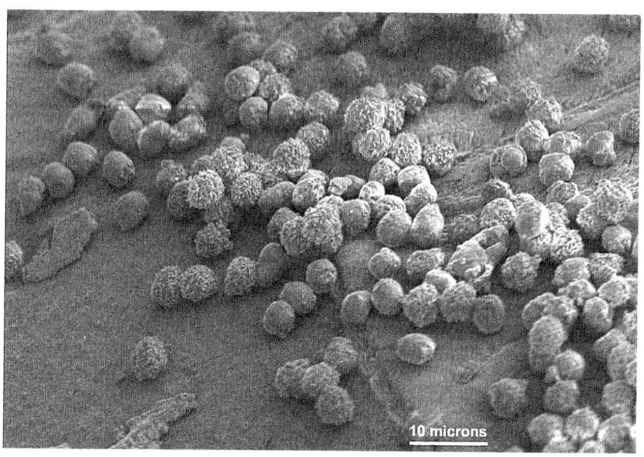

Figure 14.15 **Circulating leukaemic lymphocytes on the outside of the fibres.** This was taken after the transmigration of lymphocytes and their specific interaction with endothelial cells on the insides of the fibres.

growing evidence that lymphocyte trafficking contributes to the clinical course of chronic lymphocytic leukaemia (CLL) but to date only static *in vitro* cultures have been used to study these phenomena. To address this, a dynamic *in vitro* model was developed in which CLL cells experience shear forces equivalent to those in capillary beds and are made to flow through capillary-like hollow fibres lined with endothelial cells. CLL cells treated in this way increased their expression of CD62L, CXCR4, CD49d and CD5 directly as a result of the shear force. Furthermore, CLL cells migrated through the endothelium into the 'extravascular space' (EVS). Migrated CLL cells had significantly higher expression of CD49d, MMP-9, CD38, CD80 and CD69 compared with CLL cells that remained in the circulation. The degree of migration observed strongly correlated with CD49d expression, and treatment with the CD49d blocking antibody natalizumab resulted in significantly decreased migration. Taken together, these data provide evidence for a novel, dynamic and reproducible *in vitro* model of lymphocyte migration and cancer metastasis. This model might also prove useful in the study of stem cell homing and trafficking.

Conclusion

As the original 3D culture approach, HF systems have for decades supported the biomimetic culture of many cell types, at tissue-like densities, for extended periods of time. Novel material features supporting a renaissance in the technology include fibres composed of new materials, new surface derivatisations and porosities providing improved binding, flux and flow rates. Enhanced process control features include innovative culture monitoring approaches, automated feed and recirculation control as well as improved process understanding leading to new closed-loop control parameters. New and innovative application development has kept pace with the availability of these new technologies, expanding the scope of applicability of 3D HF cell-based assays. The HF bioreactor represents a more *in vivo*-like method by which to culture cells for assays. The new fields of metabolomics and systems biology are supporting the development of HF systems based upon multicell co-culture and defined serum-free media. The result is a robust and flexible technology with diverse applications. Such applications range from protein biological production providing ultra-clean harvest and simplified purification to improved *in vitro* viral infection models providing more accurate drug candidate pharmacokinetic/pharmacodynamic modelling.

The 3D HF-based cell assay system may not replace either animal models or human testing but does represent an important methodology for the testing of clinical compounds. It provides human kinetics and pathogen load levels before or in conjunction with human testing and bridges an important gap between animal and human testing.

References

Anderson, R. E., Meyer, W., Balch, F. *et al.* (1992) Antiviral effects of stavudine (d4T) therapy. 8th International Conference on AIDS, 19–24 July, Amsterdam, The Netherlands. Abstract WeB 1010.

Ballerman, B. & Ott, M. (1995) Adhesion and differentiation of endothelial cells by exposure to chronic shear stress: a vascular graft model. *Blood Purification*, **13**, 125–134.

Bennet, A., Liu, J., van Ryk, D. *et al.* (2007) Cryoelectron tomographic analysis of an HIV-neutralizing protein and its complex with native viral gp120. *Journal of Biological Chemistry*, **282**(38), 27754–27759.

Brown, A. N., McSharry, J. J., Weng, Q. *et al* (2011) Zanamivir, at 600 milligrams twice daily, inhibits oseltamivir-resistant 2009 pandemic H1N1 influenza virus in an in vitro hollow fiber infection model system. *Antimicrobial Agents and Chemotherapy*, **55**, 1740–1746.

Chang, Y. H., Wu, C. C., Chang, K. P. *et al.* (2009) Cell secretome analysis using hollow fiber culture system leads to the discovery of CLIC1 protein as a novel plasma marker for nasopharyngeal carcinoma. *Journal of Proteome Research*, **8**, 5465–5474.

Davis, J. M. (2007) Hollow fiber cell culture, in *Methods in Biotechnology Animal Cell Biotechnology Methods and Protocols* (ed R. Pörtner), Humana Press, New York, pp. 337–352.

Dennler, S., Goumans, M. J. & Dijke, P. (2002) Transforming growth factor beta signal transduction. *Journal of Leukocyte Biology*, **71**, 731–740.

Drusano, G. L., Bilello, P. A., Symonds, W. T. *et al.* (2002) Pharmacodynamics of Abacavir in an in vitro hollow fiber model system. *Antimicrobial Agents and Chemotherapy*, **46**(2), 464–470.

Eagle, H., Fleischman, R. & Musselman, A.D. (2004) The effective concentrations of penicillin in vitro and *in vivo* for streptococci, pneumococci, and Treponema pallidum. *Journal of Bacteriology*, **59**(5), 625–643.

Fibercell Systems (2012) Application Note. Placental Co-culture in a Hollow Fiber Bioreactor. Available at: www.fibercellsystems.com.

Gridelli, B., Vizzini, G., Pietrosi, G. *et al* (2012) Efficient human fetal liver cell isolation protocol based on vascular perfusion for liver cell-based therapy and case report on cell transplantation. *Liver Transplantation*, **18**(2), 226–237.

Hirschel, M., Gangemi, J. D., McSharry, J. & Myers, M. (2011) Novel uses for hollow fiber bioreactors. *Genetic Engineering and Biotechnology News*, **31**(12).

Hoffman, S.A., Muller-Vieira, U., Biemel, K. *et al.* (2012) Analysis of drug metabolism activities in a miniaturized liver cell bioreactor for use in pharmacological studies. *Biotechnology and Bioengineering*, **12**, 3172–3181.

Inloes, D. S., Smith, W. J., Dean P. *et al.* (2004) Hollow-fiber membrane bioreactors using immobilized E. coli for protein synthesis. *Biotechnology and Bioengineering*, **25**(11), 2653–2681.

Janke, D., Jankowski, J., Ruth, M. *et al.* (2013) The 'artificial artery' as in vitro perfusion model. *PLoS ONE*, **8**(3), e57227.

Kirstein, M. N., Brundage, R. C., Elmquist, W. F. *et al.* (2006) Characterization of an in vitro cell culture bioreactor system to evaluate anti-neoplastic drug regimens. *Breast Cancer Reasearch and Treatment*, **96**, 217–225.

Kirstein, M. N., Brundage, R. C., More, M. M. *et al* (2008) Pharmacodynamic characterization of gemcitabine cytotoxicity in an in vitro cell culture bioreactor system. *Cancer Chemotherapy Pharmacology*, **61**, 291–299.

Knazek, R. A., Gullino, P. M., Kohler, P. O. & Dedrick, R. L. (1972) Cell culture on artificial capillaries: an approach to tissue growth in vitro. *Science*, **178**(4056), 65–67.

Mancuso, A., Fernandez, E. J., Blanch, H. W. & Clark, D. S. (1990) A nuclear magnetic resonance technique for determining hybridoma cell concentration in hollow fiber bioreactors. *Biotechnology*, **8**(12), 1282–1285.

McSharry, J. J. & Drusano, G. L. (2011) Antiviral pharmacodynamics in hollow fibre bioreactors. *Antiviral Chemistry and Chemotherapy*, **21**, 183–192.

McSharry, J. J.., Deziel, M. R., Zagner, K. *et al* (2009a) Pharmacodynamics of cidofovir for vaccinia virus infection in an in vitro hollow fiber infection model system. *Antimicrobial Agents and Chemotherapy*, **53**, 129–135.

McSharry, J. J., Weng, Q., Brown, A. *et al* (2009b) Prediction of the pharmacodynamically linked variable of oseltamivir carboxylate for influenza A virus using an in vitro hollow fiber infection model system. *Antimicrobial Agents and Chemotherapy*, **53**, 2375–2381.

Preston, S. L., Piliero, P. J., Bilello, J. A. *et al.* (2003) In vitro-*in vivo* model for evaluating the antiviral activity of amprenavir in combination with ritonavir administered at 600 and 1000 milligrams, respectively, every 12 hours. *Antimicrobial Agents and Chemotherapy*, **47**(11), 3393–3399.

Redmond, E., Cahill, P. & Sitzman, J. (1995) Perfused transcapillary smooth muscle and endothelial cell co-culture- a novel in vitro model. *In Vitro Cellular and Developmental Biology. Animal*, **31**, 601–609.

Srisomsap, C., Sawangareetrakul, P., Subhasitanont, P. *et al.* (2010) Proteomic studies of cholangiocarcinoma and hepatocellular carcinoma cell secretomes. *Journal of Biomedicine and Biotechnology*, Article ID 437143.

Stanness, K. A., Westrum, L. E., Fornaciari, E. *et al.* (1997) Morphological and functional characterization of an in vitro blood-brain barrier model. *Brain Research*, **771**(2), 329–342.

Stanness, K. A., Neumaier, J. F. & Sexton, T. J. *et al* (1999) A new model of the blood–brain barrier: co-culture of neuronal, endothelial and glial cells under dynamic conditions. *Neuroreport*, **18**, 3725–3731.

Tecan Group (2012) 3D Cell Culture: Mimicking Real Tissue. Application Note. Available at: www.tecan.com.

Tharakan, J. P. & Chau, P. C. (1986) A radial flow hollow fiber bioreactor for the large-scale culture of mammalian cells. *Biotechnology and Bioengineering*, **28**, 329–342.

Usuludin, S. B. M., Cao, X. & Lim, M. (2012) Co-culture of stromal and erythroleukemia cells in a perfused hollow fiber bioreactor system as an in vitro bone marrow model for myeloid leukemia. *Biotechnology and Bioengineering*, **16**, 1248–1258.

Walsby, E., Buggins, A., Devereux, S. *et al.* (2014) Development and characterization of a physiologically relevant model of lymphocyte migration in chronic lymphocytic leukemia. *Blood*, **23**, 3607–3617.

Weeraphan, C., Diskul-Na-Ayudthaya, P., Chiablaem, K. *et al.* (2012) Effective enrichment of cholangiocarcinoma secretomes using the hollow fiber bioreactor culture system. *Talanta*, **99**, 294–301.

Whitford, W. & Cadwell, J. J. S. (2011) The potential application of hollow fiber bioreactors to large-scale production. *BioPharm International Supplements*, **24**, s21–s26.

Wu, H. Y., Chang, Y. H., Chang, Y. C. & Liao, P. C. (2009) Proteomics analysis of nasopharyngeal carcinoma cell secretome using a hollow fiber culture system and mass spectroscopy. *Journal of Proteome Research*, **8**, 380–389.

Xue, C., Kwek, K. Y., Chan, J.K. *et al.* (2014) The hollow fiber bioreactor as a stroma-supported, serum-free ex vivo expansion platform for human umbilical cord blood cells. *Biotechnology Journal*, **9**, 980–989.

CHAPTER 15

Three-dimensional engineered tissues for high-throughput compound screening: Mechanical properties of skin and ageing

Michael Conway, Ayla Annac and Tetsuro Wakatsuki
InvivoSciences, Inc., Madison, Wisconsin, USA

Introduction

Mechanical properties of skin: Health and disease

Human skin is the largest organ in the body. It is also the only organ exposed to the external environment, and as such has a number of important purposes. The major functions of skin include protecting the body from heat and cold, light damage, toxic chemicals and excess water, and it aids in body temperature regulation, as well as the retention of heat and moisture (Figure 15.1) (McLafferty *et al.*, 2012). A number of mechanical properties allow healthy skin to perform these functions. When these mechanical properties are disrupted, the skin is no longer able to play its barrier role and this has significant health consequences (Ogawa & Hsu, 2013). Children suffering from a rare genetic disorder, epidermolysis bullosa, develop skin blisters with minor mechanical stress (Cooper & Bauer, 1984) and approximately 3000–20 000 are affected (Gabrielli *et al.*, 2009). The International Scleroderma Network (Pilcher *et al.*, 1995) lists nearly 40 diseases with similar symptoms and quality of life effects. Women are more susceptible to scleroderma, with a 3–14-fold higher incidence (Gabrielli *et al.*, 2009). Investigation has identified cellular level causes for mechanical dysfunction of the skin (Reich *et al.*, 2009). Pathological analysis of scleroderma biopsies and many other skin diseases indicates that maintenance of the skin's normal mechanical properties is required for proper barrier functionality.

Underlying cellular level effects, complex subtypes of scleroderma can be classified by patterns of gene expression (Milano *et al.*, 2008). Scleroderma patient fibroblasts were found to have increased synthesis of collagen and decreased collagenase activity, consistent with the skin hardening phenotype and excessive extracellular matrix (ECM) accumulation observed in patients.

Technology Platforms for 3D Cell Culture: A User's Guide, First Edition. Edited by Stefan Przyborski.

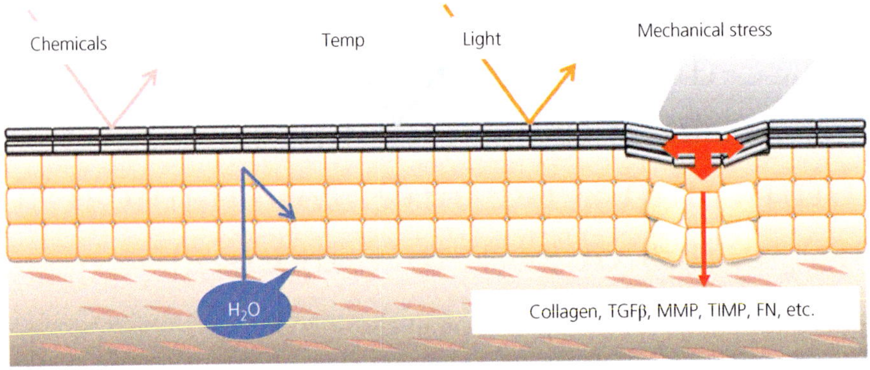

Figure 15.1 Skin functions. Human skin performs a variety of necessary homeostatic functions. It acts as a barrier to environmental chemicals, temperature, light and mechanical stress to protect internal structures. Skin also retains internal moisture and participates in maintaining body temperature. See text for more details.

Therefore, profiling ECM gene and protein expression patterns could help characterise the disease phenotypes at the molecular level and lead to new therapeutic targets.

There are nearly 400 genetic skin disorders (Bruckner-Tuderman, 2010) and the majority are rare (Grinnell *et al.*, 1987), which provides little incentive for the pharmaceutical industry to develop treatments. Establishment of an *in vitro* assay system that can analyse mechanical properties, barrier function and other biological parameters such as gene expression of *in vitro* skin tissues in high-throughput could reduce the economic barrier to screen for niche treatments.

Ageing skin and mechanical properties

The mechanical properties of skin change gradually as people age. More than 20 years ago, scientists began to investigate the effects of ageing on the mechanical properties of skin by twisting and poking it (Escoffier *et al.*, 1989; Pailler-Mattei *et al.*, 2008). These early data showed clearly that skin is an elastic material like a mechanical spring. Unlike a simple spring, however, skin takes time to recover back to its original shape after it is deformed. Therefore, healthy skin exhibits a certain degree of viscoelasticity, much like a memory foam mattress when it is depressed and allowed to return to its original form (Figure 15.2). It is the combination of ECM proteins, such as collagen and elastin, which gives skin its viscoelastic properties. Collagen provides the structure while elastin provides elasticity. Skin's viscoelastic properties can be modelled as a shock-absorbing dashpot and spring arrangement (Escoffier *et al.*, 1989). Mechanical studies of ageing skin show a longer recovery time after physical deformation compared to younger skin (Escoffier *et al.*, 1989). A number of cosmetic imperfections are the result of degradation of the skin's viscoelastic and, ultimately, mechanical properties. Examples include 'sleep lines', lines formed horizontally near the eyes,

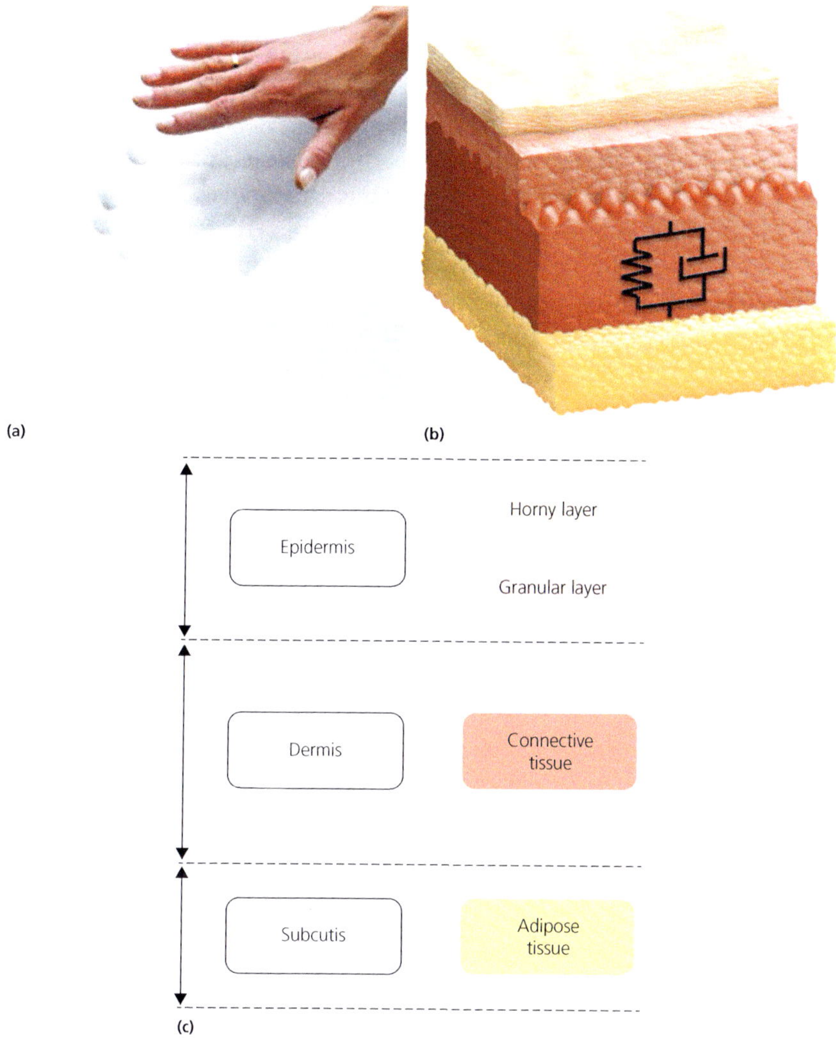

Figure 15.2 Viscoelastic properties of skin. Human skin has structural and elastic components that combine to give skin resilience and functionality. (a) The viscoelastic nature of skin acts like a memory foam mattress during deformation and reconstitution by resisting displacement and slowly returning to its original shape and position. (b) The viscoelastic nature of skin can be modelled with a dashpot and spring arrangement. (c) The layers that comprise the skin.

forehead and mouth during sleep (Sarifakioglu *et al.*, 2004), and laxity around the neck, ears and limbs. More troublesome conditions derived from the mechanical breakdown of ageing skin include various forms of dermatitis, ulcers, eczema, keratosis and thinning of the skin which can lead to physical damage (Farage *et al.*, 2013).

Key players in sun-damaged skin: Cells and molecules

Sun exposure can negatively influence the structure and mechanisms that maintain the normal physiological properties of skin. Solar ultraviolet (UV) irradiation is one well-characterised cause for skin ageing (Naylor *et al.*, 2011). UV light can penetrate through the epidermis and reach fibroblasts in the dermis as well as the ECM proteins that give skin its mechanical properties. Most of the skin itself is ECM. About 90% consists of type I collagen (Fisher *et al.*, 2008). The dermal fibroblasts deposit collagen and remodel it in part by secreting matrix metalloproteinases (MMP) which degrade collagen. Several groups have shown how UV light induces MMP expression (Quan *et al.*, 2009) which is a primary contributing factor to dermal ECM breakdown resulting in the loss of skin's normal mechanical function.

To better understand the relationship between sun exposure and the capacity for collagen recycling, fibroblasts were isolated from severely photo-damaged or photo-protected areas of the body. Both sets of fibroblasts were found to have the same collagen deposition and MMP activity (Fligiel *et al.*, 2003; Varani *et al.*, 2001, 2004). This result suggests that photo-damage does not impair the fibroblast's ability to secrete or degrade collagen but instead works through another process. So what is the difference between sun-damaged and sun-protected skin? The answer appears to be accumulated damage to the skin's collagenous ECM.

It takes about 30 years to recycle old to new collagen in human skin. During this time, collagen fibres become crosslinked to form thickened bundles (Miyachi & Ishikawa, 1998). MMPs can cut collagen and elastin fibres. As UV light induces MMP expression, the ECM is degraded, leaving thick bundles and fragmented collagen. Such collagen fragmentation introduces instability to the structure and mechanical integrity of skin. The fragmented collagen does not support healthy growth and maintenance of fibroblast physiology. In healthy young skin, non-fragmented collagen fibrils transmit mechanical signals to fibroblasts to maintain collagen production and minimise MMPs expression. In sun-damaged skin, the fibroblasts lose their mechanical connection to the ECM, resulting in low collagen deposition and increased MMP secretion. This process results in a negative loop of ECM degradation.

While scientific investigations continue to reveal complex mechanisms in premature skin ageing, a current scientific view (Fisher *et al.*, 2008) is that environmental factors such as UV irradiation accelerate the natural ageing process (Figure 15.3). While fibroblasts appear able to maintain the skin's ECM in spite of UV irradiation, ECM breakdown appears to limit the cell's ability to perform this maintenance role and drives continuous degradation. The mechanical environment of sun-damaged skin is hostile to cells and significantly different from that of sun-protected skin. Little is known about agents that can reduce the effects of UV irradiation on premature skin ageing. All-*trans* retinoic acid (RA) is one of a few agents shown to act on fibroblasts to inhibit negative biological actions induced by UV irradiation (Fisher *et al.*, 2000).

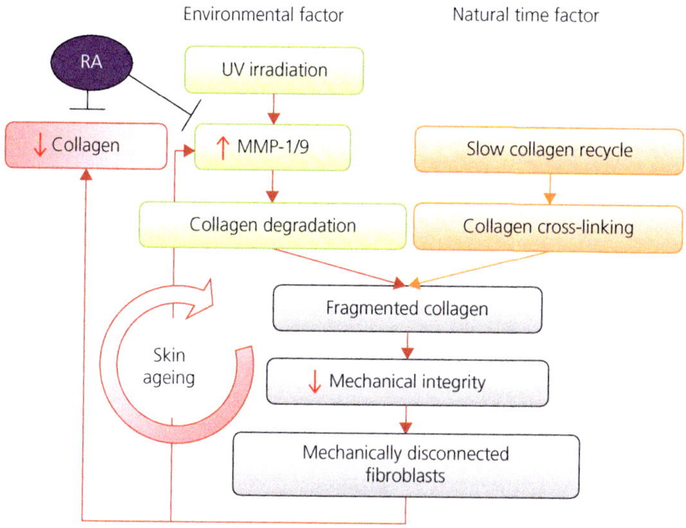

Figure 15.3 **Skin's mechanical properties and the process of ageing.** The dermal ECM is sensitive to degradation from environmental sources such as UV irradiation. Fibroblasts exposed to UV irradiation express MMPs which degrade collagen and elastin, resulting in fragmented fibrous bundles. This in turn degrades the mechanical and, ultimately, growth environment for fibroblasts, causing reduced collagen recycling and increased MMP secretion. These effects compound, resulting in UV-induced accelerated ageing.

Description of technology

Mechanical testing of connective tissue

There are well-established 3D skin *in vitro* models (e.g. Skinethic, MatTek) for testing chemical barrier function and toxicity. The ban on animal testing for cosmetic product development in the European Union (Pauwels & Rogiers, 2010) has driven the creation of acceptable *in vitro* models for this purpose. These models focus mainly on testing the safety of cosmetic ingredients and products. Therefore, cell viability remains one popular 3D skin model assay. While the mechanical properties of skin, as described above, play a pivotal role in premature skin ageing and disease processes, examination of skin mechanics has not been an integral part of cosmetic or therapeutic development. However, this is changing as progress in laboratory automation and tissue engineering is enabling mass production of miniaturised connective tissue equivalents suitable for high-throughput compound testing.

InvivoSciences: Modelling skin physiology for drug discovery

Similar to other skin models available, InvivoSciences can fabricate skin equivalents with human fibroblasts and keratinocytes but do so in a specialised engineered tissue culture chamber called the MC-8™ (Figure 15.4a,d,e). These tissue equivalents are fabricated in square microwells to form a thin membrane

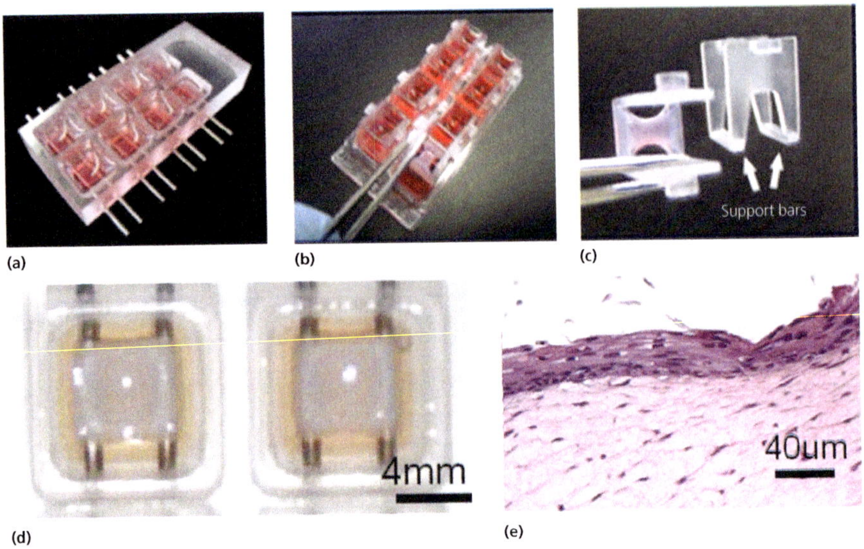

Figure 15.4 **Skin equivalents shown in the IVS MC-8 format.** (a) MC-8: the polycarbonate body has eight tissue chambers each with two stainless steel tissue support bars and optically clear well bottoms. (b,c) Alternative MC-8 format utilising a plastic scaffold system designed for manoeuvrability. (d) Close-up showing skin equivalents in the MC-8 from directly above. Note how the tissues are supported by the bars. The space between the bars is where the Palpator probe is plunged in to measure force. (e) Cross-sectional haematoxylin and eosin stain of MC-8 skin equivalents where the surface has been exposed to air to induce keratinocyte differentiation into a cutaneous epithelium.

supported horizontally by two stainless steel (or optional plastic) bars and not supported on a synthetic sheet. In this system, tissues with a monolayer of keratinocytes exposed to an air-liquid interface develop a cutaneous epithelium (Brohem *et al.*, 2011) (see Figure 15.4e). Differentiation of keratinocytes enables the 3D skin model to recapitulate morphological and biochemical features of human epidermis. Because the skin models formed in MC-8s are supported by two fixed-position bars, the mechanical properties of each tissue can be tested by plunging a force transducer into the tissue. A second important design feature of the MC-8 is optically transparent well bottoms. These permit direct observation as well as spectroscopic and fluorescent analysis. Together, the 3D skin tissue equivalents in the MC-8 format enable one to measure mechanical properties and cellular activity in real time. This system is applicable for cosmetic research and development to investigate premature skin ageing and identify compounds capable of preventing or reversing this process. The same MC-8 technology has been applied to fabricate human heart tissue equivalents with cardiomyocytes derived from human induced pluripotent stem cells (hiPSC). These engineered heart tissues (EHTs) are autonomously paced and recapitulate human cardiac physiology.

To fabricate the MC-8 skin equivalents, a solution containing collagen (rat tail type I) and the desired cells are pipetted into the wells of an MC-8 in which

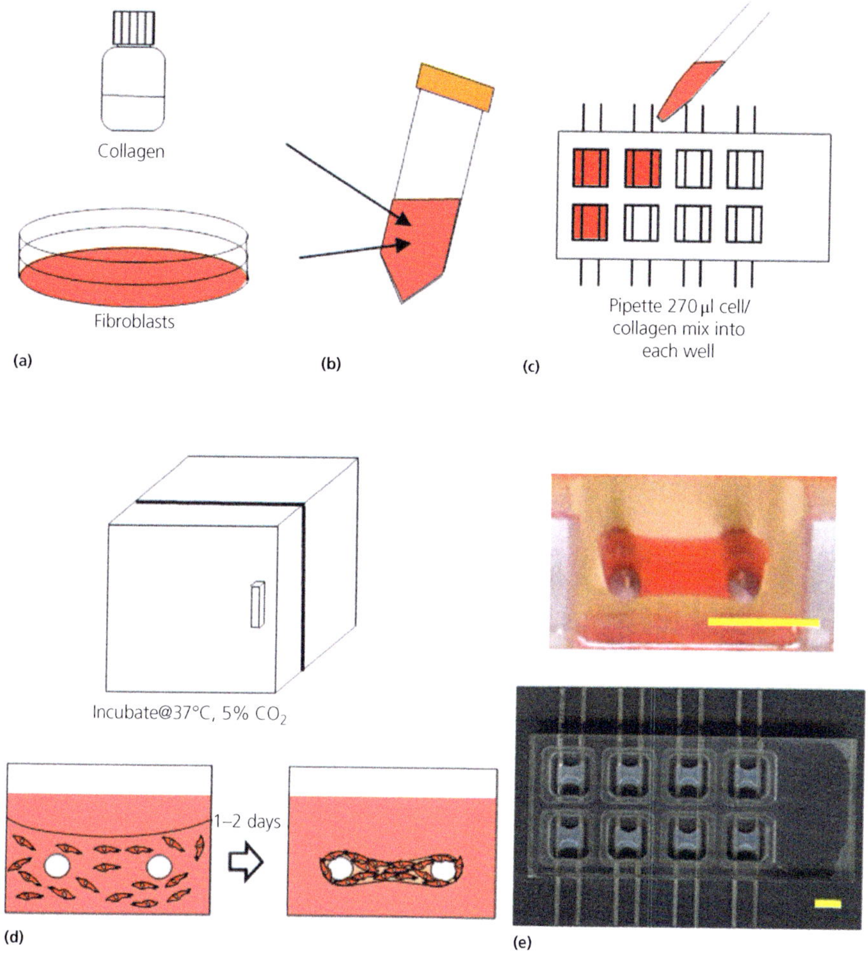

Figure 15.5 Fabrication of skin equivalents in the MC-8 format. (a) Fibroblasts are mixed with collagen to form a gel solution (b) that is pipetted into wells of the MC-8 (c). (d) Over 1–2 days the cells remodel the collagen matrix during 37 °C incubation and form tissue membranes that span the support bars (e) (cells are not drawn in real scale; yellow scale bars = 4 mm).

each contains a pair of horizontal bars to support a 4 × 4 mm tissue sheet (Figure 15.5a–c). Cells remodel the loose collagen gel to form stiff tissues within two days (Figure 15.5d,e). The total media volume required for each well is only ~500 μL, which conserves various assay reagents. Furthermore, using this system, the number of replicates can be easily increased to obtain statistically significant data. Since each construct contains at least 200 000 cells, data obtained from a single tissue represents the average of 200 000 cells. Thus, 3D tissue-based assays require significantly fewer samples (~4 tissues) than cell-based assays.

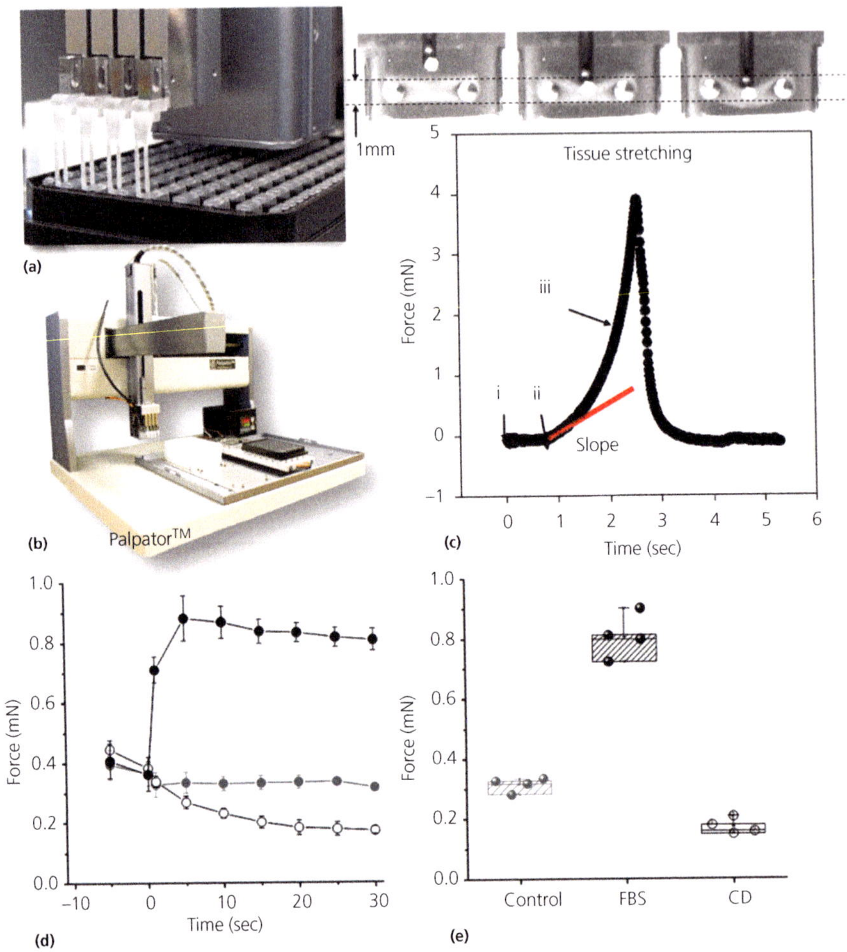

Figure 15.6 Mechanical measurements of MC-8 skin equivalents using the Palpator. (a,b) An X-Y-Z robotic device moves the Palpator head into troughs of ethanol and then water over to the MC-8 supported on a heated stage. (c) The Palpator head positions the probe directly above each tissue and measures the mechanical properties of each tissue with a probe attached to a force transducer by indenting vertically. The transducer records a profile of force response during the indentation. (d) The device can record dynamically changing contractility in response to treatments of 20% FBS or 2 μM cytochalasin D (CD). (e) The data obtained from quadruplicate samples of each condition are highly reproducible. Source: Marquez, J. P., Legant, W., Lam, V. *et al.* (2009).

As noted above, a core feature of the MC-8 or MC-96 (similar to the eight-well chamber but can grow 96 tissues for high-throughput applications) skin equivalents is the ability to collect mechanical measurements of cell force. The mechanical testing device, called the Palpator™ (Figure 15.6a), has an X-Y robotic arm that automatically places an L-shaped force-probe (Figure 15.6b,c) directly above the centre of each well and tissue. The probe indents (0.5 mm/sec) the middle of a tissue, stretching it longitudinally. The other end of the probe is connected to an isometric force transducer (see Figure 15.6c) which is moved by a vertical linear motor (±70 μm accuracy). During testing, the temperature of the tissues is kept at 37 °C using a thermocouple-controlled aluminium heating plate (see Figure 15.6a,b). The device is placed in a laminar flow hood to prevent contamination during long-term testing over the course of hours, days or weeks.

Force generated by cells interacting with the ECM in response to strain can be deconvoluted from the force exerted by the collagen matrix (Marquez *et al.*, 2009). An 'active' cellular force contribution is obtained when the 'passive' matrix mechanical contribution is controlled for. The 'total' tissue force is stimulated in response to calf serum addition whereas disrupting F-actin with cytochalasin D (CD) yields the 'passive' component (Figure 15.6d). Without the force activity of the cytoskeletal elements in the presence of CD, the force contributed by the collagenous matrix is observed and subtracted from the 'total' force to yield an 'active' force describing the cellular force contribution. The results of this type of analysis are highly reproducible (Figure 15.6e). Each engineered tissue (ET) is vertically indented eight times prior to testing, each with a resulting 20% longitudinal stretch. The recorded force-response curve gradually decreases until the third or fourth indentation, after which the measurements converge and further reductions are negligible. This convergence of tissue properties after repeated stretching is a known preconditioning effect observed in biomechanical investigations of animal tissues (Fung, 1993; Wagenseil *et al.* 2004). The repeatability and utility of this assay for high-throughput screening were confirmed to achieve a Z-factor higher than 0.5 (recommended by NIH standards – see Appendix to this chapter).

Data analysis

Recorded forces are used to determine the mechanical properties of the ETs that represent connective tissues. We have developed a mathematical description of the forces in the tissue membrane. As the probe pushes down on the tissue at a constant rate, v (0.5 mm/s), the force transducer records the resistance force, $F(t)$. Treatment of ETs with CD, a drug known to disrupt the actin cytoskeleton (Wakatsuki *et al.*, 2001), reduces $F(t)$ curves in a dose-dependent manner. These curves can be compared graphically to assess different extents of cytoskeletal disruption caused by increasing concentrations of CD. Simple parameterisation of the data curves is useful for analysing many samples in high-throughput screening (HTS) applications.

It is reasonable to assume that for small ET stretching, the ET behaves as a linear elastic material with modulus of elasticity, E. Therefore, we define $\sigma(t) = E\varepsilon(t) + \sigma_0$, where σ_0 is the pre-stress of the ET membrane (this σ_0 is a 'baseline' contractile force exerted by the cells in the ET in the absence of stretching). This simplification allows us to estimate tissue pre-stress and the modulus of elasticity through data fitting (Marquez *et al.*, 2009). By defining $\Delta Z(t) = d(t)/L_0$

and simplifying the equation by using the assumption, i.e. $d(t) << L_o$, F(t) is equal to $4\sigma_0 A\Delta Z(t\text{-}\tau) + 8EA(\Delta Z(t\text{-}\tau))^3$, ($\tau \geq \tau$ and $F(t) = 0$ for $t < \tau$) and A is a cross-sectional area of tissue. We have applied this analysis to fundamental material properties of tissues.

Example of application

Dose determinations

We previously published the effects of CD and latrunculin B (LA-B) on active cellular and tissue force using large (ring-shaped) ETs (Wakatsuki & Elson, 2003; Wakatsuki *et al.*, 2000, 2001, 2003). Here, we describe the comparison of ring ETs versus MC-8 ETs treated with CD and LA-B (Figure 15.7). For the CD experiment, samples were pretreated with 20% FBS to activate, i.e. contract, tissues (indicated as control in Figure 15.7a) before 0.02, 0.2 and 2 μM CD treatments. We plotted a single data point from each curve at 1 mm indentation against the CD dose (Figure 15.7b) and compared these results with published data. The dose-dependent inhibition of force was remarkably similar to that obtained using the larger ring system. The statistical significance of the data from the 96-well format exceeded that from the publication (eight samples), because the data were averaged over 32 samples. The magnitude of force obtained by the mini-tissue is 0.4–0.9 mN, very similar to the force measured with ring tissues (0.2–0.8 mN).

A similar study was performed using various doses of LA-B, which disrupts actin filaments by sequestering actin monomers (Figure 15.7c). We also generated a dose-dependence curve (Figure 15.7d) and compared the results with the data from the ring tissues (Wakatsuki *et al.*, 2001). The dose-dependence curve observed with the mini-tissues closely resembles that seen with the ring system. The estimated IC_{50} was ~2 μM for mini-tissues and ~1 μM for ring tissues. The small discrepancy may be due to different pretreatment of samples before taking measurements (the ring system was preincubated with 20% FBS and the mini-tissues with 10% FBS).

Toxicity assessments with multiple parameters

To assess chemical toxicity with the ET-based assay system, four ETs were treated with varying concentrations of the mitochondrial uncoupler and known metabolic poison, dinitrophenol (DNP), for 18 hours and then tissue force was measured with the Palpator (Figure 15.8). Eight concentrations of DNP were used: 0 (shown in figure as 1), 15, 50, 100, 150, 500, 1500 and 3000 μM. Tetramethylrhodamine ethyl ester (TMRE) was used to simultaneously follow mitochondrial membrane potential, an indicator of metabolism and viability. TMRE signal and ET contractility exhibited drops with increasing concentrations of DNP. By fitting a sigmoid function, EC_{50} values of 55 and 109 μM were

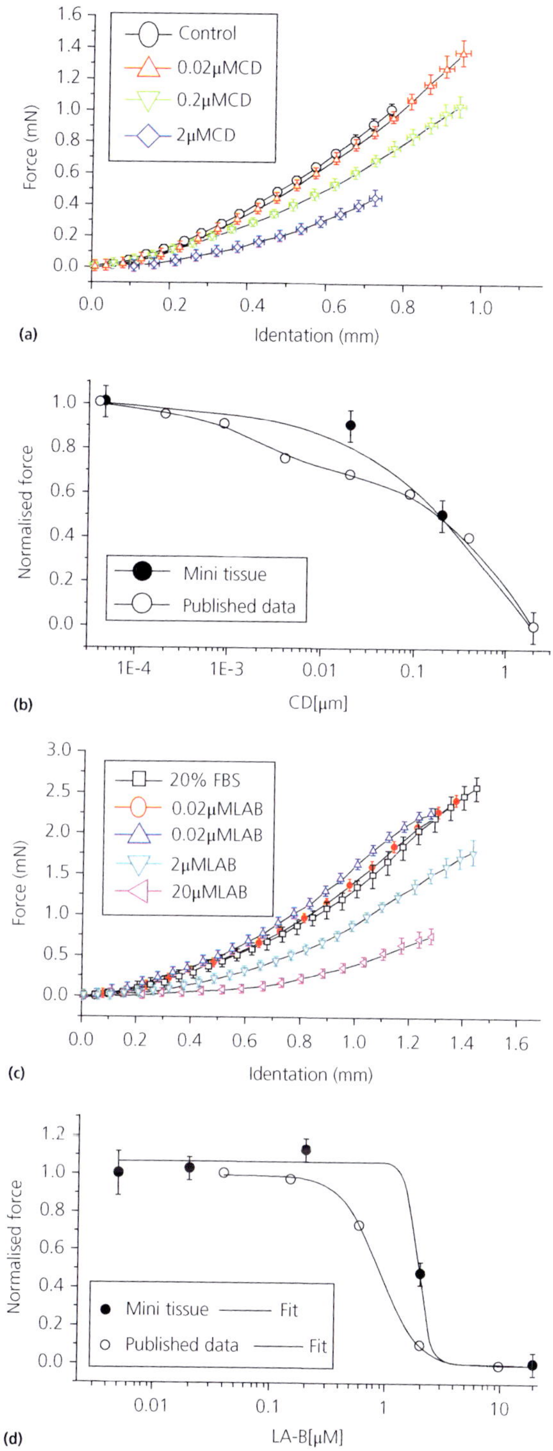

Figure 15.7 Comparison of mechanical measurements of ring and mini-tissue constructs after treatment with CD and LA-B. (a,c) Dose-dependent reduction of the force curves after CD and LA-B treatment on mini-ETs. CD (b) and LA-B (d) dose-dependent reduction of force of mini-ET were compared to published data obtained using ring constructs.

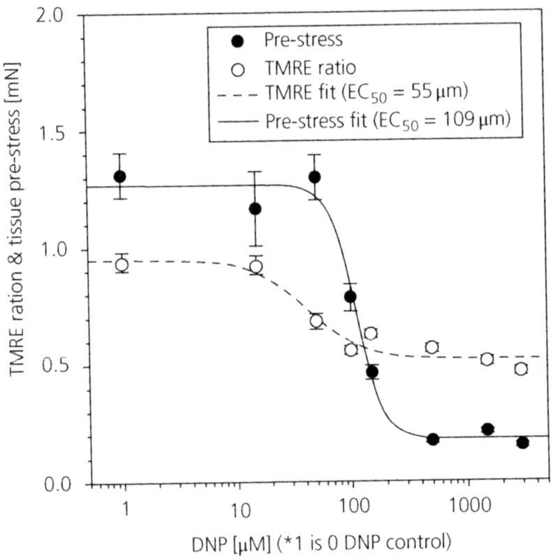

Figure 15.8 Toxicity assay for chemicals. ETs were treated with DNP (0 – shown as 1, 15, 50, 100, 150, 500, 1500 and 3000 μM) for 18 hours and then tested with the Palpator to measure force and a plated reader to measure TMRE signal representing mitochondrial membrane potential (n = 4, standard error bars shown).

estimated for TMRE fluorescence and ET contractility, respectively (see Figure 15.8). A known lethal dose, 107–270 μM, and recommended dose, 27–43 μM, as a diet pill are in a similar range as the EC_{50} to reduce TMRE fluorescence and ET contractility. This result suggests that if DNP's concentration is high enough to reduce cellular mitochondrial potential yet low enough to maintain cellular contractility, DNP could be effective to burn energy without producing ATP and yet maintain contractile tone. However, the exact mechanism by which DNP induces the loss of tissue contractility remains to be elucidated. Even though the loss in the TMRE signal, which indicates a drop in mitochondrial potential, was the expected result of DNP treatment, the loss of tissue contractility in response to DNP treatment is a novel experimental result that is unique to our ET and testing system.

Myofibroblast transformation and contractility

Fibroblasts in primary cell culture on two-dimensional (2D) surfaces transform into myofibroblasts expressing α-smooth muscle actin (Hinz, 2007). Myofibroblasts are the major cell type that appears during injury to promote healthy wound healing. However, the sustained activity of myofibroblasts can lead to fibrotic diseases in various organs and tissues (Wynn, 2007). Growth factors and hormones in serum or secreted by cells into nearby extracellular spaces can drive the myofibroblast transformation (Hinz, 2007). In addition to paracrine signalling, a mechanical signal can also promote the transformation. Because the substrate in conventional 2D culture is almost infinitely stiff

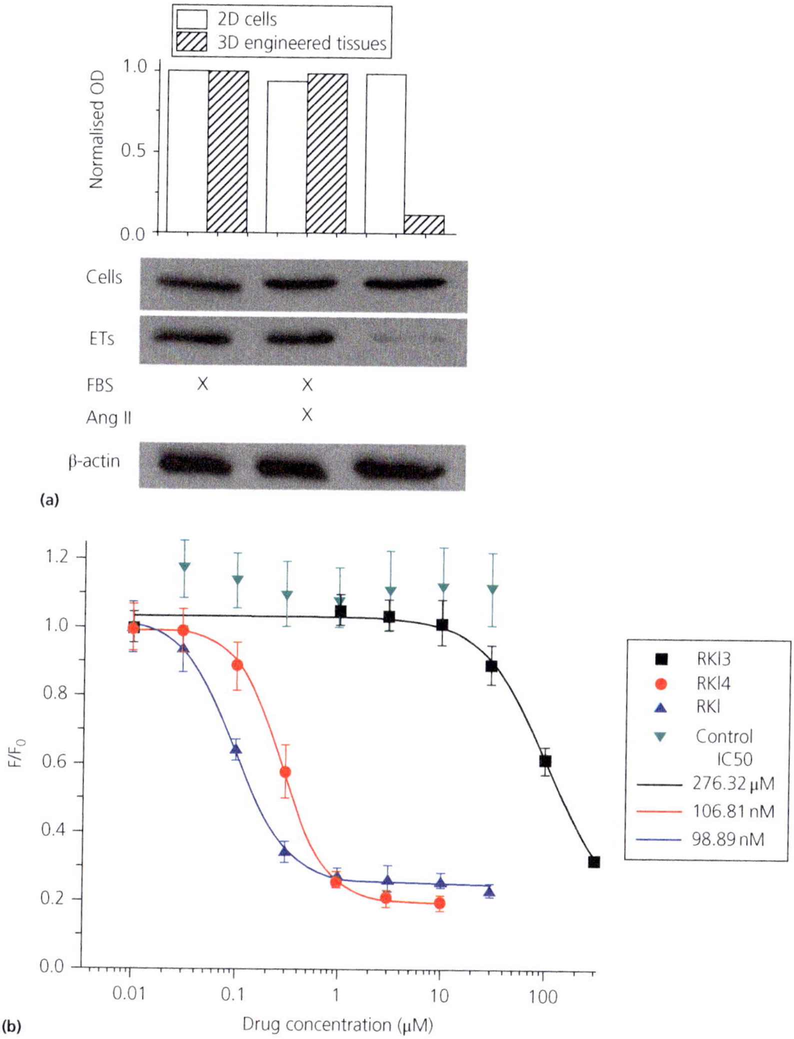

Figure 15.9 Myofibroblasts and ROCK-mediated contractility. (a) Western blot analysis showed that conventional 2D cell culture promoted an increased expression of α-smooth muscle actin, i.e. myofibroblast transformation, without any FBS or angiotensin II. Presence of FBS (10%) and angiotensin II (10 uM) promoted myofibroblast transformation in 2D and 3D culture. (b) Different types of ROCK inhibitors reduce contractility of ETs dose dependently. RKI: H1152, RKI4: Rockout, RKI3: see text. Error bars = S.E. n = 4.

compared to the ECM *in vivo*, cells cultured in 2D eventually transform into myofibroblasts. We confirmed that human dermal fibroblasts growing in conventional 2D culture transformed into myofibroblasts even in the absence of FBS. Fibroblasts cultured in 3D collagen hydrogels using the MC-8 had suppressed expression of α-smooth muscle actin and addition of FBS (10%) or angiotensin II (10 μM) promoted myofibroblast transformation (Figure 15.9a).

Rho-associated coiled-coil protein kinase (ROCK) is one of the first effectors characterised downstream of RhoA, and two isoforms are known: ROCK1 and ROCK2. Binding of either the Rho binding domain or the pleckstrin homology domain to the amino-terminal catalytic region inhibits the kinase activity (Feng *et al.*, 1999). Activated, GTP-bound Rho interacts with the Rho binding domain, activating the phosphotransferase activity of ROCK by disrupting the interaction between the C-terminal regulatory region and the N-terminal catalytic site (Amano *et al.*, 2000), releasing the catalytic site from autoinhibition. Both ROCK isoforms phosphorylate a number of downstream substrates and are involved in actin cytoskeleton regulation, cell migration, adhesion and contraction (Riento & Ridley, 2003). Small molecule inhibitors of ROCK, including derivatives of isoquinoline-sulfonamide, were developed, and fasudil is the first in this class of kinase inhibitors to be used clinically in Japan to treat cerebral vasospasm (Zhao *et al.*, 2011).

Because ROCK regulates contraction of non-muscle and smooth muscle cells, we applied our phenotypic screening system to analyse the effectiveness of various ROCK inhibitors. As described above, ETs were fabricated with human dermal fibroblasts and with FBS transformed into myofibroblasts. Using the Palpator, the dose-dependent effectiveness of ROCK inhibitors was analysed. 3-(4-Pyridyl)-1H-indole or Rockout (RKI3) (Yarrow *et al.*, 2005) was not as effective as two other derivatives of isoquinoline-sulfonamide, (S)-hexahydro-4-glycyl-2-methyl-1-(4-methylisoquinoline-5-sulfonyl)-1H-1,4-diazepine (RKI4) (Tamura *et al.*, 2005) and H-1152P (RKI) (Sasaki *et al.*, 2002), at reducing contractile force (Figure 15.9b). This example demonstrates the ability of the Palpator compound screening system to analyse modulators of myofibroblast contractility and to remodel the ECM, which contributes to the ageing of skin and fibrotic skin diseases as discussed above.

Troubleshooting

- Most types of tissues and organs can be recapitulated using 3D hydrogels and substrate materials can be natural or synthetic polymers.
- Skeletal, cardiac and smooth muscles can be reconstituted in the MC-8 and are suitable for mechanical testing using the Palpator.
- Mechanical properties of the cell and ECM can be separately analysed using published protocols (Marquez *et al.*, 2009; Wakatsuki *et al.*, 2000).
- Immunohistochemical analysis of proteins can be performed in ETs using techniques established for tissue explants (e.g. paraffin embedding, frozen tissue sectioning).

Conclusion

The key to solving a cosmetic development goal or introducing a new therapy for a skin disorder is a representative model. Instead of testing compounds on animals or even generic cell-based skin tissue models, in the near future *in vitro* tissue models will be fabricated with cells isolated from a specific person. One clear benefit is a skin tissue model that can recapitulate rare genetic disorders. For instance, 3D tissues could model skin from individuals with wrinkly skin syndrome, who develop signs of premature ageing at a young age (Reversade *et al.*, 2009). In combination with models of other maladies, high-throughput tissue screening may reveal basic properties underlying these conditions, helping to improve treatment strategies. While agents that reduce fibrosis or improve ECM integrity will be discovered with 3D tissue equivalents, individual specific models will allow for ideal combinations of therapies to be identified similar to how patient-specific cardiomyopathies have been addressed (Liang *et al.*, 2013).

We are beginning to understand the molecular mechanisms by which UV irradiation accelerates skin ageing. Recent advances in tissue engineering and human genome sequencing support development of innovative cosmetic products that can delay and possibly reverse physiological changes in UV-induced ageing skin. Automated high-throughput tissue modelling with generic, limited or even broad genetic backgrounds will hopefully produce novel cosmetics, antiageing treatments and therapies for those who currently have no satisfactory option.

Acknowledgements

This work has been supported in part by NIH grants R44 GM087784 and R01 HL109505. We appreciate the scientific collaboration with Dr Zelmira Lazarova in the Department of Dermatology at the Medical College of Wisconsin for developing skin tissue equivalents in the MC-8.

Appendix

Data reproducibility and statistical significance

To test the reproducibility of the force measurements, four identically prepared ETs grown in different wells of the same chamber were tested. Four force-curves showed slight discrepancies (Figure 15.A1a). However, once the curves were each shifted based on the time that the force probe touched each ET, the curves overlapped indistinguishably (Figure 15.A1b). This demonstrates that the forces measured in the ETs were highly reproducible and that ETs prepared similarly have almost identical mechanical properties.

The Z-factor

The Z-factor is a useful statistical parameter for comparing and evaluating the quality of high-throughput assays, and can be used to evaluate assay optimisation and validation methods (Zhang *et al.*, 1999). The Z-factor is described by the

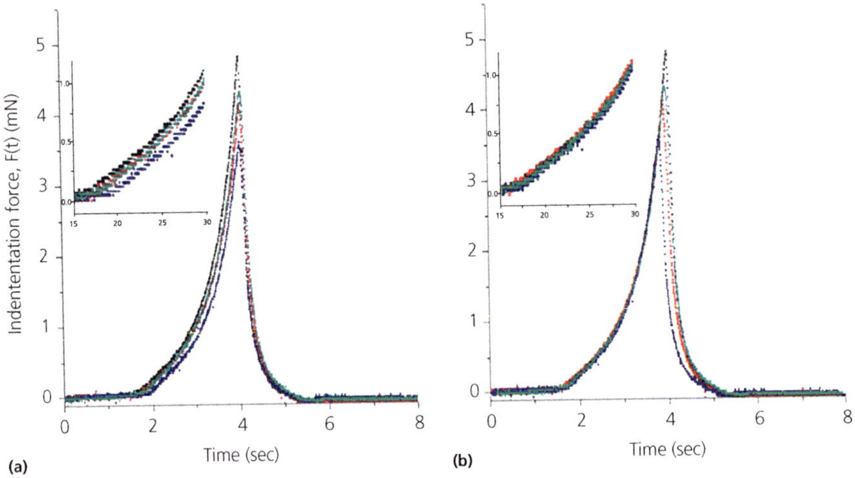

Figure 15.A1 High data reproducibility. (a) Force curves from four different constructs show slight variation in profile (inset shows initial deviation from background forces). (b) Following τ correction (force profiles are time shifted to correct for probe-tissue contact time), force profiles overlapped to show high reproducibility of data acquisition.

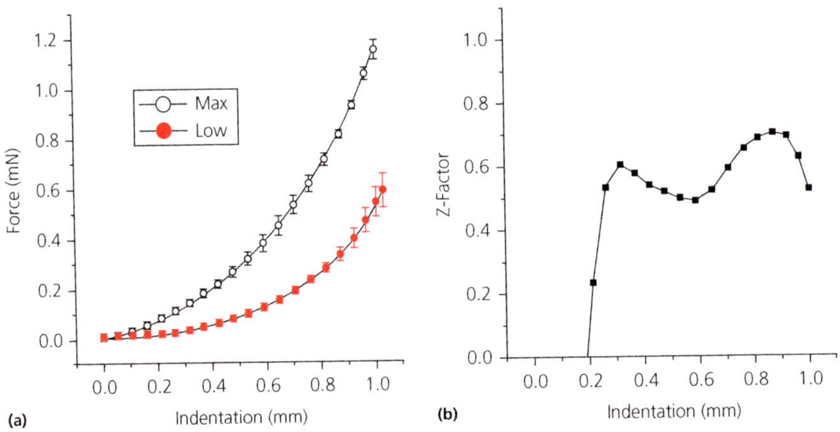

Figure 15.A2 Z-factor value of the indentation method. Maximally activated and control samples were indented and the force was recorded. (a) The Z-factor value was calculated for each indentation point. (b) Indentations between 0.3 and 1 mm have excellent Z-factor values (0.5–1.0).

difference between the means of the positive (μ_p) and negative (μ_n) controls and their standard deviations (σ_p, σ_m) as follows:

$$Zfactor = 1 - 3 \times \left(\sigma_p + \sigma_n\right) / \left|\mu_p - \mu_n\right|$$

A Z-factor value of 1 is the ideal; an assay exhibiting a Z-factor between 0.5 and 1 is considered excellent for high-throughput screening. Figure 15.A2a shows data from a test indentation assay with mini-tissues prepared from rat fibroblasts. Many of the data points obtained from maximally activated ETs (white circles) are very different from those obtained from ETs in which CD abolished active cellular force (red circles). Although the Z-factors for very small indentations are less than 0.5 (Figure 15.A2b), the indentation points between 0.3 and 1 mm have excellent Z-factors.

References

Amano, M., Fukata, Y. & Kaibuchi, K. (2000) Regulation and functions of Rho-associated kinase. *Experimental Cell Research*, **261**, 44–51.

Brohem, C. A., Cardeal, L. B., Tiago, M. *et al.* (2011) Artificial skin in perspective: concepts and applications. *Pigment Cell & Melanoma Research*, **24**, 35–50.

Bruckner-Tuderman, L. (2010) Systemic therapy for a genetic skin disease. *New England Journal of Medicine*, **363**, 680–682.

Cooper, T. W. & Bauer, E. A. (1984) Epidermolysis bullosa: a review. *Pediatric Dermatology*, **1**, 181–188.

Escoffier, C., de Rigal, J., Rochefort, A. *et al.* (1989) Age-related mechanical properties of human skin: an *in vivo* study. *Journal of Investigative Dermatology*, **93**, 353–357.

Farage, M. A., Miller, K. W., Elsner, P. & Maibach, H. I. (2013) Characteristics of the aging skin. *Advances in Wound Care*, **2**, 5–10.

Feng, J., Ito, M., Kureishi, Y. *et al.* (1999) Rho-associated kinase of chicken gizzard smooth muscle. *Journal of Biological Chemistry*, **274**, 3744–3752.

Fisher, G. J., Datta, S., Wang, Z. *et al.* (2000) c-Jun-dependent inhibition of cutaneous procollagen transcription following ultraviolet irradiation is reversed by all-trans retinoic acid. *Journal of Clinical Investigation*, **106**, 663–670.

Fisher, G. J., Varani, J. & Voorhees, J. J. (2008) Looking older: fibroblast collapse and therapeutic implications. *Archives of Dermatology*, **144**, 666–672.

Fligiel, S. E. G., Varani, J., Datta, S. C. *et al.* (2003) Collagen degradation in aged photodamaged skin *in vivo* and after exposure to matrix metalloproteinase-1 *in vitro*. *Journal of Investigative Dermatology*, **120**, 842–848.

Fung, Y. C. (1993) *Biomechanics: Mechanical Properties of Living Tissues*. Springer-Verlag, New York.

Gabrielli, A., Avvedimento, E. V. & Krieg, T. (2009) Scleroderma. *New England Journal of Medicine*, **360**, 1989–2003.

Grinnell, F., Toda, K. & Takashima, A. (1987) Activation of keratinocyte fibronectin receptor function during cutaneous wound healing. *Journal of Cell Science*, **8**, 199–209.

Hinz, B. (2007) Formation and function of the myofibroblast during tissue repair. *Journal of Investigative Dermatology*, **127**, 526–537.

Liang, P., Lan F., Lee, A. S. *et al.* (2013) Drug screening using a library of human induced pluripotent stem cell-derived cardiomyocytes reveals disease-specific patterns of cardiotoxicity. *Circulation*, **127**, 1677–1691.

Marquez, J. P., Legant, W., Lam, V. *et al.* (2009) High-throughput measurements of hydrogel tissue construct mechanics. *Tissue Engineering Part C Methods*, **15**, 181–190.

McLafferty, E., Hendry, C. & Farley, A. (2012) The integumentary system: anatomy, physiology and function of skin. *Nursing Standard*, **27**, 35–42.

Milano, A., Pendergrass, S. A., Sargent, J. L. *et al.* (2008) Molecular subsets in the gene expression signatures of scleroderma skin. *Plos One*, **3**, E2696.

Miyachi, Y. & Ishikawa, O. (1998) Dermal connective tissue metabolism in photoageing. *Australasian Journal of Dermatology*, **39**, 19–23.

Naylor, E. C., Watson, R. E. B. & Sherratt, M. J. (2011) Molecular aspects of skin ageing. *Maturitas*, **69**, 249–256.

Ogawa, R. & Hsu, C. K. (2013) Mechanobiological dysregulation of the epidermis and dermis in skin disorders and in degeneration. *Journal of Cell Molecular Medicine*, **17**, 817–822.

Pailler-Mattei, C., Bec, S. & Zahouani, H. (2008) *In vivo* measurements of the elastic mechanical properties of human skin by indentation tests. *Medical Engineering & Physics*, **30**, 599–606.

Pauwels, M. & Rogiers, V. (2010) Human health safety evaluation of cosmetics in the EU: a legally imposed challenge to science. *Toxicology and Applied Pharmacology*, **243**, 260–274.

Pilcher, B. K., Levine, N. S. & Tomasek, J. J. (1995) Thrombin promotion of isometric contraction in fibroblasts: its extracellular mechanism of action. *Plastic and Reconstructive Surgery*, **96**, 1188–1195.

Quan, T., Qin, Z., Xia, W. *et al.* (2009) Matrix-degrading metalloproteinases in photoageing. *Journal of Investigative Dermatology*, **14**, 20–24.

Reich, A., Meurer, M., Eckes, B. *et al.* (2009) Surface morphology and mechanical properties of fibroblasts from scleroderma patients. *Journal of Cellular & Molecular Medicine*, **13**, 1644–1652.

Reversade, B., Escande-Beillard, N., Dimopoulou, A. *et al.* (2009) Mutations in Pycr1 cause cutis laxa with progeroid features. *Nature Genetics*, **41**, 1016–1021.

Riento, K. & Ridley, A. J. (2003) Rocks: multifunctional kinases in cell behaviour. *Nature Reviews Molecular Cell Biology*, **4**, 446–456.

Sarifakioglu, N., Terzioğlu, A., Levent, A. & Aslan, G. (2004) A new phenomenon: 'sleep lines' on the face. *Scandinavian Journal of Plastic and Reconstructive Surgery and Hand Surgery*, **38**, 244–247.

Sasaki, Y., Suzuki, M. & Hidaka, H. (2002) The novel and specific Rho-kinase inhibitor (S)-(+)-2-methyl-1-[(4-methyl-5-isoquinoline)sulfonyl]-homopiperazine as a probing molecule for Rho-kinase-involved pathway. *Pharmacology & Therapeutics*, **93**, 225–232.

Tamura, M., Nakao, H., Yoshizaki, H. *et al.* (2005) Development of specific Rho-kinase inhibitors and their clinical application. *Biochimica et Biophysica Acta – Proteins And Proteomics*, **1754**, 245–252.

Varani, J., Spearman, D., Perone, P. *et al.* (2001) Inhibition of type I procollagen synthesis by damaged collagen in photoaged skin and by collagenase-degraded collagen *in vitro*. *American Journal of Pathology*, **158**, 931–942.

Varani, J., Schuger, L., Dame, M. K. *et al.* (2004) Reduced fibroblast interaction with intact collagen as a mechanism for depressed collagen in synthesis in photodamaged skin. *Journal of Investigative Dermatology*, **122**, 1471–1479.

Wagenseil, J. E., Elson, E. L. & Okamoto, R. J. (2004) Cell orientation influences the biaxial mechanical properties of fibroblast populated collagen vessels. *Annals of Biomedical Engineering*, **32**, 720–731.

Wakatsuki, T. & Elson, E.L. (2003) Reciprocal interactions between cells and extracellular matrix during remodeling of tissue constructs. *Biophysical Chemistry*, **100**, 593–605.

Wakatsuki, T., Kolodney, M. S., Zahalak, G. I. & Elson, E. L. (2000) Cell mechanics studied by a reconstituted model tissue. *Biophysical Journal*, **79**, 2353–2368.

Wakatsuki, T., Schwab, B., Thompson, N. C. & Elson, E. L. (2001) Effects of cytochalasin D and latrunculin B on mechanical properties of cells. *Journal of Cell Science*, **114**, 1025–1036.

Wakatsuki, T., Wysolmerski, R. B. & Elson, E. L. (2003) Mechanics of cell spreading: role of myosin Ii. *Journal of Cell Science*, **116**, 1617–1625.

Wynn, T. A. (2007) Common and unique mechanisms regulate fibrosis in various fibroproliferative diseases. *Journal of Clinical Investigation*, **117**, 524–529.

Yarrow, J. C., Totsukawa, G., Charras, G. T. & Mitchinson, T. J (2005) Screening for cell migration inhibitors via automated microscopy reveals a Rho-kinase inhibitor. *Chemistry & Biology*, **12**, 385–395.

Zhang, J. H., Chung, T. D. & Oldenburg, K. R. (1999) A simple statistical parameter for use in evaluation and validation of high throughput screening assays. *Journal of Biomolecular Screening*, **4**, 67–73.

Zhao, J., Zhou, D. Guo, J. *et al.* (2011) Efficacy and safety of fasudil in patients with subarachnoid hemorrhage: final results of a randomized trial of fasudil versus nimodipine. *Neurologia Medico-Chirurgica*, **51**, 679–683.

CHAPTER 16

Three-dimensional cell culture in the Rotary Cell Culture System™

Stephen S. Navran

Synthecon, Inc., Houston, Texas, USA

Introduction

Cell culture has been used as a tool in biological research for over 100 years (Harrison, 1907). For most of this period, cells have been grown in two-dimensional (2D) monolayers. It is only in the last 10–15 years that it has come to be appreciated that this environment is unnatural and that three-dimensional (3D) cell culture more closely resembles the conditions experienced by cells *in vivo* (Abbott, 2003; Griffith & Swartz, 2006; Schmeichel & Bissell, 2003). A variety of different approaches to 3D cell culture have been developed, some of which are discussed in other chapters of this book.

The subject of this chapter, the Rotary Cell Culture System™, was not originally intended to address the challenges of growing cells in 3D. In the mid 1980s the National Aeronautics and Space Administration (NASA) came to recognise that prolonged exposure to microgravity was producing serious adverse effects on the health of astronauts. These problems included immunosuppression, bone loss and muscle atrophy. In order for human exploration in space to continue, NASA realised that these issues had to be addressed. Consequently, a research programme was initiated to study human physiology in microgravity and develop countermeasures. One part of this programme was to create a ground-based simulation to study the effects of microgravity at the cellular level. Out of this came a new bioreactor design employing a horizontally rotating culture vessel with a coaxial tubular oxygenator. The bioreactor, known in NASA terminology as the rotating wall vessel (RWV), is completely filled with medium, leaving no head space, and upon initiation of rotation, the medium rotates as a solid body. The resolution of forces acting upon cells during the rotation induces a slow sedimentation and hence an approximation of microgravity (Wolf & Schwartz, 1991, 1992). A more detailed discussion of the fluid mechanics of the RWV and how it has facilitated 3D cell culture is outlined below.

Technology Platforms for 3D Cell Culture: A User's Guide, First Edition. Edited by Stefan Przyborski.

Whether the conditions created by the RWV are a true simulation of microgravity is questionable (Hammond & Hammond, 2001). The microgravity experienced by astronauts in orbit is due to the reduction of gravity from being at a distance from the earth, whereas the rotation of the RWV produces radial and tangential forces that oppose the force of gravity. Nevertheless, this technology is the best model available for ground-based microgravity experiments.

Apart from its use as a model for microgravity, the RWV has found its greatest benefit as a unique approach to generation and maintenance of 3D cell cultures. The particular characteristics of the RWV that differentiate it from other methods of 3D culture are the dynamic nature of the bioreactor (rotation) and the suspension of cells with negligible shear stress. Cells suspended by rotation in such a low shear stress environment can readily aggregate into 3D spheroids because the forces that would tend to pull them apart have been minimised.

The cell culture environment within the RWV can be summarised in three basic principles:

1 **Sedimentation**. As the bioreactor vessel rotates, particles (dispersed cells, cell aggregates or cells attached to scaffolds) accelerate until they reach terminal (sedimentation) velocity where the gravitational force is partially counterbalanced by centrifugal and Coriolis forces (Figure 16.1). The primary determinant of sedimentation velocity is the size of the particle which, according to the Stokes equation, increases in proportion to the square of the radius of the particle (Wolf & Schwartz, 1991). Therefore, as cell aggregates grow in size,

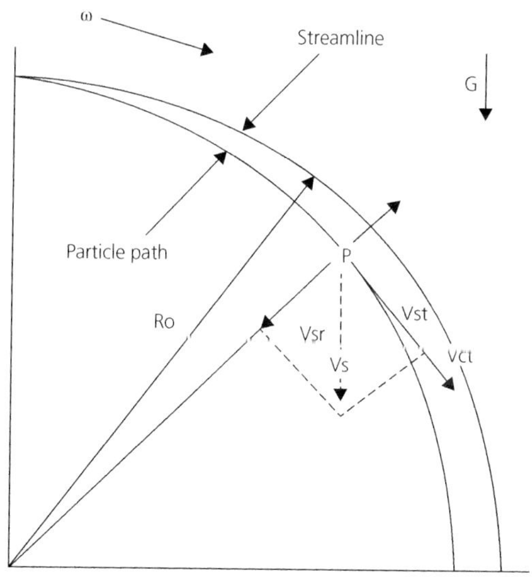

Figure 16.1 **Vector velocity diagram shows the individual velocity vectors contributing to the particle motion through the medium.** The sedimentation velocity Vs can be resolved into radial (Vsr) and tangential (Vst) components. There is an outwardly directed vector due to centrifugal force. Tangential Coriolis-induced motion leads to spiralling of the particles.

they will sediment more rapidly and it is necessary to increase the rotational speed of the RWV in order to keep the aggregates suspended and avoid damage from wall collisions.

2 **Coriolis spiralling**. As the RWV rotates, the radial force produces a secondary spiralling of the particles. The relationship between the amplitude of these spirals and the rotational speed of the RWV is shown in Figure 16.2 (see Wolf & Schwartz for the derivation of this graph). At low rotational speed, the spirals are large and result in many wall collisions. As the speed of rotation is increased, the spirals decrease in amplitude and wall collisions are reduced. At very high rotational speeds, the effect of centrifugal force becomes dominant and wall collisions rise. Therefore, an optimal rotational speed must be determined that minimises wall collisions, i.e. at the intersection of the two lines in Figure 16.2. At first glance, determination of the optimal speed might seem difficult but in practice, it is quite simple. As the aggregates increase in size to a point where adjusting the rotational speed is necessary to keep them in suspension, they become visible to the naked eye and the speed can be optimised by observing the motion of the aggregate through the clear wall of the RWV.

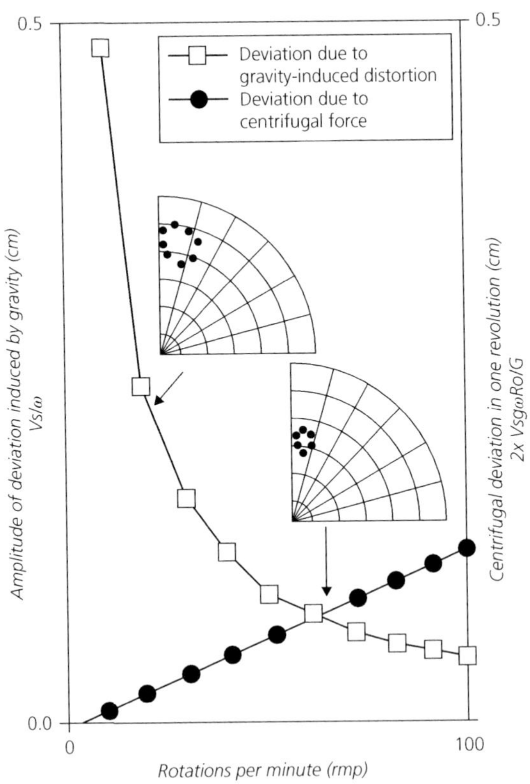

Figure 16.2 Relationship between centrifugal and gravity-induced distortions and rotational speed. The radius of Coriolis-induced spiralling is reduced by increasing rotational speed but as the speed increases, the particles will eventually collide with the vessel wall.

3 **Mass transport**. Nutrient and oxygen delivery to and metabolic waste removal from 3D cell aggregates are critically important to their survival and function in culture. The availability of nutrients and oxygen to cell aggregates in suspension is dependent on the relative contribution of diffusion versus convective bulk flow. As cell aggregates sediment through the medium in the RWV, nutrients and oxygen are continuously supplied by convective bulk flow. When sedimentation velocity is very low or approaching zero, as in the case of single cell suspensions or very small aggregates, the availability of nutrients and oxygen is limited to diffusion.

During early work on the the RWV, several members of the development team decided to start Synthecon, Inc. as a commercial entity to make the bioreactor available to the larger research community. The commercial version was renamed the Rotary Cell Culture System™ (RCCS). Although the original NASA terminology, RWV, is still used in the literature, they are equivalent. In the ensuing years, a number of variations of the basic technology have been developed. For example, there are two configurations of the bioreactor vessel (Figure 16.3): the STLV (slow turning lateral vessel), which was the original cylindrical vessel with the gas exchange membrane in the centre, and the HARV (high aspect ratio vessel), a disc-shaped vessel with the membrane on the back of the vessel, which was developed later. The HARV can accommodate larger scaffolds than the STLV. Synthecon has also developed disposable versions of the HARV. Unlike the STLV and HARV, which are batch culture systems, the Rotary Cell Culture Max is a perfusion system that can be scaled up to several litres (Figure 16.4).

Investigators in many disciplines have devised 3D cell culture applications for the RCCS. A few representative examples will be presented in this chapter. A complete list of publications can be found in the bibliography section at www.synthecon.com. One way of categorising the methods for culturing cells in the RCCS is by how the cells are organised, i.e. as spheroids, as attached to various types of scaffolds, as co-cultures or as tissue explants.

(a) (b)

Figure 16.3 The two configurations of the RCCS. (a) Slow turning lateral vessel (STLV) and (b) high aspect ratio vessel (HARV).

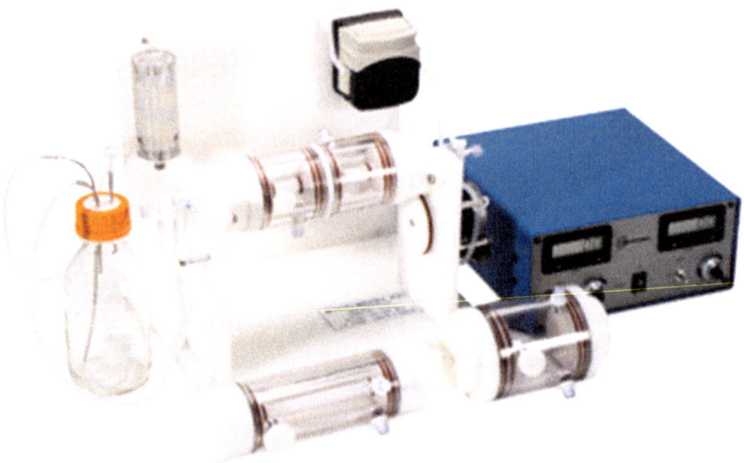

Figure 16.4 **Externally perfused RCCS.** The perfused version of the RCCS is composed of a power supply, rotator base and medium reservoir. Bioreactor vessels are available in various sizes and configurations as shown.

Spheroid assembly in the RCCS

Almost any anchorage-dependent cell will form a spheroid in the RCCS. The low shear stress environment allows cells to come in contact and adhere to each other in suspension via surface adhesion molecules. As the culture is continued, the cells will proliferate and secrete extracellular matrix, creating a more tissue-like structure. One example is RCCS-generated tumour spheroids which have been widely used in cancer research to study basic mechanisms of tumour development and drug responses (Nakamura *et al.*, 2002; Vamvakidou *et al.*, 2007).

Another example is cartilage, where the RCCS environment is highly effective in producing a construct resembling the natural tissue (Sakai *et al.*, 2009; Yu *et al.*, 2011). A third example is the production of embryoid bodies from embryonic stem cells (Gerecht-Nir *et al.*, 2004). In this study, it was demonstrated that the RCCS, in particular the STLV, was particularly efficient in producing embryoid bodies compared to static culture.

The use of scaffolds in the RCCS

Some of the earliest experiments with the RWV at the NASA Johnson Space Center involved the use of microcarrier beads. Interestingly, it was found that the microcarriers and cells could be introduced into the bioreactor vessel separately and the cells would attach to the beads while the vessel rotated. This observation was another demonstration of the benefits of the low shear, microgravity-like environment in facilitating the attachment of cells to biocompatible materials. Investigators have used many types of scaffolds, including porous, fibrous and hydrogels made from both natural and artificial materials, to culture cells in 3D in the RCCS. The only requirements are that the materials must allow cell attachment and the density and size of the scaffolds must be such that they do not sediment too rapidly. For example, the RCCS promotes the differentiation of embryonic stem cells into hepatocytes cultured on biodegradable scaffolds (Wang *et al.*, 2012). In a recent study, Lei *et al.* (2011) showed that epidermal stem cells cultured on microcarriers grew rapidly, maintained their stem cell phenotype and exhibited an epidermal structure in the RCCS, in contrast to cells in static culture which grew much more slowly, tended to become terminally differentiated and grew as a monolayer. A number of investigators have found advantages in culturing encapsulated cells in the RCCS (Lin *et al.*, 2004; Ma *et al.*, 2008; Yuan *et al.*, 2013). A detailed examination of the latter two papers will be provided later in this chapter.

Co-culture in the RCCS

Since tissues almost never contain a singe cell type, it is of great interest when culturing cells in 3D to be able to recapitulate the complexity of the naturally occurring structure. By introducing multiple cell types into the RCCS, this objective can be readily achieved. Two examples are in cancer research, where Sung *et al.* (2008) demonstrated that in co-cultures of prostate cancer and stromal cells grown in the RCCS, the growth and metastatic potential were enhanced compared to a simple monoculture, and in islet transplantation (Qureshi *et al.*, 2015), where it was shown that dispersed cells from pancreatic islets can be recombined with amniotic epithelial cells in the RCCS to make functional spheroids which exhibit improved insulin secretion and are immunoprotected by the epithelial cells.

Tissue explant culture in the RCCS

In instances where culturing intact tissues is desirable, the RCCS has been shown to be quite effective. Attempts to culture tissue explants in static culture have proven to be very challenging. The structure of the tissue tends to break down as the tissues make contact with a surface and some cells migrate out of the tissue onto the surface while other cells within the tissue may become necrotic from lack of access to nutrients and oxygen. On the other hand, in the RCCS the tissues do not come in contact with any surfaces and the dynamic environment of the bioreactor is sufficient to maintain oxygen and nutrient exchange provided that the explants are not too large. Explant culture in the RCCS has been used in cancer research to study multiple myeloma and responses to therapeutic agents (Ferrarini *et al.*, 2013), in the study of HIV infection in intact lymphoid tissue (Margolis *et al.*, 1997) and in the study of the developmental biology of zebrafish (Shimada *et al.*, 2005).

Description and operation of the RCCS

Batch culture systems

The Rotary Cell Culture System, the commercial version of the NASA-designed RWV, consists of either a cylindrical (STLV) or a disc-shaped (HARV) culture vessel, a motorised base which rotates the vessel and a power supply to drive the motor. A control to regulate the rotational speed is built into the power supply unit. The culture vessels employ a silicone membrane either in the centre (STLV) or on the back (HARV) of the vessel to enable gas transfer between the external environment and the culture medium in the vessel. The silicone membrane has no pores and transmits gas by diffusion without the formation of bubbles. The culture vessels are attached to the rotator base via a threaded shaft. The rotator base also contains a pump which directs the ambient gas through the centre of the shaft against the membrane, applying a small pressure which improves gas transport to the culture medium. The rotator base and culture vessel are placed in a humidified, CO_2 incubator. The power supply is located outside the incubator and is connected to the rotator base with a ribbon cable. The cable is thin enough so that it will not hinder closing of the door.

The bioreactor vessels have clear polycarbonate walls and Delrin endcaps and can be autoclaved many times. An exception is the disposable HARV which is γ-sterilised and intended for single use.

Cell seeding and medium changes

Cells, scaffolds and medium are added through the large fill port on the endcap of the STLV (Figure 16.5a) or on the front face of the HARV. After filling the vessel, the port is capped, leaving some dead space. It is important to completely fill the vessel, leaving no dead space or bubbles, as this will cause turbulence which will damage cultured cells (Cherry & Papoutsakis, 1998). The remaining dead space is filled by attaching two syringes to the syringe ports, one filled with medium and the other empty (Figure 16.5b). By injecting medium, the air is displaced into the empty syringe until the vessel is completely filled with liquid. The vessel is now ready to be attached to the rotator base in the incubator.

When a medium change is necessary, the vessel is removed from the rotator base and the cells and/or scaffolds allowed to settle to the bottom. Approximately three-quarters of the medium is removed through the large port, replaced with fresh medium and any remaining bubbles removed as described above.

Setting the rotational speed

If the culture consists of isolated cells with the aim of producing spheroids, the initial rotational speed need not be more than 5 rpm. As the cells aggregate and become visible, their motion can be observed through the clear wall of the vessel and the speed of rotation can be increased to compensate for the larger

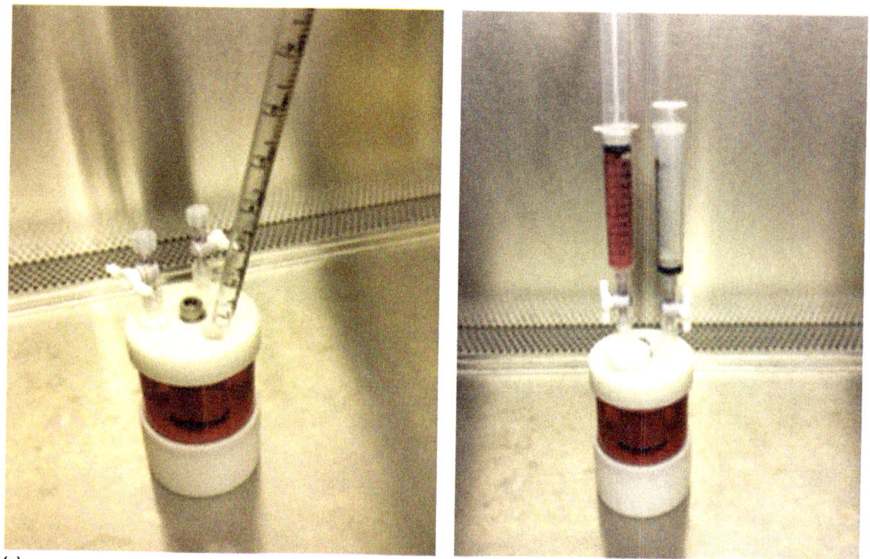

(a) (b)

Figure 16.5 Cell seeding and medium changes. (a) Loading the RCCS vessel through the large fill port. (b) Removing bubbles via the syringe ports.

size of the spheroids. If scaffolds are being used, the speed can be set to keep the scaffolds in suspension from the initiation of the culture and will not need any adjustment.

Perfusion culture RCCS

The perfused version of the RCCS (see Figure 16.4) was developed to address several deficiencies of batch culture systems. First, the culture environment of a batch system fluctuates between medium changes because of the depletion of nutrients and the accumulation of wastes. A perfusion system with a large external volume of medium tends to smooth out these variations. Second, the batch systems are restricted in scaleability. As the volume is increased by the power of 3, the gas exchange area can only increase by the square so that gas exchange becomes limiting. In the perfusion system, the flow rate can be increased to supply larger culture vessels.

The culture vessel of the perfused RCCS resembles the STLV and is rotated to create the same microgravity-like environment. Unlike the STLV, the core in the centre of the vessel is used to perfuse the medium from the external reservoir. The core is a porous polymer designed to allow the flow of medium into and out of the bioreactor vessel. The size of the pores (15–30 μ) is small enough to keep multicellular aggregates or scaffolds in the vessel. The flow of medium is maintained by a peristaltic pump which draws the medium from the reservoir through silicone tubing to a gas exchange module before it enters the vessel (see schematic, Figure 16.6). On each endcap is a disposable, γ-sterilised rotary coupling

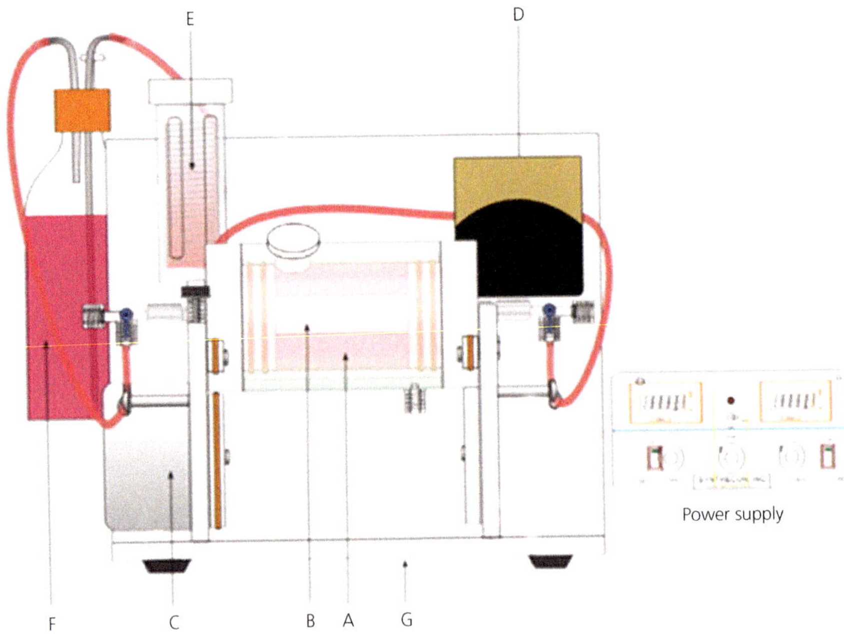

Figure 16.6 Schematic diagram of the perfused RCCS. A: Bioreactor vessel. B: Porous perfusion core. C: Drive motor. D: Peristaltic pump. E: Gas exchange module. F: Medium reservoir. G: Rotator base.

which allows simultaneous rotation and perfusion. The gas exchange module contains approximately 10 metres of spiral-wrapped, thin-walled silicone tubing which produces a surface area of 1.2 square metres. As with the batch systems, the rotator base operates in a standard humidified CO_2 incubator.

Cell seeding and medium changes

After autoclaving the bioreactor vessel, medium bottle, gas exchange module and tubing, the flow loop is assembled in a laminar flow hood with the γ-sterilised rotary couplings. Medium, cells and scaffolds, if used, are added via the large port in the vessel wall. The medium bottle is also filled with the desired volume of medium at this time. As with the batch systems, there will be some volume of bubbles left after this step. To remove these bubbles, an empty syringe is placed on the syringe port and the valve on the inlet side of the vessel is opened. The bubbles are manoeuvred under the syringe port and are removed through the empty syringe while medium from the bottle flows in to replace the bubbles. The flow loop can then be installed on the rotator base in the incubator and the culture initiated. To change the medium, it is only necessary to change the medium bottle. It is recommended that this step be performed in a hood to minimise the risk of contamination.

Example application: 3D neural tissue engineering

A number of years ago, Ma and associates at the Naval Research Laboratory developed a method for culturing neuronal stem cells in 3D collagen gels in multiwell plates (O'Connor *et al.*, 2000). The stem cell-derived neurons were able to form synapses and functional neuronal circuits (Ma *et al.*, 2004). Nevertheless, after a certain period, the cells became necrotic due to inadequate diffusion of nutrients and oxygen in static culture. When the encapsulated cells were placed in the dynamic culture environment of the RWV (Figure 16.7), they not only survived for much longer periods but formed complex brain-like tissues (Lin *et al.*, 2004; Ma *et al.*, 2008).

Methods

Neural stem and progenitor cells were dissociated from rat embryonic day 13 cerebral cortex using papain and DNAse and added to liquid type I collagen. Aliquots of the collagen-cell suspension were placed in the wells of a 24-well tissue culture plate and allowed to gel. Some cell-collagen constructs were transferred to an RCCS bioreactor vessel and some were left in the multiwell plates as static controls. Both sets of constructs were cultured in neurobasal/B27 medium with 10 ng/mL bFGF. After two weeks in culture, brain-derived neurotrophic factor (BNDF), 10 ng/mL, was added.

Cell viability was determined with calcein-AM and ethidium homodimer staining. The identity of the cells within the collagen gels was determined by immunostaining with antibodies to neuron-specific β-tubulin III (TuJ1), astrocyte-specific glial fibrillary acid protein (GFAP) and neural progenitor-specific nestin. Images of stained cells were captured by confocal laser scanning microscopy.

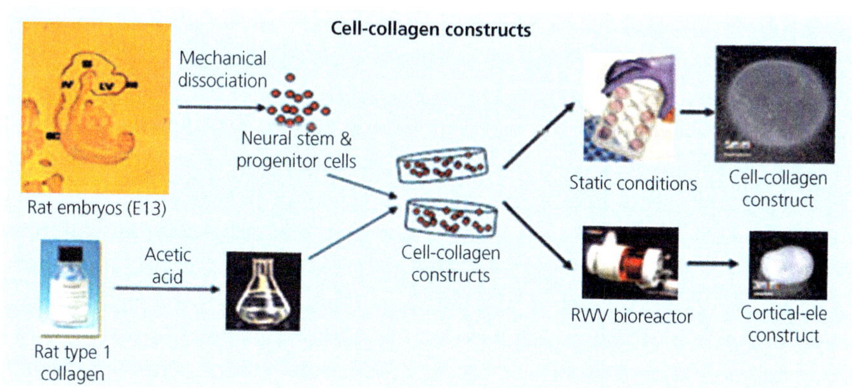

Figure 16.7 Comparison of static and dynamic culture. Experimental design for comparison of static and dynamic (RCCS) culture of neural stem cells. Source: Ma *et al.* 2008. Reproduced with permission of Mary Ann Liebert, Inc.

Functional expression of neurotransmitter receptors was determined by measuring agonist-stimulated intracellular Ca^{++} release with fluo-3 AM and an Attofluor Ratio Vision workstation.

Results

Cell clusters were rapidly formed in the gels of both static and RCCS cultures. The clusters expanded rapidly in static culture while those in the RCCS grew at a slower rate for the first three weeks. Subsequently, the static cells began to die off and the RCCS-cultured cell growth accelerated (Figure 16.8a,b). At the same time, the RCCS-cultured cells showed extensive proliferation of nestin+neural progenitors (Figure 16.8c), differentiated TuJ1+ neurons and GFAP+ astrocytes (Figure 16.8d). The RCCS-cultured neural cells showed extensive neurite outgrowth surrounded by astrocytes whereas the static cultured cells exhibited randomly scattered astrocytes and neurons with few or no extensions (Figure 16.9).

Between five and 10 weeks of culture, while the static cultured cells were dying, the RCCS-cultured cells formed into elliptically shaped tissue-like structures. The structures were organised in layers similar to the developing rat cerebral cortex (Figure 16.10). In the developing rat cerebral cortex (Figure 16.10a), the proliferating neural progenitor cells form the neuroepithelium (NE) while the differentiating neuronal and glial cells are found in the cortical plate (CP) region. In the RCCS cell-collagen constructs (Figure 16.10b), layer I contained nestin+and PCNA+ cells corresponding to the NE while layer II contained TuJ1+ neuronal cells similar to the CP region.

Not only do the RCCS-cultured constructs resemble the cerebral cortex structurally, but they also show functional Ca^{++} release responses coupled to neurotransmitter receptors found in the layers of the cerebral cortex (Figure 16.11). In layer I, composed largely of neural progenitors, the cells responded to purinergic (ATP) and cholinergic (carbachol) agonists whereas layer II, composed of differentiated neurons, responded to GABA (muscimol), glutamate and voltage-dependent Na^{+} channel (veratridine) agonists. These responses were similar to those observed in the developing rat cerebral cortex (Maric *et al.*, 2000).

These two papers, encompassing the development of an organotypic 3D model of a 3D cerebral cortex, demonstrate some of the advantages of the RCCS compared to a conventional static 3D encapsulation approach. The dynamic RCCS-generated tissues had higher rates of survival, proliferation and differentiation due to the improved availability of nutrients and oxygen. While it is only speculation, the microgravity-like environment may also contribute to the process by reducing the constraints of gravitational force to allow the encapsulated cells to organise into tissues. Studies such as these highlight the potential for tissue engineering applications of the RCCS such as replacements for injured tissues and improved *in vitro* models for drug development.

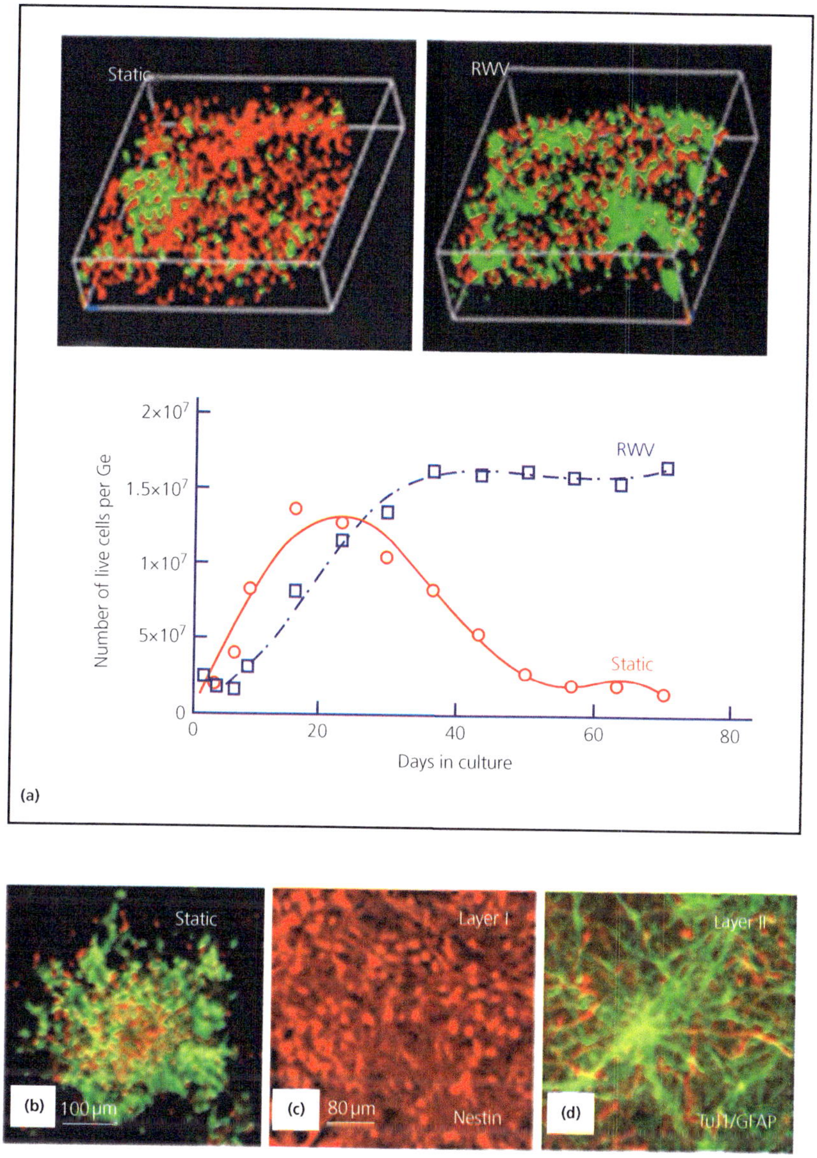

Figure 16.8 **Comparison of cell growth and differentiation in static and RCCS culture conditions.** (a) Time course of collagen-encapsulated neural stem cell growth in static or RCCS culture. Cell viability was quantitated by calcein/ethidium live/dead assay. The computerised 3D reconstruction (*top*) shows proportion of live and dead cells in the static and RCCS cultures at day 40. (b) The static culture contains a necrotic core after 14 days in culture. (c) Immunostaining of the surface layer of the gel cultured in the RCCS stained for nestin. (d) Double immunostaining of the deeper layer of the RCCS-cultured gel for the neuronal marker TuJ1 (*green*) and the astrocyte marker GFAP (*red*). Source: Ma *et al.* 2008. Reproduced with permission of Mary Ann Liebert, Inc.

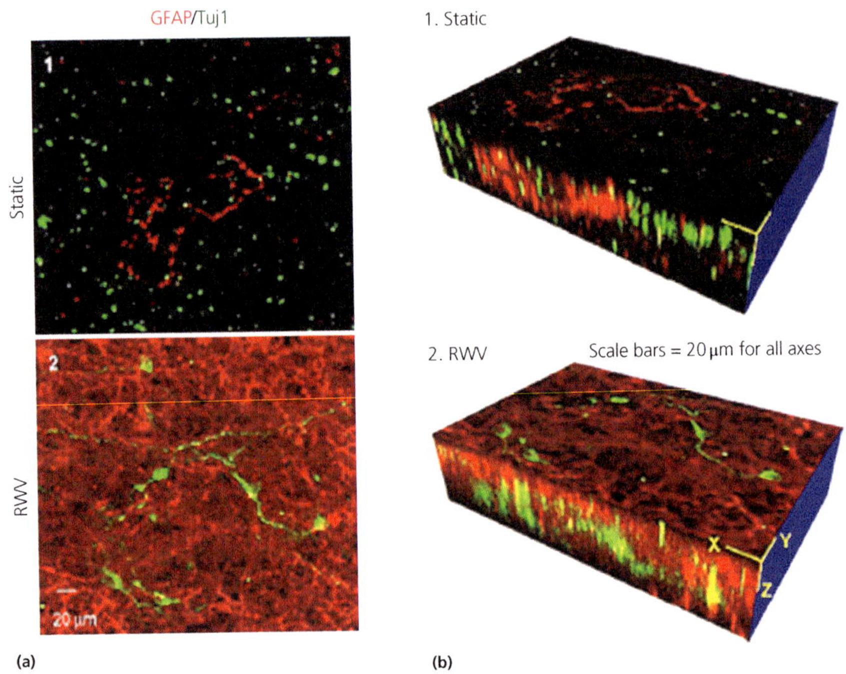

Figure 16.9 RCCS cultures maintain organotypic differentiation. (a) Confocal images of cell-collagen gels cultured for three weeks in static or RCCS culture. Neural cells, TuJ1 (*green*), astrocytes, GFAP (*red*). (b) Top-down and side projections created from confocal image stacks of cell-collagen gels under static and RCCS culture conditions. Source: Lin, H. J., O'shaughnessy, T. J., Kelly, J. & Ma, W. 2004. Reproduced with permission of Elsevier.

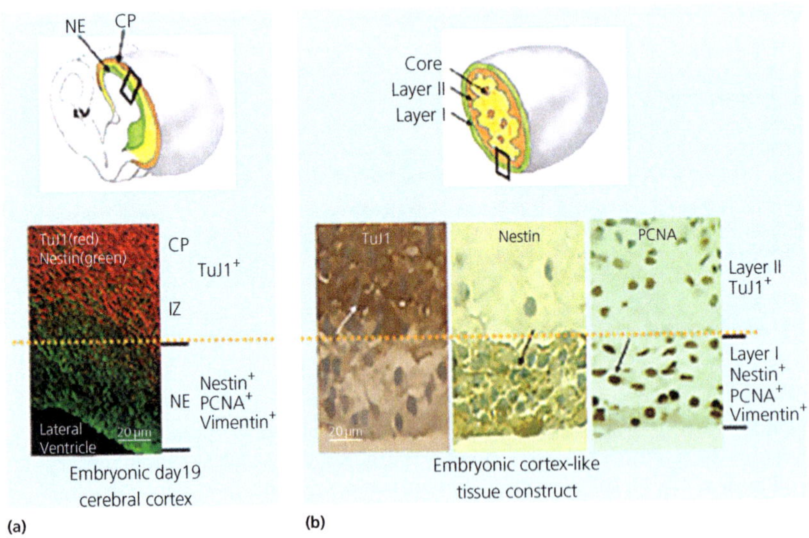

Figure 16.10 Similarity in architecture between embryonic day 19 rat cerebral cortex and RCCS-generated brain-like tissue. (a) Upper panel is a drawing of a section of an E19 rat brain. The box indicates the area shown in the lower panel, a transverse section through the E19 rat cortex immunostained for nestin (*green*) and TuJ1 (*red*). (b) Upper panel is a drawing of a transverse section of a 3D cell-collagen construct cultured in the RCCS for five weeks. The box indicates the area shown in the lower panel. The nestin+progenitors are layer I, similar to the NE, while the TuJ1+ neurons are located in layer II, similar to the CP. Source: Ma *et al.* 2008. Reproduced with permission of Mary Ann Liebert, Inc.

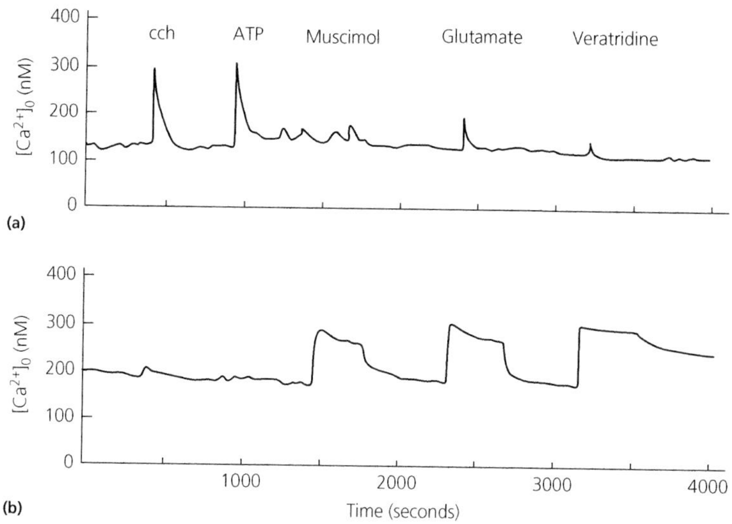

Figure 16.11 **Release of cytosolic Ca^{++} in response to pharmacological agonists to neurotransmitter receptors and voltage-dependent Na^+ channels in RCCS-cultured constructs.** (a) A representative cell recorded in layer I. Approximately 55% of the cells in layer I responded to carbachol (100 µM) and 86% of the cells responded to ATP (1 mM). (b) A representative cell recorded in layer II. Approximately 85% of the layer II cells responded to muscimol (10 µM), glutamate (50 µM) and veratridine (100 µM). Source: Ma *et al.* 2008. Reproduced with permission of Mary Ann Liebert, Inc.

Troubleshooting

That the RCCS is so fundamentally different from conventional cell culture technologies sometimes causes problems, especially with new users. For example, since the bioreactor vessels are closed, it is sometimes assumed that it is not necessary to maintain a humidified atmosphere in the incubator. Unfortunately, this assumption leads to the rapid formation of a large bubble in the vessel due to the fact that the silicone membrane used for gas exchange also permits water vapour to pass through it. Another problem that can occur is the formation of overly large spheroids with necrotic centres. Seeding too many cells when the culture is initiated usually causes this; it can be easily addressed by lowering the seeding density. The ideal seeding densities must be determined empirically depending on the characteristics of each cell type.

Conclusion

The Rotary Cell Culture System, originally developed by NASA, features a horizontally rotating cylindrical culture vessel completely filled with medium. The absence of a head space minimises turbulence which can injure cells. The rotation of cells in these conditions facilitates the assembly of 3D multicellular spheroids, allows attachment of cells to scaffolds and helps to maintain cell viability compared to static 3D culture methods. Many studies have been done using a variety of cell types in the RCCS that have clearly demonstrated the ability of this technology to produce tissue-like constructs that mimic *in vivo* structure and function.

Acknowledgements

The author wishes to acknowledge the support of the National Aeronautics and Space Administration and the National Institutes of Health through the SBIR and STTR programmes.

References

Abbott, A. (2003) Biology's new dimension. *Nature*, **421**, 870–872.

Cherry, R. S. & Papoutsakis, E. T. (1998) Physical mechanisms of cell damage in microcarrier cell culture bioreactors. *Biotechnology and Bioengineering*, **32**, 1001–1014.

Ferrarini, M., Steimberg, N., Ponzoni, M. *et al.* (2013) Ex-vivo dynamic 3-D culture of human tissues in the RCCS™ bioreactor allows the study of multiple myeloma biology and response to therapy. *PLoS One*, **8**, e71613.

Gerecht-Nir, S., Cohen, S., Itskovitz-Eldor, J. (2004) Bioreactor cultivation enhances the efficiency of human embryoid body (hEB) formation and differentiation. *Biotechnology and Bioengineering*, **86**, 493–502.

Griffith, L. G. & Swartz, M. A. (2006) Capturing complex 3D tissue physiology *in vitro*. *Nature Reviews Molecular Cell Biology*, **7**, 211–224.

Hammond, T. G. & Hammond, J. M. (2001) Optimized suspension culture: the rotating-wall vessel. *American Journal of Physiology Renal Physiology*, **281**, F12–F25.

Harrison, R. G. (1907) Observations on the living development of nerve fiber. *Proceedings of the Society for Experimental Biology and Medicine*, **4**, 140–143.

Lei, X. H., Ning, L. N., Cao, Y. J. *et al.* (2011) NASA-approved rotary bioreactor enhances proliferation of human epidermal stem cells and supports formation of 3D epidermis-like structure. *PLoS One*, **6**, e26603.

Lin, H. J., O'shaughnessy, T. J., Kelly, J. & Ma, W. (2004) Neural stem cell differentiation in a cell-collagen-bioreactor culture system. *Brain Research*, **153**,163–173.

Ma, W., Fitzgerald, W., Liu, Q. Y. *et al.* (2004) CNS stem and progenitor cells differentiation into functional neuronal circuits in three-dimensional collagen gels, *Experimental Neurology*, **190**, 276–288.

Ma, W., Tavakoli, T., Chen, S. *et al.* (2008) Reconstruction of functional cortical-like tissues from neural stem and progenitor cells. *Tissue Engineering Part A*, **14**, 1687–1697.

Margolis, L., Fitzgerald, W., Glushakova, S. *et al.* (1997) Lymphocyte trafficking and HIV infection of human lymphoid tissue in a rotating wall vessel bioreactor. *AIDS Research and Human Retroviruses*, **13**, 1411–1420.

Maric, D., Maric, I., Chang, Y. H. & Barker, J. L. (2000) Stereotypical physiological properties emerge during early neuronal and glial lineage development in the embryonic rat neocortex. *Cerebral Cortex*, **10**, 729–747.

Nakamura, K., Kuga, H., Morisaki, T. *et al.* (2002) Simulated microgravity culture system for a 3-D carcinoma. *Biotechniques*, **33**, 1068–1076.

O'Connor, S. M., Stenger, D. A., Shaffer, K.M. *et al.* (2000) Primary neural precursor cell expansion, differentiation and cytosolic Ca(2+) response in three-dimensional collagen gel. *Journal of Neuroscience Methods*, **102**, 187–195.

Qureshi, K. M., Lee, J., Paget, M. B. *et al.* (2015) Low gravity rotational culture and the integration of immunomodulatory stem cells reduce human islet allo-reactivity. *Clinical Transplantaion*, **29**, 90–98.

Sakai, S., Mishima, H., Ishi, I. T. *et al.* (2009) Rotating three-dimensional dynamic culture of adult human bone marrow-derived cells for tissue engineering of hyaline cartilage. *Journal of Orthopaedic Research*, **27**, 517–521.

Schmeichel, K. L. & Bissell, M. J. (2003) Modeling tissue-specific signaling and organ function in three dimensions. *Journal of Cell Science*, **116**, 2377–2388.

Shimada, N., Sokunbi, G. & Moorman, S. J. (2005) Changes in gravitational force affect gene expression in developing organ systems at different developmental times. *BMC Developmental Biology*, **5**, 10.

Sung, S. Y., Hsieh, C. L., Law, A. *et al.* (2008) Coevolution of prostate cancer and bone stroma in three-dimensional coculture: implications for cancer growth and metastasis. *Cancer Research*, **68**, 9996–10003.

Vamvakidou, A. P., Mondrinos. M. J., Petushi, S. P. *et al.* (2007) Heterogeneous breast tumoroids: an in vitro assay for investigating cellular heterogeneity and drug delivery. *Journal of Biomolecular Screening*, **12**, 13–20.

Wang, Y., Zhang, Y., Zhang, S. *et al.* (2012) Rotating microgravity-bioreactor cultivation enhances the hepatic differentiation of mouse embryonic stem cells on biodegradable polymer scaffolds. *Tissue Engineering Part A*, **18**, 2376–2385.

Wolf, D. A. & Schwartz, R. P. (1991) Analysis of Gravity-Induced Particle Motion and Fluid Perfusion Flow in the NASA-Designed Rotating Zero-Head-Space Tissue Culture Vessel. NASA Technical Paper 3143. NASA, Washington, DC.

Wolf, D. A. & Schwartz, R. P. (1992) Experimental Measurements of the Orbital Paths of Particles Sedimenting Within a Rotating Viscous Fluid as Influenced by Gravity. NASA Technical Paper 3200. NASA, Washington, DC.

Yu, B., Yu, D., Cao, L. *et al.* (2011) Simulated microgravity using a rotary cell culture system promotes chondrogenesis of human adipose-derived mesenchymal stem cells via the p38 MAPK pathway. *Biochemical and Biophysical Research Communications*, **414**, 412–418.

Yuan, Y., Sin, W. Y., Xue, B. *et al.* (2013) Novel alginate three-dimensional static and rotating culture systems for effective ex vivo amplification of human cord blood hematopoietic stem cells and in vivo functional analysis of amplified cells in NOD/SCID mice. *Transfusion*, **53**, 2001–2011.

Glossary

3D cell culture: The growth of cells within a biocompatible natural or synthetic extracellular matrix or physical support that creates a three-dimensional space for cell growth as opposed to using conventional culturing techniques involving a flat, two-dimensional flask or culture dish.

Adult stem cells: Multipotent cell population found in specific adult organs and/or tissues that can differentiate into specialised cell types found in their own tissue of origin.

Basement membrane: The basement membrane is a thin, cell-adherent layer of extracellular matrix that typically supports epithelia. Its function is to provide mechanical structure that separates different cell types, and has also been found to participate in cell differentiation, migration and survival by inducing important signalling cues. Major matrix proteins found in the basement membrane are laminin, collagen IV and heparan sulfate proteoglycans.

Biomimetic: Relating to synthetic methods used to mimic biochemical processes.

Bioreactor: An engineered or manufactured 3D cell culture system combining culturing techniques such as natural or synthetic substrates or suspension cultures to support a biologically active environment.

Blastocyst: An early-stage embryo containing about 150 cells made up by the inner cell mass and its surrounding cell layer referred to as the trophoblast.

Cancer-associated fibroblasts: A subpopulation of cells found in the tumour stroma that strongly resemble fibroblasts and support tumour growth. Rather than being a cell type, this is a facet of tumour-supportive fibroblasts.

Cancer stem cells: Cancerous cells that possess similar characteristics to stem cells and are believed to be able to give rise to all cells found in a cancer sample.

Carcinoma: Tumour with an epithelial origin, usually from endodermally or ectodermally derived cells.

Cardiospheres: Cardiac-derived cells growing as three-dimensional spheroids in suspension.

Cell adhesion: Binding of a cell to substrate or surface via adhesion molecules.

Cell culture: The growth of cells in a controlled setting, normally outside or dissociated from their parent tissue.

Cell density: Number of cells per cm^2 of substrate.

Technology Platforms for 3D Cell Culture: A User's Guide, First Edition. Edited by Stefan Przyborski.

Cell line: Cells that can be indefinitely proliferated in culture, often described as immortalised.

Chemically defined: Made from known constituents or with a known chemical composition.

Chondrospheres: Chondrocytes growing as three-dimensional spheroids in suspension.

Co-culture: A culture condition where two or more different cell types are grown together in order to reflect a more physiological environment.

Compliance: The inverse of stiffness. Referring to the properties of a material, the more compliant, the less stiff the material is.

Confluence: Describes the proportion of the surface covered by cells adhered to the surface of a flask or culture dish. A culture is said to be confluent when a monolayer of cells completely covers the substrate and all cells are in contact with other cells.

Cytokine: A signalling molecule or factor released by cells that elicits a receptor-mediated effect on other cells. These effects include but are not limited to differentiation, proliferation and inflammation.

Dextran: Complex branched glucan (polysaccharide consisting of glucose units) with α-linked D-glucopyranosyl repeating units.

Differentiation: The process by which a cell undergoes change. For example, in the context of stem cells, it is the process by which an undifferentiated stem cell becomes more specialised, gaining function but losing potency.

Directed differentiation: The induction of stem cell differentiation into specific lineages and cell types by modifying the culturing conditions.

Ectoderm: The ectoderm is the outermost layer of the germ cell layers of an embryo. It comes from the inner cell mass and differentiates to give rise to skin and its appendages, the nervous system and other outermost specialised tissues and organs.

Electrospinning: Technique in which fibres from a wide variety of natural and synthetic polymers are electrospun. This technique uses an electric charge.

Embryoid bodies: A suspension culture of pluripotent stem cells showing cells derived from all three germ layers in a single cell aggregate.

Embryonic stem cells: Cells with stem cell potential derived from the inner cell mass of the blastocyst. These stem cells, however, are different from adult stem cells, as they are pluripotent.

Endoderm: This germ layer is the innermost layer of an embryo. It comes from the inner cell mass of the blastocyst and gives rise to the linings of the lungs and intestines, as well as the liver and the pancreas.

Endothelium: A thin, epithelium-like cell layer lining cavities and organs, for example blood and lymphatic vessels and the heart.

Epithelial: Cells derived from epithelium; however, it often describes cells with a polygonal shape.

Epithelial-mesenchymal transition: Process by which epithelial cells undergo transformation to adopt a mesenchymal phenotype, losing cell polarity and cell-cell adhesion. This is seen in early stages of development during embryogenesis, in wound healing, during the formation of the tumour microenvironment and in early stages of metastasis.

Epithelium: A basic type of tissue often arranged in a lining of cells covering cavities and surfaces, such as the lining of the gut.

Extracellular matrix: The extracellular space supporting cellular growth, made of a complex mixture of polysaccharides, proteins and growth factors secreted by cells. It has a far-reaching role involving physical support, providing chemical cues that influence cellular behaviour and cell fate differentiation, and acting as a spatial organiser where gradients of nutrients can be established.

Feeder layer: Feeder layers are made of cells, normally mouse or human embryonic fibroblasts inactivated by γ-irradiation or mitomycin to halt cell division. Culturing certain cells at low or clonal density requires enriching the medium with feeder cells. This cell layer can maintain stem cells in an undifferentiated state without losing pluripotency, by providing an optimal environment with secreted factors and physical cues in the form of an extracellular matrix and cellular contacts for the maintenance of stem cells.

Fibroblast: Precursor cell of the differentiated fibrocyte and the responsible cell type in the synthesis of collagen and extracellular matrix.

Fluorophore: Fluorescent chemical compound used as a probe to detect protein expression.

Germ layers: The three layers present in an early embryo (ectoderm, mesoderm and endoderm) responsible for giving rise to different organs and tissues in the body.

Growth factor: A signalling molecule or factor released by cells that promotes cell proliferation by acting mostly via, but not limited to, a paracrine effect.

Growth medium: Basal medium with additives such as growth factors and serum aimed at propagating cell lines.

Haematopoietic: Relating to the making of blood or blood cells.

Haematopoietic stem cell: Multipotent stem cells able to differentiate into all other blood cell types. This cell type is found in adult bone marrow, umbilical cord blood, peripheral blood and the fetal liver.

Hanging drop technique: A well-established method for culturing cells, where a drop of fluid is suspended from a coverslip and as it hangs, it encourages aggregate formation leading to three-dimensional spheroids.

Heparan sulfate: A highly sulfated glycosaminoglycan found in the extracellular matrix.

Heterologous: A mixed cell population.

High-throughput screening: A drug discovery method requiring automation and large-scale capacity to quickly test the biological or biochemical activity of various compounds.

Hyaluronic acid: Hyaluronic acid or hyaluronan is an unsulfated glycosaminoglycan that remains unbound to a core protein. It is a key constituent of the extracellular matrix, providing both physical and chemical cues.

Hydrogel: 3D cell culturing system consisting of a network of macromolecules covalently or ionically linked together. Hydrogels are highly superabsorbent with high water content.

Hydroxypropyl cellulose: Water- and organic-soluble derivative of cellulose.

Induced pluripotent stem cells: iPSCs are adult cells reprogrammed to a pluripotent-like state to differentiate into a wide variety of cells. They are often compared to embryonic stem cells. The engineering of their potency is achieved through the overexpression of several stem cell markers.

Inner cell mass: The cluster of cells in the inner space of the blastocyst, surrounded by the trophoblast. The inner cell mass gives rise to all embryonic tissues, but not extraembryonic tissues such as placenta, which originates from the trophoblast. Cells from this mass are utilised to make human embryonic stem cell lines.

***In situ* gelation:** Process in which a drug delivery system in the form of a polymer, initially in fluid suspension, undergoes gelation in the desired site. Gelation is influenced by changes in temperature, pH, ions and UV irradiation.

***In vitro*:** Describes a biological process that occurs and/or exists outside the living organism, in an artificial environment most often in the laboratory.

***In vivo*:** Commonly used to contrast the terms *in vitro* and *ex vivo*, this term describes a state or process occurring within the body. It also describes testing conditions that take place within humans and/or animals.

Live cell imaging: Study of cells with time-lapse microscopy. This involves taking sequences of images to then view them at a greater speed in order to monitor dynamic events, for example cell motility.

Long-term self-renewal: Describes the ability of stem cells to replicate in the undifferentiated state over time.

Matrix metalloproteinase: Matrix metalloproteinases are a group of enzymes responsible for degrading proteins in the extracellular matrix.

Mechanotransduction: Process whereby cells can sense mechanical stimuli from their environment and neighbouring cells and translate it to biochemical signals. These signals can then alter cellular and extracellular structure.

Mesenchymal stem cells: Pluripotent adult stem cells found in bone marrow, which can be differentiated into connective tissue, bone, cartilage and fat.

Mesoderm: The mesoderm is the middle germ layer of an embryo. It gives rise to bone, muscle, connective and vascular tissues and the kidneys.

Microenvironment: The collection of physical (extracellular matrix, cell-cell interactions, three-dimensional space) and chemical (cytokines, metabolites, dissolved gases) cues influencing cellular behaviour and differentiation fate.

Micropatterning: The modification of biomaterials' surfaces to model tissue microenvironments where cell shape, position, multiphase tissue architecture and attachment can be controlled and studied.

Monoclonal: Cells or cell products derived from a single clone. In immunohistochemistry, monoclonal antibodies are produced by hybridomas, a fusion of B cells and myeloma cells.

Morphology: The form and structure of a cell.

Multipotent: Describing stem cells that can only differentiate into a subset of all adult tissues, for example adult stem cells.

Nanofibres: Synthetic or natural fibres that replicate the structure of tissue at the nanoscale. These fibres are often utilised as the framework of scaffolds for 3D cell culture.

Neural stem cells: Adult stem cells responsible for differentiating into nerve cells (neurons) and their supporting cells (glia (astrocytes) and oligodendrocytes).

Neurons: Nerve cells composed of a cell body and processes (axon and dendrites). These cells' function is to transmit impulses to other neurons by neurotransmitter release at synapses.

Neurospheres: Suspension culture of neural stem cells, encouraging the formation of spheroids, which can proliferate and differentiate into neurons and glia.

Organ culture: Growth and maintenance of an organ *in vitro* with the purpose of preserving the tissue architecture and allow differentiation.

Organotypic: Three-dimensional culture of one cell type aimed at mimicking the *in vivo* interactions of an organ. Differing from an organ culture, an organotypic culture involves rebuilding from dissociated cells or fragments of tissue, as distinct from organ culture, in which the structural integrity of the explanted tissue is retained.

Passage: Passaging cells is the process of maintaining a cell line by diluting them in fresh medium.

Passage number: The number of times a cell culture has been subcultured.

Phenotype: The collection of the expressed properties of a cell.

Photopolymerisation: Polymerisation induced by light. A photopolymer is a polymer that changes its properties when exposed to light, often in the ultraviolet or visible region of the electromagnetic spectrum. These changes are often manifested structurally – for example, hardening of the material occurs as a result of crosslinking when exposed to light.

Plasticity: A property of stem cells describing their ability to change the range of cell lineages they can differentiate into, depending on environmental cues.

Pluripotent: The ability to give rise to multiple lineages. Pluripotency mainly describes human embryonic stem cells, which can differentiate into any cell of an adult organism or the developing fetus. The only exception being the placenta, which origins from the trophoblast.

Porcine gelatin: Gelatin from porcine skin composed of a heterogeneous mixture of water-soluble proteins used in cell culture as a coating to enhance cell attachment.

Potency: The capacity of a stem cell or progenitor cell to form cells of different lineages. This characteristic varies along the differentiational cascade. As cells divide and become more specialised, these lose potency.

Precursor cell: Cells that are already committed to a particular cell type and lineage.

Primary culture: The initial culture originating from cells, tissues or organs extracted from an organism.

Progenitor cell: Cells with reduced potency and self-renewing ability, but at a proliferative stage without showing differentiated properties.

Proliferation: The reproduction or expansion of a cell population by mitosis.

Regenerative medicine: Like cellular therapy, this field of medicine utilises stem cell-derived differentiated cells as a means to repair damaged tissue or to create new functional tissue.

RGDS: Arginine-glycine-aspartate-serine is a tetrapeptide commonly found on proteins, fibronectin, fibrinogen α and von Willebrand factor. This peptide can interact with integrins found on cells and alter intracellular signalling.

Saturation density: Maximum number of cells per cm^2.

Scaffold: A synthetic or natural three-dimensional physical matrix for cell culture. There is a great range of methods for the design and manufacture of these highly porous platforms.

Sedimentation: Method used in biochemistry to determine properties including, but not limited to, shape, size, purity and molar mass.

Seeding efficiency: Also known as attachment efficiency. The percentage of cells plated (seeded, inoculated) which attach to the surface of the culture vessel within a specified period of time. The conditions under which such a determination is made should always be stated.

Shear elastic modulus: Coefficient of elasticity for a shearing force denoted by G, or sometimes S or μ.

Shear stress: In biology, it refers to the force exerted by the tangential drag of a fluid as it flows over a surface.

Simulated microgravity: Reproduction of an environmental factor in space influencing cellular behaviour using clinostats and simulators.

Somatic cells: Describes all cells in an organism excluding the reproductive cells: egg and sperm cells.

Spheroid: Three-dimensional aggregate of cells encouraged by suspension culture.

Stem cell: Undifferentiated cell at the beginning of a lineage with the ability to replicate and maintain its population and give rise to differentiated and specialised cells in this lineage.

Stromal cells/stroma: Refers to the part of connective tissue responsible for providing support for the parenchyma, the functional part of tissue in an organ.

Surface markers: Cell surface proteins that are cell type specific and can be targeted for detection and identification with the respective antibodies.

Suspension culture: Cell culture type where cells multiply suspended in growth medium.

Teratoma: A germ cell tumour containing cells derived from all three germ layers, usually found in the testis or ovary. The formation of teratomas after the xeno-transplantation of human embryonic stem cells into immunocompromised mice demonstrates the stem cells' pluripotency.

Thiol: Organosulfur compound that contains a carbon-bonded sulfur hydrogen bond.

Tissue engineering: A field incorporating biomaterials and cells to engineer tissues. Tissue engineering can be performed in vitro combining that platforms for 3D cell culture to build new tissues or organs from cells.

Totipotent: Describes a stem cell that can give rise to any embryonic or extraembryonic tissue.

Transdifferentiation: Stem cells from one lineage having the ability to differentiate into cells from a different lineage.

Trophoblast: Cells surrounding the inner cell mass, which give rise to the placenta. The trophoblast along with the inner cell mass form the blastocyst.

Undifferentiated: Describing a cell at a stage where its specialised structures or functions have not been yet determined.

Unipotent stem cells: Stem cells with potential to self-renew but limited to differentiating only into one specialised cell type.

Index

Technology Platforms for 3D Cell Culture: A User's Guide, First Edition. Edited by Stefan Przyborski.

Printed and bound by CPI Group (UK) Ltd, Croydon, CR0 4YY
21/09/2026
14965827-0001